An
Annotated Catalog
of Centipedes
(Chilopoda)

From the United States of America, Canada and Greenland
(1758–2008)

AN ANNOTATED CATALOG OF CENTIPEDES (CHILOPODA)

From the United States of America, Canada and Greenland (1758–2008)

Randy J. Mercurio

This catalog is dedicated to my mother and father:

Mrs. Gail A. Mercurio and Dr. Anthony R. Mercurio

for encouraging me to embrace nature

and the wonderful world of arthropods.

Author Note: Be it known that this publication does not necessarily follow the International Code of Zoological Nomenclature (ICZN). Although the author has attempted to keep the nomenclatorial content of this catalog accurate, the higher classification and overall taxonomic arrangement may not necessarily reflect the current and/or accepted classification of centipedes (Chilopoda). The information contained in this catalog is in no way meant to be a taxonomic revision, however; it is intended as a resourceful reference for the advancement of centipede taxonomy, ecology and natural history.

To order additional copies of this book, contact:
Xlibris Corporation
1-888-795-4274
www.Xlibris.com
Orders@Xlibris.com
57823

Contents

Introduction

Essentially, this catalog incorporates over 600 references on centipedes from 1758 to 2008 whose geographic coverage includes the forty-eight contiguous United States, Alaska, Hawaii, Canada and Greenland. North America will be used to refer to the covered geographic range in this catalog. Doubtless, a catalog covering all orders of centipedes for the North American region has been wanting. No effort has ever been made in the United States or the world for that matter to assemble the vast amount of pertinent information on all orders of chilopods. With much of the taxonomy for North American centipedes in disarray it is the intent of the author to have this catalog serve as a place to provide some organization. There is also a growing need for a reference source that will provide access to summarized data in order to conduct research without the labor of tracking down literature. The author has attempted to work on smaller projects regarding North American chilopod fauna and found it difficult to proceed without a catalog of this sort. Summarizing the literature seemed to be the best way of organizing the fragmented information. This catalog covers roughly 20% of the approximately 3000 known valid centipede species.

There is no need to provide an overview of the works of past contributors to chilopodology such as Rafinesque, Say, Brandt and many others as Underwood has discussed them in some detail through 1893 in Bollman's posthumous work (Bollman, 1893). To start where Underwood left off after Bollman's untimely death, Cook demonstrated a particular interest in the geophilids and published seven papers that dealt with centipedes of concern in this catalog. The relevant works of Pocock, Chamberlin, Crabill, Hoffman, Shear, Shelley and many others have been incorporated into this catalog. There are five PhD theses that are unpublished but have been combined into this catalog in hopes of assimilating disjunct data to increase the organization of centipedes in North America because they contain useful information. The doctoral thesis of Matthews (1935) compiled much data on the centipedes of Wisconsin, which also contains an ecological perspective on centipedes. Causey's (1940) doctoral thesis provided additional information on chilopods from North Carolina. Crabill's

(1952b) doctoral thesis from Cornell has been included because it contains a wealth of valuable information. Johnson's (1952) PhD thesis supplied a tremendous amount of information on the centipedes of Michigan and is highly recommended for its assimilation of ecological and natural history data on centipedes. Finally, the PhD thesis of Mundel (1981) has contributed additional information about the lithobiomorphs.

Despite that centipedes are relatively soft-bodied animals there are a few sedimentary fossils from North America. In fact, the United States contains representatives of at least three of the five extant orders in the fossil record in addition to a recently discovered extinct order. These have been included in the extinct section of the catalog. The earliest records of fossil centipedes in the United States are from the Middle Devonian Epoch (Shear et al., 1984; Shear et al., 1998). According to Almond (1985) Paleozoic chilopoda have yet to be substantiated outside of North America. Centipede fossils in the U.S. are sedimentary in nature, but elsewhere they are also known to occur in amber. Although the author is not aware of any amber centipede fossils within the U.S., one can find inclusions of *Newportia* sp. in Mexican (Grimaldi, 1996) and Dominican Republic amber (Poinar & Poinar, 1999).

The systematic classification of centipede orders are according to papers dealing with recent phylogenetic studies (see Kraus, 1998; Edgecombe, Giribet & Wheeler, 1999; Edgecombe & Giribet, 2002, 2004). The arrangement within the lithobiomorpha essentially follows that of Eason (1992). Within each order the families, subfamilies, genera, subgenera, species and subspecies follow an alphabetical listing. The type species of each genus directly follows the genus listing. Under the genus (in brackets) is a quick reference to the total number of species in the genus, the number of introduced in parentheses, and its general distribution. Each genus has a list of references that is not meant to be all-inclusive but is intended to provide the user quick reference to useful information. Species accounts start with a male and/or female symbol indicating which sexes are known and if no symbols appear it means neither sex is known. For each taxon the references follow a chronological order including the author, date of publication, page number(s), figures, tables and a summary of the reference. When specific qualitative plectrotaxic (the arrangement and nomenclature of the pedal spurs of lithobiomorph centipedes, see Crabill, 1962b) information is available it has been noted in the taxonomic history, typically as a table. Synonymic names are listed alphabetically following the valid name references and their references also follow a chronological listing. Author notes and comments follow each genus or species that particularly show a need for further attention. There may be additional notes in the references following each species listing that may provide tables of ecological information as well as taxonomic importance. The distribution does not necessarily represent the natural distribution but records all localities where each species has been found unless otherwise noted. A question mark in front of a state or province (i.e. "?California") means: 1) there is some ambiguity to as to whether the species is truly known to exist in this state; 2) the report is or may be a misidentification; 3) at one time it was reported from this state and it is not certain if it has established itself in the state (typically introduced species). For non-indigenous species there are additional references that do not necessarily address the fact that these species are found in the regions covered by the catalog, however, the author felt it was advantageous to provide them for figures or other important information they contain about the species. There are numerous cases where data such as holotype locality or the sex of specimens could not be found. Where possible holotype depositories have been included. A questionmark in front of the depository abbreviation indicates that this type is expected to be housed there, however, it has not been confirmed. For example, many times in the literature Chamberlin mentions the type(s) to be stored in his private collection and it is known that most of his private collection is housed at the NMNH. In some instances the holotype may be established as unknown and it will be stated as so usually with a reference following that explains the remark. When "?Unknown" is listed in place of a depository abbreviation the place of deposition was unknown to the author.

This catalog is not strictly taxonomic; however, it was meant to be a tool to assist people researching centipedes from many different perspectives. Therefore, when I have come across agricultural, anatomical, behavioral, ecological, molecular, physiological and toxicological or other significant papers I have included them. There is no doubt in my mind that there are other publications obscured by titles and topics that contain useful information on centipedes of the region covered that have escaped notice. As one can imagine, much effort has gone into compiling all the available data for each species in the literature and to verify depositories of type material. Museums and universities thought or known to house the material have been visited and/or contacted where possible. Otherwise, databases made available to me were used to help decipher where type material may be deposited. The information available in the literature has been compiled as thoroughly as possible without errors. If any literature is missing it is not on purpose and I apologize in advance if this has occurred. Undoubtedly, in a work of this nature there are errors and/ or omissions for which the author alone takes the full responsibility and if these are discovered please bring them to my attention so that they may be corrected/added to future editions, Email: chilopods@yahoo.com.

An Ecological Note

Chilopods represent one of the top invertebrate predators in many regions of the world and for this reason they play an important role in ecosystems. Centipedes are primarily carnivorous and typically consume other arthropods, however, larger species are known to predate on amphibians, reptiles and mammals (e.g., Wells-Cole, 1898; Okeden, 1903; Martin et al., 1961; Shugg, 1961; Easterla, 1975; Clark, 1979; McCormick & Polis, 1982; Molinari et al., 2005; Carpenter & Gillingham, 1984). Filinger (1928) has apparently observed and suggested two species of lithobiomorphs were responsible for keeping the symphylan *Scutigerella immaculata* Newport in check within greenhouses. Barker (2004, and references therein) reported of centipedes eating gastropods or being malacophagic. On the other hand, centipedes also provide an abundant food source for various arthropods. Amazingly, they are predated upon by chilopodophagus ants that specialize on eating geophilomorphous centipedes (e.g., Haskins, 1928; Traniello, 1982; Hölldobler & Wilson, 1990; Wild, 2005). Furthermore, larger vertebrate organisms such as salamanders, frogs, lizards, snakes and birds are known to include centipedes as part of their diet (e.g., Jameson, 1944; Auerbach, 1952a; Hoffman & Hubricht, 1954; Judd, 1957; Martin et al., 1961; Shelley, 1987; Burke & Mercurio, 2002). From an agricultural standpoint, centipedes of the order Geophilomorpha have been reported to feed on the roots of sugar cane (Pemberton, 1925; Williams, 1931). The reasoning for this behavior was attributed to their need for moisture, which may very well be true, however, it may also be a possibility that they are predating on an organism that could have originally created the pits in the roots. Gunn & Cherrett (1993) observed that geophilomorphs feed on earthworms, slugs, diplurans, carabid beetle adults, carrion and that lithobiomorphs eat acari while both consume "faeces/detritus". Speaking of a lithobiomorph study done by Roberts (1956), Lewis (1965) stated "There were large quantities of plant material such as fragments of dead leaves, rootlets, fungal hyphae and spores and bryophyte leaves and spores in the guts of these species and Roberts concluded that these had been taken with prey." In captivity, a personal observation of the author is that *Scutigera coleoptrata* (Linnaeus, 1758) will readily accept banana and at least some large tropical scolopendrids will happily take fleshy fruits such as mango and even relish hard-boiled eggs.

It is also interesting to note that Buchsbaum (1948) claims "Though terrestrial, such a centipede as *Lithobius* can survive many hours completely immersed in water but will die in a few hours in an uncovered dish of dry earth." Metz (1987) observed what appears to be a lithobiid submerged in an ephemeral puddle leaving a trail in the substrate. In any case, the ecology of centipedes in North America is poorly known with a small number of publications either touching upon or addressing this topic in some depth (e.g., Holmquist, 1926; Causey, 1940; Dowdy, 1944; Auerbach, 1946, 1949, 1951a; Johnson, 1952; Summers et al., 1979; Lee, 1980). Ecologists studying North American centipedes may look to research done in Europe for further information as more studies have been undertaken there and doubtless the data contained in this literature will be useful (e.g., Albert, 1979; see also Appendix V). The author hopes that this catalog will facilitate more ecological studies on centipedes, as they will reveal a wealth of information about our complex ecosystems. Furthermore, it is envisioned that this catalog will promote studies of the phylogenetic relationships through the combined use of morphology and DNA. We do have a reasonable amount of data concerning geographic distributions now summarized in this catalog, however, we are lacking in many other areas of their biology, ecology and life history. While some species of chilopods seem to be generalists, most are microhabitat specialists. Therefore, preserving these habitats are crucial to maintaining healthy localized populations of indigenous fauna and to avoid adding species of centipedes to an extinction list. For those individuals undertaking ecological studies on centipedes it will more than likely be necessary to become familiar with those of foreign origin in order to recognize these invading species. A list of these can be found in Appendix II.

Sadly, there are many introduced centipedes that originate primarily from Europe, but some come from China, Japan, Mexico and other foreign countries. The total number of known invasive and potentially invasive species is approximately seventy-eight. Including species taken at quarantine facilities, this is about fourteen percent of the known indigenous fauna, which is roughly 555 species. The extent of their establishment can be understood by looking at the distribution of each species in the proceeding pages. Recent examples of verified populations of two introduced species of centipedes are *R. longipes* and *S. morsitans* (Shelley & Edwards, 2004; Shelley, Edwards & Chagas, 2005 respectively). There is no doubt that the geographic distribution of these established species could only worsen. Urban areas show a tendency for introduced species to dominate and next to if not completely displace indigenous species (pers. obs.). Based on material from the American Museum of Natural History (AMNH) *Lithobius forficatus* (Linnaeus) seems to have been well established in the New York City area at least since the early 1900's. The author has found numerous cases of *Lithobius forficatus*, in particular, to be very abundant in mulch piles and shows an extremely high rate of reproduction here due to the moist habitat and the typical profusion of appropriate food items associated with these habitats. In some cases, introduced species can be further spread by means of using things like infested mulch from distant sources to be dispersed on paths in our State Parks (pers. obs.). These practices just exacerbate the distribution of these non-indigenous species to locations, otherwise, of relatively pristine habitat. Another abundant species is *Cryptops hortensis*, which can be found frequently in close proximity to *Lithobius forficatus*. In the halls of the AMNH *Scutigera coleoptrata* is not uncommon and has clearly been established in the building since the early 1900's as specimens in the myriapod collection demonstrates.

Notes on Techniques of Study and Preservation

Chilopods are not the easiest organisms to study as they require much patience during examination of morphological characters and haste can easily cause misidentification. Living centipedes are not at all cooperative and can typically only be examined in detail when preserved. Once preserved, however, their legs tend to be folded medially and are hardened from the typical alcohol preservative causing crucial ventral leg characters needed for classification to be obscured (applies most significantly to Lithobiomorphous centipedes). The doctoral thesis of Johnson (1952) notes how he maintained living organisms and anesthetized them to observe these characters with much success and what seems to be an excellent method worth mentioning here. Johnson used ether (presumably diethyl ether) as an anesthetizing agent which not only immobilized the specimens but the effects of the ether were such that the legs became extended allowing easy observation of the dorsal and ventral spination of each leg. Incidentally, this technique is especially useful for those wishing to maintain living organisms for other observations. Prior to reading of this technique by Johnson, the author thought of anesthetizing them for the purpose of classification while under live laboratory observation. The author and colleague Dr. Robert A. Monaco, DVM, undertook an experiment of this sort in 1999 using the more modern anesthetizing agent isoflurane (unpublished data). A geophilomorphous centipede in a petridish with substrate was placed in an induction chamber and the isoflurane was introduced. Soon enough the centipede started to exhibit peristaltic-like behavior became limp and immobile after a short time and was easily observed under a microscope. Isoflurane seems to be an appropriate anesthetic for use on centipedes and if properly dosed will most likely allow full recovery of a specimen after sedation for other future studies in the lab requiring a living specimen. In addition to observing anesthetized centipedes it is very difficult to re-examine centipedes at a later date when preserved if the legs are all flexed medially. According to Johnson, he would always anesthetize centipedes with ether prior to placing in a preserving agent, which would keep the legs extended allowing easy observation of leg spination once preserved. Furthermore, to avoid hardened/fragile specimens preserved in 70% alcohol alone, Johnson also used the following preserving agent (an acid alcohol with glycerin): 8 parts 70% ethyl alcohol, 2 parts glycerin and 1/200 parts glacial acetic acid. If this technique works as Johnson says it does, it seems to be the better choice for preservation of all centipedes intended for morphological character studies. Johnson's "Techniques" chapter is recommended for other tips on collecting, rearing, feeding, handling and preserving centipedes that is available through UMI Dissertation Services.

Acknowledgements

First and foremost, I would like to express my appreciation to all of the myriapodologists and other authors who appear in the bibliography plus those I may have inadvertently omitted. This work has been accomplished on personal time over the course of more than six years, but could not have been completed without the influence and help of numerous people. Although I have had a curiosity about arthropods since my early childhood, I thank my uncle Joseph Mercurio who was influential with regards to my interest in arthropods and nature by stirring my curiosity through gifts of field guides. I express my appreciation to my father for allowing me to, well . . . permanently borrow his laptop computer, on which the majority of this catalog was completed. My sister Laura Mercurio informed me of the ability to automate the table of contents and index, an enormous help. My parents, relatives, colleagues and friends have been very supportive of this work and without their encouragement would not have been possible. Over the years, the AMNH library and its staff have been of great assistance and in particular Mary DeJong, Tom Baione, Ingrid Lennon, Anette Springer, Barbara Rhodes, Mai Quaraman and Rebecca Molton have been of outstanding assistance. I would like to express my deepest gratitude to Louis Sorkin (AMNH) who was my mentor during the five years I volunteered at the AMNH and for the myriad ways in which he was of assistance to this catalog. This catalogue has gone through a series of evolutionary steps and would not have been as functional or intelligible if it did not have input from Lee Herman (AMNH) whose general formatting was adopted from his catalog on staphylinids. I am indebted to Rowland Shelley who initiated my long process of collecting references, which he kindly made available to me from his personal library and offered much advice through discussions. Many thanks to Jonathan Coddington, David Furth and Dana De Roche for providing access to the type collection at the Smithsonian National Museum of Natural History. Gonzalo Giribet and Laura Leibensperger of the Harvard Museum of Comparative Zoology for access to the type collection. Richard Hoffman, Alessandro Minelli and Casimir Jeekel for providing reprints of their papers. Cornell University kindly allowed me to borrow the PhD thesis of Ralph E. Crabill (a huge inspiration) through a New York University inter-library loan during my undergraduate studies that provided an enormous wealth of data in one package. A special thank you to all of those involved with Chilobase (see Minelli in bibliography). Christy Bills from the University of Utah provided assistance in looking for missing holotypes of Chamberlin. David Grimaldi (AMNH) for discussion on the fossil centipedes. I would like to thank Emeritus Curator William Emerson (AMNH) for giving me access to his personal library that provided useful reference information. Thomas Demere (SDSNH) assisted in verifying the depository of the fossil centipede *Calciphilus abboti* Chamberlin, 1949. I thank Michael S. Engel (UKNHM) for hunting down the Gunthorp types. Marla Coppolino brought literature to my attention. Suzanne Rab-Green and Jacob Mey translated German and Danish respectively and I am grateful for their help. Jay Cordeiro and the Natural Heritage Network for steering me towards useful references. Author photograph taken by Jacquelyn Barry, who has tolerated my absence during the day through long and very late nights working to finalize this catalog. I thank my publisher, Xlibris, and all involved for their help in getting this through the final steps. Last but not least, there are numerous people who have indirectly contributed to this work through inter-library loans and other assistance that I never met . . . thank you.

Type Depository Abbreviations

AMNH	American Museum of Natural History, New York, New York, USA.
ANSP	The Academy of Natural Sciences, Philadelphia, Pennsylvania, USA.
BMNH	The Natural History Museum, London, United Kingdom.
CMN	Canadian Museum of Nature, Canadian National Collection, Ottawa, Ontario, Canada.
CMNH	Cleveland Museum of Natural History, Cleveland, Ohio, USA.
FHB	Federal Horticultural Board, United States Department of Agriculture, Washington, D.C.
FMNH	Field Museum of Natural History, Chicago, Illinois, USA.
HM	Hamburg Museum, Germany.
ISM	Indiana State Museum, Indiana, USA.
LSUK	Linnaean Collection, Linnaean Society of London, United Kingdom.
MCZ	Museum of Comparative Zoology, Harvard University, Cambridge, Massachusetts, USA.
MSNV	Museo Civico Di Storia Naturale Di Verona, Ferrara, Italy.
NCSM	North Carolina State Museum of Natural Sciences, Raleigh, North Carolina, USA.
NMB	National Museum at Budapest, Hungary.
NMNH	United States National Museum of Natural History (Smithsonian Institution), Washington DC, USA.
NMW	Naturhistorisches Museum, Wien, Vienna, Austria.
SAC	Private collection of Stanely I. Auerbach.
SDSNH	San Diego Natural History Museum, San Diego, California, USA.
UC	University of California, Department of Zoology, Berkeley, California, USA.
UKNHM	University of Kansas, Natural History Museum & Biodiversity Research Center, Lawrence, Kansas, USA.
UMNH	Utah Museum of Natural History, University of Utah, Salt Lake City, Utah, USA.

UP	University of Pardova, Italy.
UPA	University of Philadelphia, Philadelphia, Pennsylvania, USA.
VMNH	Virginia Museum of Natural History, Martinsville, Virginia, USA.
ZMUC	Zoological Museum, University of Copenhagen, Denmark.

Catalog of Extant Centipedes

Class **Chilopoda** (Latrielle, 1817)
Subclass **Notostigmophora** (Pocock, 1902)
Superorder **Anamorpha** Haase, 1880
Order **Scutigeromorpha** Pocock, 1895
Family **Scutigeridae** (Gervais, 1837)

Genus *Allothereua* Verhoeff, 1905
[1 species (Introduced); USA: Hawaii]

Allothereua Verhoeff, 1905a: 101-102 (characters). TYPE SPECIES: *Cermatia maculata* Newport, 1844, fixed by original designation.
— Jeekel, 2005: 1 (list).

A. lesueurii (Lucas, 1840)
— Lucas, 1840: 538 (*Scutigera*; characters; "Nouvelle Hollande").
— Haase, 1887: 21 (*Scutigera*;). [**Note:** Reference not obtained].
— Gervais, 1897: 223 (*Scutigera*;). [**Note:** Reference not obtained].
— Attems, 1914: 48, 56 (*Scutigera*; list).
— Chamberlin, 1920c: 80, 245 (*Allothereua*; cited as unknown subgenus; list; note; Australia). [**Note:** Chamberlin notes "The position of this species must be doubtful until it is restudied."].
straba Wood, 1862: 11 (*Cermatia*; characters; USA: Hawaii). **HOLOTYPE:** NMNH, No. 252, 1 specimen, Hawaii: Oahu, on the North Pacific Exploration Expedition, by W. Stimpson, M.D.
— Chamberlin, 1920c: 82, Table pg. 245 (*Scutigera*; cited as unknown subgenus; listed as species from Australian region).
— Haase, 1887: 21 (synonym of *lesueurii*). [**Note:** synonymy taken from Chilobase, original reference not obtained].
— Nishida, 1994: 27 (*Scutigera*; checklist). [**Note:** Nishida lists this species as "adventive" or in other words as an accidental introduction].

— Nishida, 1997: 195 (*Scutigera*; checklist). [**Note:** Nishida lists this species as "adventive" or in other words as an accidental introduction].

Distribution: USA: ?Hawaii.

HOLOTYPE: ?Unknown. Australia: Queensland: Rockhampton. [**Note:** MCZ has 3 specimens in their general collection from Australia: Queensland; catalog #'s: 32576, 32577 & 32580].

Genus *Scutigera* Lamarck, 1801
[6 species (1 Introduced); Canada & USA]

Scutigera Lamarck, 1801: 182 (characters). TYPE SPECIES: *Scolopendra coleoptrata* Linnaeus, 1758, fixed by subsequent designation of Latrielle, 1804: 57. [**Note:** See Jeekel (2005: 4) for alternate type-species information].
— Bailey, 1928b: 22 (notes).
— Haupt, 1979: 398, 399 (notes on eyes, phylogenetics).
— Jeekel, 2005: 4, 6 (list; note).
Cermatia Illiger, 1807: 199 (characters). TYPE SPECIES: *Scolopendra lineata* Rossi, 1790, fixed by monotypy. [**Note:** Reference not obtained].
— Chamberlin, 1920f: 203 (note).
— Jeekel, 2005: 2, 6 (list).
Cryptomera Rafinesque, 1820: 7 (characters). TYPE SPECIES: *Cryptomera nemura* Rafinesque, 1820, fixed by subsequent designation.
— Hoffman & Crabill, 1953: 75-76, 78 (synonym of *Scutigera*; designation of type species; taxonomic discussion). [**Note:** Hoffman & Crabill have considered this genus "Isogenotypic with *Scutigera* through synonymy"].
— Jeekel, 2005: 2, 6 (list).
Selista Rafinesque, 1820: 7 (characters). TYPE SPECIES: *Selista forceps* Rafinesque, 1820, fixed by monotypy.
— Wood, 1862: 9 (synonym of *Cermatia*).
— Hoffman & Crabill, 1953: 75 (taxonomic discussion).
— Jeekel, 2005: 5, 6 (list).

♀*S. buda* Chamberlin, 1944
— Chamberlin, 1944d: 213-214, pl. 17; Fig. 25 (*Scutigera*; characters; USA: Texas).
— Reddell, 1965: 166 (*Scutigera*; checklist; note; USA: Texas).

Distribution: USA: Texas.

HOLOTYPE: FMNH, ♀, Texas: Hays Co., 10 mi. N of Buda, 23 April 1941, by K.P. Schmidt.

PARATYPES: NMNH, 2 specimens, Texas: (Stagr Co.?), 10 mi. NW of Buda, 29 April 1941, K.P. Schmidt.

♂♀*S. coleoptrata* (Linnaeus, 1758)
— Linnaeus, 1758: 638 (*Scolopendra*; characters).
— Lamarck, 1801: 182 (*Scutigera*; characters).
— Say, 1821: 109 (*Cermatia*; note; USA: Florida, Georgia).
— Newport, 1856: 7-8 (*Cermatia*; characters).
— Bollman, 1893: 130, 149 (*Cermatia*; catalogue; note).
— Baker, 1920: 93 (note; CANADA: Ontario).
— Dubuisson, 1928: 49-57; figs. 1-8 (*Scutigera*; blood circulation; tracheal ventilation). [**Note:** In French, but figures are fairly easy to understand].
— Williams & Hefner, 1928: 144 (*Scutigera*; characters; habitat; notes; USA: Ohio).
— Chamberlin, 1931b: 97 (*Scutigera*; USA: Oklahoma).
— Matthews, 1935: 108; Map: 112, pl. 9 (*Scutigera*; characters; habitat; notes; USA: Wisconsin).
— Brimley, 1938: 500 (*Scutigera*; habitat; notes; USA: North Carolina).
— Causey, 1940: 78 (*Scutigera*; characters; habitat; notes; USA: North Carolina).
— Chamberlin & Mulaik, 1940: 158 (*Scutigera*; note; USA: Texas).
— Chamberlin, 1942d: 188 (*Scutigera*; habitat; USA: Missouri).
— Chamberlin, 1943b: 107 (*Scutigera*; USA: Texas).
— Chamberlin, 1944a: 35 (*Scutigera*; USA: Georgia).
— Chamberlin, 1944d: 212 (*Scutigera*; USA: California, Illinois, Maryland, Tennessee).
— Crabill, 1952b: 395-396, pl. 9; Figs. 114-122 (*Scutigera*; characters; USA: northeast).
— Johnson, 1952: 193-200; Tab. 27-28; Map 19; Figs. 27-28 (*Scutigera*; autecology; distribution; experimental feeding diet; habitat; natural history; USA: Michigan).
— Hoffman & Crabill, 1953: 76, 78 (*Scutigera*). [**Note:** mentioned in taxonomic discussion with relation to species described by Rafinesque].
— Crabill, 1955b: 41 (*Scutigera*; note; USA: Missouri).

— Crabill, 1955h: 159 (*Scutigera*; notes; USA: Missouri).
— Crabill, 1958b: 99 (*Scutigera*; note; USA: Wisconsin).
— Branson & Batch, 1967: 86 (*Scutigera*; characters; habitat; note; USA: Kentucky).
— Wray, 1967: 156 (*Scutigera*; list; notes).
— Shelley, 1978: 221 (*Scutigera*; notes).
— Kevan, 1979: 298 (*Scutigera*; synopsis; CANADA). [**Note:** Kevan considered this a transcontinental species].
— Summers, 1979: 690, 695, 697 (*Scutigera*; checklist, key, notes).
— Lee, 1980: 5; Tab. 1 (*Scutigera*; habitat; USA: Ohio).
— Summers et al., 1980: 249 (*Scutigera*; habitat; notes; USA: Illinois).
— Summers et al., 1981: 60-61 (*Scutigera*; habitat; notes; USA: Illinois).
— Shelley 1987: 499 (*Scutigera*; note, USA: North Carolina).
— Snider, 1991: 190 (*Scutigera*; list; note).
— Behan-Pelletier, 1993: 27 (*Scutigera*; list).
— Hoffman, 1995a: 30-31 (*Scutigera*; list; notes).
— Watermolen, 1997: 7 (*Scutigera*; checklist, notes; USA Wisconsin).
— Kraus, 1998: 300; Fig. 22.4 (*Scutigera*; electronmicrograph of tracheal lung, ventral view).
— Kraus, 2001: 112; Fig. 7 (*Scutigera*; dorsal view of left mandible).
— Hilken et al., 2003a: 181-189 (*Scutigera*; hemocytes and interactions with tracheae).
— Hilken et al., 2003b: 169-173 (*Scutigera*; plasmocytes and wound healing).
— Hilken et al., 2003c: 175-184 (*Scutigera*; maxillary organ gland).
— Müller et al., 2003: 191-209 (*Scutigera*; phylogeny of mandibulata; ultrastructure of compound eye).
— Müller, Rosenberg & Meyer-Rochow, 2003: 185-197 (*Scutigera*; interommatidial exocrine glands).
— Negrisolo et al., 2004a: 413-423 (*Scutigera*; mitochondrial genome, gene order rearrangement).
— Negrisolo et al., 2004b: 770-780 (*Scutigera*; mitochondrial genome, myriapod phylogeny).
— Stoev & Geoffroy, 2004: 3 (*Scutigera*; catalogue; distribution).
— Shelley et al., 2005: 39-40 (*Scutigera*; note).
— Meyer-Rochow, et al., 2006 (*Scutigera*; spectral sensitivity of the eye through electrophysiological recordings).

floridana Newport, 1845b: 353 (*Cermatia*; characters in Latin, notes in English: USA: Florida). [**Note:** (*v. in Mus. Brit.*) appears under habitat suggesting the type of this species has a voucher in the British Museum].

— Newport, 1856: 8 (*Cermatia*; characters in Latin).

— Gervais, 1847: 225 (*Scutigera*; characters; note).

— Bollman, 1893: 130 (*Cermatia*; synonym of *forceps*).

forceps Rafinesque, 1820: 7 (*Selista*; characters; USA: "near Baltimore").

— Wood, 1862: 9-10 (*Cermatia*; synonym of *coleoptrata*, characters in Latin & English; notes; USA: District of Columbia, Missouri).

— Wood, 1865: 145-146, pl. 3; Figs. 1, 1a (*Cermatia*; synonym of *coleoptrata*, characters in Latin & English; notes; USA: east of Rocky Mountains).

— Brodie & White, 1883: 67 (*Cermatia*; CANADA).

— Cragin, 1885: 143 (*Cermatia*; habitat; notes; USA: Kansas).

— Underwood, 1885: 149 (*Scutigera*; synonym of *coleoptrata*, habitat; USA: Illinois, Pennsylvania).

— Meinert, 1886a: 171-172 (*Scutigera*; synonym of *coleoptrata*, characters in Latin; notes; USA: Massachusetts, Texas).

— Bollman, 1887b: 264-265, 266 [= 1893: 31-32, 33] (*Scutigera*; characters; list; USA: Indiana).

— Bollman, 1888b: 8 [= 1893: 80] (*Scutigera*; note; USA: Arkansas).

— Bollman, 1888e: 410 [= 1893: 111] (*Scutigera*; USA: Indiana).

— Bollman, 1893: 149 (*Scutigera*; note).

— Kenyon, 1893a: 18 (*Scutigera*; note; USA: Nebraska).

— Kenyon, 1893b: 162 (*Scutigera*; USA: Nebraska).

— Chamberlin, 1910: 366-367 (*Scutigera*; synonym of *coleoptrata*; notes; range; USA: California, New York).

— Gunthorp, 1913: 165 (*Scutigera*; note; USA: Kansas).

— Hewitt, 1914: 219 (*Scutigera*; behavior; notes; CANADA: Ontario).

— Marlatt, 1914: 1-4, Fig. 1, 2 a, b, c (*Scutigera*; adult & juvenile characters; habits; occurrence indoors; distribution notes; envenomation).

— Chamberlin, 1918a: 24 (*Scutigera*; habitat; USA: Tennessee).

— Anonymous, 1920: 35 (*Scutigera*; habitat; USA: California).

— Willey, 1920: 8 (*Cermatia*; notes; CANADA: Quebec).

— Gunthorp, 1921: 89 (*Scutigera*; habitat; USA: Kansas).

— Cameron, 1926: 169-178; tab. 1-4 (*Scutigera*; autotomy; food; regeneration; USA: Kansas, Nebraska).

— Bailey, 1928a: 1081 (*Scutigera*; list).

— Bailey, 1928b: 19, 23 (*Scutigera*; synonym of *coleoptrata*, characters; list; notes; USA: Mississippi, New York, Virginia).

— Back, 1939: 2-3, Fig. 4 (*Scutigera*; habitats; occurrence indoors; notes).
— Hatch, 1939: 189 (*Scutigera*; note; USA: Washington).
— Spencer, 1942: 24 (*Cermatia*; cited as subgenus *Scutigera*; CANADA: British Columbia).
— Auerbach, 1951a: 101-102 (*Scutigera*; characters; habitat; notes; USA: Illinois).
— Jeekel, 2005: 5 (*Selista*; list).
nemura Rafinesque, 1820: 8 (*Cryptomera*; characters; USA: Maryland).
— Hoffman & Crabill, 1953: 76 (*Cryptomera*; synonym of *coleoptrata*).
— Jeekel, 2005: 2 (*Cryptomera*; list).

[**Note:** This is considered a cosmopolitan species that is thought to have originated from Europe and can potentially be found in all states indoors. We may never know exactly when or even if this species was actually introduced. There is a probability that it is actually Holarctic but a brief summary of some known facts follows. Rafinesque described *Selista forceps* in 1820 (now considered a junior synonym of *S. coleoptrata*) and to the author's knowledge this makes him the first person to document the earliest sighting of this species in the United States near Baltimore, Maryland. Newport reported and described *Cermatia floridana* (another junior synonym) in 1845 from Florida. According to Hewitt (1914) *S. coleoptrata* was reported from Pennsylvania as early as 1849. Cragin (1885) mentions that "It seems to feel equally at home in houses, in woods, and in shady ravines" in the state of Kansas. There is no doubt that this species was well established in New York City by the early 1900's as there are ample specimens collected from within the AMNH building(s) and remain deposited in the AMNH collection. In 1939, Hatch reported it from the state of Washington and mentioned that it had been found a number of years earlier elsewhere in the state. Crabill (1955) notes *S. coleoptrata* " . . . has been found to forsake human dwellings commonly in warmer parts of the country." First recorded from Kentucky in 1967 by Branson & Batch (1967). The author has recently (2006) seen specimens collected from out of doors in Central Park, New York City].

Distribution: CANADA: British Columbia, Ontario, Quebec; USA: Arkansas, California, ?Florida, Georgia, Illinois, Indiana, Kansas, Kentucky, Massachusetts, Maryland, Michigan, Missouri, Mississippi, Nebraska, New York, North Carolina, Ohio, Oklahoma, Pennsylvania, Tennessee, Texas, Virginia, Washington, Wisconsin.

HOLOTYPE: Unknown, "Habitat in Hispania". (see Stoev & Geoffroy, 2004).

♀*S. dorothea* Chamberlin, 1943
— Chamberlin, 1943b: 108 (*Scutigera*; characters; USA: Texas).
— Würmli, 1973: 79 (*Scutigera*; synonym of *linceci*). [**Note:** Würmli synonymized this species without studying the type specimen of *linceci* or *dorothea*. Although the type specimen of *linceci* is not known to exist, no one has made the effort to collect a neotype from a plausible type locality in Texas. There is no justification for Würmli to have made the synonymy he did based on specimens of *linceci* from Costa Rica which is a locality more than 1200 miles away, as the crow flies, from the most southern point of Texas to the most northern part of Costa Rica. Therefore, the author does not recognize this synonymy until further investigation.]

Distribution: USA: Texas.

HOLOTYPE: NMNH, Texas: Brewster Co., 4 mi. W of Alpine, 2 June 1941.

♀*S. homa* Chamberlin, 1942
— Chamberlin, 1942a: 10-11; Fig. (*Scutigera*; characters; USA: Arizona).
— Würmli, 1973: 79 (*Scutigera*; synonym of *linceci*). [**Note:** Würmli synonymized this species without studying the type specimen of *linceci* or *homa*. Although the type specimen of *linceci* is not known to exist, no one has made the effort to collect a neotype from a plausible type locality in Texas. There is no justification for Würmli to have made the synonymy he did based on specimens of *linceci* from Costa Rica which is a locality more than 2000 miles away, as the crow flies, from the most southeastern point of Arizona to the most northern part of Costa Rica. Therefore, the author does not recognize this synonymy until further investigation].

Distribution: USA: Arizona. [**Note:** The author has seen photos of a specimen collected from the Superstition Mountains, Arizona].

HOLOTYPE: ?Unknown, ♀, Arizona, 22 miles SE of Ajo, 3 January 1941, by S. & D. Mulaik. [**Note:** Chamberlin noted that the type was "in the writer's collection", but has apparently not been deposited at the NMNH.

The author has taken what is believed to be a specimen of this species from under a dampened bail of hay from Vail, Arizona, which is deposited at the AMNH.]

S. linceci (Wood, 1867)
— Wood, 1867a: 42-43 (*Cermatia*; characters; habitat; USA: Texas).
— Underwood, 1885: 149 (*Scutigera*; note).
— Bollman, 1887b: 265, 266 [= 1893: 32, 33] (*Scutigera*; list; note).
— Bollman, 1893: 130 (catalogue).
— Pocock, 1895: 1-2; Tab. 1; Figs. 1, 1a, b (*Scutigera*; characters; habitat, USA: Texas).
— Gunthorp, 1920: 113 (*Cermatia*; list).
— Chamberlin, 1943b: 108 (*Scutigera*; USA: Texas).
— Würmli, 1973: 76-79; Figs. 1-5 (*Scutigera*; redescription). [**Note:** Würmli redescribed this species without studying the type specimen of *linceci*. Although the type specimen of *linceci* is not known to exist, no one has made the effort to collect a neotype from a plausible type locality in Texas. There is no justification for Würmli to have made the redescription he did based on specimens of *linceci* from Costa Rica which is a locality more than 1200 miles away, as the crow flies, from the southern most point of Texas to the most northern part of Costa Rica. Therefore, the author does not recognize this redescription until further investigation].
— Stoev & Geoffroy, 2004: 4 (*Scutigera*; catalogue). [**Note:** Stoev & Geoffroy list material from Costa Rica and Guatemala].

Distribution: USA: Arizona, Texas.

SYNTYPES: Unknown. (see Stoev & Geoffroy, 2004)

♀*S. phana* Chamberlin, 1943
— Chamberlin, 1943b: 108 (*Scutigera*; characters; USA: Texas).
— Würmli, 1973: 79 (*Scutigera*; synonym of *linceci*). [**Note:** Würmli synonymized this species without studying the type specimen of *linceci* or *phana*. Although the type specimen of *linceci* is not known to exist, no one has made the effort to collect a neotype from a plausible type locality in Texas. There is no justification for Würmli to have made the synonymy he did based on specimens of *linceci* from Costa Rica which is a locality more than 1200 miles away, as the crow flies, from the most southern point of Texas to the most northern part of Costa Rica. Therefore,

the author does not recognize this synonymy until further investigation].

Distribution: USA: Texas.

HOLOTYPE: NMNH, ♀, Texas: Hidalgo Co., Edinburg, 8 June 1941.

Subclass **Pleurostigmophora** (Pocock, 1902)
Superorder **Anamorpha** Haase, 1880
Order **Lithobiomorpha** Pocock, 1891
Family **Henicopidae** Pocock, 1901
Subfamily **Henicopinae** Pocock, 1901
Tribe **Henicopini** Pocock, 1901

Genus *Lamyctes* Meinert, 1868
[8 species (3 Introduced); Canada, Greenland & USA]

Lamyctes Meinert, 1868: 266-267 (characters; notes) Type Species: *Lamyctes fulvicornis* Meinert, 1868, [= *Lamyctes emarginatus* (Newport, 1844)] fixed by original designation. [**Note:** junior subjective synonym of *Henicops emarginatus* Newport, 1844 fide Eason, 1996 (see Hollington & Edgecombe, 2004)].
— Chamberlin, 1912g: 5-6 (characters; key to species of the United States).
— Auerbach, 1952b: 413-414 (taxonomic notes).
— Edgecombe & Giribet, 2003a: 2-3 (assigned species list including synonyms; notes).
— Jeekel, 2005: 20 (list).

♂♀*L. africanus* (Porat, 1871)
— Porat, 1871: 140 (*Henicops;*). [**Note:** Reference not obtained].
— Porat, 1894: 10 (*Henicops;*). [**Note:** Reference not obtained].
— Attems, 1928: 56-59; Fig. 11-14 (*Lamyctes;* characters; distribution). [**Note:** The text of this reference is actually in English and is useful for identifying this species].
— Zapparoli & Shelley, 2000: 35-36 (*Lamyctes;* characters; notes; USA: Hawaii).
insignis Pocock, 1891b: 154 (*Henicops;* characters; notes).
— Attems, 1928: 56 (*Henicops;* synonym of *africanus*).

Distribution: USA: ?Hawaii. [**Note:** This species is introduced to Hawaii. It is indigenous to southern Africa and Madagascar].

HOLOTYPE: ?Unknown.

♀*L. caducens* Chamberlin, 1938

— Chamberlin, 1938b: 625-626 (*Lamyctes*, characters; USA: New
Mexico).
— Chamberlin & Mulaik, 1940: 126 (*Lamyctes*, list).
— Hollington & Edgecombe, 2004: 23 (*Lamyctes*, synoptic
classification).

Distribution: USA: New Mexico.

HOLOTYPE: NMNH, ♀, New Mexico: Bear Canyon, Camp Mary White,
August 1934, by S. Mulaik.

♂♀**L. coeculus** (Brölemann, 1889)
— Brölemann, 1889: 273-276; Fig. 1 (*Lithobius*, characters (adult,
stadia); Italy). [**Note:** Interestingly, Brölemann has 3 males to 104
females; a ratio of less than 3% males to females. This is not surprising
because this species is known to be parthenogenetic].
— Silvestri, 1909: 39; Fig. 1 (*Lamyctinus*,). [**Note:** Reference not obtained].
— Chamberlin, 1920c: 72, 243 (*Lamyctinus*; USA: Hawaii).
— Attems, 1928: 62 (*Lamyctinus*; distribution; USA: Hawaii).
— Brölemann, 1932: ? (*Lamyctes*;). [**Note:** Reference not obtained].
— Attems, 1938b: 366, 369 (*Lamyctinus*; list; note).
— Chamberlin, 1930c: 68 (*Lamyctinus*; USA: Hawaii). [**Note:** The
report was of this species found in soil about plants, which originated
from Tahiti on two separate dates].
— Auerbach, 1952b: 413 (*Lamyctinus*; notes; USA: Illinois).
— Enghoff, 1975: 45-46 (*Lamyctes*; notes). [**Note:** It is interesting to
note that Enghoff suspects this species is thelytokous].
— Andersson, 1979: 73, 74, 77, 78, 79; Fig. 1; tab. 1-5 (*Lamyctinus*;
characters of larval stadia).
— Nishida, 1994: 26 (*Lamyctinus*; checklist).
— Nishida, 1997: 195 (*Lamyctinus*; checklist).
— Zapparoli & Shelley, 2000: 37 (*Lamyctes*; characters; notes; USA:
Hawaii).
— Edgecombe & Giribet, 2003a: 4-6; Figs. 1-4, 38 (*Lamyctes*;
cladograms; molecular and morphological discussion).
— Edgecombe, 2004b: 33; Fig. 3 (*Lamyctes*; scanning electron
micrograph of pretarsal claw).

Distribution: USA: Hawaii.

HOLOTYPE: ?Unknown.

♀*L. diffusus* Chamberlin & Mulaik, 1940
— Chamberlin & Mulaik, 1940: 126 (*Lamyctes*, characters; USA: Texas). [**Note:** The holotype lacks the posterior legs!]
— Hollington & Edgecombe, 2004: 23 (*Lamyctes*, synoptic classification).

Distribution: USA: Texas.

HOLOTYPE: NMNH, ♀, Texas: Edinburg, December 1939.

♂♀*L. emarginatus* (Newport, 1844)
— Newport, 1844: 96 (*Henicops*; characters).
— Hollington & Edgecombe, 2004: 23 (*Henicops*; synoptic classification; cited as senior synonym of *Lamyctes fulvicornis*).
fulvicornis Meinert, 1868: 267-268 (*Lamyctes*; characters).
— Meinert, 1872: 343-344 (*Lamyctes*; characters; distribution).
— Stuxberg, 1875a: 72 (*Lamyctes*; USA: New York).
— Stuxberg, 1875b: 18 (*Lamyctes*; list).
— Stuxberg, 1875c: 31 [=1876: 138] (*Lamyctes*; list).
— Latzel, 1880: 133; Pl. 4; Figs. 31-34 (*Henicops*; characters (adult, stadia); distribution).
— Bollman, 1887b: 265 [= 1893: 32] (*Henicops*; list).
— Bollman, 1888b: 7 [= 1893: 78] (*Henicops*; notes; USA: Arkansas). [**Note:** Bollman mentions that this species was also recorded from Mount Lebanon, New York].
— Bollman, 1893: 130, 184 (*Henicops*; catalogue, notes; USA: Minnesota).
— Chamberlin, 1901: 25. (*Henicops*; synonym of *pinampus* in part, see: Chamberlin, 1902b; USA: Arkansas, Minnesota, New York, Utah)
— Chamberlin, 1902b: 797 (*Henicops*; key to species; notes; USA: Utah).
— Chamberlin, 1903c: 335 (*Henicops*; Chamberlin notes that this species is transferred from *Henicops* to *Lamyctes* based on work of Pocock; range; USA: Idaho, Oregon, Utah).
— Chamberlin, 1909b: 191 (*Lamyctes*; USA: Nevada, Oregon, Utah).
— Chamberlin, 1910: 368; Fig. 132a (*Lamyctes*; characters; USA: Oregon, ?California). [**Note:** Chamberlin suggests it will be found in California].
— Chamberlin, 1911a: 67 (*Lamyctes*; USA: Colorado).
— Chamberlin, 1911b: 48, pl. 3; Fig. 1 (*Lamyctes*; USA: Wisconsin).

— Chamberlin, 1911c: 98 (*Lamyctes*; notes; USA: Illinois, Michigan, Minnesota, Nebraska, Wisconsin).

— Chamberlin, 1912g: 7-10, pl. 1; Figs. 1-12, pl. 2; Figs. 1-3 (*Lamyctes*; characters (adult, agenitalis); habitat; notes; USA: Arkansas, Colorado, Idaho, Illinois, Maine, Massachusetts, Michigan, Minnesota, Nebraska, New York, Oregon, Pennsylvania, Utah, Wisconsin).

— Attems, 1914: 93 (*Lamyctes*; distribution; list).

— Williams & Hefner, 1928: 138 (*Lamyctes*; characters; USA: Ohio).

— Chamberlin, 1930a: 111 (*Lamyctes*; USA: Utah).

— Chamberlin, 1930c: 68 (*Lamyctes*; USA: Hawaii). [**Note:** The report was of this species found in soil about plants, which originated from Los Angeles, California].

— Chamberlin, 1946c: 183-184 (*Lamyctes*; USA: Alaska).

— Crabill, 1952b: 264-265, pl. 6; Fig. 78 (*Lamyctes*; characters; habitat; range; subgenus *Lamyctes*; CANADA: Ontario; USA: Illinois, Maine, Massachusetts, Michigan, New York, Ohio, Pennsylvania, Wisconsin).

— Johnson, 1952: 146-147 (*Lamyctes*; autecology; distribution; notes).

— Palmén, 1954: 143-144 (*Lamyctes*; notes; CANADA: Newfoundland).

— Crabill, 1958b: 97 (*Lamyctes*; distribution; habitat; notes; USA: Wisconsin).

— Eason, 1964: 247-251; Figs. 477-487 (*Lamyctes*; characters; distribution; habitat).

— Judd, 1964: 2 (*Lamyctes*; note; CANADA: Ontario).

— Andersson, 1979: 77, 78, 79; tab. 1-5 (*Lamyctes*; characters of larval stadia).

— Kevan, 1979: 298 (*Lamyctes*; CANADA: Newfoundland).

— Summers, 1979: 696, 697 (*Lamyctes*; checklist; key).

— Summers & Uetz, 1979; Tab. 1, 2: 348-349, 351 (*Lamyctes*; microhabitat, notes, USA: Illinois).

— Summers et al., 1980: 249; Fig. 16 (*Lamyctes*; checklist; USA: Illinois).

— Summers et al., 1981: 61 (*Lamyctes*; checklist; USA: Illinois).

— Böcher & Enghoff, 1984: 49-50. (*Lamyctes*; habitat; notes; Greenland). [**Note:** This species is almost certainly introduced].

— Snider, 1991: 189 (*Lamyctes*; list; note).

— Behan-Pelletier, 1993: 26 (*Lamyctes*; list).

— Hoffman, 1995a: 27 (*Lamyctes*; list; notes). [**Note:** Hoffman suggests that this species will be found in Virginia].

— Watermolen, 1997: 7 (*Lamyctes*; checklist, notes; USA: Wisconsin).

— Edgecombe & Giribet, 2003a: 2, 3 (*Lamyctes*; synonym of *emarginatus*).

— Jensen, 2003: 68 (*Lamyctes*; Greenland).

— Hollington & Edgecombe, 2004: 23 (*Lamyctes*; cited as junior subjective synonym of *Henicops emarginatus*).

— Jeekel, 2005: 20 (*Lamyctes*; list).

hawaiiensis Silvestri, 1904: 325; Figs. 1–2 (*Lamyctes*; nominal taxon *fulvicornis*; characters; note; USA: Hawaii).

— Attems, 1909?b: 8 (*Lamyctes*; nominal taxon *fulvicornis*;). [**Note:** Reference not obtained].

— Attems, 1914: 48, 56, 93 (*Lamyctes*; nominal taxon *fulvicornis*; list).

— Chamberlin, 1920c: 71, 243; Tab. (*Lamyctes*; nominal taxon *fulvicornis*; list).

— Attems, 1928: 61 (*Lamyctes*; nominal taxon *fulvicornis*; list).

— Attems, 1938b: 366, 369 (*Lamyctes*; list; note).

— Nishida, 1994: 26 (*Lamyctes*; nominal taxon *fulvicornis*; list).

— Nishida, 1997: 195 (*Lamyctes*; nominal taxon *fulvicornis*; list).

— Zapparoli & Shelley, 2000: 36 (*Lamyctes*; nominal taxon *fulvicornis*; synonym of *emarginatus*).

Distribution: CANADA: Newfoundland, Ontario; GREENLAND; USA: Alaska, Arkansas, ?California, Colorado, Hawaii, Idaho, Illinois, Massachusetts, Maine, Michigan, Minnesota, Nebraska, Nevada, New York, Ohio, Oregon, Pennsylvania, Utah, Wisconsin.

HOLOTYPE: Unknown.

LECTOTYPE: BMNH, Hawaii: Kona, 4,000 ft., August 1892, Perkins.

PARALECTOTYPE: BMNH, Hawaii: Kona, 4,000 ft., August 1892, Perkins.

♀*L. pinampus* Chamberlin, 1910

— Chamberlin, 1910: 368; Fig. 132b (*Lamyctes*; characters; USA: California, Nevada)

— Chamberlin, 1911b: 48, pl. 3; Fig. 3 (*Lamyctes*; USA: California).

— Chamberlin, 1912g: 15-18, pl. 2; Figs. 4-6 (*Lamyctes*; characters (adult, stadia); USA: California, Nevada).

— Anonymous, 1920: 35 (*Lamyctes*; USA: California).

— Chamberlin, 1928d: 307 (*Lamyctes*; USA: Utah).
— Chamberlin, 1958c: 133 (*Lamyctes*; checklist; note).
— Hollington & Edgecombe, 2004: 24 (*Lamyctes*; synoptic classification).
fulvicornis Chamberlin (not Meinert), 1909b: 191 (*Lamyctes*; synonym of
pinampus in part; USA: Idaho, Nevada, Oregon, Utah).
— Chamberlin, 1912g: 15 (*Lamyctes*; synonym of *pinampus* in part).

Distribution: USA: California, ?Idaho, Nevada, ?Oregon, ?Utah. [**Note:**
Because some of the specimens reported by Chamberlin in 1909b as *fulvicornis*
are in part *pinampus* and he doesn't indicate which specimens and from what
states I have questioned those states never reported to have *pinampus*. Perhaps
Chamberlin's *pinampus* is actually *emarginatus*].

HOLOTYPE: ?Unknown, Nevada: Las Vegas or California: Claremont.
[**Note:** Chamberlin does not indicate which state the types are from].

L. pius Chamberlin, 1911
— Chamberlin, 1911b: 33 (*Lamyctes*; nominal taxon *tivius*; cited as
variety *pius*; characters; USA: North Carolina).
— Chamberlin, 1912g: 13-15, pl. 2; Figs. 7-8 (*Lamyctes*; characters;
notes; USA: Georgia, New Jersey, North Carolina, Pennsylvania).
— Matthews, 1935: 98-102; Map, 103 pl. 8; Fig. 1-7 (*Lamyctes*;
characters; USA: Wisconsin).
— Brimley, 1938: 500 (*Lamyctes*; nominal taxon *tivius*; cited as variety
pius; list).
— Causey, 1940: 67 (*Lamyctes*; characters; notes).
— Crabill, 1952b: 265-267 (*Lamyctes*; characters; habitat; taxonomic
notes; subgenus *Lamyctes*; USA: New York).
— Wray, 1967: 154 (*Lamyctes*; list).
— Hoffman, 1995a: 27 (*Lamyctes*; list; notes; USA: Virginia).
— Watermolen, 1997: 7 (*Lamyctes*; checklist, notes).
— Hollington & Edgecombe, 2004: 24 (*Lamyctes*; synoptic classification).

Distribution: USA: Georgia, North Carolina, New Jersey, New York,
Pennsylvania, Virginia, Wisconsin.

SYNTYPES: MCZ, 2 specimens, No. 14515, North Carolina: Madison,
Hot Springs, R.V. Chamberlin coll. [**Note:** The MCZ also has 4 specimens
from North Carolina collected 6-8-1910, catalog # 27770 and Chamberlin
refers to "length of specimens" meaning he had more than one. I suspect
these specimens are part of the type series from Chamberlin].

♀*L. tivius* Chamberlin, 1911
— Chamberlin, 1911b: 33, pl. 3; Fig. 2 (*Lamyctes*; characters; USA: Georgia, Louisiana, Mississippi, North Carolina). [**Note:** Chamberlin doesn't specify exactly which locality of the six he lists the type specimen came from, but it is assumed to be from one of the two in Mississippi as this is the first state locality listed. Furthermore, Chamberlin originally reported this species from Hot Springs, North Carolina but he later (1912g) regards this report as an error and that all of these specimens belong to *L. pius*].
— Chamberlin, 1912g: 10-13; Figs. 9-10 (*Lamyctes*; characters (adult, agenitalis I, pullus IV); notes; USA: Alabama, Louisiana, Mississippi). [**Note:** The record from New Orleans, Louisiana Chamberlin refers to as "variety a", but it is most likely just a geographic color variation. In fact, based on Chamberlin's key it is very likely that all three species, *tivius*, *pinampus* and *pius* are simply *emarginatus* introduced from Europe].
— Brimley, 1938: 500 (*Lamyctes*; list).
— Wray, 1967: 154 (*Lamyctes*; list).
— Hollington & Edgecombe, 2004: 24 (*Lamyctes*; synoptic classification).

Distribution: USA: Alabama, Georgia, Louisiana, Mississippi.

HOLOTYPE: ?Unknown, ?Mississippi: Byram or Holly Springs.

Genus *Pleotarsobius* Attems, 1909
[1 species; USA: Hawaii]

Pleotarsobius Attems, 1909: 12 (characters). **TYPE SPECIES:** *Lamyctes heterotarsus* Silvestri, 1904, fixed by monotypy.
— Attems, 1928: 63-64 (characters). [**Note:** Characters in English].
— Jeekel, 2005: 28 (list).

♀*P. heterotarsus* (Silvestri, 1904)
— Silvestri, 1904: 325-326, pl. 11; Figs. 3-4 (*Lamyctes*; characters; notes; USA: Hawaii).
— Attems, 1909b: 12 (*Pleotarsobius*;). [**Note:** Reference not obtained].
— Attems, 1914: 48, 56, 94 (*Pleotarsobius*; list).
— Chamberlin, 1920c: 73, 243 (*Pleotarsobius*; USA: Hawaii).
— Attems, 1928: 64 (*Pleotarsobius*; list).

— Attems, 1938b: 366, 369 (*Pleotarsobius*; list; note).
— Nishida, 1994: 26 (*Pleotarsobius*; checklist).
— Nishida, 1997: 195 (*Pleotarsobius*; checklist).
— Zapparoli & Shelley, 2000: 37 (*Pleotarsobius*; characters; notes).
— Hollington & Edgecombe, 2004: 24 (*Lamyctes*; synoptic classification).
— Jeekel, 2005: 28 (*Lamyctes*; list).

Distribution: USA: Hawaii.

HOLOTYPE: ?Unknown, Hawaii: Kona, (Perkins).

Subfamily **Zygethobiinae** Chamberlin, 1912
Tribe **Zygethobiini** Chamberlin, 1912

Genus *Buethobius* Chamberlin, 1911
[5 species; Eastern & Western USA]

Buethobius Chamberlin, 1911b: 34 (characters; note). TYPE SPECIES: *Buethobius oabitus* Chamberlin, 1911, fixed by original designation and monotypy.
— Chamberlin, 1912g: 18-19 (characters; key to species).
— Crabill, 1952b: 267-268 (characters; taxonomic notes).
— Jeekel, 2005: 13 (list).

♂♀*B. arizonicus* Chamberlin, 1925
— Chamberlin, 1925a: 53 (*Buethobius*; characters; USA: Arizona).
— Hollington & Edgecombe, 2004: 25 (*Buethobius*; synoptic classification).

Distribution: USA: Arizona.

HOLOTYPE: NMNH, ♀, Arizona: Santa Catalina Mts., 2 June 1924, by R. H. Peebles.

PARATYPES: NMNH, ♂♀, Arizona: Santa Catalina Mts., 2 June 1924, by R. H. Peebles.

♂♀*B. coniugans* Chamberlin, 1911
— Chamberlin, 1911f: 377, 383-384 (*Buethobius*; characters; notes; USA: California).

— Chamberlin, 1912g: 22-25, pl. 4; Fig. 3 (*Buethobius*; characters (pseudomaturus, praematurus; USA: California).

— Hollington & Edgecombe, 2004: 25 (*Buethobius*; synoptic classification).

Distribution: USA: California.

HOLOTYPE: NMNH, California: Berkeley or Mill Valley. [**Note:** Author found vial in type collection with the following data: California: 12 mi. NE Hammond, 21 March 1941. The identification had a question mark after the specific name and there are about 12 specimens. The actual type has not been located at this time].

♂♀*B. huestoni* Williams & Hefner, 1928

— Williams & Hefner, 1928: 139; Fig. 26 (*Buethobius*; characters; USA: Ohio).

— Crabill, 1952b: 269, 385 (*Buethobius*; characters; notes). [**Note:** "The affinity and identity of . . ." *Buethobius huestoni* Williams & Hefner " . . . is uncertain due to the inadequate original description[s] and the unavailablity of the type."].

— Summers, 1979: 695, 697 (*Buethobius*; checklist; key).

— Summers et al., 1980: 249; Fig. 16 (*Buethobius*; checklist; USA: Illinois).

— Summers et al., 1981: 61 (*Buethobius*; checklist; USA: Illinois).

— Hollington & Edgecombe, 2004: 25 (*Buethobius*; synoptic classification).

Distribution: USA: Illinois, Ohio.

HOLOTYPE: ?Unknown, Ohio: Hueston's Woods.

♂♀*B. oabitus* Chamberlin, 1911

— Chamberlin, 1911b: 34 (*Buethobius*; characters; USA: Mississippi).

— Chamberlin, 1912g: 20-21, pl. 3; Figs. 1-10, pl. 4; Figs. 1-2 (*Buethobius* ; characters; USA: Mississippi).

— Hollington & Edgecombe, 2004: 25 (*Buethobius*; synoptic classification).

— Jeekel, 2005: 13 (*Buethobius*; list).

Distribution: USA: Mississippi.

SYNTYPE: NMNH, Mississippi: Byram or Canton.

♀*B. translucens* Williams & Hefner, 1928
— Williams & Hefner, 1928: 139 (*Buethobius*; characters; USA: Ohio).
— Crabill, 1952b: 268-269, 385 (*Buethobius*; characters). [**Note:** "The affinity and identity of..." *Buethobius translucens* Williams & Hefner "... is uncertain due to the inadequate original description[s] and the unavailablity of the type."].
— Summers, 1979: 696, 697 (*Buethobius*; checklist; key).
— Hollington & Edgecombe, 2004: 25 (*Buethobius*; synoptic classification).

Distribution: USA: Ohio.

HOLOTYPE: ?Unknown, ♀, Ohio: Columbus, Ohio State University.

Genus *Yobius* Chamberlin, 1945
[1 species; Western USA]

Yobius Chamberlin, 1945a: 153-154 (characters). TYPE SPECIES: *Yobius haywardi* Chamberlin, 1945, fixed by original designation and monotypy.
— Jeekel, 2005: 34 (list).

♀*Y. haywardi* Chamberlin, 1945
— Chamberlin, 1945a: 154. (*Yobius*; characters; USA: Utah). [**Note:** The single female specimen Chamberlin used to describe this genus and species lacks the ultimate legs. This species has no ocelli and is similar to *Buethobius* but is different in having biarticulate tarsi on all legs].
— Hollington & Edgecombe, 2004: 25 (*Yobius*; synoptic classification).
— Jeekel, 2005: 34 (*Yobius*; list).

Distribution: USA: Utah.

HOLOTYPE: NMNH, ♀, Utah: Utah Co., on "Y" Mountain E of Provo, 12 May 1944, by C. L. Hayward.

Genus *Zygethobius* Chamberlin, 1903
[5 species; Canada and Eastern/Western USA]

Zygethobius Chamberlin, 1903c: 335 (name change from *Henicops dolichopus* Chamberlin to *Zygethobius dolichopus* (Chamberlin)). **TYPE SPECIES:** *Henicops dolichopus* Chamberlin, 1901, fixed by original designation and monotypy. [**Note:** During a visit by the author to the MCZ in August 2006, a holotype of a possible new species was found, "*Zygethobius montana* Chamb.," however the author knows no publication of this species. **HOLOTYPE:** MCZ, No. 381, Unique No. 14441, Montana: Flathead Lake, Summer 1912, C.C. Adams. Original locality/determined label reads "*Zygethobius* sp. n., Montana: Flathead Lake, Summer 1912, C.C. Adams"; Holotype vial at MCZ was found dry with a slight hint of moisture. It is not clear if the actual species is *montana* because this was only written by Crabill's label, which reads "*Zygethobius montana* Chamb., TC-201, Holotype"].

— Chamberlin, 1912g; 25-27 (characters; key to subgenera and species; notes).
— Bailey, 1928a: 1082 (list).
— Bailey, 1928b: 19, 32-33 (list; notes). [**Note:** Bailey reports of a *Zygethobius* sp. from New York at the McLean Bogs Reservation, 24 April 1924, by C. R. Crosby.]
— Crabill, 1952b: 269-270 (characters; range; taxonomic notes).
— Edgecombe & Giribet, 2003b: 21, 23, 27; Fig. 4 (systematics).
— Jeekel, 2005: 35 (list).

Zantethobius Chamberlin, 1911b: 34 (characters). **TYPE SPECIES:** *Zygethobius pontis* Chamberlin, 1911, fixed by original designation.

— Crabill, 1981c: 360 (synonym of *Zygethobius*).
— Hollington & Edgecombe, 2004: 25 (synoptic classification; synonym of *Zygethobius*).
— Jeekel, 2005: 34 (list).

♀*Z. columbiensis* Chamberlin, 1912

— Chamberlin, 1912g: 33-34 (*Zygethobius*; characters; CANADA: British Columbia).
— Hollington & Edgecombe, 2004: 25 (*Zygethobius*; synoptic classification).

Distribution: CANADA: British Columbia.

HOLOTYPE: MCZ, Unique No. 14293, ♀, CANADA: British Columbia, Powder Creek, Kaslo, [?W.P. Currie].

♂♀*Z. dolichopus* (Chamberlin, 1902)

— Chamberlin, 1902b: 797-800 (*Henicops*; characters (adult, stadia);
habitat; key to species; range; USA: Utah). [**Note:** According to
Chamberlin the specimens he reported of *Henicops fulvicornis* in
1901 were mostly *dolichopus*].

— Chamberlin, 1903c: 335 (*Zygethobius*; transferred to the new genus
Zygethobius from *Henicops*).

— Chamberlin, 1910: 368 (*Zygethobius*; habitat; notes; USA: California,
Utah).

— Chamberlin, 1911f: 383 (*Zygethobius*; notes).

— Chamberlin, 1912g: 27-30 (*Zygethobius*; characters (adult,
praematurus, agenitalis, pullus IV); habitat; USA: California, Utah).

— Hollington & Edgecombe, 2004: 25 (*Henicops*; synoptic
classification).

— Jeekel, 2005: 35 (*Henicops*; list).

fulvicornis Chamberlin (not Meinert), 1901: 25 (*Henicops*; habitat; note;
USA: Utah).

— Chamberlin, 1902b: 797 (misidentification in part).

Distribution: USA: California, Utah.

HOLOTYPE: NMNH, No. 787, Utah: Wahsatch Mountains, elevations
from 6,000 to 10,000 feet above sea level.

♀*Z. ecologus* Chamberlin, 1938

— Chamberlin, 1938b: 625 (*Zygethobius*; characters; note; USA:
Oregon).

— Hollington & Edgecombe, 2004: 25 (*Zygethobius*; synoptic
classification).

Distribution: USA: Oregon.

HOLOTYPE: NMNH, ♀, Oregon: Boyer, June or July 1937, by J. A.
Macnab.

♀*Z. pontis* Chamberlin, 1911

— Chamberlin, 1911b: 34-35, pl. 4; Fig. 2 (*Zygethobius*; subgenus
Zantethobius; characters; USA: Tennessee, Virginia).

— Chamberlin, 1912g: 35-36, pl. 4; Fig. 9, pl. 5; Fig. 9 (*Zygethobius*;
characters; USA: Tennessee, Virginia).

— Crabill, 1952b: 270-272 (*Zygethobius*; characters; taxonomic notes;
subgenus *Zantethobius*).

— Branson & Batch, 1967: 87 (*Zygethobius*; brief description; habitat; USA: Kentucky).
— Lee, 1980: 4, 5; Tab. 1 (*Zygethobius*; habitat; USA: Ohio).
— Crabill, 1981c: 360 (*Zygethobius*; notes; taxonomic discussion).
— Hoffman, 1995a: 26-27 (*Zygethobius*; list; notes).
— Edgecombe & Giribet, 2003b: 14, 23; Figs. 1-3; Tab. 1-2 (cladograms, morphological character matrix, phylogenetic relationship).
— Edgecombe, 2004a: 66, 68; Fig. 37; Tab. 1 (*Zygethobius*; cladogram; morphological codings).
— Hollington & Edgecombe, 2004: 25 (*Zygethobius*; synoptic classification; cited as subgenus *Zantethobius*).
— Jeekel, 2005: 34 (*Zygethobius*; list).

Distribution: USA: Kentucky, Ohio, Tennessee, Virginia.

SYNTYPE: MCZ, 1 specimen, No. 29900, Virginia: Natural Bridge; 11-7-1910, R.V. Chamberlin coll. [**Note:** Crabill (1981c) states the types cannot be found].

♂♀*Z. sokarienus* Chamberlin, 1911
— Chamberlin, 1911f: 382-383 (*Zygethobius*; characters; habitat; notes; USA: California).
— Chamberlin, 1912g: 30-33, pl. 4; Figs. 4-8, pl. 5; Figs. 1-8 (*Zygethobius*; characters; USA: California).
— Hollington & Edgecombe, 2004: 25 (*Zygethobius*; synoptic classification).

Distribution: USA: California.

HOLOTYPE: NMNH, ♀, California: Mill Valley.

Superfamily **Lithobioidea** Chamberlin, 1915
Family **Lithobiidae** Newport, 1844
Subfamily **Ethopolyinae** Chamberlin, 1915

Genus *Archethopolys* Chamberlin, 1925
[3 species; Southwestern USA]

Archethopolys Chamberlin, 1925c: 412 (no description, designation of type).
TYPE SPECIES: *Ethopolys integer* Chamberlin, 1919, fixed by original
designation.
— Chamberlin, 1930a: 113 (key).
— Jeekel, 2005: 11 (list; note).

♀*A. gosobius* Chamberlin, 1928
— Chamberlin, 1928d: 307-308. (*Archethopolys*; characters; USA:
Utah).
— Chamberlin, 1958c: 133 (*Archethopolys*; checklist; note).

Distribution: USA: Utah.

HOLOTYPE: NMNH, ♀, Utah: San Juan Co., at Devil's Canyon between
Blanding and Monticello, 18 April, by R. V. Chamberlin & W. J. Gertsch.

♂♀*A. kaibabus* Chamberlin, 1930
— Chamberlin, 1930a: 113 (*Archethopolys*; characters; USA: Arizona).

Distribution: USA: Arizona.

HOLOTYPE: NMNH, ♀, Arizona: Kaibab Forest, 7 July 1929, by Lowell
Woodbury.

♀*A. parowanus* (Chamberlin, 1925)
— Chamberlin, 1925e: 57 (*Ethopolys*; characters; USA: Utah).
— Chamberlin, 1930a: 113 (*Archethopolys*; note; USA: Utah).

Distribution: USA: Utah.

HOLOTYPE: NMNH, Utah: Parowan, April and May. [**Note:** NMNH
database has two types listed *parowanus* and *parawanus*].

Genus *Bothropolys* Wood, 1862
[11 species (4 Introduced); Canada & USA]

Bothropolys Wood, 1862: 15 TYPE SPECIES: *Bothropolys nobilis* Wood, 1862,
[=*Bothropolys multidentatus* (Newport, 1845)] fixed by subsequent
designation of Crabill, 1955c: 110.
— Chamberlin, 1902a: 42 (*Lithobius*; cited as subgenus *Bothropolys*; key
to American species).

— Chamberlin, 1903b: 152 (*Lithobius*; cited as subgenus *Bothropolys*; key to species of California and Oregon).
— Chamberlin, 1925c: 387-390; fig. 1 (characters; distribution; key to species; notes).
— Bailey, 1928b: 25 (note).
— Crabill, 1955c: 110 (designation of type).
— Jeekel, 1963: 193-195 (*Bothropolys*; taxonomic notes).
— Matic, 1974: 329 (*Bothropolys*; taxonomic notes).
— Jeekel, 2005: 12 (list).

Bothropolys (*Eubothropolys*) Verhoeff, 1907: 240 (characters; key). TYPE SPECIES: *Lithobius multidentatus* Newport, 1845 fixed by subsequent designation of Jeekel, 2005.
— Jeekel, 2005: 16 (list).

Bothropolys (*Poropolys*) Chamberlin, 1925c: 390, 397 (key to species; characters). TYPE SPECIES: *Lithobius permundus* Chamberlin, 1902, fixed by original designation.
— Eason, 1972b: 142 (synonym of *Bothropolys*).
— Jeekel, 2005: 29 (list).

Lithobius (*Polybothrus*) Latzel, 1880: 35 (characters). TYPE SPECIES: *Lithobius* (*Polybothrus*) *nobilis* Wood, 1862, fixed by direct substitution.
— Crabill, 1955c: 110 (synonym of *Bothropolys*).
— Jeekel, 2005: 28 (list).

Polybothrus Latzel, 1880: 35 (characters). TYPE SPECIES: *Bothropolys nobilis* Wood, 1862, [=*Bothropolys multidentatus* (Newport, 1845)] fixed by objective synonymy.
— Crabill, 1955c: 108-110 (synonym of *Bothropolys*).
— Jeekel, 1963: 193-195 (*Polybothrus*; taxonomic notes; synonym of *Bothropolys*).
— Jeekel, 2005: 12 (list).

♂♀*B. hoples* (Brölemann, 1896)
— Brölemann, 1896: 45-47 (*Lithobius*; subgenus *Bothropolys*; characters; "Washington Territory").
— Chamberlin, 1909b: 191 (*Lithobius*; USA: Oregon)
— Chamberlin, 1925c: 390, 397, pl. 2, fig 9, pl. 3, fig, 1-4 (*Bothropolys*; subgenus *Poropolys*; characters (adult, stadia); notes; CANADA: British Columbia; USA: Idaho, Montana, Oregon, Washington).
— Behan-Pelletier, 1993: 26; Tab. 2 (*Bothropolys*; distribution).

Distribution: USA: Idaho, Montana, Oregon, Washington; CANADA: British Columbia.

HOLOTYPE: ?MCZ, Washington.

♀*B. imaharensis* (Verhoeff, 1937)
— Verhoeff, 1937: 186 (*Bothropolys*; nominal taxon *asperatus*; characters; Japan).
— Chamberlin & Wang, 1952b: 184 (*Bothropolys*; nominal taxon *asperatus*; USA: New York). [**Note:** This species originated from Japan, taken in cargo at New York, 23 January 1951].
— Wang & Mauriès, 1996: 91 (*Bothropolys*; checklist).

Distribution: USA: ?New York. [**Note:** Natural distribution: China, Japan, Taiwan].

HOLOTYPE: ?Unknown, Japan.

♂*B. maluhianus* Attems, 1914
— Attems, 1914: 48, 57, 99 (*Bothropolys*; characters; USA: Hawaii).
— Attems, 1938b: 366, 369 (*Bothropolys*; distribution; list; note).
— Eason, 1977: 491 (*Bothropolys*: taxonomic notes). [**Note:** Eason suspects this species may be an example of *asperatus* and if this is the case that means *maluhianus* is a synonym of *rugosus* because *asperatus* is a synonym of *rugosus*].
— Nishida, 1994: 26 (*Bothropolys*; checklist).
— Nishida, 1997: 195 (*Bothropolys*; checklist).
— Zapparoli & Shelley, 2000: 37-38 (*Bothropolys*; characters; notes; USA: Hawaii).
asperatus Attems [not Koch], 1903a: 92 (*Lithobius*; characters; USA: Hawaii). [**Note:** Attems first reported this species as *Lithobius asperatus*).
— Silvestri, 1904: 323 (*Lithobius*; list).
— Attems, 1914: 48 (*Bothropolys*; list).
— Nishida, 1994: 26 (*Bothropolys*; checklist).
— Nishida, 1997: 195 (*Bothropolys*; checklist).
— Zapparoli & Shelley, 2000: 37 (*Lithobius* + *Bothropolys*; synonym of *maluhianus*).
oahuanus Chamberlin, 1920c: 78, 245 (*Bothropolys*; characters; notes; USA: Hawaii).
— Eason, 1977: 491 (*Bothropolys*; synonym of *maluhianus*).
— Zapparoli & Shelley, 2000: 37 (*Bothropolys*; synonym of *maluhianus*).

Distribution: USA: Hawaii.

HOLOTYPE: Unknown. 1♂, USA: Hawaii, Oahu, Maluhia (Dr. Schauinsland coll.).

♂♀***B. multidentatus*** (Newport, 1845)
— Newport, 1845b: 365 (*Lithobius*; characters; USA: "Prope Novum Eboracum" [= near or not far from New York]).
— Gervais, 1847, 236 (*Lithobius*; characters).
— Newport, 1856: 17 (*Lithobius*; characters; USA: "Prope Novum Eboracum" [= near or not far from New York]).
— Wood, 1865: 152; Fig. 7 (*Bothropolys*; characters; USA: Eastern United States).
— Stuxberg, 1875b: 10-11 (*Lithobius*; subgenus *Eulithobius*; list).
— Stuxberg, 1875c: 26, 32 [= 1876: 134, 139] (*Lithobius*; subgenus *Eulithobius*; characters; list).
— Meinert, 1886a: 175 (*Lithobius*; characters; USA: Kentucky, Massachusetts, Michigan, New Hampshire).
— Bollman, 1887b: 255, 263-264, 266 [= 1893: 22, 30-31, 33] (*Lithobius*; subgenus *Eulithobius*; characters; key; list; USA: Indiana, Michigan).
— Bollman, 1888a: 111 [= 1893: 85] (*Lithobius*; subgenus *Archilithobius*; USA: Tennessee).
— Bollman, 1888b: 8 [= 1893: 80] (*Lithobius*; note; USA: Arkansas).
— Bollman, 1888c: 342 [= 1893: 93] (*Lithobius*; notes; USA: Tennessee).
— Bollman, 1888d: 350 [= 1893: 103] (*Lithobius*; USA: Virginia).
— Bollman, 1888e: 410 [= 1893: 111] (*Lithobius*; note; USA: Indiana).
— Bollman, 1893: 130 (*Lithobius*; subgenus *Eulithobius*; catalogue).
— Morse, 1902: 187 (*Lithobius*; USA: Ohio).
— Chamberlin, 1902a: 43 (*Lithobius*; key to species; range; USA: Missouri, Illinois, Michigan, Pennsylvania)
— Chamberlin, 1909b: 190 (*Lithobius*; USA: District of Columbia, Michigan, New York, North Carolina, Virginia).
— Chamberlin, 1911b: 48 (*Bothropolys*; range; USA: Alabama, Mississippi, North Carolina, Tennessee, Virginia, West Virginia).
— Chamberlin, 1911c: 98-99 (*Bothropolys*; notes; USA: Illinois, Indiana, Michigan).
— Chamberlin, 1914b: 301 (*Bothiopolys* [misspelled]; habitat (p. 301); USA: Michigan).
— Chamberlin, 1918a: 24 (*Bothropolys*; USA: Tennessee).
— Chamberlin, 1920a: 95 (*Bothropolys*; note; USA: New York).

— Chamberlin, 1925c, pg. 390-395, pl. 1; Figs. 1-8 (*Bothropolys*; subgenus *Bothropolys*; characters (adult, stadia); notes; USA: Alabama, Arkansas, Delaware, District of Columbia, Florida, Georgia, Illinois, Indiana, Kentucky, Louisiana, Maryland, Massachusetts, Michigan, Mississippi, New Hampshire, New Jersey, New York, North Carolina, Ohio, Pennsylvania, South Carolina, Tennessee, Virginia, West Virginia).

— Holmquist, 1926: 411 (*Bothropolys*; notes on hibernation such as: activity, date collected, developmental stage, ecological association/niche, quantity found).

— Bailey, 1928a: 1081 (*Bothropolys*; list).

— Bailey, 1928b: 19, 25-26 (*Bothropolys*; characters; list; notes; USA: New York).

— Williams & Hefner, 1928: 143 (*Lithobius*; characters; USA: Ohio).

— Matthews, 1935: 86, 87, 88-91; Map: 92; Tab. [pl.] 7; Fig. 1-4 (*Bothropolys*; characters; USA: Wisconsin).

— Brimley, 1938: 500 (*Bothropolys*; USA: North Carolina).

— Causey, 1940: 68 (*Bothropolys*; characters; note; USA: North Carolina).

— Chamberlin, 1940c: 56 (*Bothropolys*; USA: North Carolina).

— Chamberlin, 1942d: 188 (*Bothropolys*; habitat; USA: Missouri).

— Chamberlin, 1944d: 211-212 (*Bothropolys*; USA: Illinois, Indiana, Missouri, Tennessee).

— Chamberlin, 1945d: 215 (*Bothropolys*; USA: Georgia).

— Auerbach, 1946: 6, 8, 14, 15, 18, 21, 22, 23, 34, 35; chart 1: 44-45 (*Bothropolys*; ecology, feeding habits; habitat; hygrotaxis; list; notes; USA: Illinois).

— Auerbach, 1949: 222, 223, 224, 225, 226; Fig. 3 (*Bothropolys*; ecology, habitat; hygrotaxis; notes; USA: Illinois).

— Crabill, 1950c: 202 (*Bothropolys*; notes; USA: South Carolina).

— Auerbach, 1951a: 100, 101, 102, 103, 106, 107, 108, 109, 118, 119, 120, 121, 122; Tab. 1, 2, 6-9, 14, 15; Fig. 1, 3-5 (*Bothropolys*; desiccation; ecology; habitat; heart beat; population percentage; list; notes; USA: Illinois).

— Auerbach, 1951b: 111, 112; Fig. 2 (*Bothropolys*; key).

— Auerbach, 1952a: 1 (*Bothropolys*; notes).

— Crabill, 1952b: 275-277, pl. 6; Figs. 79, 81-82 (*Bothropolys*; characters; habitat; range; subgenus *Bothropolys*; USA: New York).

— Johnson, 1952: 139-145, 202; Tab. 17, 28; Map 11 (*Bothropolys*; autecology; characters; distribution; experimental feeding diet; natural history; USA: Michigan).

— Crabill, 1953a: 119 (range).
— Crabill, 1955b: 40 (*Bothropolys*; habitat; notes; USA: Missouri).
— Crabill, 1955c: 108, 109, 110 (*Bothropolys*; nomenclatorial discussion).
— Crabill, 1955f: 259-260 (*Bothropolys*; checklist; USA: Kentucky).
— Crabill, 1955h: 156-157 (*Bothropolys*; habitat; notes; USA: Missouri).
— Crabill, 1958b: 97 (*Bothropolys*; habitat; notes; range; USA: Wisconsin).
— Branson & Batch, 1967: 86-87 (*Bothropolys*; characters; habitat; body measurements; USA: Kentucky).
— Wray, 1967: 154 (*Bothropolys*; list).
— Carter & Brown, 1973: 1066; Tab. 1 (*Bothropolys*; notes; CANADA: New Brunswick).
— Kevan, 1979: 298 (*Bothropolys*; synopsis; CANADA: Ontario). [**Note:** Kevan considered this species introduced, however, it is most likely indigenous].
— Summers, 1979: 691, 696, 697; Figs. 1, 2 (*Bothropolys*; checklist; key).
— Summers & Uetz, 1979: 347, 350, 351; Tab. 1 & 2; Fig. 3 (*Bothropolys*; microhabitat; notes, USA: Illinois).
— Lee, 1980: 2, 3, 5; Tab. 1 (*Bothropolys*; habitat; USA: Ohio).
— Summers et al., 1980: 249; Fig. 17 (*Bothropolys*; checklist; USA: Illinois).
— Summers et al., 1981: 61 (*Bothropolys*; checklist; USA: Illinois).
— Holsinger & Culver, 1988: 57 (*Bothropolys*; USA: Virginia). [**Note:** Reported as an accidental occurrence in a cave].
— Snider, 1991: 189 (*Bothropolys*; list).
— Behan-Pelletier, 1993: 26; Tab. 2 (*Bothropolys*; distribution).
— Hoffman, 1995a: 27 (*Bothropolys*; list; notes).
— Watermolen, 1997: 6 (*Bothropolys*; checklist, notes).
— Jeekel, 2005: 16 (*Lithobius*; list).
— Regier et al., 2005: 149; Tab. 1 (*Bothropolys*; table of taxon classification with GenBank accession numbers).
— Giribet & Edgecombe, 2006: 533, 534, 536; Fig 1,2; Tab. 1 (*Bothropolys*; cladograms; GenBank accession code).

filicium Attems, 1901: 111 (*Lithobius*; characters; habitat; North America).
TYPE: *Lithobius* (*Polybothrus*) *filicium* Attems, 1901, deposited at Zoologisches Museum, Hamburg (see Eason, 1985: 5).
— Crabill, 1952b: 275 (*Lithobius*; synonym of *multidentatus*).
— Eason, 1985: 5 (*Lithobius*; characters; synonym of *multidentatus*).

nobilis Wood, 1862: 15 (*Bothropolys*; characters; USA: Illinois, Missouri,
  Pennsylvania).
— Wood, 1865: 152 (*Bothropolys*; synonym of *multidentatus*).
— Stuxberg, 1875c: 26 [= 1876: 134] (*Bothropolys*; synonym of
  multidentatus).
— Meinert, 1886a: 175 (*Bothropolys*; synonym of *multidentatus*).
— Bollman, 1893: 130 (*Lithobius*; synonym of *multidentatus*).
— Gunthorp, 1920: 113 (*Bothropolys*; list).
— Chamberlin, 1925c: 390 (*Bothropolys*; synonym of *multidentatus*).
— Crabill, 1955c: 108, 109, 110 (*Bothropolys*; nomenclatorial
  discussion).
— Jeekel, 2005: 12, 28 (*Bothropolys*; list).
planus Newport, 1845b: 366 Pl. 33, Fig. 32 (*Lithobius*; characters; "Americâ
  Boreali"). **HOLOTYPE:** ?Unknown, ("v. in Mus. D. Hope".) (see
  Newport, 1845b; 1856).
— Gervais, 1847: 236-237 (*Lithobius*; characters).
— Newport, 1856: 18 (*Lithobius*; characters).
— Wood, 1862: 14 (*Lithobius*; characters).
— Wood, 1865: 151 (*Lithobius*; characters; note).
— Stuxberg, 1875c: 27, 32 [= 1876: 135, 139] (*Lithobius*; subgenus
  Lithobius; characters; list).
— Bollman, 1887b: 266 [= 1893: 33] (*Lithobius*; subgenus *Lithobius*;
  list).
— Bollman, 1893: 129 (*Lithobius*; subgenus *Lithobius*; catalogue).
— Eason, 1972a: 302–303 (*Lithobius*; synonym of *multidentatus*).

Distribution: CANADA: New Brunswick, Ontario; USA: Alabama,
Arkansas, Connecticut, District of Columbia, Delaware, Georgia, Illinois,
Indiana, Kentucky, Louisiana, Massachusetts, Maine, Michigan, Missouri,
Mississippi, North Carolina, New Hampshire, New Jersey, New York, Ohio,
Pennsylvania, Rhode Island, South Carolina, Tennessee, Texas, Virginia,
Vermont, Wisconsin, West Virginia.

HOLOTYPE: BMNH, New York: New York City.

♂♀***B. permundus*** (Chamberlin, 1902)
— Chamberlin, 1902a: 42 (*Lithobius*; characters (adult, stadia); habitat;
  key to species; range; USA: Utah).
— Chamberlin, 1925c: 390, 400-405, pl. 2; Fig. 1-8 (*Bothropolys*;
  subgenus *Poropolys*; characters (adult, stadia); notes; USA: Utah).
— Chamberlin, 1925e: 57 (*Bothropolys*; notes; USA: Utah).

— Chamberlin, 1930a: 112-113 (*Bothropolys*; notes; USA: Utah).
— Chamberlin, 1943b: 102-103 (*Bothropolys*; USA: Idaho, Utah).
— Jeekel, 2005: 29 (*Lithobius*; list).

Distribution: USA: Idaho, Utah.

HOLOTYPE: ?MCZ, Utah: Wahsatch Mountains. [**Note:** The MCZ has numerous specimens of this species from Chamberlin's collection and the catalog numbers with this locality are as follows: 677 (ethanol); 839-842 (slides)].

♂♀*B. rugosus* (Meinert, 1872)
— Meinert, 1872: 306-308 (*Lithobius*; characters; USA: Hawaii).
— Silvestri, 1904: 323 (*Lithobius*; list).
— Attems, 1914: 48, 57, 99 (*Bothropolys*; characters; list).
— Chamberlin, 1920c: 78-79, 245 (*Ethopolys*; notes; USA: Hawaii).
— Attems, 1938b: 366, 369 (*Ethopolys*; distribution; list; note). [**Note:** Although Attems considered this species endemic it has possibly been introduced from either Japan or the mainland of eastern Asia (see Eason, 1974b, 1977)].
— Eason, 1974b: 20-22; Tab. 1; Figs. 6-7 (*Bothropolys*; characters; notes; USA: Hawaii).
— Zapparoli & Shelley, 2000: 38 (*Bothropolys*; characters; notes; USA: Hawaii).
asperatus Koch, 1878: 788-789 (*Lithobius*; characters).
— Attems, 1903a: 92 (*Lithobius*; characters; USA: Hawaii).
— Silvestri, 1904: 323 (*Lithobius*; list).
— Attems, 1914: 48 (*Bothropolys*; characters; distribution; list).
— Nishida, 1994: 26 (*Bothropolys*; checklist).
— Nishida, 1997: 195 (*Bothropolys*; checklist).
— Eason, 1974b: 20 (*Lithobius*; synonym of *rugosus*).
?*migrans* Chamberlin, 1930c: 69 (*Bothropolys*; characters; USA: Hawaii). [**Note:** Chamberlin's report is of four specimens being imported to Hawaii on plants of the genus *Dioscorea* from China in 1928].
— Eason, 1974b: 20, 22 (cited as questionable synonym of *rugosus*; notes).
— Chamberlin & Wang, 1952b: 184. (*Bothropolys*; USA: Hawaii, Pennsylvania). [**Note:** Chamberlin & Wang reported this species from Hawaii at quarantine in Honolulu; they also noted a specimen from Japan appearing in cargo at quarantine in Philadelphia].
— Zapparoli &Shelley, 2000: 47. (*Bothropolys; note*). [**Note:** Although Zapparoli & Shelley deleted this species from the Hawaiian fauna

it is still a potential invader and is included here for future reference].

Distribution: USA: ?Hawaii, ?Pennsylvania.

LECTOTYPE: ZMUC, ♂, Hawaii, Oahu.

PARALECTOTYPES: ZMUC, 4♂, 1 ♂ imm., Hawaii, Oahu.

♀*B. victorianus* Chamberlin, 1925
— Chamberlin, 1925c: 390, 405-407, pl. 3; Fig. 5, 6 (*Bothropolys*; subgenus *Poropolys*; characters (adult, stadia); notes; CANADA: British Columbia; USA: Alaska).
— Chamberlin, 1962: 137 (*Bothropolys*; USA: Oregon).
— Behan-Pelletier, 1993: 26; Tab. 2 (*Bothropolys*; distribution).

Distribution: CANADA: British Columbia; USA: Alaska, Oregon.

HOLOTYPE: MCZ, No. 675, British Columbia Vancouver Island.

PARATYPE: NMNH.

(Subgenus ***Bothropolys*** Wood, 1862)

Bothropolys (*Bothropolys*) Wood, 1862: 15 (characters). TYPE SPECIES: *Bothropolys nobilis* Wood, 1862, [=*Bothropolys multidentatus* (Newport, 1845)] fixed by subsequent designation.
— Crabill, 1955c: 110 (designation of type).
— Matic, 1974: 334 (*Bothropolys*; discussion of subgenera).

♀*B. (B.) columbiensis* Chamberlin, 1925
— Chamberlin, 1925c: 395-397 (*Bothropolys*; subgenus *Bothropolys*; characters; notes; CANADA: British Columbia).
— Behan-Pelletier, 1993: 26; Tab. 2 (*Bothropolys*; distribution).

Distribution: CANADA: British Columbia.

HOLOTYPE: MCZ, No. 672, CANADA: British Columbia, Kaslo, female in late pseudomaturus stage. [**Note:** Original locality/determined label reads "*Bothropolys columbiensis*, type, B.C.: Kaslo, ?R.C. Currie"].

(Subgenus ***Calopolys*** Chamberlin, 1941)

Bothropolys (*Calopolys*) Chamberlin, 1941a: 17. **TYPE SPECIES:** *Bothropolys dasys* Chamberlin, 1941, fixed by monotypy.
— Matic, 1974: 329 (*Bothropolys*; discussion of subgenus *Calopolys*).
— Jeekel, 2005: 13 (*Bothropolys*; list; note).

♂♀*B.* (*C.*) *dasys* Chamberlin, 1941
— Chamberlin, 1941a: 17 (*Bothropolys*; subgenus *Calopolys*; characters; USA: California).
— Jeekel, 2005: 13 (list).

Distribution: USA: California.

HOLOTYPE: NMNH, California: Monterey Co., Hastings Reservation, 18 February 1941, by Dr. Linsdale.

(Subgenus *Oligopolys* Chamberlin, 1946)

Bothropolys (*Oligopolys*) Chamberlin, 1946c: 189. **TYPE SPECIES:** *Bothropolys ethus* Chamberlin, 1946, fixed by monotypy.
— Matic, 1974: 329 (*Bothropolys*; discussion of subgenus *Oligopolys*).
— Behan-Pelletier, 1993: 26; Tab. 2 (*Bothropolys*; list).
— Jeekel, 2005: 25 (list).

♂*B.* (*O.*) *ethus* Chamberlin, 1946
— Chamberlin, 1946c: 189 (*Bothropolys*; subgenus *Oligopolys*; characters; USA: Alaska).
— Jeekel, 2005: 25 (*Bothropolys*; list).

Distribution: USA: Alaska.

HOLOTYPE: NMNH, ♂, Alaska: U. S. Creek on Steese Highway, 62 mi. NE of Fairbanks, 20 June 1945.

(Subgenus *Synopolys* Chamberlin, 1931)

Bothropolys (*Synopolys*) Chamberlin, 1931a: 190-191 (characters). **TYPE SPECIES:** *Bothropolys epelus* Chamberlin, 1931, fixed by monotypy.
— Jeekel, 2005: 31 (list).

♀*B.* (*S.*) *epelus* Chamberlin, 1931

— Chamberlin, 1931a: 190-191 (*Bothropolys*; subgenus *Synopolys*; characters; notes; USA: Louisiana). [**Note:** Chamberlin suggested this specimen " . . . escaped from a ship at the dock" but it is not known from where it originated].

— Jeekel, 2005: 31 (*Bothropolys*; list).

Distribution: USA: ?Louisiana.

HOLOTYPE: NMNH, ♀, taken on the dock at New Orleans, Louisiana, 2 September 1930.

Genus *Ethopolys* Chamberlin, 1912
[12 species; Western Canada and USA]

Ethopolys Chamberlin, 1912e: 173-174 (characters, notes). TYPE SPECIES: *Bothropolys xanti* Wood, 1862, fixed by original designation.

— Chamberlin, 1919: 20H (notes).

— Chamberlin, 1925c: 411-414; fig. 3 (characters; distribution; key to subgenera and species).

— Jeekel, 2005: 15 (list; note) [**Note:** Jeekel states that *Ethopolys* is "A junior objective synonym of *Allobothropolys* Verhoeff, 1907"].

Bothropolys (*Allobothropolys*) Verhoeff, 1907: 240 (key). TYPE SPECIES: *Bothropolys* (*Allobothropolys*) *xanti* (Wood, 1862), fixed by monotypy.

— Chamberlin, 1925c: 389 (key; taxonomic notes). [**Note:** Chamberlin declares Verhoeff's subgeneric division of the genus artificial].

— Jeekel, 2005: 10 (list; note).

♀*E. calibius* Chamberlin, 1951

— Chamberlin, 1951c: 117-118 (*Ethopolys*; characters; note; USA: California).

Distribution: USA: California.

HOLOTYPE: NMNH, ♀, USA: California: Prairie Creek, 7-10-1946, Mulaik].

♂♀*E. californicus* (Daday, 1889)

— Daday, 1889: 153 (*Lithobius*; characters; note; USA: "California borealis").

— Chamberlin, 1925c: 432 (*Ethopolys*; subgenus uncertain; characters; notes). [**Note:** Chamberlin (1910) synonymized *Lithobius californicus* Daday with *Bothropolys monticola* (Stuxberg) but then in 1925 Chamberlin reports *Lithobius californicus* Daday as *Ethopolys* (?) *californicus* (Daday). Maybe Chamberlin forgot that he did this but the following species need to be reassessed: *L. californicus*, *B. monticola* and *E. pusio* (Stuxberg) and it may not be a bad idea to also compare *E. sierravagus* (Chamberlin) as well].

— Eason, 1985: 3-5 (*Lithobius*; characters; notes).

sierravagus (Chamberlin, 1903) : 154-155 (*Lithobius*; characters; key to species; range; USA: California, Oregon).

— Chamberlin, 1909b: 190 (*Lithobius*; USA: Washington).

— Chamberlin, 1912e: 174 (*Ethopolys*; comment regarding *sierravagus* being the same as *monticola* Stuxberg).

— Chamberlin, 1925c: 419-23, pl. 5; Figs. 1-4 (*Ethopolys*; subgenus *Archethopolys*; characters (adult, stadia); range; note; USA: California, Oregon).

— Kevan, 1983: 2947 (*Ethopolys*; subgenus *Archethopolys*; CANADA: Vancouver Island).

— Eason, 1985: 4 (*Ethopolys*; synonym of *californicus*).

— Behan-Pelletier, 1993: 26; Tab. 2 (*Ethopolys*; distribution).

Distribution: CANADA: British Columbia; USA: California, Oregon, Washington.

HOLOTYPE: NMB; USA: "California borealis", by D. Joanne Vadona.

♀*E. positivus* Chamberlin, 1941
— Chamberlin, 1941a: 17 (*Ethopolys*; characters; USA: California).

Distribution: USA: California.

SYNTYPE: NMNH, California: Sequoia National Park, "Cold Spring," 10 mi. E of Hammond, 20 March 1941, by S. & D. Mulaik.

♀*E. timpius* Chamberlin, 1951
— Chamberlin, 1951c: 116-117 (*Ethopolys*; characters; notes; USA: Utah).

Distribution: USA: Utah.

SYNTYPE: NMNH, ♀, Utah: Provo Canyon, Briday Veil Falls at upper limits.

(Subgenus *Archethopolys* Chamberlin, 1925)

Archethopolys (*Archethopolys*) Chamberlin, 1925c: 414 (key to species; species included: *bipunctatus, integer, integer alaskanus; pusio, sierravagus* [= *californicus*]). **Type Species:** *Ethopolys* (*Archethopolys*) *integer* Chamberlin, 1919, fixed by original designation.
— Chamberlin, 1930a: 113 (key to species; species included: *bipunctatus, gosobius, kaibabus, parowanus*). [**Note:** In regards to the above species in key, only *bipunctatus* belongs to *Ethopolys* (*Archethopolys*)].

♂♀*E.* (*A.*) *bipunctatus* (Wood, 1862)
— Wood, 1862:16 (*Bothropolys*; characters; USA: West of Rocky Mountains).
— Wood, 1865: 153 (*Bothropolys*; characters; USA: West of Rocky Mountains).
— Stuxberg, 1875b: 14 (*Lithobius*; subgenus *Archilithobius*; list).
— Stuxberg, 1875c: 30 (*Lithobius*; subgenus *Archilithobius*; list).
— Stuxberg, 1876: 6 (*Lithobius*; subgenus *Archilithobius*; list).
— Bollman, 1887b: 265 [= 1893: 32] (*Lithobius*; subgenus *Archilithobius*; list).
— Bollman, 1893: 128 (*Lithobius*; subgenus *Archilithobius*; catalogue).
— Chamberlin, 1901: 22 (*Lithobius*; characters, habitat; key; USA: Utah).
— Chamberlin, 1902a: 42 (*Lithobius*; key; range; USA: Utah).
— Chamberlin, 1909b: 191 (*Lithobius*; USA: Utah)
— Gunthorp, 1920: 113 (*Bothropolys*; list).
— Chamberlin, 1925c: 414-419, pl. 5; Fig. 5 (*Ethopolys*; subgenus *Archethopolys*; characters (adult, stadia); USA: Nevada, Utah).

Distribution: USA: ?California, Nevada, Utah.

HOLOTYPE: NMNH.

♂*E.* (*A.*) *bipunctatus insulatus* Chamberlin, 1955
— Chamberlin, 1955a: 180-181 (*Ethopolys*; nominal taxon *bipunctatus*; subspecies *insulatus*; characters; notes; USA: Utah).

Distribution: USA: Utah.

HOLOTYPE: ?Unknown, ♂, Utah: Stansbury Island, in "Spider Cave", 8 November 1952.

♂♀*E. (A.) integer* Chamberlin, 1919
— Chamberlin, 1919: 20H (*Ethopolys*; characters; notes; USA: Oregon, Washington).
— Chamberlin, 1925c: 414, 423-426, pl. 5; Fig. 6-8, pl. 6; Fig. 1 (*Ethopolys*; subgenus *Archethopolys*; characters; notes; USA: Oregon, Washington).
— Chamberlin, 1962: 137 (*Ethopolys*; USA: Oregon).
— Jeekel, 2005: 11 (*Ethopolys*; list).

Distribution: USA: Oregon, Washington.

TYPES: MCZ, No. 737, 4 specimens, Washington: Pullman, 3-28-'08.

♀*E. (A.) integer alaskanus* Chamberlin, 1919
— Chamberlin, 1919: 15H, 21H (*Ethopolys*; characters; list; USA: Alaska).
— Chamberlin, 1925c: 414, 426, pl. 6; Fig. 2 (*Ethopolys*; subgenus *Archethopolys*; nominal taxon *integer*; cited as subspecies *alaskanus*; characters; USA: Alaska).
— Chamberlin, 1946c: 188 (*Ethopolys*; nominal taxon *integer*; cited as subspecies *alaskanus*; USA: Alaska).
— Behan-Pelletier, 1993: 26; Tab. 2 (*Ethopolys*; distribution).

Distribution: USA: Alaska.

COTYPES: MCZ, No. 743, 2 specimens, Alaska: Forrester Island, May 1913, (Ronald and Prof. H. Heath) or Sitka.

♀*E. (A.) monticola* (Stuxberg, 1875)
— Stuxberg, 1875a: 65-66 (*Lithobius*; characters; USA: ?Nevada). [**Note:** The type locality is "Sierra Nevada at Summit Station" and it is assumed this is in the state of Nevada and not California].
— Stuxberg, 1875b: 14 (*Lithobius*; subgenus *Archilithobius*; list).
— Stuxberg, 1875c: 30 [=1876: 137] (*Lithobius*; subgenus *Archilithobius*; list).
— Stuxberg, 1875d: 188 (*Lithobius*; characters).
— Stuxberg, 1876: 6 (*Lithobius*; subgenus *Archilithobius*; list).

— Bollman, 1887b: 265 [= 1893: 32] (*Lithobius*; subgenus *Archilithobius*; list).

— Bollman, 1893: 128 (*Lithobius*; subgenus *Archilithobius*; catalogue).

— Chamberlin, 1902a: 42 (*Lithobius*; key to species; USA: "Sierra Nevada Mountains").

— Chamberlin, 1909b: 190 (*Lithobius*; USA: California).

— Chamberlin, 1910: 369 (*Bothropolys*; characters; notes).

— Chamberlin, 1911f: 377 (*Bothropolys*; range).

— Chamberlin, 1912e: 174 (*Ethopolys*; comment regarding *monticola* and it being the same as *sierravagus* Chamberlin).

— Chamberlin, 1925c: 430-431 (*Ethopolys*; subgenus *Archethopolys*; characters; notes). [**Note:** Chamberlin (1910) synonymized *Lithobius californicus* Daday with *Bothropolys monticola* (Stuxberg), but then in 1925 Chamberlin reports *Lithobius californicus* Daday as *Ethopolys* (?) *californicus* (Daday). Maybe Chamberlin forgot that he did this but the following species need to be reassessed as they are now placed in *Ethopolys*: *E. californicus*, *E. monticola* and *E. pusio* (Stuxberg). It may not be a bad idea to also compare *E. sierravagus* (Chamberlin) as well, however, revision of the entire genus seems appropriate].

Distribution: USA: "Sierra Nevada Mountains", California, ?Nevada, Oregon, Washington.

HOLOTYPE: ?Unknown, ?California or ?Nevada, "Sierra Nevada ad Summit Station (G. Eisen)".

♂♀*E. (A.) pusio* (Stuxberg, 1875)
— Stuxberg, 1875a: 66-67 (*Lithobius*; characters; USA: California).
— Stuxberg, 1875b: 16 (*Lithobius*; subgenus *Archilithobius*; list).
— Stuxberg, 1875c: 30 (*Lithobius*; subgenus *Archilithobius*; list).
— Stuxberg, 1875d: 188-189 (*Lithobius*; characters).
— Stuxberg, 1876: 6 (*Lithobius*; subgenus *Archilithobius*; list).
— Bollman, 1887b: 265 [= 1893: 32] (*Lithobius*; subgenus *Archilithobius*; list).
— Bollman, 1893: 129 (*Lithobius*; subgenus *Archilithobius*; catalogue).
— Chamberlin, 1909b: 187 (*Lithobius*; USA: California).
— Chamberlin, 1910: 373 (*Lithobius*; key to species; notes). [**Note:** Chamberlin doubts the validity of this species and suggests the

original description by Stuxberg was based on immature specimens of *Bothropolys monticola* (Stuxberg)].
— Chamberlin, 1911f: 378 (*Bothropolys*; characters; notes; USA: California).
— Chamberlin, 1925c: 427-430, pl. 6; Figs. 3-6 (*Ethopolys*; subgenus *Archethopolys*; characters (adult, stadia); notes; USA: California).
— Chamberlin, 1944c: 79 (*Ethopolys*; USA: California).

Distribution: USA: California.

HOLOTYPE: ?Unknown, California: San Francisco, by Eisen. [**Note:** MCZ has numerous specimens from various localities in California from the Chamberlin collection. The catalog numbers are as follows: 688-690 (ethanol), 692-695 (ethanol), 854-856 (slides)].

♀*E. (A.) spectans* Chamberlin, 1951
— Chamberlin, 1951c: 118 (*Ethopolys*; characters; CANADA: British Columbia).
— Kevan, 1979: 298 (*Ethopolys*; synopsis; CANADA: British Columbia).
— Kevan, 1983: 2947 (*Ethopolys*; note). [**Note:** Kevan mentions a new species near *spectans* from Vancouver Island, British Columbia].
— Behan-Pelletier, 1993: 26; Tab. 2 (*Ethopolys*; distribution). [**Note:** The new species Kevan mentions in 1983 is also included in Behan-Pelletier's distribution table].

Distribution: CANADA: British Columbia.

HOLOTYPE: NMNH, ♀, CANADA: British Columbia, Vancouver Id., Spectacle Lake, 12 May, by Dr. G. C. Carl.

(Subgenus *Ethopolys* Chamberlin, 1912)

Ethopolys (*Ethopolys*) Chamberlin, 1912e: 173 (characters). TYPE SPECIES: *Bothropolys xanti* Wood, 1862, fixed by original designation.
— Matic, 1974: 329 (notes).

♂♀*E. (E.) xanti* (Wood, 1862)
— Wood, 1862:15-16 (*Bothropolys*; characters; notes; USA: California, Oregon).
— Wood, 1865: 152 (*Bothropolys*; characters; USA: California).

— Wood, 1867b: 128 (*Bothropolys*; USA: California).
— Stuxberg, 1875b: 10-11 (*Lithobius*; subgenus *Lithobius*; list).
— Stuxberg, 1875c: 27 (*Lithobius*; subgenus *Lithobius*; list).
— Stuxberg, 1876: 135 (*Lithobius*; list).
— Bollman, 1887b: 255, 261, 266 [= 1893: 22, 28, 33] (*Lithobius*; subgenus *Lithobius*; characters; key; list; USA: California).
— Bollman, 1893: 129 (*Lithobius*; subgenus *Lithobius*; catalogue).
— Chamberlin, 1902a: 42 (*Lithobius*; key to species; USA: California, Oregon).
— Silvestri, 1904: 323 (*Lithobius*; list).
— Chamberlin, 1909b: 190 (*Lithobius*; USA: California).
— Chamberlin, 1910: 369 (*Bothropolys*; characters; range; USA: California).
— Chamberlin, 1911f: 378 (*Bothropolys*; notes; USA: California).
— Chamberlin, 1912e: 173 (*Ethopolys*; cited as type species of *Ethopolys*).
— Anonymous, 1920: 35 (*Ethopolys*; USA: California).
— Gunthorp, 1920: 113 (*Bothropolys*; list).
— Chamberlin, 1925c: 432-437, pl. 4; Figs. 1-6 (*Ethopolys*; subgenus *Ethopolys*; characters (adult, stadia); range; notes; USA: California, Utah).
— Attems, 1938b: 366, 369 (*Lithobius*; distribution; list; note).
— Chamberlin, 1940a: 4 (*Ethopolys*; USA: California).
— Eason, 1974b: 21 (*Ethopolys*; note).
— Jeekel, 2005: 15 (*Lithobius*; list).

Distribution: USA: California, Oregon, Utah.

HOLOTYPE: ?MCZ, California, Fort Tejon.

Genus *Zygethopolys* Chamberlin, 1925
[4 species; Western CANADA, Eastern/Western USA]

Zygethopolys Chamberlin, 1925c: 408 (characters; notes). Type Species: *Zygethopolys nothus* Chamberlin, 1925, fixed by original designation and monotypy.
— Crabill, 1953a: 120 (key to species).
— Jeekel, 2005: 35 (list).

♂ *Z. atrox* Crabill, 1953
— Crabill, 1953a: 119-120 (*Zygethopolys*; characters (plectrotaxy); notes; USA: Kentucky).
— Crabill, 1955f: 260 (*Zygethopolys*; checklist; USA: Kentucky).

— Hoffman, 1995a: 27 (*Zygethopolys*; list; notes). [**Note:** Hoffman suggests this species will likely be found in Virginia].

Distribution: USA: Kentucky.

HOLOTYPE: NMNH, Crabill collection No. C-1446, ♂, Kentucky: Whitely Co., Cumberland Falls State Park, 14 April 1952, by Theodore J. Spilman, under a rock.

♀**Z. nothus** Chamberlin, 1925
— Chamberlin, 1925c: 409-411; Fig. 2; pl. 3 fig. 7, 8 (*Zygethopolys*; characters; USA: Alaska).
— Behan-Pelletier, 1993: 26 (*Zygethopolys*; list).
— Jeekel, 2005: 35 (*Zygethopolys*; list).

Distribution: USA: Alaska

HOLOTYPE: MCZ, No. 687, Alaska, Forrester Island, May 1913, H. & R. Heath.

PARATYPE: MCZ, No. 686, Alaska, Forrester Island, July 1913, H. & R. Heath.

♂♀**Z. pugetensis** Chamberlin, 1928
— Chamberlin, 1928c: 85 (*Zygethopolys*; characters; USA: Washington).

Distribution: USA: Washington.

HOLOTYPE: NMNH, ♀, Washington: region of Puget Sound, by E. E. Smith.

ALLOTYPE: ?NMNH, ♂, Washington: region of Puget Sound, by E. E. Smith.

PARATYPES: ?NMNH, 6 specimens, Washington: region of Puget Sound, by E. E. Smith.

♂♀**Z. pugetensis tiganus** Chamberlin & Wang, 1952
— Chamberlin & Wang, 1952a: 61 (*Zygethopolys*; nominal taxon *pugetensis*; characters; CANADA: British Columbia).
— Kevan, 1979: 298 (*Zygethopolys*; synopsis; CANADA: British Columbia).

— Behan-Pelletier, 1993: 27 (*Zygethopolys*; nominal taxon *pugetensis*; list).

Distribution: CANADA: British Columbia.

SYNTYPE: NMNH, CANADA: British Columbia, Vancouver, 4 April 1933, by H. Leech.

Subfamily **Lithobiinae** Newport, 1844

Genus *Alaskobius* Chamberlin, 1946
[3 species; USA: Alaska]

Alaskobius Chamberlin, 1946c: 187 (characters). Type Species: *Alaskobius josephus* Chamberlin, 1946, fixed by original designation.
— Jeekel, 2005: 9 (list).

♂*A. adlatus* Chamberlin, 1946
— Chamberlin, 1946c: 188, pl. 2; Fig. 11 (*Alaskobius*; characters; USA: Alaska).
— Behan-Pelletier, 1993: 26; Tab. 2 (*Alaskobius*; distribution).

Distribution: USA: Alaska.

HOLOTYPE: NMNH, ♂, Alaska: Richardson Highway, 5 mi. S of Rapids.

♂♀*A. josephus* Chamberlin, 1946
— Chamberlin, 1946c: 187-188, pl. 2; Fig. 10 (*Alaskobius*; characters; USA: Alaska).
— Chamberlin, 1952c: 209 (*Alaskobius*; USA: Alaska).
— Behan-Pelletier, 1993: 26; Tab. 2 (*Alaskobius*; distribution).
— Jeekel, 2005: 9 (list).

Distribution: USA: Alaska.

HOLOTYPE: NMNH, ♀, Alaska: Matnauska, 7 May 1945. [**Note:** It is not clear why Chamberlin would have designated a female as the allotype if the holotype was a female. If indeed this is what Chamberlin did one may consider making the paratype the allotype].

ALLOTYPE: NMNH, ♀, Alaska: Matnauska, 7 May 1945.

PARATYPE: NMNH, ♂, Alaska: Circle City, October 1943.

♀A. parvior Chamberlin, 1946
— Chamberlin, 1946c: 188 (*Alaskobius*; characters; USA: Alaska).
— Behan-Pelletier, 1993: 26; Tab. 2 (*Alaskobius*; distribution).

Distribution: USA: Alaska.

HOLOTYPE: NMNH, ♀, Alaska: College, near Fairbanks, 22 September 1943.

Genus *Arebius* Chamberlin, 1916
[22 species; Western USA]

Arebius Chamberlin, 1916: 171-174; Fig. 4 (characters; habitat (p. 115-116); key to species; notes). TYPE SPECIES: *Arebius medius* Chamberlin, 1916, fixed by original designation. [**Note:** Although not clearly defined in the literature the following species of *Arebius* belong in the subgenus *Pagobius* based on the female genital forceps being tripartite: *cherosus, crenius, montivagus, navajo, petrovius, platypus, sequens, sequoius, tetanus, tridens*. The remaining species not assigned to a subgenus seem to belong to the subgenus *Arebius*].
— Attems, 1926: 382, 384 (*Arebius*;). [**Note:** reference not obtained].
— Attems, 1938a: 344 (synonym of *Lithobius* (*Monotarsobius*)).
— Jeekel, 2005: 12 (list).
Hesperobius Chamberlin, 1913: 39 (list). TYPE SPECIES: *Lithobius obesus* Stuxberg, 1875, fixed by monotypy.
— Jeekel, 2005: 19 (list; notes). [**Note:** It seems Chamberlin (1916) changed the name *Hesperobius* into *Arebius*].

♂♀A. agamus Chamberlin, 1941
— Chamberlin, 1941a: 9 (*Arebius*; characters; notes; USA: California).

Distribution: USA: California.

SYNTYPES: NMNH, California: Sequoia National Park, "Cold Spring," 10 mi. E of Hammond, 3,500 ft., 20 March 1941. [**Note:** 3 male specimens].

ALLOTYPE: NMNH, California: 12 mi. E of Hammond, 21-22 March 1941.

PARATYPE: NMNH, ♂, California: 12 mi. E of Hammond, 21-22 March 1941.

♀*A. cherosus* Chamberlin, 1941
— Chamberlin, 1941a: 9-10 (*Arebius*; characters; note; USA: California).

Distribution: USA: California.

HOLOTYPE: NMNH, ♀, California: Sequoia National Park, "Cold Spring," 10 mi. E of Hammond, 3,500 ft., 20 March 1941.

♀*A. convergens* Chamberlin, 1943
— Chamberlin, 1943b: 105-106 (*Arebius*; characters; USA: New Mexico). [**Note:** Chamberlin notes that this species is close to *A. epelus* and it appears that the described *A. convergens* may just be a more mature specimen from *epelus* making it a junior synonym].

Distribution: USA: New Mexico.

HOLOTYPE: NMNH, New Mexico: N of Glencoe, 31 May 1941.

PARATYPES: NMNH, New Mexico: Tijaras, 30 May 1941, Mulaik. NMNH, numerous specimens, New Mexico: 6 mi. S Mountainair, 31 May 1941, S.-D. Mulaik.

♂♀*A. crenius* Chamberlin, 1941
— Chamberlin, 1941a: 6-7 (*Arebius*; characters; notes; USA: California). [**Note:** Chamberlin mentions that this species represents *A. dolius*, and it appears he has described more mature specimens of *dolius*. Therefore, this may be a junior synonym of *dolius*].

Distribution: USA: California.

SYNTYPES: NMNH, 7 specimens, California: Mountain Springs, 8 January 1941, S. & D. Mulaik.

♂♀*A. elysianus* Chamberlin, 1916
— Chamberlin, 1916: 179-180, pl. 7; Fig. 2 (*Arebius*; characters; USA: California).
— Anonymous, 1920: 35 (*Arebius*; USA: California).

Distribution: USA: California.

TYPES: MCZ, No. 444, 18 specimens, California: Los Angeles, Elysian Park, June 1909, R. V. Chamberlin.

PARATYPES: MCZ, No. 355, 442, California: Pasadena; MCZ, No. 441, San Ysidro. W. M. Wheeler; MCZ, No. 443, Claremont, R. V. Chamberlin.

♂♀*A. epelus* Chamberlin, 1941
— Chamberlin, 1941a: 7 (*Arebius*; characters; notes; USA: California).

Distribution: USA: California.

HOLOTYPE: NMNH, ♀, California: Monterey Co., Carmel Valley, Hastings Reservation, 1 May 1940.

PARATYPE: NMNH.

♀*A. fremontus* Chamberlin, 1943
— Chamberlin, 1943b: 106-107 (*Arebius*; characters; USA: Montana, Wyoming).

Distribution: USA: Montana, Wyoming.

SYNTYPE: NMNH, Wyoming: Fremont Co. Brooks Lake Falls, 12 August 1941.

♂♀*A. integrior* Chamberlin, 1949
— Chamberlin, 1949a: 13-14 (*Arebius*; characters; notes; USA: Alaska).
— Behan-Pelletier, 1993: 26; Tab. 2(*Arebius*; distribution).

Distribution: USA: Alaska.

HOLOTYPE: NMNH, ♀, Alaska: north front of the Brooks Range, 68°20' N. lat. and 151°30' W. long., 20 August 1948, by Neal A. Weber.

ALLOTYPE: ?NMNH, No. 2319, ♂, Alaska: north front of the Brooks Range, 68°20' N. lat. and 151°30' W. long., 20 August 1948, by Neal A. Weber. [**Note:** The allotype and paratype specimens are missing their posterior legs].

PARATYPE: ?NMNH, No. 2309, ♀, Alaska: north front of the Brooks Range, 68°20' N. lat. and 151°30' W. long., 20 August 1948, by Neal A. Weber.

♀*A. medius* Chamberlin, 1916
- Chamberlin, 1916: 177-179, pl. 7; Fig. 1 (*Arebius*; characters; USA: California).
- Jeekel, 2005: 12 (*Arebius*; list).

Distribution: USA: California.

HOLOTYPE: MCZ, No. 356, California: Brookdale, March, 1913. R. V. Chamberlin.

PARATYPE: MCZ, No. 438, California: Brookdale, March, 1913. R. V. Chamberlin.

♀*A. montivagus* Chamberlin, 1943
- Chamberlin, 1943b: 106 (*Arebius*; characters; USA: Montana).

Distribution: USA: Montana.

SYNTYPE: NMNH, Montana: 6 mi. W of Belgrade on Gallatin River, 17 August 1941.

♀*A. navajo* Chamberlin, 1943
- Chamberlin, 1943b: 107 (*Arebius*; characters; USA: New Mexico).

Distribution: USA: New Mexico.

HOLOTYPE: NMNH, New Mexico: Escabosa, 30 May 1941.

♂♀*A. obesus* (Stuxberg, 1875)
- Stuxberg, 1875a: 67-68 [=1875d: 189-190] (*Lithobius*; characters; USA: California).
- Stuxberg, 1875b: 18 (*Lithobius*; subgenus *Archilithobius*; list).
- Stuxberg, 1875c: 31 [=1877: 138] (*Lithobius*; subgenus *Archilithobius*; list).
- Bollman, 1887b: 265 [= 1893: 32] (*Lithobius*; subgenus *Archilithobius*; list).

— Bollman, 1893: 129 (*Lithobius*; subgenus *Archilithobius*; catalogue).
— Chamberlin, 1909b: 187 (*Lithobius*; USA: California).
— Chamberlin, 1910: 373; Fig. 131 c-e (*Lithobius*; characters; key to species; USA: California).
— Chamberlin, 1911f: 379 (*Lithobius*; range; USA: California).
— Chamberlin, 1916: 175-177, pl. 7; Fig. 3-4; pl. 8; Fig. 1 (*Arebius*; characters (adult, stadia); USA: California. [**Note:** Other specimens collected by Chamberlin in California of this species are in the MCZ as follows: MCZ, No. 354, 445, 447, Stanford; MCZ, No. 449, Santa Barbara; MCZ, No. 448, Fresno].
— Jeekel, 2005: 19 (*Lithobius*; list).

Distribution: USA: California, Utah.

HOLOTYPE: ?Unknown, California: Sausalito. [**Note:** MCZ has numerous specimens from California with the following catalog numbers: 354 (slide), 445-449 (ethanol)].

♀*A. petrovius* Chamberlin, 1947
— Chamberlin, 1947a: 39 (*Arebius*; characters; USA: California). [**Note:** There were four specimens collected and Chamberlin compares one to a female of *A. diplonyx* so it is assumed that, in the least, a female was collected].

Distribution: USA: California.

HOLOTYPE: NMNH, California: Monterey Co., Finch Creek, 24 March 1945, under a rock, by Dr. J. M. Linsdale.

♂♀*A. platypus* Chamberlin, 1941
— Chamberlin, 1941a: 8 (*Arebius*; characters; USA: Washington).

Distribution: USA: Washington.

HOLOTYPE: NMNH, ♀, Washington: Naches River, 15 mi. below summit, W. 121°7' N. 46°59', 5 July 1938, by W. Ivie.

♂♀*A. sequens* Chamberlin, 1947
— Chamberlin, 1947a: 39 (*Arebius*; characters; USA: California).

Distribution: USA: California.

HOLOTYPE: NMNH, California: Monterey Co., Hastings Reservation, 10 June 1943, by Dr. J. M. Linsdale.

♂♀*A. sequoius* Chamberlin, 1941
 — Chamberlin, 1941a: 8-9 (*Arebius*; characters; USA: California).

Distribution: USA: California.

HOLOTYPE: NMNH, California: Sequoia National Park, "Cold Spring," 10 mi. E of Hammond, 3,500 ft., 20 March 1941.

♂♀*A. tetonus* Chamberlin, 1943
 — Chamberlin, 1943b: 107 (*Arebius*; characters; USA: Wyoming).

Distribution: USA: Wyoming.
3**HOLOTYPE:** NMNH, Wyoming: Teton Co., Leeks Camp, 13 September 1941.

♀*A. tridens* Chamberlin, 1941
 — Chamberlin, 1941a: 8 (*Arebius*; characters; note; USA: California).

Distribution: USA: California.

HOLOTYPE: NMNH, ♀, California: Monterey Co., Hastings Reservation, 30 January 1940.

(Subgenus *Arebius* Chamberlin, 1916)

Arebius Chamberlin, 1916: 171-174; Fig. 4 (characters; habitat (p. 115-116); key to species; notes). Type Species: *Arebius medius* Chamberlin, 1916, fixed by original designation.
 — Attems, 1926: 382, 384 (*Arebius*;). [**Note:** reference not obtained].
 — Attems, 1938a: 344 (synonym of *Lithobius* (*Monotarsobius*) Verhoeff, 1905).

♀*A.* (*A.*) *oregonensis* Chamberlin, 1916
 — Chamberlin, 1916: 181-182, pl. 7; Fig. 6 (*Arebius*; characters; USA: Oregon).

Distribution: USA: Oregon.

HOLOTYPE: MCZ, No. 353, Oregon: Portland, 3 August, 1902. R. V. Chamberlin.

PARATYPE: MCZ, No. 439, Oregon: Portland, 3 August, 1902. R. V. Chamberlin.

(Subgenus Pagobius Chamberlin, 1916)

Arebius (*Pagobius*) Chamberlin, 1916: 172–173 (characters; key). TYPE SPECIES: *Arebius diplonyx* Chamberlin, 1916, fixed by original designation.
— Attems, 1926: 382, 384 (synonym of *Lithobius* (*Monotarsobius*)). [**Note:** Reference not obtained].
— Jeekel, 2005: 26 (list).

♂♀*A.* (*P.*) *diplonyx* Chamberlin, 1916d
— Chamberlin, 1916: 184-186, pl. 5; Fig. 5; pl. 6; Fig. 7-8 (*Arebius*; subgenus *Pagobius*; characters; note; USA: California).
— Chamberlin, 1943b: 105 (*Arebius*; USA: New Mexico).
— Jeekel, 2005: 26 (*Arebius*; subgenus *Pagobius*; list).

Distribution: USA: California, New Mexico.

HOLOTYPE: MCZ, No. 342, California: Santa Barbara, R. V. Chamberlin. [**Note:** During the author's visit to the MCZ in August of 2006 the vial 342 was not found and the MCZ database remarks field said see MCZ 435 which was originally designated as a paratype vial, see below].

PARATYPE: MCZ, No. 341, 434-436, California: Santa Barbara, R. V. Chamberlin. [**Note:** Vial 435, collecting date of March 1913, originally designated as a paratype vial, has 3 specimens and was sorted at the MCZ as a type (Holotype); vials 434 & 436 were also found during author's visit in August 2006].

♀*A.* (*P.*) *dolius* Chamberlin, 1916
— Chamberlin, 1916: 186-188, pl. 6; Fig. 4-6 (*Arebius*; subgenus *Pagobius*; characters; USA: California).

Distribution: USA: California.

HOLOTYPE: MCZ, No. 343, California: Friant, March 1913, R. V. & S. C. Chamberlin.

PARATYPES: MCZ, No. 431, California: Santa Barbara, R. V. Chamberlin; MCZ, No. 432, -433, 450 California: Friant, March 1913, R. V. & S. C. Chamberlin.

♂♀*A.* (*P.*) *kochii* (Stuxberg, 1875)
— Stuxberg, 1875a: 68-69 (*Lithobius*; characters; USA: California).
— Stuxberg, 1875b: 18 (*Lithobius*; subgenus *Archilithobius*; list).
— Stuxberg, 1875c: 6-7 [=1877: 137-138] (*Lithobius*; subgenus *Archilithobius*; habitat; list).
— Stuxberg, 1875d: 190 (*Lithobius*; characters; USA: California).
— Bollman, 1893: 128 (*Lithobius*; subgenus *Archilithobius*; catalogue).
— Chamberlin (in part), 1909b: 187 (*Lithobius*; notes; USA: California, Colorado).
— Chamberlin, 1910: 371, 374 (*Lithobius*; characters; USA: California).
— Chamberlin, 1911f: 379 (*Lithobius*; USA: California).
— Chamberlin, 1916: 182-184, pl. 5; Fig. 6; pl. 6; Fig. 1-3 (*Arebius*; subgenus *Pagobius*; characters; note; USA: California, Oregon).

Distribution: USA: California, Colorado, Oregon.

HOLOTYPE: ?Unknown, California: Sausalito.

TOPOTYPES: MCZ, No. 344 (slide), 347, California: Marin: Sausalito, 4-1911, R. V. Chamberlin.

Genus *Arenobius* Chamberlin, 1912
[1 species; Southeastern USA]

Arenobius Chamberlin, 1912e: 177-178 (characters; distribution; notes).
TYPE SPECIES: *Lithobius manegitus* Chamberlin, 1911, fixed by original designation.
— Chamberlin, 1917: 247-248 (characters; notes).
— Attems, 1938a: 344 (synonym of *Lithobius* (*Pokabius*)).
— Jeekel, 2005: 12 (list).

♂♀*A. manegitus* (Chamberlin, 1911)
— Chamberlin, 1911b: 43-44, pl. 4; Figs. 4-7 (*Lithobius*; characters; notes; USA: North Carolina, Tennessee).
— Chamberlin, 1912e: 178 (*Arenobius*; designated as type species).

— Chamberlin, 1917: 248-255; pl. 6; Figs. 3-6 (*Arenobius*; characters (adult, stadia); USA: North Carolina, Tennessee).
— Brimley, 1938: 500 (*Lithobius*; USA: North Carolina).
— Causey, 1940: 69 (*Arenobius*; characters).
— Chamberlin, 1944d: 200 (*Arenobius*; USA: Tennessee).
— Crabill, 1952b: 292-293; 417, pl. 6; Fig. 80 (*Arenobius*; characters; plectrotaxy, USA: North Carolina, South Carolina, Virginia).
— Branson & Batch, 1967: 87-88 (*Arenobius*; characters; habitat; USA: Kentucky).
— Wray, 1967: 154 (*Arenobius*; list).
— Hoffman, 1995a: 28 (*Arenobius*; list; notes).
— Jeekel, 2005: 12 (*Lithobius*; list).

Distribution: USA: Kentucky, North Carolina, South Carolina, Tennessee, Virginia.

TYPES: MCZ, No. 596, 5 specimens, North Carolina: Hot Springs, 8-7-1910.

PARATYPES: MCZ, No. 588, North Carolina: Hot Springs; MCZ, No. 597, North Carolina: Catawba; MCZ, No. 594, North Carolina: Saluda; MCZ, No. 591, North Carolina: Linville Falls; MCZ, No. 590, Tennessee: Johnson City; MCZ, No. 593, Tennessee: Unaka Springs; MCZ, No. 595, Tennessee: Altapass.

Genus *Arkansobius* Chamberlin, 1938
[1 species; Southern Central USA]

Arkansobius Chamberlin, 1938b: 626 (characters). TYPE SPECIES: *Arkansobius lamprus* Chamberlin, 1938, fixed by original designation and monotypy.
— Jeekel, 2005: 12, 36 (list).

♂♀*A. lamprus* Chamberlin, 1938
— Chamberlin, 1938b: 626-627 (*Arkansobius*; characters; USA: Arkansas).
— Jeekel, 2005: 12, 39 (*Arkansobius*; list).

Distribution: USA: Arkansas.

HOLOTYPE: NMNH, Arkansas: Pike Co., 7 April 1937.

Genus *Australobius* Chamberlin, 1920
[1 species (Introduced); USA: Hawaii]

Australobius Chamberlin, 1920c: 75-76 (characters). **Type Species:**
Australobius scabrior Chamberlin, 1920, fixed by original designation.
Australobius (*Malayobius*) Chamberlin, 1938: 628 (characters). **Type Species:**
Australobius vians Chamberlin, 1938, fixed by monotypy.
— Eason, 1978: 22 (synonym of *Australobius*).
— Jeekel, 2005: 22 (list).

♀*A. vians* Chamberlin, 1938
— Chamberlin, 1938b: 628 (*Australobius*; subgenus *Maylayobius*;
characters; USA: Hawaii). [**Note:** This genus does not occur in North
America or Hawaii naturally and has been introduced. According
to Chamberlin the type may be based on a subadult female lacking
one moult].
— Zapparoli & Shelley, 2000: 47 (*Australobius*; subgenus *Malayobius*;
list; notes).

Distribution: USA: ?Hawaii.

HOLOTYPE: NMNH, ♀, Hawaii, 8 May 1931 (Origin: Malay States).

Genus *Banobius* Chamberlin, 1938
[1 species; Northwestern USA]

Banobius Chamberlin, 1938b: 628 (characters). **Type Species:** *Banobius tener*
Chamberlin, 1938, fixed by original designation and monotypy.
— Jeekel, 2005: 12 (*Banobius*; list).

♂*B. tener* Chamberlin, 1938
— Chamberlin, 1938b: 629 (*Banobius*; characters; habitat; USA:
Oregon).
— Jeekel, 2005: 12 (list).

Distribution: USA: Oregon.

HOLOTYPE: NMNH, ♂, Oregon: Boyer, 13 May 1933, by J. A. Macnab.

Genus *Calcibius* Chamberlin & Wang, 1952
[1 species; Northwestern USA]

Calcibius Chamberlin & Wang, 1952a: 55 (characters). **TYPE SPECIES:** *Calcibius calcarifer* Chamberlin & Wang, 1952, fixed by original designation and monotypy.
— Jeekel, 2005: 13 (*Calcibius*; list).

♂*C. calcarifer* Chamberlin & Wang, 1952
— Chamberlin & Wang, 1952a: 55 (*Calcibius*; characters; USA: Washington).
— Jeekel, 2005: 13 (list).

Distribution: USA: Washington.

HOLOTYPE: NMNH, ♂, Washington: Ellensberg, 29 August 1929, by R. V. Chamberlin. [**Note:** The specific name on the holotype label reads *calcifer*. Furthermore, the holotype vial at the NMNH has the 19th of August as the day collected as opposed to the 29th from the literature].

Genus *Cruzobius* Chamberlin, 1942
[1 species (Introduced); Southern Central USA]

Cruzobius Chamberlin, 1942c: 21 (characters). **TYPE SPECIES:** *Cruzobius verus* Chamberlin, 1942, fixed by original designation and monotypy.
— Jeekel, 2005: 14 (list).

♂*C. verus* Chamberlin, 1942
— Chamberlin, 1942c: 22 (*Cruzobius*; characters; USA: Texas).
— Jeekel, 2005: 14 (*Cruzobius*; list).

Distribution: USA: ?Texas.

HOLOTYPE: NMNH, Texas: Laredo, at quarantine, 5 June 1941, with orchids brought from Vera Cruz, Mexico.

Genus *Elattobius* Chamberlin, 1941
[1 species (Introduced); Southeastern USA]

Elattobius Chamberlin, 1941a: 3 (characters). **TYPE SPECIES:** *Elattobius simplex* Chamberlin, 1941, fixed by monotypy.
— Jeekel, 2005: 15 (list).

♀*E. simplex* Chamberlin, 1941

— Chamberlin, 1941a: 3 (*Elattobius*; characters; USA: Louisiana).
[**Note:** This is a species that originated from Mexico which may
have the potential of becoming established in the United States.
Chamberlin states that this genus is similar to *Arkansobius*, however,
the antennae have more than 19 articles].

— Jeekel, 2005: 15 (*Elattobius*; list).

Distribution: USA: ?Louisiana.

HOLOTYPE: NMNH, ♀, Louisiana: New Orleans, 12 July 193?, taken
on banana from Mexico.

Genus *Enarthrobius* Chamberlin, 1926
[6 species; Southeastern and Western USA]

Enarthrobius Chamberlin, 1926a: 9 (characters; note). TYPE SPECIES:
Enarthrobius bullifer Chamberlin, 1926, fixed by original designation
and monotypy.

— Chamberlin, 1944d: 204-205 (key to subgenera and species;
notes).

— Crabill, 1952b: 295 (characters; notes).

— Jeekel, 2005: 15 (list).

(Subgenus *Capnobius* Chamberlin, 1944)

Enarthrobius (*Capnobius*) Chamberlin, 1944d: 205, 206 (characters; key).
TYPE SPECIES: *Enarthrobius covenus* Chamberlin, 1944, fixed by original
designation.

— Jeekel, 2005: 13 (*Enarthrobius*; list; note).

♀(imm.) *E.* (*C.*) *covenus* Chamberlin, 1944

— Chamberlin, 1944d: 206-207 (*Enarthrobius*; subgenus *Capnobius*;
characters; notes; USA: Tennessee). [**Note:** This species has been
based on two immature female specimens].

— Crabill, 1952b: 295 (*Enarthrobius*; subgenus *Capnobius*; list).

— Jeekel, 2005: 13 (list).

Distribution: USA: Tennessee.

HOLOTYPE: FMNH, ♀, Tennessee: Great Smoky Mountains National
Park, Greenbrier Cove, 14-19 June 1942, by H.S. Dybas.

PARATYPE: FMNH, ♀, Tennessee: Great Smoky Mountains National Park, Greenbrier Cove, 14-19 June 1942, by H.S. Dybas. NMNH, 2 specimens labeled as types, Tennessee: Great Smoky Mountains National Park, Greenbrier Cove, 13-19 June 1942, by H.S. Dybas.

♂*E.* (*C.*) *dybasi* Chamberlin, 1944
— Chamberlin, 1944d: 207 (*Enarthrobius*; subgenus *Capnobius*; characters; USA: Tennessee).
— Crabill, 1952b: 295, pl. 6; Fig. 84 (*Enarthrobius*; subgenus *Capnobius*; list).

Distribution: USA: Tennessee.

HOLOTYPE: FMNH, ♂, Tennessee: Great Smoky Mountains National Park, Greenbrier Cove, 13-14 June 1942, by H.S. Dybas.

(Subgenus *Enarthrobius* Chamberlin, 1926)

Enarthrobius (*Enarthrobius*) Chamberlin, 1926a: 9 (characters). **TYPE SPECIES:** *Enarthrobius* (*Enarthrobius*) *bullifer* Chamberlin, 1926, fixed by original designation.
— Chamberlin, 1944d: 204–205 (notes; key)

♂♀*E.* (*E.*) *bullifer* Chamberlin, 1926
— Chamberlin, 1926a: 9-10 (*Enarthrobius*; characters; habitat; USA: South Carolina). [**Note:** A young female was collected with the holotype specimen. Closely related to *E. oblitus*].
— Crabill, 1952b: 295 (*Enarthrobius*; subgenus *Enarthrobius*; list).
— Jeekel, 2005: 15 (*Enarthrobius*; list).

Distribution: USA: South Carolina.

HOLOTYPE: MCZ, No. 2233, ♂, South Carolina: Charleston, taken in a heap of rubbish in a garbage by J. T. Rogers. [**Note:** NMNH also has a vial labeled as a holotype].

♂♀*E.* (*E.*) *fumans* Chamberlin, 1944
— Chamberlin, 1944d, p. 206 (*Enarthrobius*; subgenus *Enarthrobius*; characters; USA: Tennessee).
— Crabill, 1952b: 295 (*Enarthrobius*; subgenus *Enarthrobius*; list).

Distribution: USA: Tennessee.

HOLOTYPE: FMNH, ♂, Tennessee: Great Smoky Mountains National Park, Greenbrier Cove, 14-19 June 1942, by H.S. Dybas.

ALLOTYPE: FMNH, ♀, Tennessee: Great Smoky Mountains National Park, Greenbrier Cove, 14-19 June 1942, by H.S. Dybas.

♂*E. (E.) litus* Chamberlin, 1944
— Chamberlin, 1944d: 205 (*Enarthrobius*; subgenus *Enarthrobius*; characters; USA: Tennessee).
— Crabill, 1952b: 295 (*Enarthrobius*; subgenus *Enarthrobius*; list).

Distribution: USA: Tennessee.

HOLOTYPE: FMNH, ♂, Tennessee: Great Smoky Mountains National Park, Greenbrier Cove, 14-19 June 1942, by H.S. Dybas. [**Note:** NMNH has a vial labeled as a holotype].

PARATYPE: FMNH, ♂, Tennessee: Great Smoky Mountains National Park, Greenbrier Cove, 14-19 June 1942, by H.S. Dybas.

♂♀*E. (E.) oblitus* Chamberlin & Wang, 1952
— Chamberlin & Wang, 1952a: 55-56 (*Enarthrobius*; characters; notes; USA: California).

Distribution: USA: California.

SYNTYPE: NMNH, California: Claremont.

Genus *Escimobius* Chamberlin, 1949
[1 species; USA: Alaska]

Escimobius Chamberlin, 1949a: 12-13 (characters). Type Species: *Escimobius cryophilus* Chamberlin, 1949, fixed by original designation and monotypy.
— Jeekel, 2005: 15 (list).

♂♀*E. cryophilus* Chamberlin, 1949
— Chamberlin, 1949a: 13 (*Escimobius*; characters; USA: Alaska). [**Note:** The female specimen collected is said to be mutilated but agrees in general characters with the male holotype which is missing the tibia and tarsi from each ultimate leg, as well as the claws from the penultimate legs. Chamberlin distinguishes this genus from *Oabius* by " . . . a definite dorsal keel at the distal end

of the fifth joint of the penultimate legs, with the succeeding two joints abruptly and considerably thinner". Otherwise, *Escimobius* seems to be described from a less than mature *Oabius*].
— Behan-Pelletier, 1993: 26; Tab. 2 (*Escimobius*; distribution).
— Jeekel, 2005: 15 (*Escimobius*; list).

Distribution: USA: Alaska.

HOLOTYPE: NMNH, ♂, Alaska: north front of the Brooks Range, 68°20' N. lat. and 151°30' W. long., 20 August 1948, by Neal A. Weber. [**Note:** Date seems to be 26ᵗʰ of August not the 20ᵗʰ].

Genus *Eulithobius* Stuxberg, 1875
[3 species; Central and Southeastern USA]

Eulithobius Stuxberg, 1875c [=1877: 134]: 26 (cited as subgenus; list). **Type Species:** *Lithobius* (*Lithobius*) *punctulatus* C. L. Koch, 1847 [= *Lithobius validus* Meinert, 1872], fixed by subsequent designation of Jeekel, 2005.
— Crabill, 1958d: 260, 262 (elevated to genus level, key to American species; notes).
— Eason, 1974a: 71 (notes). [**Note:** The author has left these three species under *Eulithobius* but based on the type species designation and according to Eason they belong in the genus *Lithobius*].
— Jeekel, 2005: 16 (list; notes; designation of type).

♀*E. fattigi* Chamberlin, 1945
— Chamberlin, 1945d: 216 (*Eulithobius*; characters; USA: Georgia).

Distribution: USA: Georgia.

HOLOTYPE: NMNH, ♀, Georgia: Boston, 14 April 1939.

♂♀*E. hypogeus* Chamberlin, 1940
— Chamberlin, 1940b: 48-49 (*Eulithobius*; characters; USA: Florida). [**Note:** This species is apparently associated with the Florida Pocket Gopher (*Geomys floridanus*) making it a unique Eulithobiid. *Pholobius goffi*, a monotypic genus, is also a cohabitant].
— Crabill, 1958d: 262 (key).

Distribution: USA: Florida.

HOLOTYPE: NMNH, Florida: ?Putnam Co., ?Melrose, 1939

♂♀*E. sphactes* Crabill, 1958
— Crabill, 1958d: 260-262 (*Eulithobius*; characters (plectrotaxy); key; USA: Oklahoma).

Distribution: USA: Oklahoma.

HOLOTYPE: NMNH, No. 2467, ♂, Oklahoma, Muskogee Co., Fort Gibson, Dresser's Cave, 28 January 1958, by Thomas C. Barr.

ALLOTYPE: ?NMNH, ♀, Oklahoma, Muskogee Co., Fort Gibson, Dresser's Cave, 28 January 1958, by Thomas C. Barr.

PARATYPE: ?NMNH, ♂, Oklahoma, Muskogee Co., Fort Gibson, Dresser's Cave, 28 January 1958, by Thomas C. Barr.

Genus *Garibius* Chamberlin, 1913
[9 species; Eastern USA]

Garibius Chamberlin, 1913: 61-63 (characters; distribution; habitat; key to species; note). TYPE SPECIES: *Garibius monticolens* Chamberlin, 1913, fixed by original designation.
— Attems, 1938a: 344 (synonym of *Lithobius* (*Pokabius*)).
— Crabill, 1952b: 374 (characters; notes; species list).
— Palmén 1954: 143 (Palmén reports a *Lithobius* sp. from Newfoundland, Canada and suggests that it may belong to the genus *Garibius*. The only species of *Garibius* that is known to exists so far north is *Garibius opicolens* (Chamberlin) and this may be worth investigating).
— Crabill, 1957b: 375-376 (notes).
— Jeekel, 2005: 17 (list).

♂*G. alabamae* Chamberlin, 1913
— Chamberlin, 1913: 78-79, pl. 4; Fig. 8-9 (*Garibius*; characters; USA: Alabama).
— Crabill, 1952b: 375 (*Garibius*; list).

Distribution: USA: Alabama.

HOLOTYPE: MCZ, No. 455, Alabama: Morgan, 7-27-10.

♂♀*G. branneri* (Bollman, 1888)
- Bollman, 1888a: 107-108, 111, 112 [= 1893: 82, 84, 85] (*Lithobius*; subgenus *Archilithobius*; characters; USA: Tennessee).
- Bollman, 1888b: 7 [= 1893: 79] (*Lithobius*; note; USA: Arkansas).
- Bollman, 1888c: 342 [= 1893: 93] (*Lithobius*; note; USA: Tennessee).
- Bollman, 1893: 128 (*Lithobius*; subgenus *Archilithobius*; catalogue).
- Chamberlin, 1911b: 39 (*Lithobius*; notes; USA: Alabama, Mississippi, North Carolina, Tennessee).
- Chamberlin, 1913: 80 (*Garibius*; characters; USA: Tennessee).
- Brimley, 1938: 500 (*Lithobius*; USA: North Carolina).
- Causey, 1940: 69 (*Garibius*; characters).
- Crabill, 1952b: 385 (*Garibius*). [**Note:** According to Crabill, "The affinity and identity of . . ." *Garibius branneri* (Bollman) " . . . is uncertain due to the inadequate original description[s] and the unavailablity of the type]."
- Crabill, 1957b: 377. (*Garibius*; notes). [**Note:** Crabill regards this species as a species inquirendae].
- Wray, 1967: 154 (*Garibius*; list).
- Hoffman, 1995a: 28 (*Garibius*; list; notes). [**Note:** Hoffman believes this species will be found in Virginia].

Distribution: USA: Alabama, Arkansas, Georgia, Mississippi, North Carolina, Tennessee.

HOLOTYPE: ISM, Tennessee: Knox co., Knoxville, in the woods ca. 1 mi. S. of the river at Knoxville, near Maryville road, May 21, 1887.

COTYPES: NMNH, (see Crabill, 1957b: 377).

♂*G. catawbae* Chamberlin, 1913
- Chamberlin, 1913: 63-64, pl. 4; Fig. 8-9 (*Garibius*; characters; USA: North Carolina).
- Brimley, 1938: 500 (*Lithobius*; note; USA: North Carolina).
- Causey, 1940: 70 (*Garibius*; characters).
- Crabill, 1952b: 375 (*Garibius*; list).
- Wray, 1967: 154 (*Garibius*; list).

Distribution: USA: North Carolina.

HOLOTYPE: MCZ, No. 460, North Carolina: Catawba, 12/8/40.

♂♀ *G. georgiae* Chamberlin, 1913
 — Chamberlin, 1913: 76-78 (*Garibius*; characters; USA: Georgia).
 — Chamberlin, 1951b: 32 (*Garibius*; USA: Georgia).
 — Crabill, 1952b: 375 (*Garibius*; list).
 — Wray, 1967: 154 (*Garibius*; list).
 — Hoffman, 1995a: 28 (*Garibius*; list; notes).

Distribution: USA: Georgia, North Carolina.

TYPES: MCZ, No. 453, 4 specimens, Georgia: Atlanta, 7-30-1910, R.V.C.

♂ *G. mississippiensis* Chamberlin, 1913
 — Chamberlin, 1913: 74-76 (*Garibius*; characters; USA: Mississippi).
 — Crabill, 1952b: 375 (*Garibius*; list).

Distribution: USA: Mississippi.

HOLOTYPE: MCZ, No. 454, Mississippi: Brookhaven, 7-19-'10. [**Note:** Holotype vial found dry at MCZ during author's visit in August 2006].

♂ *G. monticolens* Chamberlin, 1913
 — Chamberlin, 1913: 65-66, pl. 4; Fig. 1-4, 6-7, pl. 5; Fig. 1 (*Garibius*; characters; USA: North Carolina, Tennessee).
 — Brimley, 1938: 500 (*Lithobius*; note; USA: North Carolina).
 — Causey, 1940: 70 (*Garibius*; characters).
 — Crabill, 1950c: 202 (*Garibius*; USA: South Carolina).
 — Crabill, 1952b: 375 (*Garibius*; list).
 — Wray, 1967: 154 (*Garibius*; list).
 — Lee, 1980: 5; Tab. 1 (*Garibius*; habitat; USA: Ohio).
 — Hoffman, 1995a: 28 (*Garibius*; list; notes). [**Note:** Hoffman believes this species will be found in Virginia].
 — Jeekel, 2005: 17 (*Garibius*; list).

Distribution: USA: North Carolina, Ohio, South Carolina, Tennessee.

HOLOTYPE: MCZ, No. 461, Tennessee: Unaka Springs, Aug. 11, 1910. [**Note:** Holotype vial found dry during author's visit in August 2006].

♂♀ *G. opicolens* Chamberlin, 1913

— Chamberlin, 1913: 67-70 (*Garibius*; characters (adult, immaturus, agenitalis I); habitat; notes; USA: New Jersey; Pennsylvania). [**Note:** Chamberlin states that *G. pagoketes*, is a close ally to *G. monticolens* and *G. dendrophilus*, now a synonym of *Lithobius crassipes*].

— Crabill, 1952b: 375-378, pl. 8; Fig. 100 (*Garibius*; characters; range; taxonomic notes; USA: Kentucky, Massachusetts, New Jersey, New York, Pennsylvania, Virginia, West Virginia).

— Crabill, 1955f: 260 (*Garibius*; checklist; USA: Kentucky).

— Crabill, 1957b: 376-377 (*Garibius*; synonomy of *pagoketes* with *opicolens*).

— Lee, 1980: 5; Tab. 1 (*Garibius*; habitat; USA: Ohio).

— Summers et al., 1980: 249; Fig. 18 (*Garibius*; checklist; USA: Illinois).

— Summers et al., 1981: 61 (*Garibius*; checklist; USA: Illinois).

— Hoffman, 1995a: 28 (*Garibius*; list; habitat; notes).

— Mercurio, 2005: 232 (*Garibius*; habitat; USA: New York).

pagoketes Chamberlin, 1913: 72-74 (*Garibius*; characters; notes; USA: Massachusetts). **Type Locality:** Massachusetts: Blue Hills.

— Crabill, 1952b: 375 (*Garibius*; synonym of *opicolens*).

— Crabill, 1957b: 377 (*Garibius*; synonym of *opicolens*).

— Branson & Batch, 1967: 88 (*Garibius*; brief description; habitat; USA: Kentucky).

— Lee, 1980: 5; Tab. 1 (*Garibius*; habitat; USA: Ohio).

— Hoffman, 1995a: 28 (*Garibius*; list; notes; USA: Virginia). [**Note:** Hoffman has kept *pagoketes* as a valid species until further scrutiny].

Distribution: USA: Illinois, Kentucky, Massachusetts, New Jersey, New York, Ohio, Pennsylvania, Virginia, West Virginia.

HOLOTYPE: MCZ, New Jersey: Masonville.

♀*G. pagoketes* Chamberlin, 1913

— Chamberlin, 1913: 72-74 (*Garibius*; characters; notes; USA: Massachusetts).

— Crabill, 1952b: 375 (*Garibius*; synonym of *opicolens*).

— Crabill, 1957b: 377 (*Garibius*; synonym of *opicolens*).

— Branson & Batch, 1967: 88 (*Garibius*; brief description; habitat; USA: Kentucky).

— Lee, 1980: 5; Tab. 1 (*Garibius*; habitat; USA: Ohio).

— Hoffman, 1995a: 28 (*Garibius*; list; notes; USA: Virginia).
[**Note:** Hoffman has kept *pagoketes* as a valid species until further
scrutiny].

Distribution: USA: Kentucky, Massachusetts, Ohio, Virginia.

HOLOTYPE: ?NMNH; Massachusetts: Blue Hills.

♂ *G. psychrophilus* Crabill, 1957
— Crabill, 1957b: 376 (*Garibius*; characters (plectrotaxy); USA:
Virginia).
— Hoffman, 1995a: 28 (*Garibius*; list; notes; USA: Virginia).

Distribution: USA: Virginia.

HOLOTYPE: NMNH; No. 2382; ♂, Virginia: Montgomery Co.,
Blacksburg, February 1957, by R. L. Hoffman.

Genus *Georgibius* Chamberlin, 1944
[1 species; Southeastern USA]

Georgibius Chamberlin, 1944a: 34 (characters). **TYPE SPECIES:** *Georgibius
georgiae* Chamberlin, 1944, fixed by original designation and monotypy.
[**Note:** Chamberlin erected the genus *Georgibius* based on one male
specimen and compares *Georgibius* to *Sonibius* and *Enarthrobius*. but never
mentions *Paitobius*. The NMNH has a holotype labeled as *Georgibius
pulblanus* Chamberlin in the database].
Crabill, 1952b: 304 (synonym of *Paitobius*; key to species of
northeastern USA). [**Note:** Crabill seemed to suspect that *Georgibius*
was a junior synonym of *Paitobius* Chamberlin, but Crabill never
stated of what *Paitobius* species *G. georgiae* is a synonym. There
are four possible species it could belong to as follows: *P. atlantae*
Chamberlin, *naiwatus* (Chamberlin), *watsuitus* (Chamberlin) and *P.
zinus* (Meinert) that are known to exist in Georgia. The author finds
georgiae to be very close to *naiwatus* based on the general description
but the holotype should be studied in detail and compared thoroughly
with *Paitobius* species from Georgia. In this case, because the evidence
is not substantial and this information is not from a peer reviewed
publication, the author is not accepting the synonymy of Crabill until
there is further evidence the genus *Georgibius* is indeed *Paitobius* or
some other genus].

— Jeekel, 2005: 17 (list).

♂G. georgiae Chamberlin, 1944
   — Chamberlin, 1944a: 34-35 (*Georgibius*; characters; USA: Georgia).
   — Jeekel, 2005: 17 (*Georgibius*; list).

Distribution: USA: Georgia.

HOLOTYPE: ?NMNH, ♂, Georgia: Brier Creek, 7 miles north of Sylvania, April 12, 1943.

Genus *Gonibius* Chamberlin, 1925
[2 species; Southern USA]

Gonibius Chamberlin, 1925: 445-447; fig. 2 (characters; distribution; key to species). TYPE SPECIES: *Lithobius rex* Bollman, 1888, by subsequent designation of Chamberlin, 1925: 445.
   — Attems, 1938a: 344 (synonym of *Lithobius* (*Australobius*)).
   — Crabill, 1952b: 294 (characters; notes).
   — Jeekel, 2005: 17 (list).

♂♀G. rex (Bollman, 1888)
   — Bollman, 1888d: 350 [= 1893: 102-103] (*Lithobius*; characters; notes; USA: Georgia).
   — Bollman, 1893: 130 (*Lithobius*; subgenus *Eulithobius*; catalogue).
   — Chamberlin, 1911b: 47 (*Lithobius*; USA: Georgia).
   — Chamberlin, 1921a: 230 (*Sonibius*; note; USA: Tennessee).
   — Chamberlin, 1925d: 447-448, pl. 1; Fig. 4, 5 (*Gonibius*; subgenus *Gonibius*; characters; USA: Georgia).
   — Chamberlin, 1944d: 209 (*Gonibius*; notes; USA: Tennessee).
   — Crabill, 1952b: 294-295 (*Gonibius*; characters; notes).
   — Jeekel, 2005: 17 (*Lithobius*; list).

Distribution: USA: Georgia, Tennessee.

HOLOTYPE: NMNH, Acc. 19542, 21, ♀, Georgia: Tallulah.

(Subgenus *Tambius* Chamberlin, 1925)

Gonibius (*Tambius*) Chamberlin, 1925d: 448–450 (characters). TYPE SPECIES: *Lithobius glyptocephalus* Chamberlin, 1903, fixed by original designation.

— Attems, 1938a: 344 (synonym of *Lithobius* (*Australobius*)).
— Eason, 1992: 3 (Eason questions Attems' synonymy).
— Jeekel, 2005: 32 (list).

♂♀*G.* (*T.*) *glyptocephalus* (Chamberlin, 1903)
— Chamberlin, 1903a: 35-36 (*Lithobius*; characters; Chamberlin
remarks: this species seems to be nearest *Lithobius howei* Bollman;
USA: New Mexico).
— Chamberlin, 1925d: 448-450 (*Gonibius*; subgenus *Tambius*; characters;
USA: New Mexico).
— Jeekel, 2005: 31 (*Lithobius*; list).

Distribution: USA: New Mexico.

HOLOTYPE: ?Unknown, New Mexico: Beulah, ca. 8,000 ft.

Genus *Gosibius* Chamberlin, 1912
[21 species; Western & Southern USA]

Gosibius Chamberlin, 1912e: 204 (characters; note). Type Species: *Gosibius
monicus* Chamberlin, 1912, fixed by monotypy. [**Note:** The original
designation of the type species *Lithobius paucidens* Wood, 1862, by
Chamberlin for *Gosibius* is invalid due to a literature date priority of
Chamberlin, 1912d over Chamberlin, 1912e].
— Attems, 1938a: 344 (synonym of *Lithobius* (*Alokobius*)).
— Crabill, 1952b: 294 (characters; notes).
— Jeekel, 2005: 17 (list; notes).

♂♀*G. aberrantus* Chamberlin, 1943
— Chamberlin, 1943b: 103 (*Gosibius*; characters; note; USA: New
Mexico).

Distribution: USA: New Mexico.

HOLOTYPE: NMNH, ♀, New Mexico: West Ruidosa Junction, 1 June
1941. [**Note:** The type determined label *Gosibius aberrantus ruidosus*].

PARATYPES: NMNH, New Mexico. [**Note:** 5 vials from various localities
with numerous specimens].

♀*G. ameles* Chamberlin, 1940

89

— Chamberlin, 1940d: 54-55 (*Gosibius*; characters; notes; USA: Utah). [**Note:** According to Chamberlin this species is close in form to *G. monicus*].

Distribution: USA: Utah.

HOLOTYPE: NMNH, ♀, Utah: near Verdure, 18 April 1928, R. V. Chamberlin.

♂ *G. atopops* Chamberlin, 1941
— Chamberlin, 1941a: 20 (*Gosibius*; characters; USA: California).

Distribution: USA: California.

HOLOTYPE: NMNH, ♂, California: Kern Co., 7 mi. N of Glenville, 19 March 1941, S. & D. Mulaik. [**Note:** Type is labeled as *Guambius* not *Gosibius*].

♀ *G. benespinosus* Chamberlin, 1941
— Chamberlin, 1941a: 19-20 (*Gosibius*; characters; USA: California).
— Chamberlin, 1944c: 79 (*Gosibius*; USA: California).

Distribution: USA: California.

HOLOTYPE: NMNH, ♀, California: 6 mi. W of Bishop, and at Benton Station, 16 March 1941, by S. & D. Mulaik. [**Note:** The NMNH has two vials that each has a designated female holotype specimen. The vial with one specimen is from 6 mi. W of Bishop, however, the second vial with a female holotype is from Benton Station].

PARATYPES: NMNH, numerous specimens, California: 11 mi. W of Bishop, 16 March 1941, by S. & D. Mulaik. NMNH, Benton Station, 16 March 1941, by S. & D. Mulaik.

♂♀ *G. escabosanus* Chamberlin, 1943
— Chamberlin, 1943b: 103-104 (*Gosibius*; characters; USA: New Mexico).

Distribution: USA: New Mexico.

HOLOTYPE: NMNH, New Mexico: Escobosa, 30 May 1941.

♂♀ *G. louisianus* Chamberlin, 1942
— Chamberlin, 1942d: 186-187 (*Gosibius*; characters; habitat; USA: Louisiana).

Distribution: USA: Louisiana.

HOLOTYPE: NMNH, ♂, Louisiana: 5 mi. NW of Shreveport, 13 April 1936, under log, by Leslie Hubricht.

ALLOTYPE: NMNH, ♀, Louisiana: 5 mi. NW of Shreveport, 13 April 1936, under log, by Leslie Hubricht.

♂♀ *G. mulaiki* Chamberlin, 1938
— Chamberlin, 1938b: 634-635 (*Gosibius*; characters; USA: New Mexico).
— Chamberlin & Mulaik, 1940: 126-127 (*Gosibius*; list).

Distribution: USA: New Mexico.

HOLOTYPE: NMNH, New Mexico: Bear Canyon, Camp Mary White, 15 mi. SE of Cloudcroft, 20 August 1934, by S. Mulaik.

♂♀ *G. saccharogeus* Chamberlin, 1941
— Chamberlin, 1941a: 21-22 (*Gosibius*; characters; notes; USA: Texas). [**Note:** According to Chamberlin close in form to *G. monicus*].
— Chamberlin, 1951b: 33 (*Gosibius*; USA: Louisiana).

Distribution: USA: Louisiana, Texas.

HOLOTYPE: NMNH, Texas: Sugarland, 26 February 1938.

♂ *G. texicolens* Chamberlin, 1938
— Chamberlin, 1938b: 635 (*Gosibius*; characters; USA: Texas).
— Chamberlin & Mulaik, 1940: 126 (*Gosibius*; USA: Texas).
— Chamberlin, 1943b: 103 (*Gosibius*; USA: Texas).

Distribution: USA: Texas.

HOLOTYPE: NMNH, ♂, Texas: Edinburg, 29 November 1936, by S. Mulaik.

PARATYPE: NMNH, ♂, Texas, 1934.

(Subgenus *Abatobius* Chamberlin, 1917)

Gosibius (*Abatobius*) Chamberlin, 1917: 208 (key). TYPE SPECIES: *Gosibius arizonensis* Chamberlin, 1917, fixed by subsequent designation.
— Attems, 1926: 382, 383 (synonym of *Lithobius* (*Alokobius*)). [**Note:** Reference not obtained].
— Jeekel, 2005: 9 (list; note; designation of type).

♂♀ *G.* (*A.*) *angelicus* Chamberlin, 1944
— Chamberlin, 1944d: 200 (*Gosibius*; subgenus *Abatobius*; characters; USA: California).

Distribution: USA: California.

HOLOTYPE: FMNH, ♂, California: Los Angeles, 2 February 1936, by Gordon Grant.

ALLOTYPE: FMNH, ♀, California: Los Angeles, 2 February 1936, by Gordon Grant.

PARATYPES: FMNH, six specimens, California: Los Angeles, 2 February 1936, by Gordon Grant. NMNH, 2 specimens, California: Los Angeles, 2 February 1936, by Gordon Grant.

♂♀ *G.* (*A.*) *arizonensis* Chamberlin, 1917
— Chamberlin, 1917: 208, 223-225, pl. 2; Fig. 5; pl. 3; Fig. 1 (*Gosibius*; subgenus *Abatobius*; characters; key; USA: Arizona).
— Chamberlin, 1920d: 44 (*Gosibius*; USA: Arizona).
— Chamberlin, 1925e: 56-57 (*Gosibius*; note; USA: Utah).
— Chamberlin, 1928d: 307 (*Gosibius*; note; USA: Utah).
— Chamberlin, 1943b: 103 (*Gosibius*; USA: New Mexico).
— Crabill, 1952b, pl. 6; Fig. 83 (*Gosibius*; female gonopods).
— Chamberlin, 1958c: 133 (*Gosibius*; checklist; note). [**Note:** specific name cited as *arizonicus*].

Distribution: USA: Arizona, New Mexico, Utah.

SYNTYPES: MCZ, No. 581, 3 specimens, Arizona: Fort Williams, 7000 ft., 28 January 1911, W. M. Wheeler.

PARATYPES: MCZ, No. 612, Arizona: Fort Williams, W. M. Wheeler; MCZ, No. 611, Grand Canyon, W. M. Wheeler; MCZ, No. 610, Huachucha

Mountains, W. M. Wheeler; MCZ, No. 608, 613, Arizona, H. K. Morrison; and NMNH.

♀ *G.* (*A.*) *auxodontus* Chamberlin & Wang, 1952
— Chamberlin & Wang, 1952a: 60 (*Gosibius*; subgenus *Abatobius*; characters; note; USA: probably Utah). [**Note:** Apparently Chamberlin didn't know exactly where this specimen came from but he indicated " . . . probably Utah or an adjacent area"].

Distribution: USA: ?Utah.

HOLOTYPE: NMNH, ♀, ?Utah.

♀ *G.* (*A.*) *montereus* Chamberlin, 1917
— Chamberlin, 1917: 225-227, pl. 1; Fig. 5 (*Gosibius*; subgenus *Abatobius*; characters; USA: California).

Distribution: USA: California.

TYPES: MCZ, No. 582, 3 specimens, [**Note:** Chamberlin designated a ♀ as the holotype], California: Pacific Grove, March, 1913.

(Subgenus *Amplobius* Chamberlin, 1941)

Amplobius Chamberlin, 1941a: 20 (characters). **TYPE SPECIES:** *Gosibius fusatus* Chamberlin, 1941, fixed by original designation and monotypy.
— Jeekel, 2005: 11 (list; note).

♂♀ *G.* (*A.*) *fusatus* Chamberlin, 1941
— Chamberlin, 1941a: 21 (*Gosibius*; subgenus *Amplobius*; characters; USA: California).
— Jeekel, 2005: 11 (*Gosibius*; list).

Distribution: USA: California.

HOLOTYPE: NMNH, California: 11 mi. W of Glenville, 5, 000 ft., 19 March 1941

PARATYPE: NMNH.

(Subgenus *Gosibius* Chamberlin, 1912)

Gosibius (*Gosibius*) Chamberlin, 1912d: 146–147 (characters). TYPE SPECIES: *Gosibius monicus* Chamberlin, 1912, fixed by monotypy.
— Attems, 1938: 344 (synonym of *Lithobius* (*Alokobius*)).
— Jeekel, 2005: 17 (notes).

♂♀*G.* (*G.*) *brevicornis* Chamberlin, 1917
— Chamberlin, 1917: 215-219, pl. 2.; Figs. 1-4 (*Gosibius*; subgenus *Gosibius*; characters (adult, stadia); habitat; note; USA: California).

Distribution: USA: California.

HOLOTYPE: MCZ, No. 552, California: Friant, March, 1913. [**Note:** MCZ type vial has a label written by Crabill with the MCZ No. 352 not 552].

PARATYPES: MCZ, No. 545-548, 550-551, 553, California: Friant; MCZ, No. 549, California: Fresno. R. V. & S. C. Chamberlin.

♀*G.* (*G.*) *monicus* Chamberlin, 1912
— Chamberlin, 1912d: 146-147 (*Gosibius*; characters; USA: California).
— Chamberlin, 1917: 219-221 (*Gosibius*; subgenus *Gosibius*; characters; notes).
— Jeekel, 2005: 17 (*Gosibius*; list).

Distribution: USA: California.

COTYPES: MCZ, No. 578, 2 specimens, California: Santa Monica, June 1909, R. V. Chamberlin.

♂♀*G.* (*G.*) *paucidens* (Wood, 1862)
— Wood, 1862: 14 (*Lithobius*; characters; note; USA: California, ?Missouri).
— Wood, 1865: 151 (*Lithobius*; characters; USA: California).
— Stuxberg, 1875b: 12 (*Lithobius*; chart).
— Stuxberg, 1875c: 29 [=1877: 137] (*Lithobius*; list).
— Bollman, 1887b: 266 [= 1893: 33] (*Lithobius*; subgenus *Lithobius*; list).
— Bollman, 1893: 129 (*Lithobius*; subgenus *Lithobius*; catalogue).
— Chamberlin, 1909b: 191 (*Lithobius*; USA: California).
— Chamberlin, 1910: 372 (*Lithobius*; characters; key to species; USA: California).

— Chamberlin, 1912e: 204 (*Gosibius*; cited as type species).
— Chamberlin, 1917: 210-215, pl. 1; Figs. 1-4 (*Gosibius*; subgenus *Gosibius*; characters (adult, stadia); note; USA: California).
— Anonymous, 1920: 35 (*Gosibius*; USA: California).
— Gunthorp, 1920: 113 (*Lithobius*; list).
— Chamberlin, 1944d: 200 (*Gosibius*; USA: California).
— Crabill, 1952b: 294 (*Gosibius*; listed as type species; note).
— Jeekel, 2005: 17 (*Lithobius*; note).

Distribution: USA: California, ?Missouri.

HOLOTYPE: ?MCZ, California: Fort Tejon.

♂*G.* (*G.*) *submarginis* Chamberlin & Wang, 1952
— Chamberlin & Wang, 1952a: 60 (*Gosibius*; subgenus *Gosibius*; characters; notes; USA: Washington).

Distribution: USA: Washington.

HOLOTYPE: NMNH, ♂, Washington: between Seattle and Everett, August.

(Subgenus *Timiobius* Chamberlin, 1917)

Gosibius (*Timiobius*) Chamberlin, 1917: 221-223 (characters). Type Species: *Gosibius intermedius* Chamberlin, 1917, fixed by monotypy.
— Attems, 1926: 382, 383 (?).[**Note:** Reference not obtained].
— Attems, 1938: 344 (synonym of *Lithobius* (*Alokobius*)).
— Jeekel, 2005: 33 (list).

♂*G.* (*T.*) *intermedius* Chamberlin, 1917
— Chamberlin, 1917: 221-223, pl. 3; Figs. 2-3 (*Gosibius*; subgenus *Timiobius*; characters; note; USA: California).
— Jeekel, 2005: 33 (*Gosibius*; list).

Distribution: USA: California.

HOLOTYPE: MCZ, No. 580, ♂, California: Santa Barbara, March 1913. [**Note:** MCZ vial 580 is sorted as the paratype vial, containing 2 specimens, apparently based on a label from Crabill. Another vial (No. 579) was seemingly labeled by Crabill as the holotype, which contains one specimen. Chamberlin originally designated the vial No. 580 as the holotype vial, which contains 2 specimens. At this time it is unclear if there was an error by Chamberlin in the literature or an error based on the labels created by Crabill. Assuming Crabill examined the sex of the specimens and that he is right then the MCZ specimen No. 579 is actually the holotype unless the number labels were accidentally switched].

♀*G. (T.) sequens* Chamberlin, 1941
— Chamberlin, 1941a: 19 (*Gosibius*; subgenus *Timiobius*; characters; USA: California).

Distribution: USA: California.

HOLOTYPE: NMNH, ♀, California: Monterey Co., Hastings Reservation, 8 April 1941, by Dr. J. M. Linsdale.

Genus *Guambius* Chamberlin, 1912
[9 species; Western & Southern USA]

Guambius Chamberlin, 1912d: 144 (note). TYPE SPECIES: *Lithobius euthus* Chamberlin, 1904, fixed by original designation.
— Chamberlin, 1917: 233-236; fig. 3 (characters; distribution; key to subgenera and species; notes).
— Attems, 1938a: 344 (synonym of *Lithobius* (*Alokobius*)).
— Chamberlin, 1941a: 18-19 (key to species).
— Crabill, 1952b: 293 (characters; notes).
— Jeekel, 2005: 18 (list).

♂♀*G. hesperus* Chamberlin, 1941
— Chamberlin, 1941a: 18 (*Guambius*; characters; notes; USA: California).

Distribution: USA: California.

HOLOTYPE: NMNH, California: Monterey Co., Hastings Reservation.

(Subgenus *Guambius* Chamberlin, 1912)

Guambius (*Guambius*) Chamberlin, 1912: 144 (characters). TYPE SPECIES: *Lithobius euthus* Chamberlin, 1904, fixed by original designation.

♂*G. (G.) curtior* Chamberlin, 1917
— Chamberlin, 1917: 236-237, pl. 5; Fig. 6; pl. 6; Figs. 1-2 (*Guambius*; subgenus *Guambius*; characters; USA: Mississippi).

Distribution: USA: Mississippi.

HOLOTYPE: MCZ, No. 605, ♂, Mississippi: Gulfport, 7-20,21-1910.

♂♀ *G.* (*G.*) *euthus* (Chamberlin, 1904)
— Chamberlin, 1904: 652-653 (*Lithobius*; characters; USA: Texas).
— Chamberlin, 1911b: 40 (*Lithobius*; USA: Mississippi).
— Chamberlin, 1917: 235, 240-241 (*Guambius*; subgenus *Guambius*; cited as type species; characters; note; USA: Texas).
— Chamberlin, 1944d: 201 (*Guambius*; note; USA: Texas).
— Crabill, 1952b: 293 (*Guambius*; listed as type species; note).
— Jeekel, 2005: 18 (*Guambius*; list).

Distribution: USA: Mississippi, Texas.

HOLOTYPE: MCZ, No. 606, ♀, Texas: Austin, Mar. 12-18, 1903, J.H. Comstock. [**Note:** The holotype vial, containing 4 or 5 specimens, at the MCZ was found dry with the slightest hint of moisture].

♂♀ *G.* (*G.*) *mississippiensis* (Chamberlin, 1912)
— Chamberlin, 1912d: 144-146 (*Guambius*; characters; note; USA: Mississippi).
— Chamberlin, 1917: 237-240, pl. 5; Figs. 4-5 (*Guambius*; subgenus *Guambius*; characters; USA: Mississippi).

Distribution: USA: Mississippi.

HOLOTYPE: MCZ, No. 604, ♂, Mississippi: Byram, July 18, 1910, R. V. Chamberlin. [**Note:** original label reads "see mount" referring to the slide].

PARATYPE: MCZ, No. 603, Mississippi: Byram, July 1910, R. V. Chamberlin.

♀ *G.* (*G.*) *pinguis* (Bollman, 1888)
— Bollman, 1888b: 7 [= 1893: 79] (*Lithobius*; characters; note; USA: Arkansas).
— Bollman, 1893: 129 (*Lithobius*; subgenus *Archilithobius*; catalogue).
— Chamberlin, 1911b: 40 (*Lithobius*; notes; USA: Mississippi).
— Chamberlin, 1917: 241-242 (*Guambius*; subgenus *Guambius*; characters; note; USA: Arkansas, Mississippi). [**Note:** It should

be pointed out that the record for Mississippi was based on a "much rubbed specimen"].

Distribution: USA: Arkansas, Mississippi.

HOLOTYPE: ISM, Arkansas: Little Rock, by C.H. Bollman. [**Note:** NMNH has a type specimen].

(Subgenus *Sibibius* Chamberlin, 1912)

Guambius (*Sibibius*) Chamberlin, 1912e: 177, 178 (some characters but no description; originally under the genus *Arenobius*). TYPE SPECIES: *Arenobius coloradanus* Chamberlin, 1912, fixed by subsequent designation of Jeekel, 2005.
— Chamberlin, 1917: 235-236 (transferred from *Arenobius* to *Guambius*; key).
— Attems, 1938: 344 (synonym of *Lithobius* (*Alokobius*)).
— Jeekel, 2005: 30 (list; notes; designation of type species).

♂♀ *G.* (*S.*) *christianus* Chamberlin, 1946
— Chamberlin, 1946b: 194-195 (*Guambius*; subgenus *Sibibius*; characters; notes; USA: Mississippi).

Distribution: USA: Mississippi.

HOLOTYPE: NMNH, ♀, Mississippi: Pass Christian, 15 February 1946, by J. & W. Rapp.

ALLOTYPE: NMNH, ♂, Mississippi: Pass Christian, 15 February 1946.

♂♀ *G.* (*S.*) *coloradanus* (Chamberlin, 1912)
— Chamberlin, 1912d: 141-143 (*Arenobius*; characters; USA: Colorado).
— Chamberlin, 1917: 242-245, pl. 4; Figs. 3-5; pl. 5; Fig. 1 (*Guambius*; subgenus *Sibibius*; characters; transferred to *Guambius*; USA: Colorado).
— Chamberlin, 1943b: 104 (*Guambius*; USA: Colorado).
— Jeekel, 2005: 30 (*Arenobius*; list).

Distribution: USA: Colorado.

TYPES: MCZ, No. 599, 8 specimens, Colorado: Manitou, 8-21-1910, R. V. Chamberlin.

PARATYPE: MCZ, No. 600, Colorado: Durango, 1894, C. F. Baker.

♂♀ *G. (S.) hubrichti* Chamberlin, 1942
— Chamberlin, 1942d: 187 (*Guambius*; characters; habitat; USA: Louisiana). [**Note:** The female is immature that is known for this species].

Distribution: USA: Louisiana.

HOLOTYPE: NMNH, ♂, Louisiana: Natchatoches Parish, 2 mi. S of Saline, under logs, by Leslie Hubricht.

♂♀ *G. (S.) oedipes* (Bollman, 1888)
— Bollman, 1888b: 8 [= 1893: 80] (*Lithobius*; characters; note; USA: Arkansas).
— Bollman, 1893: 129 (*Lithobius*; subgenus *Lithobius*; catalogue).
— Chamberlin, 1911a: 68 (*Lithobius*; key to species; USA: Colorado).
— Chamberlin, 1911b: 43 (*Lithobius*; USA: Mississippi).
— Chamberlin, 1917: 245-247, pl. 5; Figs. 2-3 (*Guambius*; subgenus *Sibibius*; characters; notes).

Distribution: USA: Arkansas, Colorado, Louisiana, Mississippi.

SYNTYPE: NMNH, Arkansas: Little Rock.

Genus *Helembius* Chamberlin, 1918
[1 species; Southeastern USA]

Helembius Chamberlin, 1918b: 377-378 (characters). Type Species: *Helembius nannus* Chamberlin, 1918, fixed by original designation and monotypy.
— Attems, 1938a: 344 (synonym of *Lithobius* (*Monotarsobius*)).
— Crabill, 1952b: 366 (characters; note).
— Eason, 1992: 4 (taxonomic discussion). [**Note:** Eason suggests that *Helembius* Chamberlin may be a synonym of *Monotarsobius* but it has been left as a separate genus until further study. Eason also incorrectly states that *Helembius* is from Louisiana when it is only known from Georgia and possibly South Carolina].
— Jeekel, 2005: 18 (list).

♂♀ *H. nannus* Chamberlin, 1918

— Chamberlin, 1918b: 378-379 (*Helembius*; characters; USA: Georgia).
— Crabill, 1950c: 202 (*Helembius*; possible report from USA: South Carolina). [**Note:** Crabill hesitatingly reports of a *Helembius* sp. from South Carolina and does not refer the specimens to *nannus* due to a U-shaped diastema instead of a V-shaped].
— Crabill, 1952b: 366 (*Helembius*; characters).
— Jeekel, 2005: 18 (*Helembius*; list).

Distribution: USA: Georgia, ?South Carolina.

HOLOTYPE: MCZ, Unique No. 14453, Georgia: Okefenokee Swamp, Billy's Id., June 1912, C. U. Exped.

Genus *Juanobius* Chamberlin, 1928
[1 species; Western USA]

Juanobius Chamberlin, 1928d: 309 (characters). **TYPE SPECIES:** *Juanobius eremus* Chamberlin, 1928, fixed by original designation and monotypy.
— Jeekel, 2005: 19 (list).

♂*J. eremus* Chamberlin, 1928
— Chamberlin, 1928d: 309 (*Juanobius*; characters; USA: Utah).
— Chamberlin, 1958c: 133 (*Juanobius*; checklist; note).
— Jeekel, 2005: 19 (*Juanobius*; list).

Distribution: USA: Utah.

HOLOTYPE: NMNH, ♂, Utah: Devil's Canyon, San Juan, 18 April, by W. J. Gertsch. [**Note:** NMNH has a holotype of this genus labeled as *gertschi*, being that this is a monotypic genus and it was collected by Gertsch this is most likely the holotype for *eremus*. Holotype specimen lacking ultimate legs!].

Genus *Kiberbius* Chamberlin, 1916
[7 species; Western USA]

Kiberbius Chamberlin, 1916: 153-154; Fig. 2 (characters; key to species; notes). **TYPE SPECIES:** *Kiberbius ogmopus* Chamberlin, 1916, fixed by original designation.
— Attems, 1938a: 344 (subgenus of *Lithobius*; key).
— Chamberlin, 1941a: 13-14 (key to species; notes).

— Jeekel, 2005: 20 (list).

♂*K. cayoteus* Chamberlin, 1941
— Chamberlin, 1941a: 14 (*Kiberbius*; characters; note; USA: California).

Distribution: USA: California.

HOLOTYPE: NMNH, ♂, California: Coyote Well, 8 January 1941, by S.
& D. Mulaik.

♂*K. dyscritus* Chamberlin, 1941
— Chamberlin, 1941a: 15 (*Kiberbius*; characters; note; USA: Arizona).

Distribution: USA: Arizona.

HOLOTYPE: NMNH, ♂, Arizona: South Mountain View, 28 December
1940, by S. & D. Mulaik.

♂♀*K. gosobius* Chamberlin, 1941
— Chamberlin, 1941a: 16 (*Kiberbius*; characters; USA: Arizona).

Distribution: USA: Arizona.

HOLOTYPE: ?Unknown, Arizona: Covered Wells, 3 January 1941, by S.
& D. Mulaik.

♀*K. nannus* Chamberlin, 1916
— Chamberlin, 1916: 157-158, pl. 4; Fig. 7; pl. 5; Fig. 4 (*Kiberbius*;
characters; notes; USA: California).

Distribution: USA: California.

HOLOTYPE: MCZ, No. 340, California: Eaton's Canyon, near Altadena,
April, 1913. R. V. Chamberlin.

♀*K. ogmopus* Chamberlin, 1916
— Chamberlin, 1916: 154-157, pl. 4; Fig. 5-6; pl. 5; Fig. 1-3 (*Kiberbius*;
characters; note; USA: California).
— Jeekel, 2005: 20 (*Kiberbius*; list).

Distribution: USA: California.

HOLOTYPE: MCZ, No. 339, California: Eaton's Canyon, near Altadena. R.V. Chamberlin.

PARATYPES: MCZ, No. 419, California: Eaton's Canyon, near Altadena. R.V. Chamberlin.

♂*K. remex* (Chamberlin, 1903)
— Chamberlin, 1903b: 156 (*Lithobius*; characters; key to species; USA: California).
— Chamberlin, 1910: 374 (*Lithobius*; characters; key to species; USA: California).
— Chamberlin, 1916: 159-160, pl. 4; Fig. 5-6; pl. 5; Fig. 1-3 (*Kiberbius*; characters; note; USA: California).

Distribution: USA: California.

HOLOTYPE: MCZ, No. 14534, ♂, California: Shasta Springs, 1902. R.V. Chamberlin.

♀*K. robles* Chamberlin, 1941
— Chamberlin, 1941a: 15 (*Kiberbius*; characters; note; USA: Arizona).

Distribution: USA: Arizona.

HOLOTYPE: NMNH, ♀, Arizona: 15 mi. W of Robles, 2 January 1941, by S. & D. Mulaik.

Genus *Labrobius* Chamberlin, 1915
[1 species (Introduced); Western USA]

Labrobius Chamberlin, 1915: 536 (characters). **TYPE SPECIES:** *Labrobius minor* Chamberlin, 1915, fixed by original designation. [**Note:** The holotype of *minor* is deposited at the MCZ, No. 1737].
— Attems, 1938a: 344 (synonym of *Lithobius* (*Alokobius*)).

♀*L. investigans* Chamberlin, 1938
— Chamberlin, 1938b: 634 (*Labrobius*; characters; USA: California). [**Note:** An introduced species from Mexico taken at quarantine in San Francisco and if it is established in California is not known].
— Mundel, 1981: 118 (*Labrobius*; notes).

Distribution: USA: ?California.

HOLOTYPE: NMNH, #15-203, ♀, California: San Francisco, 23 November 1936 (Origin: Mexico: Hidalgo).

Genus *Liobius* Chamberlin & Mulaik, 1940
[1 species; Southern Central USA]

Liobius Chamberlin & Mulaik, 1940: 127 (characters). **Type Species:** *Liobius mimus* Chamberlin & Mulaik, 1940, fixed by original designation and monotypy. [**Note:** Chamberlin compares this genus to *Oabius*].
— Mundel, 1981: 188 (list).

♂*L. mimus* Chamberlin & Mulaik, 1940
— Chamberlin & Mulaik, 1940: 127-128 (*Liobius*; characters; USA: Texas).

Distribution: USA: Texas.

HOLOTYPE: NMNH, Texas: 37 mi. W of Brady, December, 1939.

Genus *Lithobius* Leach, 1815
[38 species (15 Introduced); CANADA, USA, GREENLAND]

Lithobius Leach, 1815: 381 (characters). **Type Species:** *Scolopendra forficata* Linnaeus, 1758, fixed by subsequent designation of Latrielle, 1831.
— Chamberlin, 1901: 21-22 (key to species of Utah).
— Chamberlin, 1903b: 152-153 (key to species of California and Oregon).
— Chamberlin, 1910: 370-372 (key to species of California).
— Chamberlin, 1911a: 67-68 (key to species of Colorado).
— Chamberlin, 1911c: 98-101 (key to species of Illinois, Iowa, Nebraska, Michigan and Wisconsin).
— Chamberlin, 1925d: 450-452 (characters; distribution; key to species).
— Holmquist, 1926: 411 (notes on hibernation such as: activity, date collected, developmental stage, ecological association/niche, quantity found).
— Bailey, 1928b: 28 (characters).

— Crabill, 1955a: 134-136 (official designation of *Lithobius* as the type genus of Lithobiidae).
— Eason, 1974a: 71 (notes).
— Haupt, 1979: 394, 397, 398, 399; figs. 2d, 2e, 3d (chemotactile sensilla, temporal organ, eyes).
— Jeekel, 2005: 21 (list; notes).

Ezembius Chamberlin, 1919: 19H (notes). TYPE SPECIES: *Lithobius stejnegeri* Bollman, 1893, fixed by original designation. [**Note:** Chamberlin erected this genus with no indication of it being a new genus and there were no characters given until 1923].
— Chamberlin, 1923a: 240-241 (characters; notes).
— Eason, 1974b: 30 (note). [**Note:** Eason states that *"Ezembius . . . was erected by Chamberlin (1919) as a monotypic genus to receive L. stejnegeri* Bollman so that *Ezembius* is the correct name for the group of Asiatic species which can be arranged around *L. rapax*; but it should be regarded only as a subgenus of *Lithobius"*].
— Jeekel, 2005: 17 (list).

Haplolithobius Verhoeff, 1925: 153 (characters). TYPE SPECIES: *Lithobius olivarum* Verhoeff, 1925, fixed by monotypy.
— Matic, 1966: 101 (synonym of *Lithobius*).

Kauabius Chamberlin, 1920c: 77 (characters). TYPE SPECIES: *Lithobius hawaiiensis* Silvestri, 1904, fixed by original designation and monotypy.
— Eason, 1977: 485 (Eason regards Chamberlin's monotypic genus of *Kauabius* as a synonym of *Lithobius* but leaves the validity of it open to question).
— Jeekel, 2005: 19 (list).

Lithobius (*Archilithobius*) Stuxberg, 1875b: 8 (characters). TYPE SPECIES: *Lithobius erythrocephalus* C.L. Koch, 1847, fixed by subsequent designation of Chamberlin, 1952.
— Brölemann, 1930: 240 (synonym of *Lithobius*).
— Jeekel, 2005: 12, 35 (list; note).

Lithobius (*Hemilithobius*) Stuxberg, 1875b: 8 (characters). TYPE SPECIES: *Lithobius borealis* Meinert, 1868, fixed by subsequent designation of Jeekel, 2005.
— Brölemann, 1930: 240 (synonym of *Lithobius*).
— Jeekel, 2005: 18 (list; note; designation of type species).

Lithobius (*Oligobothrus*) Latzel, 1880: 36 (characters). TYPE SPECIES: *Scolopendra forficata* Linnaeus, 1758, fixed by subsequent designation of Jeekel, 2005.
— Jeekel, 2005: 25 (list; notes; designation of type species; junior objective synonym of *Lithobius*).

Mesobius Chamberlin, 1951c: 115-116 (characters; note). **TYPE SPECIES:** *Mesobius danianus* Chamberlin, 1951, fixed by original designation and monotypy.
— Matic, 1970: 10 (synonym of *Lithobius*).
— Eason, 1974a: 70 (generic classification).
— Jeekel, 2005: 22 (list; note).
Stenomera Rafinesque, 1820: 8 (characters). **TYPE SPECIES:** *Stenomera interrupta* Rafinesque, 1820, fixed by monotypy.
— Hoffman & Crabill, 1953: 76-77 (synonym of *Lithobius*).
— Jeekel, 2005: 31 (list).

♀*L. apheles* Chamberlin, 1940
— Chamberlin, 1940f: 77 (*Lithobius*; characters; notes; USA: North Carolina).
— Causey, 1940: 78 (*Lithobius*; note).
— Wray, 1967: 155 (*Lithobius*; list).

Distribution: USA: North Carolina.

HOLOTYPE: NMNH, ♀, North Carolina: Linville, 14 October 1939, by Nelle B. Causey. [**Note:** Chamberlin states that the type specimen is a male, however, the actual type specimen at the NMNH is a female].

♂♀*L. aureus* McNeill, 1887
— McNeill, 1887a: 327 (*Lithobius*; characters; USA: Florida). [**Note:** McNeill notes that the two specimens upon which he described this species lack the anal legs].
— Bollman, 1887b: 266 [= 1893: 33] (*Lithobius*; subgenus *Lithobius*; list).
— Bollman, 1893: 129 (*Lithobius*; subgenus *Lithobius*; catalogue).
— Verhoeff, 1925: 690 (*Lithobius*; list).

Distribution: USA: Florida.

SYNTYPES: MCZ, No. 14247, 2 specimens, Florida: Pensacola.

♀*L. bellulus* Chamberlin, 1903
— Chamberlin, 1903b: 155-156 (*Lithobius*; characters; key to species; USA: Oregon).

Distribution: USA: Oregon.

HOLOTYPE: ?Unknown, ♀, Oregon: Portland.

♂*L. beulae* Chamberlin, 1903
— Chamberlin, 1903a: 36-37 (*Lithobius*; characters; Chamberlin remarks: the two known male specimens of this species have lost their anal legs and that the general appearance is not unlike that of *L. utahensis* Chamberlin; USA: New Mexico).

Distribution: USA: New Mexico.

HOLOTYPE: ?Unknown, ♂, New Mexico: Beulah, ca. 8,000 ft.

L. cantabrigensis Meinert, 1886
— Meinert, 1886a: 177 (*Lithobius*; characters; USA: Massachusetts).
— Bollman, 1887b: 266 [= 1893: 33] (*Lithobius*; subgenus *Hemilithobius*; list).
— Bollman, 1888a: 111 [= 1893: 85] (*Lithobius*; USA: Tennessee).
— Bollman, 1888c: 342 [= 1893: 93] (*Lithobius*; note; USA: Tennessee).
— Bollman, 1893: 129 (*Lithobius*; subgenus *Hemilithobius*; catalogue).
— Chamberlin, 1911b: 41 (*Lithobius*; notes; USA: North Carolina, South Carolina, Tennessee, Virginia). [**Note:** See note under *Paitobius zinus* Chamberlin].
— Brimley, 1938: 500 (*Lithobius*; USA: North Carolina).
— Wray, 1967: 155 (*Lithobius*; list).
— Kevan, 1983: 2949 (*Nampabius*; list) [**Note:** Kevan lists this as a *Nampabius* species because *Hemilithobius* is listed as a synonym of *Nampabius* but it has been left in *Lithobius* until further investigation].

Distribution: USA: Massachusetts, North Carolina, South Carolina, Tennessee, Virginia. [**Note:** Chilobase (see Minelli in bibliography) lists Alabama & Georgia as a locality, however, the author was not able to locate literature stating this].

HOLOTYPE: MCZ, No. 818 (ethanol/slide), 1 specimen, Massachusetts: Cambridge, by Mr. H.H. James.

♂♀*L. celer* Bollman, 1888
— Bollman, 1888b: 7-8 [= 1893: 79] (*Lithobius*; characters; notes; USA: Arkansas).
— Bollman, 1893: 129 (*Lithobius*; subgenus *Lithobius*; catalogue).
— Kenyon, 1893a: 17 (*Lithobius*; note; USA: Nebraska).

— Kenyon, 1893b: 162 (*Lithobius*; USA: Nebraska).
— Chamberlin, 1909b: 190 (*Lithobius*; USA: Michigan).
— Chamberlin, 1911b: 43 (*Lithobius*; USA: Kentucky).
— Chamberlin, 1911c: 99, 102 (*Lithobius*; key to species; notes; USA: Michigan, Wisconsin).
— Chamberlin, 1925d: 463-464 (*Lithobius*; characters; notes; USA: Arkansas). [**Note:** Type material is said to be in bad condition].
— Crabill, 1955f: 260 (*Lithobius*; checklist; USA: Kentucky).
— Summers, 1979: 696, 698 (*Lithobius*; checklist; key).
— Watermolen, 1997: 4 (*Lithobius*; checklist, notes).

Distribution: USA: Arkansas, ?Illinois, ?Indiana, ?Iowa, Kentucky, Michigan, ?Missouri, Nebraska, Ohio, Wisconsin. [**Note:** Based on the distribution it is presumed to be found in Iowa].

SYNTYPE: NMNH, Arkansas: ?Little Rock.

♂♀*L. chumasanus* Chamberlin, 1903
— Chamberlin, 1903b: 154 (*Lithobius*; characters; USA: California).
— Chamberlin, 1910: 372 (*Lithobius*; characters; USA: California).
— Chamberlin, 1925d: 464-466 (*Lithobius*; characters; notes; USA: California).

Distribution: USA: California.

TYPES: MCZ, No. 2427, 2 specimens, California: Santa Barbara, 1902. [**Note:** A label in vial reads "Also see vial among reg. material for 3rd type"].

♀*L. cockerelli* Chamberlin, 1904
— Chamberlin, 1904: 651-652 (*Lithobius*; characters; USA: New Mexico).

Distribution: USA: New Mexico.

TYPES: MCZ, Unique No. 14287, 4 specimens, [**Note:** Chamberlin designated a ♀ as the holotype], New Mexico, July 28, Prof. T.D.A. Cockerell.

L. hardyi Chamberlin, 1946
— Chamberlin, 1946f: 21 (*Lithobius*; characters; USA: Texas).

Distribution: USA: Texas.

HOLOTYPE: NMNH, Texas: Laguna Madre, 23 mi. SE Harlingen.

L. intermontanus Chamberlin, 1901
— Chamberlin, 1901: 24 (*Lithobius*; characters; key to species; USA: Utah; according to Chamberlin all eight specimens obtained have lost their posterior pairs of legs).

Distribution: USA: Utah.

SYNTYPE: NMNH, No. 785, Utah: Branch of Mill Creek Canyon.

♂♀ *L. lindrothi* Palmén, 1954
— Palmén, 1954: 141-143; Fig. 2(a), 16-17 (*Lithobius*; characters; plectrotaxy; notes; CANADA: Newfoundland). [**Note:** Palmén suggests this species belongs to the *Sonibius* group of Chamberlin].
— Kevan, 1979: 298 (*Lithobius*; synopsis; CANADA: Newfoundland).
— Behan-Pelletier, 1993: 26 (*Lithobius*; list).

Distribution: CANADA: Newfoundland.

HOLOTYPE: Canadian National Collection: ♂, CANADA: Newfoundland: Cow Head, 8 August 1949.

PARATYPES: Canadian National Collection: 3 ♂, 2 ♀, CANADA: Newfoundland: Cow Head, 8 August 1949.

♀*L. minnesotae* Bollman, 1887
— Bollman, 1887a: 81 [= 1893: 19] (*Lithobius*; characters; USA: Minnesota). [**Note:** Bollman originally described this species inadequately and elaborated by redescribing it as a new species again shortly after (see Bollman, 1887b)].
— Bollman, 1887b: 254, 255-256, 265 [= 1893: 21, 22-23, 32] (*Lithobius*; subgenus *Archilithobius*; characters; key; list; notes; USA: Minnesota).
— Bollman, 1893: 128 (*Lithobius*; subgenus *Archilithobius*; catalogue).
— Chamberlin, 1911c: 101, 104 (*Lithobius*; key to species; USA: Minnesota, Wisconsin).
— Kevan, 1983: 2949 (*Nadabius*; checklist; questioned synonym of *Nadabius jowensis*). [**Note:** Kevan lists this as a *Nadabius* species because *Archilithobius* is considered a synonym of *Nadabius*. It

has been left in *Lithobius* until further investigation because there is
no supporting evidence for this synonymy. Kevan (1983: 2943) even
states "In the following list, which does not claim to be infallible
insofar as synonymy . . .".].

Distribution: USA: Minnesota, Wisconsin.

HOLOTYPE: NMNH, Minnesota: Fort Snelling, by W.D. Howe.

♂*L. paradoxus* Stuxberg, 1875
— Stuxberg, 1875a: 67 [=1875d: 189] (*Lithobius*; characters; USA:
  California).
— Stuxberg, 1875b: 18 (*Lithobius*; subgenus *Archilithobius*; list).
— Stuxberg, 1875c: 31 [=1877: 138] (*Lithobius*; subgenus *Archilithobius*;
  list).
— Bollman, 1887b: 265 [= 1893: 32] (*Lithobius*; subgenus *Archilithobius*;
  list).
— Bollman, 1893: 129 (*Lithobius*; subgenus *Archilithobius*; catalogue).
— Chamberlin, 1909b: 187 (*Lithobius*; USA: California).
— Chamberlin, 1910: 374 (*Lithobius*; characters; key to species;
  Chamberlin notes this species may have been based on a young
  specimen; USA: California).

Distribution: USA: California.

HOLOTYPE: ?Unknown, ♂, California: San Pedro, by G. Eisen.

L. spinipes Say, 1821
— Say, 1821: 108-109 (*Lithobius*; characters; habitat; USA: State
  Unknown). [**Note:** Say claims that the species is very common under
  stones].
— Lucas, 1840: 543 (*Lithobius*; characters; distribution).
— Newport, 1845b: 365, 366 (*Lithobius*; notes). [**Note:** Newport
  was inclined to believe that his description of the new species *L.
  americanus* was conspecific with *spinipes*].
— Gervais, 1847: 235-236 (*Lithobius*; characters; notes).
— Provancher, 1873: 415 (*Lithobius*; characters).
— Brodie & White, 1883: 67 (*Lithobius*; CANADA).
— Bollman, 1893: 146-147 (*Lithobius*; notes). [**Note:** Bollman
  regarded *N. transmarinus* and *N. mordax* as synonyms of
  spinipes, and he argues that Say's specimen was from the

"Southern United States" and that they couldn't be *forficatus* because *forficatus* was not known from south of Virginia].

— Kevan, 1983: 2949 (notes that *N. mordax*, *N. howei* and *N. transmarinus* are " . . . all probably junior synonyms of *spinipes* (Say) s. str., . . .").

mordax Bollman (not Koch), 1893: 146 (*Lithobius*; cited as synonym of *spinipes*; notes).

transmarinus Bollman (not Koch), 1893: 146-147 (Lithobius; notes; cited as synonym of spinipes).

Distribution: CANADA; USA: Southern. [**Note:** This species is not listed in any state list for the United States because it is unclear in which state specimens were found].

HOLOTYPE: ?Unknown; USA.

♂*L. watovius* Chamberlin, 1911
— Chamberlin, 1911b: 37 (*Lithobius*; characters; USA: Mississippi).

Distribution: USA: Mississippi.

HOLOTYPE: ?MCZ, ♂, Mississippi: Byram.

(Subgenus *Ezembius* Chamberlin, 1919)

Lithobius (*Ezembius*) Chamberlin, 1919: 19 (characters). **TYPE SPECIES:** *Lithobius stejnegeri* Bollman, 1893, fixed by subsequent designation.
— Chamberlin, 1923a: 240-241 (characters; cited as genus).
— Eason, 1976a: 123-125 (*Ezembius* designated as subgenus of *Lithobius*).
— Jeekel, 2005: 17 (list; note).

♂♀*L.* (*E.*) *rapax* Meinert, 1872
— Meinert, 1872: 325-326 (*Lithobius*; characters; distribution).
— Eason, 1974b: 27-30; Figs. 13-15 (*Lithobius*; subgenus *Ezembius*; characters; notes).
— Eason, 1996: 118, 120; Figs. 3-4 (*Lithobius*; subgenus *Ezembius*; characters; notes).

arcticus Attems, 1909a: 19 (*Monotarsobius*; see *stejnegeri* Bollman, 1893)
— Chamberlin, 1946c: 186 (*Monotarsobius*; synonym of *rapax*).

— Eason, 1974b: 27 (*Monotarsobius*; synonym of *rapax*).

haasei Attems, 1909a: 22 (*Lithobius*; subgenus *Archilithobius*; characters;
"Behring-Insel").

— Chamberlin, 1946c: 186 (*Lithobius*; subgenus *Archilithobius*; synonym
of *stejnegeri*).

— Eason, 1974b: 27 (*Lithobius*; subgenus *Archilithobius*; synonym of
rapax).

sibericus Haase, 1880: 223 (*Lithobius*; characters).

— Attems, 1909a: 22 (*Lithobius*; notes; synonym of *hassei*).

sulcipes Stuxberg, 1876: 20-22 (*Lithobius*; characters).

— Cook, 1904: 72 (*Lithobius*; USA: Alaska).

— Attems, 1909a: 9 (*Lithobius*; list).

— Chamberlin, 1911e: 260 (*Lithobius*; USA: Alaska).

— Eason, 1974b: 27 (*Lithobius*; synonym of *rapax*).

— Behan-Pelletier, 1993: 26; Tab. 2 (*Ezembius*; distribution).

yamashinai Verhoeff, 1938: 103, Fig. 5 (*Archilithobius*; subgenus *Archilithobius*;
characters).

— Chamberlin & Wang, 1952b: 183 (*Ezembius*; Japan).

— Eason, 1974b: 27 (*Ezembius*; synonym of *rapax*).

Distribution: USA: Alaska.

LECTOTYPE: ZMUC, ♂, U.S.S.R., Sartung, Sakhalin.

PARALECTOTYPES: ZMUC, 2♂, U.S.S.R., Sartung, Sakhalin.

♂♀*L. (E.) stejnegeri* Bollman, 1893

— Bollman, 1893: 199-200 (*Lithobius*; characters; Bering Island).

— Cook, 1904: 71-72 (*Lithobius*; subgenus *Archilithobius*; notes; USA:
Alaska).

— Attems, 1909a: 9 (*Lithobius*; list).

— Chamberlin, 1911e: 260 (*Lithobius*; taxonomic notes).

— Chamberlin, 1919: 15H, 19H (*Ezembius*; list; notes).

— Chamberlin, 1920d: 44 (*Ezembius*; USA: Alaska).

— Chamberlin, 1923a: 241-242 (*Ezembius*; notes; USA: Alaska).

— Chamberlin, 1946c: 186-187 (*Ezembius*; notes; USA: Alaska).

— Eason, 1974b: 27 (*Ezembius*; synonym of *rapax*). [**Note:** Eason (1996)
later explains his incorrect synonomy of *rapax* with *stejnegeri*].

— Eason, 1996: 118; Figs. 1-2 (*Lithobius*; subgenus *Ezembius*; characters;
notes).

— Jeekel, 2005: 17 (*Lithobius*; list).

arcticus Attems, 1909a: 19–20 (*Monotarsobius*; characters; notes; "Behring-Insel").
— Chamberlin, 1946c: 186 (*Monotarsobius*; synonym of *rapax*).
— Eason, 1974b: 27 (*Monotarsobius*; synonym of *rapax*).
— Eason, 1996: 118 (*Monotarsobius*; synonym of *stejnegeri*).

Distribution: USA: Alaska.

HOLOTYPE: NMNH (No. 92 & 93), Alaska: Bering Island, Topor Rof Island, and Gavaus Kaya Topka, Copper Island; Stejneger.

(Subgenus *Lithobius* Leach, 1814)

Lithobius (*Lithobius*) Leach, 1814: 387 (?characters). [**Note:** Reference not obtained]. **TYPE SPECIES:** *Scolopendra forficata* Linnaeus, 1758, fixed by subsequent designation.

♂♀*L.* (*L.*) *atkinsoni* Bollman, 1887
— Bollman, 1887c: 625 [= 1893: 42] (*Lithobius*; subgenus *Lithobius*; characters; note; USA: North Carolina).
— Bollman, 1888d: 349 [= 1893: 101] (*Lithobius*; characters; USA: Georgia).
— Bollman, 1893: 129 (*Lithobius*; subgenus *Lithobius*; catalogue).
— Chamberlin, 1911b: 42 (*Lithobius*; USA: Georgia, North Carolina, South Carolina).
— Chamberlin, 1925d: 466-470, pl. 1; Fig. 6 (*Lithobius*; characters (adult, stadia); USA: Georgia, North Carolina, South Carolina).
— Brimley, 1938: 500 (*Lithobius*; USA: North Carolina).
— Causey, 1940: 78 (*Lithobius*; characters).
— Crabill, 1950c: 202 (*Lithobius*; USA: South Carolina).
— Chamberlin, 1951b: 32 (*Lithobius*; USA: Georgia).
— Wray, 1967: 155 (*Lithobius*; list).
— Hoffman, 1995a: 27 (*Lithobius*; list; notes). [**Note:** Hoffman believes this species could be found in Virginia].

Distribution: USA: Georgia, North Carolina, South Carolina.

HOLOTYPE: ?Unknown, North Carolina: Jackson Co., Balsam, by Prof. George F. Atkinson. [**Note:** MCZ has numerous specimens with the following catalog numbers: 27900, 27903-27905, 27908, 27911, 27939, 27941-27943, ?27949].

♂♀*L.* (*L.*) *borealis* Meinert, 1868

— Meinert, 1868: 263 [non Meinert, 1872: 322] (*Lithobius*; characters; note).
— Stuxberg, 1875b: 14 (*Lithobius*; subgenus *Hemilithobius*; list).
— Chamberlin, 1930c: 68 (*Lithobius*; USA: Hawaii). [**Note:** Chamberlin reported this European species was taken at Hawaii in packing of orchids from England on November 6, 1929].
— Eason, 1964: 210-211; Tab. 6 (*Lithobius*; notes).
— Eason, 1974b: 5-7 (*Lithobius*; subgenus *Lithobius*; characters; notes). [**Note:** According to Eason "All that can be said with certainty is that most records of *L. lapidicola* in the literature refer to *L. borealis*, whereas most of those of *L. borealis* refer either to *L. lusitanus* or to some other species"].
— Zapparoli & Shelley, 2000: 47 (*Lithobius*; list; notes).
— Jeekel, 2005: 18 (*Lithobius*; list).
lapidicola Latzel, 1880: 106 (non Meinert, 1872: 328-329) [**Note:** According to Eason (1974b) this is a synonym of *borealis* and most subsequent authors who cited Latzel's *lapidicola* were actually referring to *borealis*].
— Verhoeff (not Latzel), 1925: 146 (*Lithobius*; nominal taxon *erythrocephalus*; characters).
— Eason, 1974b: 6 (*Lithobius*; synonym of *borealis*).
saalachiensis Verhoeff, 1937: 214, 227; fig. 51 (*Lithobius*; characters; key).
— Eason, 1974b: 6 (*Lithobius*; synonym of *borealis*).
würmanus Verhoeff, 1937: 212, 224; fig. 54 (*Lithobius*; nominal species *lusitanus*; characters).
— Eason, 1974b: 6 (*Lithobius*; synonym of *borealis*).

Distribution: USA: ?Hawaii.

SYNTYPES: Destroyed (see Eason, 1974b: 6).

NEOTYPE: ZMUC, ♂, Denmark: Faeroe Islands, "Thorshavn, Uldspinderiet, under sten, 9.4.1925, Krygr".

♂♀*L. (L.) calcaratus* C. L. Koch, 1844
— Koch, 1844: 23 (*Lithobius*; characters; Germany). [**Note:** reference not obtained].
— Eason, 1972b: 109 (*Lithobius*; characters; notes).
danianus Chamberlin, 1951c: 115-116 (*Mesobius*; characters; note; USA: Florida).
— Matic, 1970: 10 (*Mesobius* transferred to *Lithobius*).
— Eason, 1974: 70 (probable synonym of *calcaratus*).

113

Distribution: USA: Florida. [Introduced].

HOLOTYPE: Germany.

NEOTYPE: BMNH; Nuremburg, #13.6.18.18. (see Eason, 1972).

L. (*L.*) *eucnemis* Stuxberg, 1875
— Stuxberg, 1875a: 70-71 (*Lithobius*; characters; notes; USA: New York).
— Stuxberg, 1875b: 14 (*Lithobius*; subgenus *Hemilithobius*; list).
— Stuxberg, 1876: 6 (*Lithobius*; subgenus *Hemilithobius*; list).
— Bollman, 1887b: 266 [= 1893: 33] (*Lithobius*; subgenus *Hemilithobius*; list).
— Bollman, 1893: 129 (*Lithobius*; subgenus *Hemilithobius*; catalogue).
— Verhoeff, 1925: 690 (*Lithobius*; list).
— Brölemann, 1930: 30 (transferred to *Lithobius* (*Lithobius*) [= *Lithobius* (*Hemilithobius*)]).

Distribution: USA: New York.

HOLOTYPE: ?Unknown, New York: Mount Lebanon, by G. Eisen, 1873.

♂♀*L.* (*L.*) *forficatus* (Linnaeus, 1758)
— Linnaeus, 1758: 638 (*Scolopendra*; characters; distribution).
— Leach, 1817: 39-40; Tab. 137 (*Lithobius*; characters).
— Newport, 1845b: 367 (*Lithobius*; characters; notes; Europe).
— Gervais, 1847: 230 (*Lithobius*; characters).
— Newport, 1856: 18-19 (*Lithobius*; characters; Europe).
— Meinert, 1872: 315-317 (*Lithobius*; characters; distribution; notes).
— Stuxberg, 1875b: 10-11 (*Lithobius*; subgenus *Lithobius*; list).
— Stuxberg, 1876: 4-5 (*Lithobius*; list).
— Bollman, 1887b: 255, 260, 266 [= 1893: 22, 27, 33] (*Lithobius*; subgenus *Lithobius*; characters (adult, stadium); key; list; note; USA: Indiana, Michigan).
— Bollman, 1888e: 409 [= 1893: 110] (*Lithobius*; note; USA: Indiana).
— Bollman, 1893: 129, 185 (*Lithobius*; subgenus *Lithobius*; catalogue; note; USA: Minnesota).
— Kenyon, 1893a: 17 (*Lithobius*; note; USA: Nebraska).
— Kenyon, 1893b: 162 (*Lithobius*; USA: Nebraska).

— Chamberlin, 1901: 24 (*Lithobius*; characters, habitat; key to species; USA: Utah).
— Chamberlin, 1909b: 190-191 (*Lithobius*; note; USA: Idaho, Utah).
— Chamberlin, 1911a: 69 (*Lithobius*; key to species; notes; USA: Colorado).
— Chamberlin, 1911b: 42-43 (*Lithobius*; range; USA: Kentucky, North Carolina, South Carolina, Virginia, West Virginia).
— Chamberlin, 1911c: 99, 101-102 (*Lithobius*; key to species; notes; USA: Illinois, Indiana, Iowa, Michigan, Minnesota, Wisconsin).
— Gunthorp, 1913: 165 (*Lithobius*; notes; USA: Kansas).
— Chamberlin, 1920a: 95 (*Lithobius*; notes; CANADA: Ontario).
— Chamberlin, 1920b: 166 (*Lithobius*; CANADA: Ontario).
— Gunthorp, 1921: 89 (*Lithobius*; notes; USA: Kansas).
— Chamberlin, 1925d: 453-462; fig. 4; pl. 2; Fig. 1-3 (*Lithobius*; characters (adult, stadia); CANADA: Newfoundland, Nova Scotia, Ontario, Quebec; USA: Colorado, Connecticut, District of Columbia, Idaho, Illinois, Indiana, Iowa, Kansas, Kentucky, Maine, Maryland, Massachusetts, Minnesota, Missouri, Nebraska, New Hampshire, New Jersey, New York, North Carolina, Ohio, Pennsylvania, Rhode Island, South Carolina, Vermont, Virginia, West Virginia, Wisconsin).
— Chamberlin, 1925e: 56 (*Lithobius*; USA: Utah).
— Holmquist, 1926: 411 (*Lithobius*; notes on hibernation such as: activity, date collected, developmental stage, ecological association/niche, quantity found).
— Bailey, 1928a: 1081–1082 (*Lithobius*; list).
— Bailey, 1928b: 19, 28 (*Lithobius*; list; notes; USA: New York).
— Filinger, 1928: 359 (*Lithobius*; note).
— Williams & Hefner, 1928: 142 (*Lithobius*; characters; USA: Ohio).
— Matthews, 1935: 37-41; map 42; pl. 3; Fig. 1-7 (*Lithobius*; characters; USA: Wisconsin).
— Attems, 1938b: 366, 369 (*Lithobius*; distribution; list; note; USA: Hawaii).
— Brimley, 1938: 500 (*Lithobius*; USA: North Carolina).
— Back, 1939: 1-2 (*Lithobius*; note).
— Causey, 1940: 77; fig. 37 (*Lithobius*; characters; USA: North Carolina).
— Chamberlin, 1942d: 187 (*Lithobius*; USA: Missouri).
— Chamberlin, 1943b: 104 (*Lithobius*; USA: Utah, Wyoming).
— Chamberlin, 1944d: 204 (*Lithobius*; USA: Illinois, New York).

— Cloudsley-Thompson, 1945: 537-538 (*Lithobius*; behavior; notes).

— Cole, 1946: 60-61 (*Lithobius*; behavior; habitat; notes; USA: Illinois).

— Rapp, 1946: 667 (*Lithobius*; USA: Illinois).

— Auerbach, 1951a: 100, 101, 102, 103, 107, 108, 109, 115, 118, 119, 120, 121, 122; tab. 1, 6-8, 14; fig. 1, 3-5 (*Lithobius*; ecology; desiccation; food habits; habitat; heart beat; hygrotaxis; list; population percentage; USA: Illinois).

— Chamberlin, 1951b: 31-32 (*Lithobius*; USA: Maryland, Nebraska, Nevada, New Jersey, New York, Vermont).

— Crabill, 1952b: 315-317; pl. 6; figs. 91-92 (*Lithobius*; characters; distribution; note; USA: New York).

— Johnson, 1952: 147-164; Tab. 18, 28; map 12; figs. 24-26 (*Lithobius*; autecology; characters; distribution; experimental feeding diet; natural history; USA: Michigan).

— Hoffman & Crabill, 1953: 76, 77 (*Lithobius*; taxonomical discussion of *Cryptomera* and *Stenomera*).

— Palmén, 1954: 139-140; fig. 3 (*Lithobius*; notes; CANADA: Newfoundland).

— Crabill, 1955a: 134, 135, 136 (*Scolopendra*; nomenclature).

— Crabill, 1955f: 260 (*Lithobius*; checklist; USA: Kentucky).

— Crabill, 1955h: 156 (*Lithobius*; notes; USA: Missouri). [**Note:** This species has almost certainly been introduced form Europe more than once. Although, Crabill (1955h) notes "one specimen of this species has been recorded from Missouri" Stuxberg (1876) has already identified that this species occurs there. Furthermore, the Greenland distribution is known only from one specimen found in a greenhouse, presumably introduced and the status of its establishment there is presently not known (see Jensen, 2003)].

— Crabill, 1958b: 97-98 (*Lithobius*; notes; USA: Wisconsin).

— Crabill, 1962b: 408; Tab. 1 (*Lithobius*; plectrotaxy).

— Eason, 1964: 194-198; tab. 2; figs. 326-337 (*Lithobius*; characters; distribution; habitat).

— Judd, 1964: 2 (*Lithobius*; note; CANADA: Ontario).

— Wray, 1967: 155 (*Lithobius*; list).

— Rilling, 1968: 1-136 (*Lithobius*; internal anatomy; musculature; nervous system; respiratory system) [**Note:** In German with excellent illustrations].

— Wood & Wheeler, 1972: 1363 (*Lithobius*; North American report of European parasitic Tachinid fly on centipede; USA: New York).

— Andersson, 1978: 63-70; figs. 1-2; tab. 1-4 (post-embryonic development).

— Kevan & Vickery, 1978: 194 (*Lithobius*; CANADA: Quebec).

— Andersson, 1979: 77, 78, 79; tab. 1, 3-5 (*Lithobius*; characters of larval stadia).

— Haupt, 1979: 396; figs. 2d, e (*Lithobius*; olfactory sensilla).

— Kevan, 1979: 298 (*Lithobius*; synopsis; note; CANADA: "?Trans."). [**Note:** Kevan thought that this species may be transcontinental].

— Summers, 1979: 693, 696, 698; Fig. 4 (*Lithobius*; checklist, key).

— Lee, 1980: 5; Tab. 1 (*Lithobius*; habitat; USA: Ohio).

— Summers et al., 1980: 251; Fig. 19 (*Lithobius*; checklist; USA: Illinois).

— Summers et al., 1981: 61 (*Lithobius*; checklist; USA: Illinois).

— Kevan, 1983: 2948 (*Lithobius*; note; CANADA).

— Snider, 1991: 189 (*Lithobius*; list).

— Behan-Pelletier, 1993: 26; Tab. 2 (*Lithobius*; distribution).

— Hoffman, 1995a: 27 (*Lithobius*; list; notes).

— Watermolen, 1997: 4-5 (*Lithobius*; checklist, notes; USA: Wisconsin).

— Zapparoli & Shelley, 2000: 39–40 (*Lithobius*; characters; notes).

— Jensen, 2003: 68 (*Lithobius*; GREENLAND). [**Note:** Reported as an introduced species].

— Müller, Rosenberg & Meyer-Rochow, 2003: 185-197 (*Lithobius*; interommatidial exocrine glands).

— Jeekel, 2005: 21 (*Scolopendra*; list; notes).

americanus Newport, 1845b: 365; Tab. 33; Fig. 29 (*Lithobius*; characters; "America Boreali").

— Gervais, 1847: 236 (*Lithobius*; characters; note).

— Newport, 1856: 17 (*Lithobius*; characters; USA).

— Wood, 1862: 14 (*Lithobius*; characters; USA: "between Pike Lake and Ft. Union").

— Wood, 1865: 148 (*Lithobius*; characters; USA: "Eastern United States").

— Brodie & White, 1883: 67 (*Lithobius*; CANADA).

— Bollman, 1893: 129 (*Lithobius*; subgenus *Lithobius*; synonym of *forficatus*).

interrupta Rafinesque, 1820: 8 (*Stenomera*; characters; USA: Massachusetts).

— Hoffman & Crabill, 1953: 76-77 (*Stenomera*; synonym of *forficatus*).

— Jeekel, 2005: 31 (*Stenomera*; list).

multidentatus Wood [not Newport], 1862: 13 (*Lithobius*; characters; notes; USA: Arkansas, Illinois, Missouri and "En route from N. Orleans to Galveston").
— Crabill, 1952b: 315 (*Lithobius*; synonym of *forficatus*).

Distribution: CANADA: Newfoundland, Nova Scotia, Ontario, Quebec; GREENLAND; USA: Arkansas, Colorado, Connecticut, District of Colombia, Hawaii, Idaho, Illinois, Indiana, Iowa, Kansas, Kentucky, Maryland, Maine, Massachusetts, Michigan, Minnesota, Missouri, North Carolina, Nebraska, New Hampshire, New Jersey, Nevada, New York, Ohio, Pennsylvania, Rhode Island, South Carolina, Utah, Vermont, Virginia, Wisconsin, West Virginia, Wyoming.

HOLOTYPE: ?Unknown, Europe.

♂♀*L. (L.) hawaiiensis* Silvestri, 1904
— Silvestri, 1904: 324-325 (*Lithobius*; characters; USA: Hawaii).
— Attems, 1914: 48, 57, 95 (*Archilithobius*; list).
— Chamberlin, 1920c: 78 (*Kauabius*; USA: Hawaii).
— Attems, 1938b: 366, 369 (*Lithobius*; list).
— Eason, 1977: 485-490; figs. 1-6 (*Lithobius*; description of female lectotype and male paralectotype, notes on this species and Hawaiian Lithobiidae in general). [**Note:** Eason points out that this is most likely not an indigenous species and is more likely an immigrant from eastern Asia].
— Nishida, 1994: 26 (*Lithobius*; checklist).
— Nishida, 1997: 195 (*Lithobius*; checklist).
— Zapparoli & Shelley, 2000: 39 (*Lithobius*; subgenus *Lithobius*; characters; notes; USA: Hawaii).
— Jeekel, 2005: 19 (*Lithobius*; list).

Distribution: USA: ?Hawaii (see Eason, 1977).

LECTOTYPE: BMNH, Reg. No. 1904.10.28.2, ♀, Hawaii: Kauai, Makaweli, 3000 ft., 1897 (Perkins).

PARALECTOTYPE: BMNH, Reg. No. 1904.10.28.1, ♂, Hawaii: Kauai, Koholuamano, (Perkins, 1895).

♂♀*L. (L.) melanops* Newport, 1845
— Newport, 1845b: 371 (*Lithobius*; characters; notes; England).
— Palmén, 1954: 140; Fig. 1(2) (*Lithobius*; notes; CANADA: Newfoundland).

— Eason, 1964: 202-206; Figs. 350-362 (*Lithobius*; characters; distribution; habitat).
— Eason, 1972a: 304 (*Lithobius*; notes).
— Kevan, 1979: 298 (*Lithobius*; synopsis; CANADA: Newfoundland).
— Behan-Pelletier, 1993: 26 (*Lithobius*; list).
— Watermolen, 1997: 5 (*Lithobius*; checklist, notes).

glabratus C. Koch, 1847: 149 (*Lithobius*; characters; USA: California).
— Stuxberg, 1875b: 12-13 (*Lithobius*; subgenus *Lithobius*; list).
— Pocock, 1890: 62 (*Lithobius*; synonym of *melanops*).
— Eason, 1972b: 113 (*Lithobius*; synonym of *melanops*).

harrietae — Chamberlin, 1906: 3-4 (*Lithobius*; characters; types consist of four adults; USA: Colorado). **TYPES:** MCZ, No. 143, 3 specimens, Colorado: east of Glenwood Springs, 1904. [**Note:** Original locality/determined label reads "*Taiyubius harrielae* Ch., types, Colorado, near Glenwood Sps., 1904"].
— Chamberlin, 1909b: 191 (*Lithobius*; USA: Colorado).
— Chamberlin, 1911a: 69 (*Lithobius*; key to species; USA: Colorado).
— Chamberlin, 1912e: 176 (*Taiyubius*; misspelled *harrielae*; transferred to *Taiyubius*).
— Chamberlin, 1922b: 309-311, pl. 5; Fig. 6 (*Taiyubius*; misspelled *harrielae*; characters; USA: Colorado, Massachusetts).
— Chamberlin, 1951b: 32 (*Taiyubius*; USA: Nevada).
— Crabill, 1952b: 325-326, 430, pl. 7; Figs. 93-96 (*Taiyubius*; misspelled *harrielae*; characters; plectrotaxy; taxonomic notes; USA: Massachusetts, New York).
— Crabill, 1958b: 93, 98-99 (*Taiyubius*; misspelled *harrielae*; habitat; notes; USA: Wisconsin).
— Crabill, 1962b: 410; Tab. 7 (*Taiyubius*; misspelled *harrielae*; plectrotaxy).
— Eason, 1976b: 65–66 (*Taiyubius*; notes; synonym of *melanops*).
— Watermolen, 1997: 7 (note).

Distribution: CANADA: Newfoundland; USA: California, Colorado, Massachusetts, Nevada, New York, Wisconsin.

HOLOTYPE: BMNH, England.

♂ *L. (L.) obscurus* Meinert, 1872
— Meinert, 1872: 300-301 (*Lithobius*; characters; Spain).
— Eason, 1974b: 18-20 (*Lithobius*; characters; notes).

— Zapparoli & Shelley, 2000: 40 (*Lithobius*; subgenus *Lithobius*; characters; notes; USA: Hawaii). [**Note:** Zapparoli & Shelley reported it for the first time in the Hamakua Forest Reserve indicating that it has probably established itself. This species is probably introduced].

Distribution: USA: Hawaii.

LECTOTYPE: ZMUC, Near Granada; Caratraca near Malaga, Spain. (see Eason, 1974b).

PARALECTOTYPE: ZMUC, Near Granada; Caratraca near Malaga, Spain. (see Eason, 1974b).

♂♀*L. (L.) peregrinus* Latzel, 1880
 — Latzel, 1880: 63-64 (*Lithobius*; characters; Italy).
 — Attems, 1928: 79 (*Lithobius*; characters; notes; South Africa).
 — Brölemann, 1930: 254 (*Lithobius*; characters).
 — Matic, 1957: 12; figs. 5-8 (*Lithobius*;). [**Note:** Reference not obtained].
 — Matic & Darabantu, 1968: 111; fig. 4a-e (*Lithobius*;). [**Note:** Reference not obtained].
 — Zalesskaya, 1978: 76; pl. 32; Figs. 1-6 (*Lithobius*;). [**Note:** Reference not obtained].
 — Barber & Eason, 1986: 431, 432-436; figs. 1-4; tab. 1 (*Lithobius*; characters (adult, stadia); habitat; notes; plectrotaxy; redescription).
 — Eason, 1991: 23 (*Lithobius*; noted as being recorded from New Orleans, Louisiana).
ethochaetus Chamberlin, 1938b: 627-628 (*Lithobius*; characters; notes; USA: Louisiana). [**Note:** HOLOTYPE: NMNH, ♀, Louisiana: New Orleans, taken at quarantine in cargo from Panama, August 10, 1936].
 — Barber & Eason, 1986: 431, 432 (*Lithobius*; synonym of *peregrinus*).
provocator Pocock, 1891: 152-153 (*Lithobius*; characters; notes). Type locality: Bermuda.
 — Chamberlin, 1920g: 280-283; pl. 29; Figs. 6-8 (*Lithobius*; characters (adult, stadia); distribution; notes).
 — Chamberlin, 1925c: 408 (*Lithobius*; note).
 — Eason, 1975: 59 (*Lithobius*; notes).
 — Barber & Eason, 1986: 432 (*Lithobius*; synonym of *peregrinus*).

Distribution: USA: ?Louisiana.

LECTOTYPE: ?Unknown.

L. (L.) pinetorum Harger, 1872
— Harger, 1872: 118 (*Lithobius*; characters; habitat; USA: Oregon).
— Stuxberg, 1875b: 18 (*Lithobius*; subgenus *Lithobius*; list).
— Stuxberg, 1875c: 29 [=1877: 136-137] (*Lithobius*; subgenus *Lithobius*; list).
— Bollman, 1887b: 266 [= 1893: 33] (*Lithobius*; subgenus *Lithobius*; list).
— Bollman, 1893: 129 (*Lithobius*; subgenus *Lithobius*; catalogue).
— Kevan, 1983: 2948 (*Lithobius*; checklist).

Distribution: USA: Oregon.

HOLOTYPE: ?Yale Peabody Museum, Oregon: "in the valley of the John Day River", October, 1871, by Professor G.H. Collier, under bark of decaying pine logs.

(Subgenus *Metalithobius* Chamberlin, 1910)

Lithobius (*Metalithobius*) Chamberlin, 1910: 370 (characters; key). Type Species: *Lithobius* (*Metalithobius*) *obesus* (Stuxberg, 1875), fixed by subsequent designation of Jeekel, 2005.
— Jeekel, 2005: 23 (list; notes).

♀*L. (M.) cardinalis* (Bollman, 1887)
— Bollman, 1887a: 82 [= 1893: 20] (*Lithobius*; characters; USA: Indiana).
— Bollman, 1887b: 254, 258-259, 265 [= 1893: 21, 25-26, 32] (*Lithobius*; subgenus *Archilithobius*; characters; habitat; key; list; note; USA: Indiana).
— Bollman, 1888e: 409 [= 1893: 110] (*Lithobius*; note; USA: Indiana).
— Bollman, 1893: 128, 134 (*Lithobius*; subgenus *Archilithobius*; catalogue; characters; habitat; note).
— Chamberlin, 1911c: 101, 103 (*Lithobius*; key to species; USA: Indiana).
— Williams & Hefner, 1928: 140 (*Lithobius*; characters; USA: Ohio).
— Crabill, 1952b: 385 (*Lithobius*; [**Note:** Bollman (1887a) originally described this species very inadequately and then elaborated more

by redescribing it as a new species a second time shortly after (1887b), but the type is apparently unknown and Crabill states "The affinity and identity of . . ." *Lithobius cardinalis* Bollman " . . . is uncertain due to the inadequate original description[s] and the unavailablity of the type." Crabill seemed to be unaware of Bollman's redescription, therefore this species deserves further investigation].

— Kevan, 1983: 2948 (*Metalithobius*; list). [**Note:** Kevan lists this as a *Metalithobius* species because *Archilithobius* is listed as a synonym of *Metalithobius*. The *Lithobius* mentioned by Chamberlin (1910) that are listed in his key need further investigation but the author has listed this species as belonging to the subgenus Metalithobius according to Kevan, however, even Kevan states on page 2943 "In the following list, which does not claim to be infallible insofar as synonymy and distribution are concerned . . ."].

Distribution: USA: Indiana, Ohio.

COTYPES: NMNH, 7 specimens, Indiana: Bloomington.

(Subgenus *Monotarsobius* Verhoeff, 1905)

Lithobius (*Monotarsobius*) Verhoeff, 1905b: 249 (?). [**Note:** Reference not obtained]. TYPE SPECIES: *Lithobius curtipes* C. L. Koch, 1847, fixed by subsequent designation.

— Chamberlin & Wang, 1952b: 183 (*Monotarsobius*; USA: Hawaii). [**Note:** Originally cited as a *Monotarsobius* species that is now considered a subgenus of *Lithobius*. Apparently a female specimen of this subgenus was intercepted at quarantine in Honolulu Hawaii, 25 June 1938].

— Eason, 1976a: 123 (taxonomic note). [**Note:** Eason points out that Verhoeff originally erected *Monotarsobius* as a genus and has been regarded by many subsequent authors as either a subgenus or a synonym of *Lithobius*. The author is following Eason and regarding *Monotarsobius* as a subgenus of *Lithobius*].

— Jeekel, 2005: 23 (list).

Nipponobius Chamberlin, 1929: 37 (characters). TYPE SPECIES: *Nipponobius migrans* Chamberlin, 1929, fixed by original designation.

— Eason, 1976a: 123 (synonym of *Lithobius* (*Monotarsobius*)). [**Note:** Eason states that *Nipponobius* Chamberlin should be

regarded as a synonym of *Lithobius* (*Monotarsobius*) Verhoeff.
Accordingly all species of *Nipponobius* are transferred].
— Jeekel, 2005: 24 (list).
Onebius Chamberlin, 1926b: 92 (characters). **Type Species:** *Onebius
moananus* Chamberlin, 1926, fixed by original designation.
— Eason, 1992: 4 (*Onebius*; notes).
— Zapparoli & Shelley, 2000: 44 (synonym of *Lithobius* (*Monotarsobius*)).
— Jeekel, 2005: 25 (list).

♀*L.* (*M.*) *annectus* Chamberlin, 1941
— Chamberlin, 1941a: 4-5 (*Nipponobius*; characters; notes; USA:
California).
— Eason, 1976a: 123 (*Nipponobius* transferred to *Lithobius* (*Monotarsobius*)).

Distribution: USA: California.

HOLOTYPE: NMNH, ♀, California: 11 mi. E of Glenville, 5,000 ft., 19
March 1941, by S. & D. Mulaik.

♀ *L.* (*M.*) *australis* (Chamberlin, 1944)
— Chamberlin, 1944b: 64 (*Nipponobius*; characters; notes; USA:
Hawaii).
— Eason, 1976a: 123 (*Nipponobius* synonym of *Lithobius*
(*Monotarsobius*)).
— Zapparoli & Shelley, 2000: 47 (*Nipponobius*; list; notes).

Distribution: USA: ?Hawaii.

HOLOTYPE: NMNH, ♀, Hawaii: Honolulu, from packing about a plant
of *Epidendrum* sp. imported from Australia.

♀ *L.* (*M.*) *cepeus* (Chamberlin, 1940)
— Chamberlin, 1940d: 50 (*Nipponobius*; characters; note; USA: Hawaii).
[**Note:** This species was found in packing material about *Vandatteres
grandiflora* from Japan by a plant quarantine inspector. It is not likely
to be established in Hawaii].
— Eason, 1976a: 123 (*Nipponobius*; synonym of and transferred to
Lithobius (*Monotarsobius*)).
— Zapparoli & Shelley, 2000: 47 (*Nipponobius*; list; notes).

Distribution: USA: ?Hawaii.

HOLOTYPE: NMNH, ♀, Hawaii: Honolulu, 11 April 1938.

♂♀*L.* (*M.*) *crassipes* (L. Koch, 1862)
— L. Koch, 1862: 71-72; Tab. 2, Fig. 31 (*Lithobius*; characters; Germany).
— Latzel, 1880: 128-129 (*Lithobius*; characters).
— Brölemann, 1930: 325; figs. 449-452 (*Lithobius*; characters).
— Verhoeff, 1937: 188 (*Monotarsobius*; key).
— Crabill, 1952b: 380-384, 450 (*Monotarsobius*; characters; distribution; habitat; plectrotaxy; range; taxonomic notes; introduced; USA: Illinois, Massachusetts, New York, Pennsylvania, Vermont). [**Note:** Crabill reports that specimens have been intercepted in Chicago, Illinois on imported bulbs from Germany].
— Eason, 1964: 234-236; figs. 440-451 (*Lithobius*; characters; distribution; habitat).
— Eason, 1982: 24-25, 28 (*Lithobius*; subgenus *Monotarsobius*; systematic review; key).
dendrophilus Chamberlin, 1913: 70-72 (*Garibius*; characters; notes; USA: Pennsylvania). [**Note:** Holotype deposited at the MCZ, No. 459, original locality label reads "Phil., Oct. 28-'12, near 69th & [?Machet St.], under boards in the open"; synonymy taken from Crabill, 1952b: 380].
— Crabill, 1952b: 380 (*Garibius*; synonym of *Monotarsobius crassipes*).
?*vagrans* Chamberlin, 1916: 169-170 (*Paobius*; characters; note; USA: Vermont). [**Note:** synonymy taken from Crabill, 1952b: 380; holotype at MCZ, No. 352; paratypes at MCZ, No. 412, 414].
— Chamberlin, 1951b: 32 (*Paobius*; USA: Vermont).
— Crabill, 1952b: 380 (*Paobius*; cited as new synonym of *Monotarsobius crassipes*).

Distribution: USA: Illinois, Massachusetts, New York, Pennsylvania, Vermont.

HOLOTYPE: ?Unknown. Germany.

♂*L.* (*M.*) *migrans* (Chamberlin, 1929)
— Chamberlin, 1929: 37 (*Nipponobius*; characters; USA: Pennsylvania). [**Note:** Taken from quarantine in Philadelphia from lily bulbs imported form Japan].
— Chamberlin, 1930c: 69 (*Nipponobius*; notes).

— Eason, 1976a: 120, 123 (*Nipponobius* transferred to *Lithobius*
(*Monotarsobius*); note).
— Jeekel, 2005: 23 (*Nipponobius*; list).

Distribution: USA: ?Pennsylvania.

HOLOTYPE: NMNH, ♂, Pennsylvania: Philadelphia. (Origin: Japan:
Saitama),

♀*L. (M.) moananus* (Chamberlin, 1926)
— Chamberlin, 1926b: 92 (*Onebius*; characters; USA: Hawaii).
— Attems, 1938b: 366, 369 (*Onebius*; list).
— Eason, 1977: 491 (*Onebius*; notes).
— Butler & Usinger, 1963: 239 (**Note:** Reference not obtained).
— Nishida, 1994: 26 (*Onebius*; list).
— Nishida, 1997: 195 (*Onebius*; list).
— Zapparoli & Shelley, 2000: 44 (*Lithobius*; subgenus *Monotarsobius*;
characters; USA: Hawaii).
— Jeekel, 2005: 25 (*Onebius*; list).

Distribution: USA: Hawaii (Origin probably Japan, see Zapparoli & Shelley,
2000).

HOLOTYPE: ?Unknown; Hawaii, Apr. 18, 1923 (Fullaway): 6 females.

♂*L. (M.) sinensis* (Chamberlin, 1930)
— Chamberlin, 1930c: 68-69 (*Nipponobius*; characters; USA: Hawaii).
[**Note:** Taken at Hawaii in 1928 on *Lillium* species imported from
China].
— Eason, 1976a: 123 (*Nipponobius*; synonym of *Lithobius*
(*Monotarsobius*)).
— Zapparoli & Shelley, 2000: 47 (*Nipponobius*; list; notes).

Distribution: USA: ?Hawaii.

HOLOTYPE: NMNH, Hawaii.

L. (M.) tricalcaratus (Attems, 1909)
— Attems, 1909a: 9, 20-21, pl. 1; Fig. 5 (*Monotarsobius*; characters; list;
USA: Alaska).
— Chamberlin, 1919: 15H (*Monotarsobius*; list).

125

— Chamberlin, 1946c: 186 (*Monotarsobius*; list).
— Eason, 1976a: 125 (*Monotarsobius* regarded as subgenus of *Lithobius*).
— Kevan, 1983: 2949 (*Monotarsobius*; list). [**Note:** Kevan (1983, see footnote 29) suggests that species of " . . . *Monotarsobius* may eventually be found on the Alaskan side of the Bering Straight."].

Distribution: USA: Alaska.

SYNTYPE: MCZ, No. 14576, 2 specimens, Vega Expedition, Alaska: Port Clarence.

(Subgenus *Sigibius* Chamberlin, 1913)

♂♀*L. (S.) bullatus* Eason, 1993
— Eason, 1993: 192-195; figs. 15-19; tab. 5 (*Lithobius*; subgenus *Sigibius*; characters; notes; China).
— Zapparoli & Shelley, 2000: 42-44; Figs. 5-8; Tab. 2 (*Lithobius*; subgenus *Sigibius*; characters; plectrotaxy; USA: Hawaii).

Distribution: USA: Hawaii.

HOLOTYPE: MSNV; ♂, China: Hong Kong: Castle Park, 25.I.1981; leg. Cottarelli & Osella.

PARATYPES: MSNV; 2♀; China: Hong Kong: Tai Po, 26.I.1981; leg. Osella.

♂♀*L. (S.) microps* Meinert, 1868
— Meinert, 1868: 265 (*Lithobius*; characters; notes).
— Meinert, 1872: 330-331 (*Lithobius*; characters; distribution; [**Note:** see Eason, 1974b]).
— Eason, 1974b: 7-9 (*Lithobius*; subgenus *Monotarsobius*; characters; notes).
— Andersson, 1979: tab. 1, 3-5 (*Lithobius*; characters of larval stadia).
— Kevan, 1979: 298 (*Lithobius*; synopsis; CANADA: Newfoundland).
— Behan-Pelletier, 1993: 26 (*Sigibius*; list).
— Mercurio, 2005: 232 (*Lithobius*; habitat; USA: New York).
duboscqui Brölemann, 1896: 116. (*Lithobius*;). [**Note:** reference not obtained].
— Verhoeff, 1937: 189 (*Monotarsobius*; key).

— Eason, 1964: 242-246; Tab. 15; Figs. 463-476 (*Lithobius*; characters; distribution; habitat; notes).

— Kevan, 1979: 298 (*Lithobius*; synopsis; CANADA: Newfoundland).

olivarum Verhoeff. 1925: 153-155 (*Lithobius*; subgenus *Haplolithobius*; characters).

puritanus Chamberlin, 1913: 102-104 (*Sigibius*; characters; USA: Maine, Massachusetts).

— Chamberlin, 1920a: 95 (*Sigibius*; note; CANADA: Quebec).

— Crabill, 1952b: 336, 338, 339, 340-341, 434, pl. 7; Fig. 103 (*Sigibius*; characters; habitat; plectrotaxy; range; synonym of *microps*; taxonomic notes; USA: Connecticut, Maine, Massachusetts, Michigan, New York).

— Palmén, 1954: 141 (*Sigibius*; note).

— Judd, 1964: 2 (*Sigibius*; note; CANADA: Ontario).

— Kevan, 1979: 298 (*Sigibius*; synopsis; CANADA: Ontario, Quebec).

— Summers, 1979: 696, 698 (*Sigibius*; checklist, key).

— Jeekel, 2005: 30 (*Sigibius*; list).

— Mercurio, 2005: 232 (synonym of *microps*; habitat; USA: New York).

Distribution: CANADA: Newfoundland, Ontario, Quebec; USA: Connecticut, Massachusetts, Maine, Michigan, New York.

HOLOTYPE: ZMUC, ♀imm., Palm-House, Copenhagen Botanic Gardens (Denmark).

Genus *Llanobius* Chamberlin & Mulaik, 1940
[3 species; South Central and Southeastern USA]

Llanobius Chamberlin & Mulaik, 1940: 128 (characters). **Type Species:** *Llanobius paucispinus* Chamberlin & Mulaik, 1940, fixed by original designation.

— Jeekel, 2005: 21 (*Llanobius*; list).

♂*L. chamberlini* Causey, 1942

— Causey, 1942: 83 (*Llanobius*; characters; USA: North Carolina).

— Wray, 1967: 155 (*Llanobius*; list).

— Crabill, 1952b: 385 (*Llanobius*). [**Note:** According to Crabill "The affinity and identity of . . ." *Llanobius chamberlini* Causey " . . . is uncertain due to the inadequate original description[s] and the unavailablity of the type."].

— Hoffman, 1995a: 28 (*Llanobius*; list). [**Note:** Hoffman questions if *Taiyubius dux* Chamberlin, 1940 is a senior synonym of *chamberlini*. It is peculiar that *Taiyubius dux* is the only species of the genus that is on the east coast. The issue needs further taxonomic scrutiny].

Distribution: USA: North Carolina.

HOLOTYPE: ANSP, ♂, North Carolina: Durham, Duke Forest, by Mr. James H. Starling.

♂*L. paucispinus* Chamberlin & Mulaik, 1940
— Chamberlin & Mulaik, 1940: 128 (*Llanobius*; characters; USA: Texas). [**Note:** Chamberlin states that the type may be one moult away from maturity].
— Jeekel, 2005: 21 (*Llanobius*; list).

Distribution: USA: Texas.

HOLOTYPE: NMNH, ♂, Texas: Kerr Co., Raven Ranch, December 1939.

♂*L. santus* Chamberlin & Mulaik, 1940
— Chamberlin & Mulaik, 1940: 156 (*Llanobius*; characters; USA: Texas).

Distribution: USA: Texas.

HOLOTYPE: NMNH, Texas: Tom Green Co., 11 mi. E of San Angelo, 15 December 1939.

Genus *Lophobius* Chamberlin, 1922
[6 species; Western CANADA and USA]

Lophobius Chamberlin, 1922b: 343, 367, 368; fig. 9 (characters of subgenus within *Pokabius*; distribution; key to subgenera and species within *Pokabius*). **TYPE SPECIES:** *Lithobius socius* Chamberlin, 1901, fixed by subsequent designation. [**Note:** Chamberlin originally created the subgenus *Lophobius* within the genus *Pokabius*]
— Chamberlin, 1925e: 55, 56 [**Note:** Chamberlin's first use of *Lophobius* as a genus and he apparently elevated the subgenus to genus status with the description of two new species *loganus* and *francisae*].

128

— Chamberlin, 1928a: 94-95 (key to species). [**Note:** Chamberlin provides a key to the species of *Lophobius* with the description of the new species *lasalanus*. In this key he has seemingly transferred *castellopes*, *collium*, *eremus*, *helenae*, *pungonius* and *socius* from *Pokabius* to *Lophobius* (did he mean subspecies?) but apparently forgot to include *loganus* in his key to species].

— Chamberlin, 1940d: 53-54 (key to species).

— Jeekel, 2005: 21 (list; notes).

♂*L. apachus* Chamberlin, 1940
— Chamberlin, 1940d: 53 (*Lophobius*; characters; notes; USA: Arizona).

Distribution: USA: Arizona.

HOLOTYPE: NMNH, ♂, Arizona: Duncan, 5 September 1939.

♂*L. carinipes* (Daday, 1889)
— Daday, 1889: 153, pl. 5; Fig. 31 (*Lithobius*; characters; USA: "California borealis".

— Chamberlin, 1910: 372 (*Lithobius*; characters; key to species; USA: California).

— Chamberlin, 1922b: 367 (*Pokabius*; characters; USA: California).

— Eason, 1985: 1-3; Figs. 1, 2 (*Lophobius*; characters; notes).

Distribution: USA: California.

HOLOTYPE: NMB, ♂, California: Northern California, by D. J. Vadona.

♂♀*L. collium* (Chamberlin, 1901)
— Chamberlin, 1901: 23 (*Lithobius*; characters; habitat; key to species; USA: Utah)

— Chamberlin, 1913: 62 (*Nadabius*; note).

— Chamberlin, 1922b: 370-372, pl. 11; Fig. 4-6 (*Pokabius*; subgenus *Lophobius*; characters (adult, stadia); habitat; USA: Utah).

— Chamberlin, 1928a: 94 (*Lophobius*; key). [**Note:** transferred from *Pokabius* being placed in key to species of *Lophobius*].

— Chamberlin, 1928d: 308 (*Lophobius*; USA: Utah).

— Chamberlin, 1930a: 112 (*Lophobius*; habitat; note; USA: Utah).

— Chamberlin, 1958c: 133 (*Lophobius*; checklist; note).

Distribution: USA: Utah.

SYNTYPES: NMNH, No. 783, Utah: foot hills near Salt Lake City, 3/5, 1901, R.V. Chamberlin.

♂♀*L. francisae* Chamberlin, 1925
— Chamberlin, 1925e: 56 (*Lophobius*; characters; USA: Utah).
— Chamberlin, 1928a: 95 (*Lophobius*; key).
— Chamberlin, 1930a: 112 (*Lophobius*; notes; USA: Utah).

Distribution: USA: Utah.

HOLOTYPE: NMNH, Utah: Cedar City, May 1924.

♂♀*L. lasalanus* Chamberlin, 1928
— Chamberlin, 1928a: 94 (*Lophobius*; characters; USA: Utah).

Distribution: USA: Utah.

HOLOTYPE: NMNH, Utah: San Juan Co., La Sal Mountains, July, 1927, by Dr. V. M. Tanner.

♂♀*L. sororis* Chamberlin, 1940
— Chamberlin, 1940d: 52 (*Lophobius*; characters; habitat; notes; USA: Nevada).

Distribution: USA: Nevada.

HOLOTYPE: NMNH, Nevada: Ruby Valley, 15 June 1935, under leaves on ground under growth of willow, R.V. Chamberlin.

Genus *Nadabius* Chamberlin, 1913
[17 species; CANADA; USA]

Nadabius Chamberlin, 1913: 62 (no description just mentioned in text).
TYPE SPECIES: *Lithobius jowensis* Meinert, 1886, fixed by subsequent designation by Chamberlin, 1922: 320.
— Chamberlin, 1922b: 319-322; fig. 7 (characters; distribution; key to species).
— Bailey, 1928b: 29 (characters; note).
— Attems, 1938a: 344 (synonym of *Lithobius* (*Pokabius*)).
— Crabill, 1952b: 356-358 (characters; key to species of northeastern USA; notes).

— Crabill, 1962b: 410; Tab. 8 & 9 (anterior and posterior limits of spurs including *aristeus*, *jowensis* and *pullus*).
— Carter & Brown, 1973: 1066, Tab. 1 (data from study on soil arthropods; CANADA: New Brunswick).
— Watermolen, 1997: 5 (notes).
— Jeekel, 2005: 23 (list; note).

♂*N. ameles* Chamberlin, 1944

— Chamberlin, 1944d: 210 (*Nadabius*; characters; USA: Indiana).
— Crabill, 1952b: 364-365, 444, pl. 8; Fig. 104 (*Nadabius*; characters; plectrotaxy; taxonomic notes; USA: Connecticut, Illinois, Indiana).
— Crabill, 1955b:41 (*Nadabius*; notes; USA: Missouri).
— Summers, 1979: 696, 698, (*Nadabius*; checklist, key).
— Summers & Uetz, 1979: 350; Fig. 3 (*Nadabius*; microhabitat; USA: Illinois).
— Lee, 1980: 5; Tab. 1 (*Nadabius*; habitat; USA: Ohio).
— Summers et al., 1980: 251; Fig. 20 (*Nadabius*; checklist; USA: Illinois).
— Summers et al., 1981: 61 (*Nadabius*; checklist; USA: Illinois).
— Snider, 1991: 190 (*Nadabius*; list; habitat; USA: Michigan).
— Watermolen, 1997: 7 (*Nadabius*; notes).

Distribution: USA: Connecticut, Illinois, Indiana, Michigan, Missouri, Ohio.

HOLOTYPE: NMNH, Indiana: Porter Co., Dune Acres (=Mineral springs), 18 April 1942, by H.S. Dybas.

♂♀*N. aristeus* Chamberlin, 1922

— Chamberlin, 1922b: 322-324, pl. 6; Fig. 3, 4 (*Nadabius*; characters (adult, stadia); USA: Massachusetts, New Jersey, New York).
— Bailey, 1928a: 1082 (*Nadabius*; list).
— Bailey, 1928b: 19, 29 (*Nadabius*; characters; list; USA: New York).
— Crabill, 1952b: 360-362, 441, pl. 8; Figs. 105-107 (*Nadabius*; characters; habitat; plectrotaxy; range; CANADA: Ontario; USA: Massachusetts, New Jersey, New York, Pennsylvania, Virginia).
— Crabill, 1962b: 408; Tab. 3; 410; Tab. 8, 9; 411; Tab. 11 (*Nadabius*; plectrotaxy characters, anterior and posterior limits of spurs and age correlation).
— Lee, 1980: 5; Tab. 1 (*Nadabius*; habitat; USA: Ohio).
— Hoffman, 1995a: 28 (*Nadabius*; list; notes).

Distribution: CANADA: Ontario; USA: Massachusetts, New Jersey, New York, Ohio, Pennsylvania, Virginia.

TYPES: MCZ, No. 247, 2 specimens, New Jersey: Macapin, November 10, 1913, J.H. Emerton.

♂♀(imm.)*N. caducipes* Chamberlin, 1946
— Chamberlin, 1946c: 187 (*Nadabius*; characters; USA: Alaska). [**Note:** Except for the male holotype the several immature females are missing their posterior legs].
— Behan-Pelletier, 1993: 26 (*Nadabius*; list).

Distribution: USA: Alaska.

HOLOTYPE: NMNH, ♂, Alaska: Beluga Flats, 3-4 mi. SW Beluga River on Cook Inlet, 1-3 September 1945, J.C. Chamberlin.

PARATYPES: NMNH, 4 specimens, Alaska: Beluga Flats, 3-4 mi. SW of Beluga River on Cook Inlet, 1-3 September 1945, J.C. Chamberlin

♂*N. cherokeenus* Chamberlin, 1947
— Chamberlin, 1947a: 38; fig. p. 39 (*Nadabius*; characters; notes; USA: Georgia).
— Crabill, 1952b: 358 (*Nadabius*; note).

Distribution: USA: Georgia.

HOLOTYPE: NMNH, ♂, Georgia: Atlanta, 17 February 1946, by P.W. Fattig.

♂♀*N. coloradensis* (Cockerell, 1893)
— Cockerell, 1893: 370 (*Lithobius*; nominal taxon *kochii*; cited as variety *coloradensis*; note; USA: Colorado).
— Chamberlin, 1911a: 70 (*Lithobius*; notes).
— Chamberlin, 1922b: 327-329, pl. 7; Fig. 4 (*Nadabius*; characters (adult, stadia); USA: Colorado).
— Chamberlin, 1943b: 104-105 (*Nadabius*; USA: Colorado, Montana, Wyoming).
dopaintus Chamberlin, 1911a: 69 (*Lithobius*; characters; key to species; USA: Colorado). [**Note:** Chamberlin states that this species is most closely related to *L. socius*. Two vials of types deposited at the MCZ, No. 218, 11 specimens, original locality/determined label reads "*Pokabius coloradensis* (Ckll.), types of *dopaintus*

Ch., Col: Manitou, 8-21-1910"; the second vial, No. 220, 6 specimens, original locality/determined label reads "*N. coloradensis* (Ckll.), [= type of *dopaintus* Ch.] Col: Tollau[?n]d, Coll. Aug. 1911"].

— Chamberlin, 1922b: 327 (synonym of *coloradensis*).

kochii Bollman (not Stuxberg), 1888: 349 [= 1893: 101] (*Lithobius*; characters; notes: USA: Colorado). [**Note:** Chamberlin (1911a: 70) cited that this report is in part *coloradensis*].

— Chamberlin (not Stuxberg, in part), 1909: 187 (*Lithobius*; notes). [**Note:** Chamberlin (1911a: 70) cited that this report is in part *coloradensis*. Chamberlin reported specimens from California and Colorado therefore those from the latter are *coloradensis*].

Distribution: USA: Colorado, Montana, Wyoming.

HOLOTYPE: ?Unknown, Colorado: West Cliff, T.D.A. Cockerell. [**Note:** MCZ has numerous specimens with the following catalog numbers: 206-207 (slides), 219, 221-222, 28715, 28721-28732, 28739-28741, 28744, 28746-28747].

♂♀*N. eigenmanni* (Bollman, 1887)
— Bollman, 1887c: 625 [= 1893: 42] (*Lithobius*; characters; note; CANADA: British Columbia).
— Bollman, 1893: 128 (*Lithobius*; subgenus *Archilithobius*; catalogue).
— Chamberlin, 1922b: 336-337, pl. 7; Fig. 5 (*Nadabius*; characters; notes: CANADA: British Columbia; USA: Oregon).
— Matthews, 1935: 58, 62-67; Map: 68 (*Nadabius*; characters; USA: Wisconsin). [**Note:** Matthews explains that based on all of his *Nadabius* collected he would consider *eigenmanni* a synonym of *jowensis* Meinert (see *holzingeri*)].
— Crabill, 1952b: 365 (*Nadabius*; systematic note).
— Kevan, 1979: 298 (*Nadabius*; synopsis; CANADA: British Columbia).
— Behan-Pelletier, 1993: 26 (*Nadabius*; list).

Distribution: CANADA: British Columbia; USA: Oregon, Wisconsin.

SYNTYPE: NMNH, CANADA: British Columbia, Glacier (C. H. Eigenmann)

♂*N. eremites* Chamberlin, 1944
— Chamberlin, 1944d: 210 (*Nadabius*; characters; USA: Tennessee).

— Crabill, 1952b: 357-358 (*Nadabius*; note).

Distribution: USA: Tennessee.

HOLOTYPE: NMNH, ♂, Tennessee: Great Smoky Mountain National Park, Greenbrier Cove, 13-19 June 1942, by H.S. Dybas.

♂♀***N. holzingeri*** (Bollman, 1887)
— Bollman, 1887e: 83 [= 1893: 72] (*Lithobius*; subgenus *Archilithobius*; characters; notes; USA: Minnesota).
— Bollman, 1893: 128, 185 (*Lithobius*; subgenus *Archilithobius*; catalogue; note).
— Chamberlin, 1911c: 101, 104 (*Lithobius*; key to species; USA: Minnesota, Wisconsin).
— Chamberlin, 1922b: 338-339 (*Nadabius*; characters; notes; USA: Minnesota).
— Verhoeff, 1925: 690 (*Lithobius*; list).
— Matthews, 1935: 58, 66 (*Nadabius*; characters; USA: Wisconsin). [**Note:** Matthews explains that based on all of his *Nadabius* collected he would consider *holzingeri* a synonym of *jowensis* Meinert (see *eigenmanni*)].
— Crabill, 1952b: 362-363, 442 (*Nadabius*; characters; plectrotaxy; range).
— Crabill, 1958b: 98 (*Nadabius*; notes; USA: Wisconsin).
— Summers, 1979: 696, 698, (*Nadabius*; checklist, key).
— Summers et al., 1980: 251; Fig. 21 (*Nadabius*; checklist; USA: Illinois).
— Summers et al., 1981: 61 (*Nadabius*; checklist; USA: Illinois).
— Watermolen, 1997: 5 (*Nadabius*; checklist, notes).

Distribution: USA: Illinois, Minnesota, Wisconsin.

SYNTYPE: NMNH, Minnesota: Winona (J.M. Holzinger)

♂♀***N. jowensis*** (Meinert, 1886)
— Meinert, 1886a: 177 (*Lithobius*; characters; [**Note:** Meinert does not give a locality]).
— Bollman, 1887b: 265 [= 1893: 32] (*Lithobius*; subgenus *Archilithobius*; list).
— Bollman, 1888e: 409 [= 1893: 110] (*Lithobius*; notes; USA: Indiana).
— Bollman, 1893: 128 (*Lithobius*; subgenus *Archilithobius*; catalogue).

— Chamberlin, 1911a: 70 (*Lithobius*; characters; key to species; USA: Colorado).

— Chamberlin, 1911c: 100, 103 (*Lithobius*; key; notes; USA: Illinois, Indiana, Michigan, Nebraska, Wisconsin).

— Gunthorp, 1913: 165-166 (*Lithobius*; USA: Kansas).

— Chamberlin, 1914b: 302 (*Nadabius*; habitat (p. 301); USA: Michigan).

— Chamberlin, 1922b: 330-336, pl. 7, fig 6; pl. 8; Fig. 1 (*Nadabius*; characters (adult, stadia); notes; USA: Colorado, Illinois, Indiana, Iowa, Kansas, Michigan, Nebraska, Wisconsin, Wyoming).

— Verhoeff, 1925: 690 (*Lithobius*; list).

— Holmquist, 1926: 411 (*Nadabius*; notes on hibernation such as: activity, date collected, developmental stage, ecological association/ niche, quantity found).

— Chamberlin, 1928b: 154 (*Nadabius*; USA: Missouri).

— Holmquist, 1928: 83, Tab. 3 (*Nadabius*; co-inhabitant of *F. uleki* ant nests).

— Williams & Hefner, 1928: 141 (*Lithobius*; characters; USA: Ohio).

— Matthews, 1935: 59-62, 64, 64, 66; Map: 68, pl. 5; Fig. 1-5. (*Nadabius*; characters; USA: Wisconsin). [**Note:** Matthews suggested that *N. jowensis* (Meinert) is the senior synonym for *N. eigenmanni* (Bollman) and *N. holzingeri* (Bollman). The author agrees with Matthews' suspicion but has no solid evidence to justify placing them under *jowensis* until further study is undertaken of this genus as a whole].

— Chamberlin, 1942b: 15 (*Nadabius*; USA: Iowa).

— Chamberlin, 1942d: 187 (*Nadabius*; USA: Missouri).

— Chamberlin, 1943b: 105 (*Nadabius*; USA: Colorado, Wyoming).

— Chamberlin, 1944d: 209-210 (*Nadabius*; USA: Illinois).

Auerbach, 1946: 6, 14, 16, 35 (*Nadabius*; habitat, list; notes; USA: Illinois).

— Rapp, 1946: 667 (*Nadabius*; habitat; USA: Illinois).

— Auerbach, 1949: 226 (*Nadabius*; notes; USA: Illinois).

— Auerbach, 1951a: 101, 104, 112, 115, 118; Tab. 1, 8, 9; Fig. 1, 4 (*Nadabius*; desiccation; habitat; length to weight ratio; list; notes; population percentage; USA: Illinois).

— Crabill, 1952b: 363-364, 443 (*Nadabius*; characters; plectrotaxy; range; USA: Indiana, Kentucky, Michigan, Ohio, Wisconsin). [**Note:** Crabill thought that *N. ameles* Chamberlin resembled *jowensis* more so than *eigenmanni* as suggested by Chamberlin. It seems that with all of Chamberlin's single reports of new species within this genus needs to be thoroughly investigated to decipher intra-specific variation from legitimate species].

— Johnson, 1952: 165-172, 201; tabs. 19-20, 28; Map 13 (*Nadabius*; autecology; characters; distribution; experimental feeding diet; natural history; USA: Michigan).

— Crabill, 1955b:41 (*Nadabius*; notes; USA: Missouri).

— Crabill, 1955f:260 (*Nadabius*; checklist; USA: Kentucky).

— Crabill, 1955h: 156 (*Nadabius*; notes; USA: Missouri).

— Judd, 1957: 79 (*Nadabius*; cited as being eaten by a salamander; CANADA: Ontario).

— Crabill, 1958b: 98 (*Nadabius*; notes; USA: Wisconsin).

— Crabill, 1962b: 409; Tab. 4; 410; Tab. 8 & 9 (*Nadabius*; plectrotaxy characters, including anterior and posterior limits of spurs).

— Summers, 1979: 696, 698, (*Nadabius*; checklist, key).

— Summers & Uetz, 1979: 347, 350, 351; Tab. 1 & 2; Fig. 3 (*Nadabius*; microhabitat; USA: Illinois).

— Summers et al., 1980: 251; Fig. 22 (*Nadabius*; checklist; USA: Illinois).

— Summers et al., 1981: 61 (*Nadabius*; checklist; USA: Illinois).

— Kevan, 1983: 2949 (*Nadabius*; list; USA: Idaho).

— Snider, 1991: 189 (*Nadabius*; list; habitat; USA: Michigan).

— Behan-Pelletier, 1993: 26 (*Nadabius*; list).

— Watermolen, 1997: 5 (*Nadabius*; checklist, notes).

— Jeekel, 2005: 23 (*Lithobius*; list).

?*bilabiatus* Bollman [not Wood], 1887b: 254, 256, 265 [= 1893: 21, 23, 32] (*Lithobius*; subgenus *Archilithobius*; characters; note; list; USA: Indiana, Michigan).

— Chamberlin, 1911a: 70 (*Lithobius*; synonym of *jowensis*).

— Crabill, 1952b: 363 (*Lithobius*; synonym of *jowensis*).

bruneri Kenyon, 1893a: 17 (*Archilithobius*; characters; USA: Nebraska).

— Kenyon, 1893b: 162 (*Archilithobius*; characters; USA: Nebraska).

— Chamberlin, 1922b: 331 (synonym of *jowensis*).

sexdentatus Kenyon, 1893a: 17 (*Lithobius*; characters; note; USA: Nebraska).

— Kenyon, 1893b: 162 (*Lithobius*; characters; USA: Nebraska).

— Chamberlin, 1911c: 99, 103 (*Lithobius*; USA: Nebraska).

— Chamberlin, 1922b: 331 (synonym of *jowensis*).

similis Bollman, 1888a: 112 [= 1893: 85] (*Lithobius*; subgenus *Archilithobius*; characters; note; USA: Tennessee).

— Bollman, 1888c: 341 [= 1893: 92] (*Lithobius*; synonym of *trilobus*).

— Bollman, 1893: 129 (*Lithobius*; subgenus *Archilithobius*; catalogue).
trilobus Bollman, 1887a: 81 [= 1893: 19] (*Lithobius*; characters; USA: Indiana). **HOLOTYPE:** NMNH.
— Bollman, 1887b: 254, 258, 265 [= 1893: 21, 25, 32] (*Lithobius*; subgenus *Archilithobius*; characters; key; list; notes; USA: Indiana).
— Bollman, 1888c: 341 [= 1893: 92] (*Lithobius*; notes).
— Bollman [in part], 1888e: 409 [= 1893: 110] (*Lithobius*; note; USA: Indiana).
— Bollman, 1893: 129, 134-135 (*Lithobius*; subgenus *Archilithobius*; catalogue; characters).
— Chamberlin, 1911c: 101, 104 (key; localities).
— Chamberlin, 1922b: 331 (*Lithobius*; synonym of *jowensis* in part).
[**Note:** Chamberlin explains that most specimens labeled by Bollman as *L. trilobus* are referable to *jowensis*, however, some specimens were said to be *T. pullus*].
— Williams & Hefner, 1928: 141 (*Lithobius*; characters; USA: Ohio)
— Crabill, 1952b: 363 (*Lithobius*; synonym of *jowensis*, in part).
— Kevan, 1983: 2949 (*Nadabius*; checklist).

Distribution: CANADA: Ontario; USA: Colorado, Idaho, Illinois, Indiana, Iowa, Kansas, Kentucky, Michigan, Missouri, Nebraska, Ohio, Tennessee, Wisconsin, Wyoming.

HOLOTYPE: MCZ, no. 266, 1 specimen, Iowa: McGregor. [**Note:** MCZ database indicates "Davis" as the collector].

[**Note:** For some reason the only authors who used the original specific name for this species were Bollman, followed by Chamberlin (only in part), Gunthorp, Verhoeff, and Williams & Hefner who were aware of Chamberlin's genus *Nadabius* but seemingly decided to disregard it for their paper, and finally Johnson (1952) who obviously did his research making him the latest author to publish this species with the originally designated scientific name. It appears Chamberlin (1914b) started the misuse of the specific name *iowensis* when he transferred this species to his genus *Nadabius*, probably from never checking the original work by Meinert, or perhaps it was just because he assumed there was a spelling error and that Meinert must have surely meant *iowensis* because it was discovered in Iowa. In any case, Crabill and Summers took after Chamberlin's seemingly unjustified name change, which has carried through up until the present time, but appears in its original spelling here].

♀*N. mesechinus* (Chamberlin, 1903)
— Chamberlin, 1903b: 158 (*Lithobius*; characters; key to species; USA: Oregon).
— Chamberlin, 1922b: 329-330, pl. 6; Fig. 2 (*Nadabius*; characters; USA: Oregon).
— Chamberlin, 1943b: 105 (*Nadabius*; USA: Idaho, Montana).

Distribution: USA: Idaho, Montana, Oregon.

TYPES: MCZ, No. 270, 2 specimens, [**Note:** Chamberlin designated a ♀ as the holotype], Oregon: Meacham, August, 1902.

♀*N. oreinus* Chamberlin, 1922
— Chamberlin, 1922b: 325-327, pl. 5; Fig. 7; pl. 6; Fig. 1 (*Nadabius*; characters (adult, stadia); USA: California).

Distribution: USA: California.

HOLOTYPE: MCZ, No. 271, ♀, California: Shasta Springs, August, 1909.

♂*N. phanus* Chamberlin, 1941
— Chamberlin, 1941a: 16 (*Nadabius*; characters; USA: California).

Distribution: USA: California.

HOLOTYPE: NMNH, ♂, California: Monterey Co., Hastings Reservation, 25 March 1941, by Dr. J. M. Linsdale.

♂♀*N. pluto* Chamberlin & Wang, 1952
— Chamberlin & Wang, 1952a: 56 (*Nadabius*; characters; note; USA: Montana).

Distribution: USA: Montana.

SYNTYPE: NMNH, Montana: Hell Gate River, 13 August 1929, by R. V. Chamberlin & Edith S. Chamberlin.

♂♀*N. pullus* (Bollman, 1887)
— Bollman, 1887a: 81 [= 1893: 19] (*Lithobius*; characters; USA: Indiana).

— Bollman, 1887b: 254, 257-258, 265 [= 1893: 21, 24-25, 32] (*Lithobius*; subgenus *Archilithobius*; characters; key; list; USA: Indiana).
— Bollman, 1888e: 409 [= 1893: 110] (*Lithobius*; note; USA: Indiana).
— Bollman, 1893: 129, 134 (*Lithobius*; subgenus *Archilithobius*; catalogue; characters; note).
— Chamberlin, 1911c: 100, 103 (*Lithobius*; key to species; USA: Illinois, Indiana, Nebraska).
— Chamberlin, 1922b: 324 (*Nadabius*; characters; USA: District of Columbia, Illinois, Indiana, Tennessee, Virginia, West Virginia).
— Verhoeff, 1925: 690 (*Lithobius*; list).
— Williams & Hefner, 1928: 141 (*Lithobius*; characters; USA: Ohio).
— Brimley, 1938: 501 (*Lithobius*; note; USA: North Carolina).
— Crabill, 1952b: 358-360, 440 (*Nadabius*; characters; plectrotaxy; range; USA: District of Columbia, Illinois, Indiana, Kentucky, New York, Ohio, Virginia, West Virginia).
— Johnson, 1952: 172-175; Tab. 21; Map. 14 (*Nadabius*; autecology; characters; distribution; USA: Michigan).
— Crabill, 1955f: 260 (*Nadabius*; checklist; USA: Kentucky).
— Crabill, 1962b: 408; Tab. 2: 411; Tab. 10 (*Nadabius*; plectrotaxy characters, including anterior and posterior limits of spurs and age correlation).
— Branson & Batch, 1967: 88 (*Nadabius*; characters; habitat; USA: Kentucky).
— Wray, 1967: 155 (*Sozibius*; list).
— Summers, 1979: 696, 698, (*Nadabius*; checklist, key).
— Lee, 1980: 2, 3, 5; Tab. 1 (*Nadabius*; habitat; USA: Ohio).
— Summers et al., 1980: 251; Fig. 23 (*Nadabius*; checklist; USA: Illinois).
— Summers et al., 1981: 61 (*Nadabius*; checklist; USA: Illinois).
— Snider, 1991: 189 (*Nadabius*; list).
— Hoffman, 1995a: 28 (*Garibius*; list; notes).
— Watermolen, 1997: 7 (*Nadabius*; notes).

dorsospinorum Kenyon, 1893a: 17 (*Archilithobius*; characters; note; USA: Nebraska). [**Note:** Kenyon (1893b) described the species *Archilithobius dorsospinorum* and compared it to Bollman's *Lithobius elattus* [sic] which is now *N. pullus*. According to Chamberlin (1911c) *A. dorsospinorum* is a questionable synonym of *N. pullus*].

— Kenyon, 1893b: 162 (*Archilithobius*; characters; note; USA: Nebraska).

— Chamberlin, 1911c: 103 (*Lithobius*; questionable synonym of *pullus*).

elattus Bollman, 1888d: 348 [= 1893: 100-101] (*Lithobius*; characters; notes; USA: District of Columbia, Virginia).

— Bollman, 1893: 128 (*Lithobius*; subgenus *Archilithobius*; catalogue).

— Chamberlin, 1911b: 40 (*Lithobius*; notes; synonym of *pullus*; USA: Alabama, Kentucky, Mississippi, Tennessee, Virginia).

— Chamberlin, 1922b: 324 (synonym of *pullus*).

Distribution: USA: Alabama, District of Columbia, Illinois, Indiana, Kentucky, Michigan, North Carolina, Nebraska, New York, Ohio, ?South Carolina, Tennessee, Virginia, West Virginia. [**Note:** Expected to be found in South Carolina].

HOLOTYPE: ?Unknown, Indiana: Bloomington. [**Note:** MCZ has numerous specimens with the following catalog numbers: 214, 231-233, 235-242].

♂*N. saphes* Chamberlin, 1940

— Chamberlin, 1940f: 76 (*Nadabius*; characters; USA: North Carolina).

— Crabill, 1952b: 357 (*Nadabius*; note).

— Wray, 1967: 155 (*Nadabius*; list).

— Hoffman, 1995a: 28 (*Nadabius*; list; notes). [**Note:** Hoffman believes this species will be found in southern Virginia].

Distribution: USA: North Carolina.

HOLOTYPE: NMNH, ♂, North Carolina: Duke Forest, 18 October 1939, by Nelle B. Causey.

♂*N. vaquens* Chamberlin & Wang, 1952

— Chamberlin & Wang, 1952a: 56-57 (*Nadabius*; characters; USA: Wyoming).

Distribution: USA: Wyoming.

HOLOTYPE: ?Unknown, Wyoming: Yellowstone Park at Mt. Washburn, 13 August 1940.

♀*N. waccamanus* Chamberlin, 1940

— Chamberlin, 1940f: 77 (*Nadabius*; characters; notes; USA: North
Carolina).
— Causey, 1940: 72 (*Nadabius*; list).
— Wray, 1967: 155 (*Nadabius*; list).

Distribution: USA: North Carolina.

HOLOTYPE: NMNH, ♀, North Carolina: Lake Waccamaw, 24 September
1939, by Nelle B. Causey.

Genus *Nampabius* Chamberlin, 1913
[16 species; Eastern CANADA and USA]

Nampabius Chamberlin, 1913: TYPE SPECIES: *Nampabius virginiensis*
Chamberlin, 1913, fixed by original designation.
— Chamberlin, 1913: 40-42 (characters; distribution; habitat; key to
species; note).
— Bailey, 1928b: 29 (characters; notes).
— Attems, 1938a: 344 (synonym of *Lithobius* (*Pokabius*)).
— Crabill, 1952b: 366-368 (characters; key to species of northeastern
USA; notes).
— Crabill, 1952c: 206 (key to species of northeastern United States).
— Eason, 1974: 68 (note).
— Kevan, 1983: 2949 ("*Nampabius* Chamberlin = *Hemilithobius*
Stuxberg. *auctt.*, *partim*").
— Holsinger & Culver, 1988: 57 (*Nampabius*; cave species; USA: Virginia).
[**Note:** Holsinger & Culver report a *Nampabius* sp. from Virginia:
Montgomery Co.: Erharts Cave but do not indicate why they could not
identify it to species; maybe due to immaturity or possibly a new species].
— Jeckel, 2005: 23 (list).

♂♀*N. carolinensis* Chamberlin, 1913
— Chamberlin, 1913: 54-55, pl. 2; Fig. 4-5 (*Nampabius*; characters;
USA: South Carolina).
— Causey, 1940: 71 (*Nampabius*; characters; USA: North Carolina).
— Chamberlin, 1940c: 56 (*Nampabius*; USA: North Carolina).
— Wray, 1967: 155 (*Nampabius*; list).
— Hoffman, 1995a: 28 (*Nampabius*; list). [**Note:** Hoffman believes this
species will be found in Virginia].

Distribution: USA: North Carolina, South Carolina.

HOLOTYPE: MCZ, No. 495, South Carolina: Landrum, August 4, 1910. [**Note:** Label in vial reads "see mount"].

♀*N. embius* Chamberlin, 1913
— Chamberlin, 1913: 49-50, pl. 1; Fig. 3 (*Nampabius*; characters; USA: South Carolina).
— Crabill, 1952b: 367 (*Nampabius*; note).

Distribution: USA: South Carolina.

HOLOTYPE: MCZ, No. 490, South Carolina: Taylors, 8-3-10.

♂♀*N. fungiferopes* (Chamberlin, 1904)
— Chamberlin, 1904: 652 (*Lithobius*; habitat; USA: New York).
— Chamberlin, 1913: 43-45, pl. 1; Fig. 6-7 (*Nampabius*; characters; USA: New York, Vermont).
— Bailey, 1928a: 1082 (*Nampabius*; list).
— Bailey, 1928b: 19, 30 (*Nampabius*; characters; list; USA: New York).
— Crabill, 1952b: 369, 445 (*Nampabius*; characters; plectrotaxy; USA: New York, Vermont).
— Johnson, 1952: 180-181 (*Nampabius*; characters; distribution; habitat; USA: Michigan).
— Crabill, 1962b: 409; Tab. 5 (*Nampabius*; plectrotaxy).
— Snider, 1991: 190 (*Nampabius*; list).

Distribution: USA: Michigan, New York, Vermont.

TYPES: MCZ, No. 489, 2 specimens, New York: Ithaca.

♂♀*N. georgianus* Chamberlin, 1913
— Chamberlin, 1913: 52-53, pl. 2; Fig. 1-3, pl. 3; Fig. 8 (*Nampabius*; characters; USA: Georgia).
— Chamberlin, 1944a: 33 (*Nampabius*; USA: Georgia).
— Crabill, 1952b: 367 (*Nampabius*; note).

Distribution: USA: Georgia.

TYPE: MCZ, No. 493, Georgia: Tallulah Falls, 8-1-11. [**Note:** Original label reads "(see Mt.)" after the date indicating there is a slide].

COTYPE: MCZ, No. 500, Georgia: Tallulah Falls, 8-1-10.

♂*N. inimicus* Chamberlin, 1913
— Chamberlin, 1913: 50-52, pl. 3; Fig. 5-6 (*Nampabius*; characters; USA: Tennessee).
— Crabill, 1952b: 367 (*Nampabius*; note).
— Hoffman, 1995a: 28 (*Nampabius*; list; notes). [**Note:** Hoffman believes this species will be found in the Virginia counties of Lee and Scott].

Distribution: USA: Tennessee.

HOLOTYPE: MCZ, No. 492, Tennessee: Russellville, 8-8-10.

♀*N. longiceps* Chamberlin, 1913
— Chamberlin, 1913: 57-58 (*Nampabius*; characters; USA: North Carolina).
— Brimley, 1938: 500 (*Lithobius*; note; USA: North Carolina).
— Causey, 1940: 71 (*Nampabius*; characters).
— Wray, 1967: 155 (*Nampabius*; list).

Distribution: USA: North Carolina.

HOLOTYPE: MCZ, No. 491, North Carolina: Asheville, 8-5-10.

♂*N. lulae* Chamberlin, 1913
— Chamberlin, 1913: 58-60 (*Nampabius*; characters; USA: Georgia, Tennessee).

Distribution: USA: Georgia, Tennessee.

HOLOTYPE: MCZ, No. 486, Georgia: Lula, 7-31-1910.

♂♀*N. lundii* (Meinert, 1886)
— Meinert, 1886b: 111 (*Lithobius*; characters; USA: near New York).
— Bollman, 1887b: 265 [= 1893: 32] (*Lithobius*; subgenus *Archilithobius*; list).
— Bollman, 1888a: 111 [= 1893: 85] (*Lithobius*; note; USA: Tennessee).
— Bollman, 1888c: 341 [= 1893: 92] (*Lithobius*; note; USA: Tennessee).
— Bollman, 1893: 128 (*Lithobius*; subgenus *Archilithobius*; catalogue).

— Chamberlin, 1911b: 39 (*Lithobius*; range; USA: Georgia, North Carolina, South Carolina, Tennessee, Virginia). [**Note:** non lundii?].

— Chamberlin, 1913: 60-61 (*Nampabius*; characters (Latin); USA: "near New York City").

— Bailey, 1928a: 1082 (*Nampabius*; list).

— Bailey, 1928b: 19, 29-30 (*Nampabius*; characters; list; notes; USA: New York).

— Williams & Hefner, 1928: 142 (*Lithobius*; characters; USA: Ohio).

— Brimley, 1938: 500 (*Lithobius*; USA: North Carolina).

— Causey, 1940: 72 (*Nampabius*; characters; note).

— Crabill, 1952b: 385 (*Nampabius*; note). [**Note:** According to Crabill "The affinity and identity of . . ." *Nampabius lundii* (Meinert) " . . . is uncertain due to the inadequate original description[s] and the unavailablity of the type."].

— Crabill, 1952c: 206, (*Nampabius*; footnote; *species inquirenda*).

— Wray, 1967: 155 (*Lithobius*; list).

— Eason, 1974b: 47-48; Figs. 31-33 (*Nampabius*; characters; notes).

michiganensis Chamberlin, 1914b: 302 (*Nampabius*; characters; habitat (p. 301); notes; USA: Michigan).

— Crabill, 1952b: 373-374, 448, pl. 8; Figs. 110-111 (*Nampabius*; characters; plectrotaxy; CANADA: Ontario; USA: Kentucky, Michigan, New York).

— Johnson, 1952: 181-183; Tab. 23 (*Nampabius*; autecology; characters; distribution; natural history; USA: Michigan).

— Crabill, 1955f: 260 (*Nampabius*; checklist; USA: Kentucky).

— Eason, 1974b: 4, 47 (*Nampabius*; synonym of *lundii*).

— Summers, 1979: 696, 698, (*Nampabius*; checklist, key).

— Snider, 1991: 190 (*Nampabius*; list).

Distribution: CANADA: Ontario; USA: ?Georgia, Kentucky, Michigan, ?North Carolina, New York, Ohio, ?South Carolina, ?Tennessee, ?Virginia.

HOLOTYPE: ZMUC, ♂, New York: Near New York City (L. Lund). [**Note:** The question marks in front of Georgia, North Carolina, South Carolina, Tennessee, and Virginia are based on Eason (1974b) who stated some of Chamberlin's (1911) records of *L. lundii* from the Southern States of U.S.A. were almost certainly based on misidentified specimens (see Eason, 1974b)].

♂♀*N. major* Chamberlin, 1925

— Chamberlin, 1925f: 291 (*Nampabius*; characters; notes; USA: Tennessee).

— Crabill, 1952b: 368 (*Nampabius*; note).

Distribution: USA: Tennessee.

HOLOTYPE: NMNH, Tennessee: Jefferson City, by Mr. J. D. Ives. [A letter accompanied the specimens and stated the "animals were living in a cave of a little over a mile in length. Within the cave the animals had for their habitat a mound of bat faeces, situated about one fourth of a mile from the entrance in absolute darkness." Chamberlin notes that they have no special adaptation to cave life and that the eyes are unusually well developed for this genus].

♂♀*N. mycophor* Chamberlin, 1940
— Chamberlin, 1940f: 75 (*Nampabius*; characters; notes; USA: North Carolina).
— Causey, 1940: 72 (*Nampabius*; note).
— Crabill, 1952b: 368 (*Nampabius*; note).
— Wray, 1967: 154 (*Nampabius*; list).
— Hoffman, 1995a: 28-29 (*Garibius*; list; notes; USA: Virginia).

Distribution: USA: North Carolina, Virginia.

HOLOTYPE: NMNH, North Carolina: Duke Forest, 15 September 1939, by Nelle B. Causey.

♂*N. parienus* Chamberlin, 1913
— Chamberlin, 1913: 47-48, pl. 3; Fig. 1-2 (*Nampabius*; characters; USA: North Carolina).
— Causcy, 1940: 70 (*Nampabius*; characters; note).
— Crabill, 1952b: 368 (*Nampabius*; note).
— Wray, 1967: 155 (*Nampabius*; list). [**Note:** Specific name misspelled *parieus*].
— Holsinger & Culver, 1988: 57 (*Nampabius*; USA: Virginia). [**Note:** Reported as either a trogloxene or accidental occurrence in a cave].
— Hoffman, 1995a: 29 (*Garibius*; list; note).

Distribution: USA: North Carolina, South Carolina, Virginia.

HOLOTYPE: ?MCZ, North Carolina: Hot Springs.

♂♀*N. perspinosus* Chamberlin, 1928
— Chamberlin, 1928c: 85-86 (*Nampabius*; characters; USA: Washington).

Distribution: USA: Washington.

HOLOTYPE: NMNH, ♂, Washington: Puget Sound, by F.E. Smith.

ALLOTYPE: ?Unknown, ♀, Washington: Puget Sound, by F.E. Smith.

PARATYPES: ?Unknown, ♂♀ many specimens, Washington: Puget Sound, by F.E. Smith.

♂*N. tennesseensis* Chamberlin, 1913
— Chamberlin, 1913: 55-57, pl. 3; Fig. 3-4 (*Nampabius*; characters; USA: Tennessee).
— Crabill, 1952b: 368 (*Nampabius*; note).
— Hoffman, 1995a: 28 (*Nampabius*; list; note). [**Note:** Hoffman believes this species will be found in Virginia].

Distribution: USA: Tennessee.

HOLOTYPE: ?MCZ, Tennessee: Russellville.

♂♀*N. turbator* Crabill, 1952
— Crabill, 1952b: 369-372, 446 (*Nampabius*; characters; notes; plectrotaxy; USA: Virginia). [**Note:** This is from Crabill's PhD thesis, not a peer reviewed publication, but he later publishes this species in a peer reviewed journal (see Crabill, 1952c)].
— Crabill, 1952c: 203-206 (*Nampabius*; characters; notes on lithobiomorph terminology; USA: Virginia). [**Note:** This species is strongly indicative of being cavernicolous].
— Holsinger & Culver, 1988: 57 (*Nampabius*; USA: Virginia). [**Note:** Reported as a questioned troglobite].
— Hoffman, 1995a: 29 (*Nampabius*; list; note).

Distribution: USA: Virginia.

HOLOTYPE: Crabill's collection; No. C-1398-619; ♂, Virginia: Alleghany Co., Lowmoor, April 1950, 0.5 mi. within a cave, by R.L. Hoffman.

ALLOTYPE: Crabill's collection; No. C-1398-619; ♀, Virginia: Alleghany Co., Lowmoor, April 1950, 0.5 mi. within a cave, by R.L. Hoffman.

(Subgenus *Carolobius* Causey, 1942)

Nampabius (*Carolobius*) Causey, 1942: 82 (characters). **Type Species:**
Nampabius pinus Causey, 1942, fixed by monotypy.
— Jeekel, 2005: 13 (*Nampabius*; list).

♂*N.* (*C.*) *pinus* Causey, 1942
— Causey, 1942: 82 (*Nampabius*; characters; USA: North Carolina).
— Crabill, 1952b: 368 (*Nampabius*; list).
— Wray, 1967: 155 (*Nampabius*; list).
— Hoffman, 1995a: 29 (*Nampabius*; list; note).
— Jeekel, 2005: 13 (list).

Distribution: USA: North Carolina.

HOLOTYPE: ANSP; ♂, North Carolina: Durham, Duke Forest, by Mr.
James H. Starling.

COTYPE: ANSP; ♂, North Carolina: Durham, Duke Forest, by Mr. James
H. Starling.

(Subgenus *Nampabius* Chamberlin, 1913)

Nampabius (*Nampabius*) Chamberlin, 1913: 40 (characters). **Type Species:**
Nampabius (*Nampabius*) *virginiensis* Chamberlin, 1913, fixed by original
designation.
— Attems, 1938: 344 (synonym of *Lithobius* (*Pokabius*) Chamberlin,
1912).

♂♀*N.* (*N.*) *virginiensis* Chamberlin, 1913
— Chamberlin, 1913: 45-47, pl. 1; Fig. 1-2, 4-5, pl. 2. fig. 6 (*Nampabius*;
characters; USA: Tennessee, Virginia).
— Causey, 1940: 71 (*Nampabius*; characters; habitat; USA: North Carolina).
— Chamberlin, 1944d: 211 (*Nampabius*; USA: Tennessee).
— Crabill, 1952b: 372-373, 447, pl. 8; Figs. 108-109, 112-113
(*Nampabius*; characters; plectrotaxy; USA: Virginia).
— Crabill, 1962b: 409; Tab. 6 (*Nampabius*; plectrotaxy).
— Wray, 1967: 155 (*Nampabius*; list).
— Summers, 1979: 696, 698, (*Nampabius*; checklist, key).
— Hoffman, 1995a: 29 (*Nampabius*; list; note; USA: Virginia).
— Watermolen, 1997: 5 (*Nampabius*; notes).

— Jeekel, 2005: 23 (*Nampabius*; list).

Distribution: USA: North Carolina, Tennessee, Virginia. [**Note:** See Watermolen (1997: 5) for possible identification of this species in Wisconsin but it may be a misidentification. The author has left this state out of the distribution list as it is not a definitive identification].

HOLOTYPE: MCZ, Virginia: Natural Bridge.

Genus *Neolithobius* Stuxberg, 1875
[15 species; Central and Southeastern USA]

Neolithobius Stuxberg, 1875b: 8 (characters; cited as new subgenus). TYPE SPECIES: *Lithobius vorax* Meinert, 1872, fixed by subsequent designation of Chamberlin, 1925d: 471.
— Stuxberg, 1877: 135
— Latzel, 1880: 35 (in part).
— Bollman (in part), 1887b: 262 [= 1893: 29] (cited as subgenus).
— Bollman, 1887c: 626-627 [= 1893: 43-44] (cited as subgenus; key to species; species included: *clarus*; *juventus*, *latzeli*, *mordax*, *terreus*, *transmarinus*, *tyrannus* (emended), *vorax*).
— Bollman, 1893: 129, 164 (catalogue; cited as subgenus; key).
— Chamberlin, 1920d: 44 (notes; possible occurrence of this genus in Montana). [**Note:** Chamberlin reports two specimens of a *Neolithobius* sp. from St. Xavier, Montana, on the 31 of May, 1917, that was collected from the crop of an eared grebe, *Colymbus nigricollis californicus* (Brehm). This is the only report of this genus occurring in Montana so it is worth mentioning. Further collecting efforts should be made in this state to establish the actual species occurring here because it is a considerable distance from the known biogeographic distribution of the genus].
— Chamberlin, 1925d: 470-473; fig. 5 (characters; distribution, key to species; note).
— Chamberlin, 1945d: 216 (key to species having both anal and penult legs with a single claw).
— Crabill, 1952b: 296-299 (characters; key to species of northeastern USA; notes; range).
— Reddell, 1965: 166 (checklist of a *Neolithobius* sp. from cave; USA: Texas).
— Reddell, 1994: ? (notes on *Neolithobius* sp. from cave; USA: Texas).
— Frank et al., 2004 (an immature specimen referred to this genus was reported from a bromeliad in Florida).

— Jeekel, 2005: 24 (list).

Eulithobius Verhoeff (not Stuxberg), 1907: 240 (key).

— Chamberlin, 1925d: 470 (synonym of *Neolithobius*, in part).

♂♀*N. arkansensis* Chamberlin, 1944
   — Chamberlin, 1944d: 208 (*Neolithobius*; characters; notes; USA: Arkansas).
   — Crabill, 1952b: 298 (*Neolithobius*; note).

Distribution: USA: Arkansas.

HOLOTYPE: FMNH, ♂, Arkansas: Polk Co., Rich Mountain, 21 March 1938, by K.P. Schmidt.

PARATYPES: FMNH, 10 specimens, Arkansas: Polk Co., Rich Mountain, 21-22 March 1938, by K.P. Schmidt; 3 specimens; Arkansas: Garland Co., 11 mi. W of Hot Springs, 21 March 1938; 2 specimens; Arkansas: Logan Co., Mount Magazine, 2,500 ft., 24 March 1938; 1 ♂; Arkansas: Clark Co., 7 mi. N of Arkidelphia, 4 April 1937; 3 specimens (1 ♂); Arkansas: Sevier Co., 2 mi. E of Ben Lomond, 16 April 1941. NMNH, 2 specimens, Arkansas: Garland Co., 11 mi. W of Hot Springs, 21 March 1938, K.P. Schmidt.

♀*N. aztecus* (Humbert & Saussure, 1869)
   — Humbert & Saussure, 1869: 156 (*Lithobius*; characters; Mexico).
   — Bollman, 1887b: 255, 259-260, 266 [= 1893: 22, 26-27, 33] (*Lithobius*; subgenus *Lithobius* (questioned); characters; key; list; USA: California). [**Note:** Bollman was uncertain of his identification of this specimen from Ukiah, California, but it is originally described from Mexico].
   — Bollman, 1893: 129 (*Lithobius*; subgenus *Lithobius*; catalogue).
   — Chamberlin, 1910: 372 (*Lithobius*; characters; key to species; USA: ?California).
   — Chamberlin, 1943: 27 (transferred to *Neolithobius*).

Distribution: USA: ?California.

HOLOTYPE: ?Unknown, Mexico: Cordiliera orientalis. [**Note:** MCZ has a single specimen from Guatemala; catalog # 27876].

♂♀*N. devorans* (Chamberlin, 1912)
   — Chamberlin, 1912d: 147-150 (*Lithobius*; characters; notes; USA: Alabama).

— Chamberlin, 1925d: 488-492 (*Neolithobius*; characters (adult, stadia); notes; USA: Alabama).
— Crabill, 1952b: 299 (*Neolithobius*; note). [**Note:** Crabill's footnote regarding the reference to Chamberlin's 1912 paper lists Florida as a locality, but this seems to be an error as there is no such record by Chamberlin to my knowledge; unless Crabill has seen specimens from Florida that may reside at Cornell].

Distribution: USA: Alabama.

HOLOTYPE: ?Unknown, Alabama: Jackson, 1910, R. V. Chamberlin. [**Note:** MCZ has numerous specimens, which probably contain the original specimens Chamberlin used to describe the species, that have the following catalog numbers: 29506-?29509].

♂*N. entonus* Chamberlin, 1942
— Chamberlin, 1942d: 188 (*Neolithobius*; characters; USA: Arkansas, Oklahoma).
— Crabill, 1952b: 298 (*Neolithobius*; note).

Distribution: USA: Arkansas, Oklahoma.

HOLOTYPE: NMNH, ?Oklahoma: Latimer Co., 2 mi. E of Gowen, 26 April 1936, by Leslie Hubricht.

♂♀*N. ethopus* Chamberlin, 1945
— Chamberlin, 1945d: 215 (*Neolithobius*; characters; USA: Georgia).
— Chamberlin, 1951b: 32 (*Neolithobius*; USA: Florida).
— Crabill, 1952b: 298 (*Neolithobius*; note).

Distribution: USA: Florida, Georgia.

HOLOTYPE: NMNH, ♂, Georgia: Camilla, 15 April 1939.

ALLOTYPE: NMNH, ♀, Georgia: Camilla, 15 April 1939.

♂♀*N. helius* Chamberlin, 1918
— Chamberlin, 1918b: 379-380 (*Neolithobius*; characters; USA: Georgia). [**Note:** Chamberlin lists this species as coming from Louisiana (p. 369), however, the localities listed for the species are from Georgia].

— Chamberlin, 1925d: 498-499 (*Neolithobius*; characters; USA: Georgia).
— Chamberlin, 1945d: 215 (*Neolithobius*; USA: Georgia).
— Crabill, 1952b: 298 (*Neolithobius*; note).

Distribution: USA: Georgia.

HOLOTYPE: MCZ, Unique No. 14363, ♂, Georgia: Okefenokee Swamp, Billy's Island, June 1912, C. U. Exped.

♂♀*N. latzelii* (Meinert, 1886)
— Meinert, 1886a: 175-176 (*Lithobius*; characters; USA: Virginia).
— Bollman, 1887b: 266 [= 1893: 33] (*Lithobius*; subgenus *Neolithobius*; list).
— Bollman, 1887c: 626 [= 1893: 43] (*Lithobius*; subgenus *Neolithobius*; distribution; key).
— Bollman, 1888d: 349 [= 1893: 102] (*Lithobius*; characters; USA: Virginia).
— Bollman, 1893: 129 (*Lithobius*; subgenus *Neolithobius*; catalogue).
— Chamberlin, 1925d: 484-486 (*Neolithobius*; characters (adult, stadia); USA: North Carolina, Virginia).
— Verhoeff, 1925: 690 (*Lithobius*; list).
— Brimley, 1938: 500 (*Lithobius*; USA: North Carolina).
— Causey, 1940: 77 (*Neolithobius*; characters; USA: North Carolina).
— Crabill, 1952b: 303-304, 421 (*Neolithobius*; characters; plectrotaxy; USA: Virginia).
— Wray, 1967: 155 (*Neolithobius*; list). [**Note:** Specific name misspelled *latzeli*].
— Hoffman, 1995a: 27-28 (*Neolithobius*; list; notes).
— Hoffman, 1995b: 30-32 (*Neolithobius*; taxonomic and type locality notes; USA: Virginia).

Distribution: USA: North Carolina, Virginia.

HOLOTYPE: MCZ, No. 789, 1 specimen, Virginia: Crandall, Meinert.

♂♀*N. mordax* (L. Koch, 1862)
— L. Koch, 1862: 34, Tab. 1, Fig. 6 (*Lithobius*; characters; USA: Louisiana).
— Wood, 1865: 149-150 (*Lithobius*; characters; notes).
— Meinert, 1872: 294-295 (*Lithobius*; characters; notes).

— Stuxberg, 1875b: 10-11 (*Lithobius*; subgenus *Neolithobius*; list).
— Stuxberg, 1875c: 26-27 [=1877: 134] (*Lithobius*; subgenus *Neolithobius*; list).
— Cragin, 1885: 143 (*Neolithobius*; habitat; USA: Kansas).
— Bollman, 1887b: 255, 262, 266 [= 1893: 22, 29, 32] (*Lithobius*; subgenus *Neolithobius*; characters; list; key; notes; USA: Indiana, Florida).
— Bollman, 1887c: 627 [= 1893: 44] (*Lithobius*; subgenus *Neolithobius*; distribution; key).
— McNeill, 1887a: 326 (*Lithobius*; note; USA: Florida). [**Note:** Misspelled *Litnobius*].
— Bollman, 1888b: 8 [= 1893: 80] (*Lithobius*; note; USA: Arkansas).
— Bollman, 1893: 129, 185 (*Lithobius*; subgenus *Neolithobius*; catalogue; note; USA: Minnesota).
— Bollman, 1893: 146 (*Lithobius*; synonym of *spinipes*).
— Kenyon, 1893a: 17 (*Neolithobius*; notes; USA: Nebraska).
— Kenyon, 1893b: 162 (*Neolithobius*; USA: Nebraska).
— Brölemann, 1896: 47-48 (*Lithobius*; notes).
— Chamberlin, 1909b: 190 (*Lithobius*; USA: Louisiana).
— Chamberlin, 1911a: 68 (*Lithobius*; key to species; notes; USA: Colorado).
— Chamberlin, 1911c: 99, 101 (*Lithobius*; key to species; notes; USA: Iowa, Minnesota, Nebraska, Wisconsin).
— Gunthorp, 1913: 167 (*Neolithobius*; synonym of *transmarinus*).
— Chamberlin, 1918a: 24 (*Lithobius*; USA: Tennessee).
— Chamberlin, 1918b: 380 (*Neolithobius*; note; USA: Louisiana).
— Chamberlin, 1920d: 44 (*Neolithobius*; USA: Texas).
— Chamberlin, 1925d: 473-478, pl. 3; Figs. 1-2 (*Neolithobius*; characters (adult, stadia); notes; USA: Alabama, Arkansas, Iowa, Kansas, Louisiana, Minnesota, Mississippi, Nebraska, Tennessee).
— Verhoeff, 1925: 690 (*Lithobius*; list).
— Williams & Hefner, 1928: 143 (*Lithobius*; characters; USA: Ohio).
— Brimley, 1938: 501 (*Lithobius*; note; USA: North Carolina).
— Causey, 1940: 77 (*Neolithobius*; characters; USA: North Carolina).
— Chamberlin, 1946b: 194 (*Neolithobius*; USA: Mississippi).
— Chamberlin, 1951b: 32 (*Neolithobius*; USA: Nebraska).
— Crabill, 1952b: 299-301, 418 (*Neolithobius*; characters; plectrotaxy; range; USA: Ohio).
— Wray, 1967: 155 (*Neolithobius*; list).

- Summers, 1979: 696, 698, (*Neolithobius*; checklist, key).
- Summers et al., 1980: 251 (*Neolithobius*; checklist; notes; USA: Illinois).
- Summers et al., 1981: 61 (*Neolithobius*; checklist; notes; USA: Illinois).
- Kevan, 1983: 2949 (*Neolithobius*; checklist).
- Watermolen, 1997: 5 (*Neolithobius*; checklist, notes).

howei Bollman, 1887a: 81 [= 1893: 19] (*Lithobius*; characters; USA: Minnesota). [**Note:** NMNH has the holotype].
- Bollman, 1887b: 255, 259, 266 [= 1893: 22, 26, 33] (*Lithobius*; subgenus *Lithobius*; characters; key; list; note; USA: Minnesota).
- Bollman, 1888e: 409 [= 1893: 110] (*Lithobius*; note; USA: Indiana).
- Bollman, 1893: 129, 133, 185 (*Lithobius*; subgenus *Lithobius*; catalogue; characters; note).
- Chamberlin, 1911c: 99, 101 (*Lithobius*; key to species; USA: Indiana, Minnesota).
- Chamberlin, 1925d: 473 (*Lithobius*; cited as questioned synonym of *mordax*).
- Kevan, 1983: 2949 (*Neolithobius*; synonym of *mordax*). [**Note:** Kevan notes that *N. howei* is " ... probably junior synonyms of *spinipes* (Say) s. str., ... "].

louisianae Brölemann, 1896: 48 (*Lithobius*; nominal taxon *mordax*; characters; USA: Louisiana).
- Chamberlin, 1911b: 45 (*Lithobius*; nominal taxon mordax; synonym of *transmarnius*).
- Chamberlin, 1925d: 473 (synonym of *mordax*).

perarmatus Brölemann, 1896: 48-49 (*Lithobius*; nominal taxon *transmarinus*; characters; USA: Louisiana).
- Chamberlin, 1911b: 45 (*Lithobius*; nominal taxon mordax; synonym of *transmarnius*). [**Note:** specific name misspelled *permatus*].
- Chamberlin, 1925d: 473 (synonym of *mordax*).

spinipes Say, 1821: 108-109 (*Lithobius*; characters; habitat; USA).
- Bollman (in part), 1893: 146-147 (*Lithobius*; notes).

transmarinus Chamberlin (not Koch), 1911b: 45 (*Lithobius*; synonym of *mordax*, in part).
- Bollman, 1893: 146 (*Lithobius*; synonym of *spinipes*).
- Gunthorp, 1913: 166-167 (*Lithobius*; habitat; notes; USA: Kansas).
- Gunthorp, 1921: 89 (*Lithobius*; notes).
- Chamberlin, 1925d: 473 (cited as synonym of *mordax*, in part).

— Kevan, 1983: 2949 (*Neolithobius*; questioned synonym of *mordax*).

Distribution: USA: Alabama, Arkansas, Colorado, Florida, Illinois, Indiana, Iowa, Kansas, Louisiana, Minnesota, Mississippi, North Carolina, Nebraska, Ohio, Tennessee, Texas, Wisconsin.

HOLOTYPE: ?Unknown, Louisiana: New Orleans. [**Note:** Type specimen may be deposited in Munich, Germany; however, the MCZ has numerous specimens with the following catalog numbers: 790, 2083, 2407, 29448-29450, 29452-29455, 29458-29459].

♂♀*N. suprenans* Chamberlin, 1925
  — Chamberlin, 1925d: 500 (*Neolithobius*; characters; USA: Colorado, New Mexico, Texas).
  — Chamberlin, 1931b: 97 (*Neolithobius*; USA: Oklahoma).
  — Chamberlin & Mulaik, 1940, 158 (*Neolithobius*; USA: Texas).
  — Chamberlin, 1942b: 15 (*Neolithobius*; USA: Iowa).
  — Chamberlin, 1943b: 104 (*Neolithobius*; USA: Colorado, Texas).
  — Chamberlin, 1944d: 207-208 (*Neolithobius*; USA: Arkansas).
  — Crabill, 1952b: 298 (*Neolithobius*; note; USA: Kansas, Nebraska).
  — Crabill, 1955h: 155-156 (*Nadabius*; notes; USA: Missouri).
  — Reddell, 1965: 166 (*Neolithobius*; notes; checklist; USA: Texas).

Distribution: USA: Arkansas, Colorado, Iowa, Kansas, Missouri, Nebraska, New Mexico, Oklahoma, Texas.

HOLOTYPE: MCZ, No. 14563, 1 specimen, New Mexico: Las Vegas Hot Springs, Cockerell.

PARATYPES: MCZ, No. 29514, 29515, New Mexico & Colorado. [**Note:** MCZ also has other material with the following catalog numbers: 29510, 29511].

♂♀*N. transmarinus* (L. Koch, 1862)
  — Koch, 1862: 33-34; Tab. 1, Fig. 5 (*Lithobius*; characters; USA: Louisiana).
  — Stuxberg, 1875b: 10-11 (*Lithobius*; subgenus *Neolithobius*; list).
  — Stuxberg, 1875c: 26, 32 (*Lithobius*; subgenus *Neolithobius*; list).
  — Stuxberg, 1876: 3 (*Neolithobius*; subgenus *Neolithobius*; list).
  — Bollman, 1887b: 266 [= 1893: 33] (*Lithobius*; subgenus *Neolithobius*; list).

— Bollman (in part), 1887c: 626 [= 1893: 43] (*Lithobius*; subgenus *Neolithobius*; key).

— Bollman, 1888b: 8 [= 1893: 80] (*Lithobius*; note; USA: Arkansas).

— Bollman, 1893: 130, 146 (*Lithobius*; subgenus *Neolithobius*; catalogue; synonym of *spinipes*).

— Brölemann, 1896: 48 (*Lithobius*; notes).

— Chamberlin (in part), 1911b: 45 (*Lithobius*; notes; USA: Alabama, Louisiana, Mississippi).

— Gunthorp (in part), 1913: 166–167 (*Lithobius*; habitat; notes; USA: Kansas).

— Chamberlin, 1918a: 24 (*Lithobius*; USA: Tennessee).

— Chamberlin, 1918b: 380 (*Neolithobius*; USA: Louisiana).

— Chamberlin, 1925d: 478-480, pl. 3; Figs. 3-4 (*Neolithobius*; characters; USA: Arkansas, Louisiana, Mississippi, Texas).

— Chamberlin, 1942d: 188 (*Neolithobius*; USA: Louisiana, Texas).

— Chamberlin, 1944d: 207 (*Neolithobius*; USA: Texas).

— Chamberlin, 1951b: 32 (*Neolithobius*; USA: Louisiana).

— Crabill, 1952b: 298 (*Neolithobius*; note; USA: Alabama).

— Kevan, 1983: 2949 (*Neolithobius*; questioned synonym of *mordax*).

louisianae Brölemann, 1896: 48 (*Lithobius*; nominal taxon *mordax*; characters; USA: Louisiana).

— Chamberlin, 1911b: 45 (*Lithobius*; nominal taxon mordax; synonym of *transmarnius*).

?*mordax* Chamberlin (not Koch), 1911b (*Lithobius*; synonym of *transmarinus*).

spinipes Bollman (not Say, in part), 1893: 146-147 (*Lithobius*; notes).

— Brölemann, 1896: 48 (synonym of *transmarinus*).

— Kevan, 1983: 2949 (notes that *N. mordax*, *N. howei* and *N. transmarinus* are " . . . all probably junior synonyms of *spinipes* (Say) s. str., . . .").

Distribution: USA: Alabama, Arkansas, Kansas, Louisiana, Mississippi, Tennessee, Texas.

HOLOTYPE: ?Unknown; Louisiana: New Orleans. [**Note:** Type specimen may be deposited in Munich, Germany; however, the MCZ has numerous specimens with the following catalog numbers: 2212, 29456, 29457, 29461-29470, 29472-29474, 31051].

♂♀*N. tyrannus* (Bollman, 1887)

— Bollman, 1887c: 626, 627 [= 1893: 43, 44] (*Lithobius*; subgenus *Neolithobius*; characters; key; note; USA: Indiana). [**Note:** Original specific name spelling under the description was *tyrranicus* and in the key it was *tyranicus*].

— Bollman, 1888e: 409 [= 1893: 110] (*Lithobius*; note; USA: Indiana).

— Bollman, 1893: 130, 201 (*Lithobius*; catalogue). [**Note:** The original spelling of the specific name was *tyrannicus*. Underwood presents an emendation and explains that it is evidently a misprint because it appears from Bollman's manuscripts he intended the spelling as *tyrannus*].

— Chamberlin, 1911c: 99, 101 (*Lithobius*; USA: Indiana).

— Chamberlin, 1925d: 486-488 (*Neolithobius*; characters; notes; USA: Indiana).

— Verhoeff, 1925: 690 (*Lithobius*; list).

— Williams & Hefner, 1928: 143 (*Lithobius*; characters; USA: Ohio).

— Chamberlin, 1944d: 208 (*Neolithobius*; USA: Indiana).

— Auerbach, 1946: 14 (*Neolithobius*; list; USA: Illinois).

— Crabill, 1952b: 302-303, 420 (*Neolithobius*; characters; plectrotaxy; notes; USA: Illinois, Indiana, Ohio).

— Summers, 1979: 693, 696, 698; Fig. 3 (*Neolithobius*; checklist, key).

— Summers & Uetz, 1979: 347; Tab. 1; Fig. 3 (*Neolithobius*; USA: Illinois).

— Summers et al., 1980: 251; Fig. 25 (*Neolithobius*; checklist; USA: Illinois).

— Summers et al., 1981: 61 (*Neolithobius*; checklist; USA: Illinois).

— Watermolen, 1997: 7 (*Neolithobius*; notes).

mordax Bollman (not Koch, in part), 1887b: 262 (*Lithobius*; characters).

— Chamberlin (not Koch, in part), 1911: 45 (*Lithobius*; cited as synonym, in part).

Distribution: USA: Illinois, Indiana, Ohio.

HOLOTYPE: NMNH; Indiana: Bloomington.

♂♀*N. underwoodi* (Bollman, 1888)

— Bollman, 1888d: 350 [= 1893: 102] (*Lithobius*; characters; notes; USA: Georgia).

— Bollman, 1893: 130 (*Lithobius*; subgenus *Neolithobius*; catalogue).

— Chamberlin, 1911b: 46 (*Lithobius*; notes; USA: Alabama, Georgia,
  South Carolina).
— Chamberlin, 1925d: 501-504, pl. 3; Fig. 6 (*Neolithobius*; characters
  (adult, stadia); USA: Alabama, Georgia, South Carolina).
— Crabill, 1950c: 202 (*Neolithobius*; morphological notes; USA: South
  Carolina).
— Chamberlin, 1951b: 32 (*Neolithobius*; USA: Georgia).
— Crabill, 1952b: 298 (*Neolithobius*; note).

Distribution: USA: Alabama, Georgia, South Carolina.

HOLOTYPE: ?NMNH, Acc. 19542, 22, ♀, Georgia: Macon, by L. M.
Underwood. [**Note:** Although Bollman stated the types were deposited in
the U.S. National Museum along with their accession numbers they are not
seen in the database at this time; however, the MCZ has numerous specimens
with the following catalog numbers: 2397-2406, 2423-2425].

♂♀*N. voracior* (Chamberlin, 1912)
— Chamberlin, 1912d: 150-152 (*Lithobius*; characters; notes; USA:
  Mississippi).
— Chamberlin, 1925d: 492-497 (*Neolithobius*; characters (adult, stadia);
  notes; USA: Mississippi).
— Chamberlin, 1944d: 209 (*Neolithobius*; USA: Illinois, Missouri).
— Auerbach, 1946: 14, 15, 16, 21, 22, 35; chart 2: 46-47 (*Neolithobius*;
  ecology, habitat; hygrotaxis; list; notes; USA: Illinois).
— Auerbach, 1949: 222, 224, 225; Fig. 4 (*Neolithobius*; ecology;
  hygrotaxis; notes; USA: Illinois).
— Auerbach, 1951a: 100, 101, 102, 103, 109, 118, 119, 120, 121, 122;
  Tab. 1,2, 6-8, 14; Fig. 1, 3-5 (*Neolithobius*; desiccation; food habits;
  habitat; heart beat; hygrotaxis; list; population percentage; USA:
  Illinois).
— Auerbach, 1951b: 111, 112, 113; Figs. 1, 4 (*Neolithobius*; key).
— Crabill, 1952b: 301-302, 419 (*Neolithobius*; characters; plectrotaxy;
  taxonomic notes; USA: Illinois).
— Crabill, 1955b:41 (*Neolithobius*; notes; USA: Missouri).
— Crabill, 1955h:155 (*Neolithobius*; habitat; notes; USA: Missouri).
— Summers, 1979: 693, 696, 698; Fig. 5 (*Neolithobius*; checklist,
  key).
— Summers & Uetz, 1979: 347, 350, 351; Tab. 1 & 2; Fig. 3 (*Neolithobius*;
  microhabitat; USA: Illinois).

— Summers et al., 1980: 251; Fig. 26 (*Neolithobius*; checklist; USA: Illinois).

— Summers et al., 1981: 61-62 (*Neolithobius*; checklist; USA: Illinois).

— Watermolen, 1997: 5 (*Neolithobius*; checklist, notes; USA: Wisconsin).

Distribution: USA: Illinois, Missouri, Mississippi, Wisconsin.

SYNTYPES: MCZ, No. 29484, 9 specimens, Mississippi: Fernwood, 7-19-1910, R.V. Chamberlin Coll. [**Note:** Additional specimens at the MCZ are as follows: 29483, 29485-29487].

♂♀*N. vorax* (Meinert, 1872)
— Meinert, 1872: 292-294 (*Lithobius*; characters; USA: Louisiana).
— Stuxberg, 1875b: 10-11 (*Lithobius*; subgenus *Neolithobius*; list).
— Stuxberg, 1875c: 26, 32 [= 1876: 134, 139] (*Lithobius*; subgenus *Neolithobius*; characters, list).
— Stuxberg, 1877: 135 (*Lithobius*; subgenus *Neolithobius*;.
— Bollman, 1887b: 266 [= 1893: 33] (*Lithobius*; subgenus *Neolithobius*; list).
— Bollman, 1887c: 627 [= 1893: 44] (*Lithobius*; subgenus *Neolithobius*; distribution; key).
— Bollman, 1888b: 8 [= 1893: 80] (*Lithobius*; note; USA: Arkansas).
— Bollman, 1893: 130 (*Lithobius*; subgenus *Neolithobius*; catalogue).
— Chamberlin (in part), 1911b: 45-46 (*Lithobius*; character notes; USA: Alabama, Florida, Mississippi, North Carolina, Virginia).
— Chamberlin, 1925d: 480-484, pl. 2; Figs. 4-6 (*Neolithobius*; characters (adult, stadia); USA: Alabama, Florida, Mississippi, Tennessee).
— Chamberlin, 1951b: 32 (*Neolithobius*; USA: Florida).
— Crabill, 1952b: 299 (*Neolithobius*; note).
— Eason, 1974b: 12-13 (*Neolithobius*; characters; notes).
— Jeekel, 2005: 24 (*Neolithobius*; list).
clarus McNeill, 1887a: 326-327 (*Lithobius*; characters; note; USA: Florida). [**Note:** NMNH has 2 vials one labeled as cotypes and the other with 2 specimens considered types].
— Bollman, 1887b: 255, 262-263, 266 [= 1893: 22, 29-30, 33] (*Lithobius*; subgenus *Neolithobius*; characters (adult, stadium); key; list; note; USA: Florida).

— Bollman, 1893: 129 (*Lithobius*; subgenus *Neolithobius*; catalogue).

— Chamberlin, 1911b: 45 (*Lithobius*; synonym of *vorax*).

— Chamberlin, 1925d: 484 (note). [**Note:** According to Chamberlin the type for *Neolithobius clarus* McNeill is a pseudomaturus *N. vorax*].

latzeli Meinert, 1886a: 175 (*Lithobius*; characters; USA: Virginia).

— Chamberlin, 1911b: 45 (*Lithobius*; synonym of *vorax*).

tyrranicus Bollman, 1887c: 626, 627 (*Lithobius*; subgenus *Neolithobius*; characters; key; note; USA: Indiana). [**Note:** Indiana has been left out of the distribution until further scrutiny].

— Chamberlin, 1911b: 45 (*Lithobius*; synonym of *vorax*).

Distribution: USA: Alabama, Arkansas, Florida, Louisiana, Mississippi, North Carolina, Tennessee, Virginia.

LECTOTYPE: ZMUC; ♂, Mississippi: Beloxi (Biloxi?), near New Orleans.

PARALECTOTYPE: ZMUC; ♀, Mississippi: Beloxi (Biloxi?), near New Orleans.

♂♀***N. xenopus*** (Bollman, 1888)

— Bollman, 1888d: 349 [=1893: 101-102] (*Lithobius*; notes; USA: Georgia).

— Bollman, 1893: 130 (*Lithobius*; subgenus *Neolithobius*; catalogue).

— Chamberlin, 1911b: 45 (*Lithobius*; USA: Georgia).

— Chamberlin, 1925d: 497-498, pl. 3; Fig. 5 (*Neolithobius*; characters; notes; USA: Georgia).

— Chamberlin, 1944a: 35 (*Neolithobius*; note; USA: Georgia).

— Chamberlin, 1951b: 32 (*Neolithobius*; USA: Florida, Georgia).

— Crabill, 1952b: 299 (*Neolithobius*; note).

Distribution: USA: Florida, Georgia.

HOLOTYPE: NMNH, Acc. 19542, 22, ♂, Georgia: Macon, by L.M. Underwood.

Genus ***Nothembius*** Chamberlin, 1916
[4 species; Western USA]

Nothembius Chamberlin, 1916: 188-190; Fig. 5 (characters; distribution; key to species). TYPE SPECIES: *Nothembius insulae* Chamberlin, 1916, fixed by original designation.
— Attems, 1938a: 343 (subgenus of *Lithobius*; key).
— Jeekel, 2005: 24 (list).

♂♀*N. aberrans* Chamberlin, 1916
— Chamberlin, 1916: 196-198, pl. 8; Fig. 5; pl. 9; Fig. 4,5 (*Nothembius*; characters; notes; USA: California).

Distribution: USA: California.

HOLOTYPE: MCZ, No. 357, California: Eaton's Canyon, near Altadena. R. V. Chamberlin.

PARATYPES: MCZ, No. 365, 430, California: Eaton's Canyon, near Altadena. R. V. Chamberlin.

♂♀*N. amplus* Chamberlin, 1941
— Chamberlin, 1941a: 13 (*Nothembius*; characters; note; USA California).

Distribution: USA: California.

SYNTYPES: NMNH, ♂, 2♀, California: Mountain Springs, 8 January 1941, by S. & D. Mulaik.

♂♀*N. insulae* Chamberlin, 1916
— Chamberlin, 1916: 190-194, pl. 8; Fig. 6; pl. 9; Fig. 1-3 (*Nothembius*; characters (adult, stadia); notes; USA: California).
— Jeekel, 2005: 24 (*Nothembius*; list).

Distribution: USA: California.

TYPES: MCZ, No. 429, 11 specimens, California: Santa Cruz Island, La Playa Canyon, April 1913, R. V. Chamberlin. [**Note:** The total of 11 specimens are divided in 2 micro vials, one with 5 and the other with 6 specimens; some legs are free floating in the main vial].

PARATYPES: MCZ, No. 359-364, 427, 428, California: Santa Cruz Island, La Playa Canyon. R. V. Chamberlin.

♂♀*N. nampus* Chamberlin, 1916

— Chamberlin, 1916: 194-196, pl. 8; Fig. 4; pl. 9; Fig. 6,7 (*Nothembius*; characters; USA: California).
— Anonymous, 1920, p. 35 (*Nothembius*; USA: California).
— Chamberlin, 1925a: 54 (*Nothembius*; notes; USA: Arizona).
— Chamberlin, 1944d: 211 (*Nothembius*; USA: California).

Distribution: USA: Arizona, California.

TYPES: MCZ, No. 426, 10 specimens, California: Claremont. R. V. Chamberlin.

PARATYPES: MCZ, No. 420, 424, California: Claremont; MCZ, No. 358, 425, Eaton's Canyon; MCZ, No. 421-423, Santa Cruz Island. R. V. Chamberlin.

Genus *Nuevobius* Chamberlin, 1941
[1 species; Southeastern USA]

Nuevobius Chamberlin, 1941c: 188 (characters). **TYPE SPECIES:** *Nuevobius cavicolens* Chamberlin, 1941, fixed by original designation and monotypy. [**Note:** The type species is from Mexico and was taken from "dung of bat cave"].
— Crabill, 1960f: 127-130 (affinities; notes).

♂*N. cottus* Crabill, 1960
— Crabill, 1960f: 122-127; Figs. 1-4, 6 (*Nuevobius*; characters; key to genera of eastern North America with the following characteristics in combination: prosternal teeth more numerous than 2 + 2; antennal articles more numerous than 28; ultimate leg of male with pronounced sexual modifications; USA: Tennessee). [**Note:** This species is cavernicolous and shows some resemblance to *Sozibius*].

Distribution: USA: Tennessee.

HOLOTYPE: NMNH, Myriapod Catalogue Number 2673, Myriapod Collection Number 148, ♂, Tennessee: Blount Co., Tuckaleechee Caverns, near Townsend, 18 April 1959, by Thomas C. & Catherine Barr.

PARATYPE: NMNH, Myriapod Catalogue Number 2673, Myriapod Collection Number 148, ♂ (pseudomaturus stadium), Tennessee: Blount Co., Tuckaleechee Caverns, near Townsend, 18 April 1959, by Thomas C. & Catherine Barr.

Genus *Oabius* Chamberlin, 1916
[29 species; Western USA & Hawaii]

Oabius Chamberlin, 1916: 121-124; Fig. 1 (characters; distribution; habitat (p. 116); key to species). TYPE SPECIES: *Oabius utahensis* Chamberlin, var. *tiganus* Chamberlin 1910, fixed by monotypy. [**Note:** The designation by Chamberlin, 1916 of *Oabius pylorus* Chamberlin, 1916 as type-species is invalid (see Jeekel, 2005)].
— Attems, 1938a: 344 (synonym of *Lithobius* (*Monotarsobius*)).
— Chamberlin, 1949a: 14 [**Note:** Chamberlin reported that an *Oabius* sp. was taken in Canada: Whitehorse, Yukon Territory, near airport, 18 August 1948, by Neal A. Weber, under loosely buried wood in sandy soil, with second growth pine beside the airport (?NMNH No. 2280). The identification to species made difficult owing to the absence of all the posterior legs].
— Kevan, 1979: 298 [**Note:** Kevan reports of an *Oabius* species from Yukon that is probably taken from Chamberlin (1949a)].
— Jeekel, 2005: 25 (list; notes).

♀ *O. adjacens* Chamberlin, 1946
— Chamberlin, 1946c: 184, 186 (*Oabius*; characters; notes; USA: Alaska).
— Behan-Pelletier, 1993: 26 (*Oabius*; list).

Distribution: USA: Alaska.

SYNTYPE: NMNH, Alaska: Juneau, 28-29 April 1945.

♂♀ *O. ajonus* Chamberlin, 1941
— Chamberlin, 1941a: 5 (*Oabius*; characters; note; USA: Arizona).

Distribution: USA: Arizona.

SYNTYPE: NMNH, Arizona: 20 mi. S of Ajo, 4 January 1941, by S. & D. Mulaik.

♀ *O. alaskanus* Chamberlin, 1946
— Chamberlin, 1946c: 184 (*Oabius*; characters; note; USA: Alaska).
— Behan-Pelletier, 1993: 26 (*Oabius*; list).

Distribution: USA: Alaska.

SYNTYPES: NMNH, Alaska: ?Haines. [**Note:** The author was not able to find a vial labeled as the holotype, however, two vials were from Haines].

PARATYPES: NMNH, Alaska: 4 vials from Haines, Homer and Juneau.

♂♀ *O. arktaus* Chamberlin, 1946
— Chamberlin, 1946c: 186 (*Oabius*; characters; USA: Alaska).
— Behan-Pelletier, 1993: 26 (*Oabius*; list).

Distribution: USA: Alaska.

SYNTYPE: NMNH, Alaska: Fairbanks, 21-22 September 1943.

♂♀ *O. dissimulans* Chamberlin, 1916
— Chamberlin, 1916: 140-144, pl. 3; Fig. 1-3 (*Oabius*; characters (adult, stadia); notes; USA: California).

Distribution: USA: California.

TYPES: MCZ, No. 380, 30 specimens, California: Pacific Grove, April, 1911, R. V. Chamberlin.

PARATYPES: MCZ, No. 322, 375, California: Brookdale; MCZ, No. 323, 373, Sausalito; MCZ, No. 369, Santa Cruz Island; MCZ, No. 370, 376, Pacific Grove; MCZ, No. 371, Felton Big Trees; MCZ, No. 372, Mill Valley; MCZ, No. 379, Berkeley. R. V. Chamberlin. [**Note:** NMNH also has paratypes].

♀ *O. eugenus* Chamberlin, 1928
— Chamberlin, 1928c: 86 (*Oabius*; characters; USA: Oregon).

Distribution: USA: Oregon.

HOLOTYPE: NMNH, ♀, Oregon: Eugene, Hendrick's Park, 1 October 1927, by David T. Jones.

♂♀ *O. fratris* Chamberlin, 1916
— Chamberlin, 1916: 130-132, pl. 1; Fig. 8, pl. 2; Fig. 10 (*Oabius*; characters (adult, stadia); USA: California).

Distribution: USA: California.

HOLOTYPE: MCZ, No. 313, California: Friant, March, 1913. R. V. & S. C. Chamberlin. [**Note:** The holotype vial at the MCZ, labeled as No. 394 (Unique No. 29663), containing 8 specimens, was found dry with the slightest hint of moisture within a bail jar; the original locality/determined label reads "*Oabius fratris* sp. nov., type, Cal.: Friant, Mch., 1913"].

PARATYPES: MCZ, No. 314, 315, California: Friant, March, 1913. R. V. & S. C. Chamberlin.

♀*O. kernensis* Chamberlin, 1941
— Chamberlin, 1941a: 6 (*Oabius*; characters; USA: California).

Distribution: USA: California.

HOLOTYPE: NMNH, ♀, California: Kern Co., 7 mi. S of Glenville, 19 March 1941, by S. & D. Mulaik.

PARATYPE: NMNH, California.

♀*O. mercurialis* Chamberlin, 1962
— Chamberlin, 1962: 137-138 (*Oabius*; characters; USA: Nevada).

Distribution: USA: Nevada.

HOLOTYPE: ?Unknown, ♀, Nevada: Clark Co., Mercury, 26 January 1961.

♂♀*O. mimosus* Chamberlin, 1938
— Chamberlin, 1938b: 631 (*Oabius*; characters; habitat; USA: Oregon).
— Chamberlin, 1940d: 51 (*Oabius*; characters; note; USA: Oregon).
— Chamberlin, 1962: 137 (*Oabius*; USA: Oregon).

Distribution: USA: Oregon.

HOLOTYPE: NMNH, Oregon: Boyer, 10 &13 August 1933, in rotten wood, by C. J. J.

PARATYPE: NMNH.

♂♀*O. oreinus* Chamberlin, 1925
— Chamberlin, 1925a: 53 (*Oabius*; characters; USA: Arizona).

Distribution: USA: Arizona.

HOLOTYPE: NMNH, Arizona: Santa Catalina Mts., by R. H. Peebles.

♂♀ *O. paiutus* Chamberlin, 1925
— Chamberlin, 1925e: 55 (*Oabius*; characters; USA: Utah).
O. piutus Chamberlin - Chamberlin [sic], 1930a: 111.

Distribution: USA: Utah.

HOLOTYPE: MCZ, No. 14492, 1 specimen, Utah: Iron Co., Parowan Canyon, May 1924.

PARATYPE: MCZ, No. 29851, 2 specimens, Utah: Parowan Canyon, May 1924.

♂ *O. parvior* Chamberlin, 1938
— Chamberlin, 1938b: 631-632 (*Oabius*; characters; USA: New Mexico).
— Chamberlin & Mulaik, 1940: 127 (*Oabius*; list).

Distribution: USA: New Mexico.

HOLOTYPE: NMNH, ♂, New Mexico: Bear Canyon, Camp Mary White, August 1934.

♂♀ *O. patonius* (Chamberlin, 1911)
— Chamberlin, 1911f: 380 (*Lithobius*; characters; habitat; notes; USA: California).
— Chamberlin, 1916: 134-136, pl. 1; Fig. 1-4 (*Oabius*; characters (adult, stadia); USA: California).

Distribution: USA: California.

HOLOTYPE: MCZ, No. 324, California: Mill Valley. R.V. Chamberlin.

PARATYPES: MCZ, No. 401, California: Berkeley; MCZ No. 402, California: Mill Valley. R.V. Chamberlin.

♀ *O. patonius flavus* Chamberlin, 1916
— Chamberlin, 1916: 136-137, pl. 1; Fig. 5, pl. 2. fig. 7 (*Oabius*; nominal taxon *patonius*; cited as subspecies *flavus*; characters; note; USA: California).

Distribution: USA: California.

SYNTYPE: MCZ, ♀ No. 395, California: Santa Barbara. R.V. Chamberlin.

PARATYPE: MCZ, No. 327, California: Santa Barbara. R.V. Chamberlin.

♀*O. patonius micrus* Chamberlin, 1916
— Chamberlin, 1916: 137-138, pl. 1; Fig. 10, pl. 2. fig. 6 (*Oabius*; characters; notes; USA: California).

Distribution: USA: California.

HOLOTYPE: MCZ, No. 397, California: Pacific Grove. R.V. Chamberlin. [**Note:** MCZ No. 396 is labeled as the holotype with 2 specimens, but was originally designated as a paratype vial according to the literature].

PARATYPES: MCZ, No. 325, 326, California: Pacific Grove; MCZ, No. 396, Capitola, March 1913, R.V. Chamberlin.

♂♀*O. pelotes* Chamberlin, 1941
— Chamberlin, 1941a: 5 (*Oabius*; characters; note; USA: California). [**Note:** It is not clear from the literature which sex or collecting event was designated as the holotype specimen].

Distribution: USA: California.

HOLOTYPE: NMNH, California: Monterey Co., Hastings Reservation, 29 January 1940 (♀) or 12 June 1940 (♂, ♀).

♂♀*O. pylorus* Chamberlin, 1916
— Chamberlin, 1916: 125-130, pl. 1; Fig. 6, pl. 2; Fig. 1-2 (*Oabius*; characters (adult, stadia); note; USA: California).
— Chamberlin, 1930c: 68 (*Oabius*; USA: Hawaii). [**Note:** This species has been taken at Hawaii April 22, 1929, in soil about chives from San Francisco, California].
— Zapparoli & Shelley, 2000: 47 (*Oabius*; list; notes).

Distribution: USA: California, ?Hawaii.

HOLOTYPE: MCZ; No. 318; California: Capitola, R.V. Chamberlin.

PARATYPES: MCZ; No. 316, 317; California: Berkeley; MCZ; No. 384; Sausalito; MCZ; No. 381; Santa Cruz; MCZ; No. 382; Brookdale; MCZ; No. 383; Felton; Big Trees; MCZ; No. 385; Capitola. R. V. Chamberlin.

♂♀*O. rodocki* Chamberlin, 1940
   — Chamberlin, 1940d: 50 (*Oabius*; characters; habitat; USA: Idaho).

Distribution: USA: Idaho.

HOLOTYPE: NMNH, ♂, Idaho: Spaulding, 18 November 1939, taken under moist leaves in woods by R. E. Rodock.

ALLOTYPE: NMNH, ♀, Idaho: Spaulding, 18 November 1939, taken under moist leaves in woods by R.E. Rodock.

♂*O. sanjuanus* Chamberlin, 1928
   — Chamberlin, 1928d: 308-309 (*Oabius*; characters; USA: Utah).
   — Chamberlin, 1958c: 133 (*Oabius*; checklist; note).

Distribution: USA: Utah.

HOLOTYPE: NMNH; Utah: ♂, San Juan Co., Bluff, 16 April, by H.P. Critchlow.

♂♀*O. sastianus* (Chamberlin, 1903)
   — Chamberlin, 1903b: 157-158 (*Lithobius*; characters; key to species; USA: California).
   — Chamberlin, 1910: 374 (*Lithobius*; characters).
   — Chamberlin, 1916: 144 145 (*Oabius*; characters; note; USA: California).

Distribution: USA: California.

HOLOTYPE: MCZ, No. 336, California: Shasta Springs. R.V. Chamberlin.

♂♀*O. tabiphilus* Chamberlin, 1916
   — Chamberlin, 1916: 145-149, pl. 1; Fig. 9, pl. 2; Fig. 8 (*Oabius*; characters (adult, stadia); note; USA: California).

Distribution: USA: California.

HOLOTYPE: MCZ, No. 411, California: Santa Barbara, R.V. Chamberlin.

♀*O. tiganus* (Chamberlin, 1910)
  — Chamberlin, 1910: 374 (*Lithobius*; cited as subspecies of *utahensis*; characters; USA: California). [**Note:** Chamberlin's original description of the anal leg spines was 1,3,2,9. This is clearly a typographical error to have the tibial spines numbering 9, so this is supposed to be a 0 as is stated in Chamberlin's 1916 more detailed description of the species].
  — Chamberlin, 1911f: 379-380 (*Lithobius*; habitat; USA: California).
  — Chamberlin, 1916: 132-134, pl. 1; Fig. 7, pl. 2; Fig. 5 (*Oabius*; characters (adult, stadia); USA: California).
  — Jeekel, 2005: 25 (*Lithobius*; nominal taxon *utahensis*; var. *tiganus*; list).
utahensis Chamberlin, 1909b: 187 (*Lithobius*; synonym of *tiganus* in part; USA: California, Oregon). [**Note:** Some of this material reported belonged to *Pokabius tiganus* (Chamberlin); it is not clear if any specimens were taken in Oregon so for the time being this state has been left out of the distribution].

Distribution: USA: California.

HOLOTYPE: MCZ, ♀ No. 393, California: Pacific Grove, R.V. Chamberlin.

PARATYPES: MCZ, No. 319, 391, 392, California: Pacific Grove. R.V. Chamberlin.

♀*O. uleorus* Chamberlin, 1916
  — Chamberlin, 1916: 138-140, pl. 2; Fig. 9 (*Oabius*; characters; note; USA: Alaska).
  — Chamberlin, 1919: 15H (*Oabius*; list).
  — Behan-Pelletier, 1993: 26 (*Oabius*; list).

Distribution: USA: Alaska.

HOLOTYPE: MCZ, ♀ No. 337, Alaska: Forrester Island, May-June, 1913. Harold & R.W. Heath.

PARATYPES: MCZ, No. 403, 404, Alaska: Forrester Island, May-June, 1913. Harold & R.W. Heath.

♂♀*O. wamus* Chamberlin, 1962
  — Chamberlin, 1962: 137 (*Oabius*; characters; USA: Oregon).

Distribution: USA: Oregon.

HOLOTYPE: ?University of Utah, ♀, Oregon: Saddleback Mt.

(Subgenus *Nyctobius* Chamberlin, 1916)

Oabius (*Nyctobius*) Chamberlin, 1916: 121, 123 (characters; key). TYPE
SPECIES: *Oabius* (*Nyctobius*) *decipiens* Chamberlin, 1916, fixed by
subsequent designation.
— Attems, 1938a: 344 (synonym of *Lithobius* (*Monotarsobius*)).
— Jeekel, 2005: 24 (list; designation of type).

♂*O.* (*N.*) *boyeranus* Chamberlin, 1940
— Chamberlin, 1940d: 50-51 (*Oabius*; subgenus *Nyctobius*; characters;
note; USA: Oregon).

Distribution: USA: Oregon.

HOLOTYPE: NMNH, ♂, Oregon, Boyer, August 1933.

♀*O.* (*N.*) *decipiens* Chamberlin, 1916
— Chamberlin, 1916: 149-151, pl. 3; Fig. 5 (*Oabius*; subgenus *Nyctobius*;
characters; USA: California).
— Jeekel, 2005: 24 (*Oabius*; subgenus *Nyctobius*; list).

Distribution: USA: California.

TYPES: MCZ, No. 405, 3 specimens, California: Pacific Grove, June, 1902,
R.V. Chamberlin.

PARATYPES: MCZ, No. 334, California: Pacific Grove, R.V.
Chamberlin.

♀*O.* (*N.*) *ineptus* Chamberlin, 1916
— Chamberlin, 1916: 151-153, pl. 2; Fig. 3-4, pl. 3; Fig. 4 (*Oabius*;
subgenus *Nyctobius*; characters; USA: Oregon).

Distribution: USA: Oregon.

HOLOTYPE: MCZ, No. 335, Oregon: Portland, August, 1902. R.V.
Chamberlin.

PARATYPE: MCZ, No. 406, Portland, August, 1902. R.V. Chamberlin. [**Note:** This vial found labeled as holotype by Crabill].

(Subgenus *Zanobius* Chamberlin, 1938)

Oabius (*Zanobius*) Chamberlin, 1938b: 630 (characters). **TYPE SPECIES:** *Oabius* (*Zanobius*) *aiolus* Chamberlin, 1938, fixed by monotypy.
— Jeekel, 2005: 34 (list).

♂♀ *O.* (*Z.*) *aiolus* Chamberlin, 1938
— Chamberlin, 1938b: 630-631 (*Oabius*; subgenus *Zanobius*; characters; note; USA: Oregon).
— Jeekel, 2005: 34 (*Oabius*; list).

Distribution: USA: Oregon.

HOLOTYPE: NMNH, Oregon, Boyer, 23 July 1937, by J. A., Macnab.

ALLOTYPE: NMNH, Oregon, Boyer, 23 July 1937, by J. A., Macnab.

PARATYPES: NMNH, Oregon, Boyer. [**Note:** about 20 specimens in 5 vials].

Genus *Paitobius* Chamberlin, 1912
[14 species; Eastern & Western USA]

Paitobius Chamberlin, 1912e: 175-176 (characters; distribution; note). **TYPE SPECIES:** *Lithobius carolinae* Chamberlin, 1911, fixed by original designation.
— Chamberlin, 1922b: 279-283; fig. 4 (characters; key to species; notes).
— Bailey, 1928b: 32 (characters).
— Crabill, 1952b: 304 (key to species of northeastern USA).
— Jeekel, 2005: 26 (list).

♂ *P. arienus* (Chamberlin, 1911)
— Chamberlin, 1911b: 47-48 (*Lithobius*; characters; notes; USA: North Carolina).
— Chamberlin, 1922b: 283-285 (*Paitobius*; characters; note; USA: North Carolina; South Carolina).
— Brimley, 1938: 500 (*Lithobius*; USA: North Carolina).
— Causey, 1940: 75 (*Lithobius*; characters; note).
— Wray, 1967: 155 (*Paitobius*; list).

— Hoffman, 1995a: 29 (*Paitobius*; list; note). [**Note:** Hoffman believes this species will be found in southwestern Virginia].

Distribution: USA: North Carolina, South Carolina.

HOLOTYPE: MCZ, No. 116, North Carolina: Hot Springs, 6 August 1910.

♂*P. eutypus* Chamberlin, 1940
— Chamberlin, 1940f: 75-76 (*Paitobius*; characters; notes; USA: North Carolina).
— Causey, 1940: 75 (*Paitobius*; note).
— Wray, 1967: 155 (*Paitobius*; list).
— Hoffman, 1995a: 29 (*Paitobius*; list; note). [**Note:** Hoffman suggests this species will be found in Virginia].

Distribution: USA: North Carolina.

HOLOTYPE: NMNH, ♂, North Carolina: Linville, 14 October 1939, by Nelle B. Causey.

♂*P. exceptus* Chamberlin, 1922
— Chamberlin, 1922b: 300 (*Paitobius*; subgenus *Tunabius*; characters; USA: Alabama).
— Hoffman, 1995a: 29 (*Paitobius*; list; note; USA: Virginia). [**Note:** Hoffman suspects "The large geographic distance between these localities would seem, however, to argue against this identification."].

Distribution: USA: Alabama, ?Virginia.

TYPES: MCZ, No. 129, 2 specimens, [**Note:** Chamberlin designated a ♂ as the holotype], Alabama: Jackson, 7-23-1910.

♂♀*P. exiguus* (Meinert, 1886)
— Meinert, 1886b: 110 (*Lithobius*; characters; USA: New York).
— Bollman, 1887b: 265 [= 1893: 32] (*Lithobius*; subgenus *Archilithobius*; list).
— Bollman, 1893: 128 (*Lithobius*; subgenus *Archilithobius*; catalogue).
— Chamberlin, 1911b: 40 (*Lithobius*; notes; USA: Alabama, Kentucky, Mississippi, Tennessee, Virginia).
— Chamberlin, 1911c: 101, 104 (*Lithobius*; key to species; USA: Illinois, Iowa, Nebraska, Wisconsin).

— Chamberlin, 1922b: 302-303 (*Paitobius*; original characters repeated with the locality as "New York City. (L. Lund)"). [**Note:** Chamberlin states that *Paitobius exiguus* (= *Lithobius exiguus* Meinert) is probably a close ally of *Paitobius adelus* Chamberlin (see p. 295)].

— Bailey, 1928a: 1082 (*Paitobius*; list).

— Bailey, 1928b: 19, 32 (*Paitobius*; list; note).

— Crabill, 1952b: 310, 311 (*Paitobius*; taxonomic notes). [**Note:** Crabill considers this species a *species inquirenda*]. [**Note:** Crabill (1952b) notes that this species "bears more than a superficial resemblance to Chamberlin's *zinus*"].

— Crabill, 1955f: 260 (*Paitobius*; checklist; cited as *species inquirenda*; USA: Kentucky).

— Eason, 1974b: 45-47; Figs. 28-30 (*Paitobius*; lectotype description; taxonomical notes).

— Watermolen, 1997: 6 (*Paitobius*; checklist, notes).

?*zinus* Chamberlin, 1911b: 41-42 (*Lithobius*; characters; notes; nominal taxon *cantabrigensis*; cited as variety *zinus*; USA: Alabama, North Carolina).

— Brimley, 1938: 500 (*Lithobius*; nominal taxon *cantabrigensis*; cited as variety *zinus*; USA: North Carolina).

— Eason, 1974b: 45, 46 (*Lithobius*; cited as questioned synonym of *exiguus*).

Distribution: USA: Alabama, Illinois, Iowa, Kentucky, Mississippi, Nebraska, ?North Carolina, New York, Tennessee, Virginia, Wisconsin.

LECTOTYPE: ZMUC, ♀, New York City (L. Lund).

PARALECTOTYPE: ZMUC, ♂, unnamed locality but not conspecific with lectotype (see Eason, 1974b).

♂♀*P. naiwatus* (Chamberlin, 1911)

— Chamberlin, 1911b: 42 (*Lithobius*; characters; USA: Georgia, Kentucky, North Carolina, South Carolina, Tennessee).

— Chamberlin, 1912e: 175 (*Paitobius*; list).

— Chamberlin, 1922b: 289-292, pl. 4; Fig. 3,4 (*Paitobius*; characters (adult, stadia); USA: Georgia, North Carolina, South Carolina).

— Brimley, 1938: 501 (*Lithobius*; USA: North Carolina).

— Causey, 1940: 76 (*Paitobius*; characters; note).

— Crabill, 1952b: 307-308, 422 (*Paitobius*; characters; plectrotaxy; taxonomic note; USA: Kentucky).

— Crabill, 1955f: 260 (*Paitobius*; checklist; USA: Kentucky).
— Wray, 1967: 155 (*Paitobius*; list).
— Hoffman, 1995a: 29 (*Paitobius*; list; notes; USA: Virginia).

Distribution: USA: Georgia, Kentucky, North Carolina, South Carolina, Tennessee, Virginia.

TYPES: MCZ, No. 122, 5 specimens, North Carolina: Saluda, Aug, 4 & 5, 1910.

♀*P. simitus* (Chamberlin, 1911)
— Chamberlin, 1911b: 44 (*Lithobius*; characters; USA: Mississippi).
— Chamberlin, 1912e: 175 (*Paitobius*; list).
— Chamberlin, 1922b: 296-297 (*Paitobius*; characters; USA: Mississippi).

Distribution: USA: Mississippi.

HOLOTYPE: MCZ, No. 118, ♀, Mississippi: Grenada, 15 July 1910.

PARATYPE: MCZ, No. 183 (slide), Mississippi: Grenada, July 1910.

♂♀*P. tabius* (Chamberlin, 1911)
— Chamberlin, 1911b: 44 (*Lithobius*; characters; notes; USA: Tennessee).
— Chamberlin, 1912e: 175 (*Paitobius*; list).
— Chamberlin, 1922b: 292-294, pl. 5; Fig. 1 (*Paitobius*; characters; USA: Tennessee).
— Hoffman, 1995a: 29 (*Paitobius*; list; notes; USA: Virginia).

Distribution: USA: Tennessee, Virginia.

SYNTYPE: MCZ, No. 115, Tennessee: Johnson City, 9 August 1910.

PARATYPE: MCZ, No. 182 (slide), Tennessee: Johnson City, August, 1910.

(Subgenus *Paitobius* Chamberlin, 1912)

Paitobius (*Paitobius*) Chamberlin, 1912e: 175. TYPE SPECIES: *Paitobius* (*Paitobius*) *carolinae* Chamberlin, 1911, fixed by original designation.
— Attems, 1938a: 344 (subgenus of *Lithobius*; key).

— Crabill, 1960d: 157 (note; see footnote).

♂♀*P. (P.) adelus* Chamberlin, 1922
— Chamberlin, 1922b: 294-295, pl. 5; Fig. 3 (*Paitobius*; characters; habitat; taxonomic note; USA: Pennsylvania).
— Crabill, 1952b: 311-312, 425, pl. 6; Figs. 87-89 (*Paitobius*; characters; plectrotaxy; range; USA: New York, Pennsylvania, Virginia).

Distribution: USA: New York, Pennsylvania, Virginia.

TYPES: MCZ, No. 128, 5 specimens, Pennsylvania: Upsal, Oct. 20, 1912. [**Note:** Needs holotype designation].

♂♀*P. (P.) carolinae* (Chamberlin, 1911)
— Chamberlin, 1911b: 47 (*Lithobius*; characters; notes; USA: North Carolina, South Carolina, Tennessee).
— Chamberlin, 1912e: 175 (*Paitobius*; cited as type species).
— Chamberlin, 1922b: 286-289, pl. 4; Fig. 5,6 (*Paitobius*; characters (adult, stadia); note; USA: North Carolina, South Carolina, Tennessee).
— Brimley, 1938: 500 (*Lithobius*; USA: North Carolina).
— Causey, 1940: 76 (*Paitobius*; characters; note).
— Wray, 1967: 155 (*Paitobius*; list). [**Note:** Specific name misspelled *caroline*].
— Hoffman, 1995a: 29 (*Paitobius*; list; note). [**Note:** Hoffman believes this species will be found in southwestern Virginia].
— Jeekel, 2005: 26 (*Paitobius*; list).

Distribution: USA: North Carolina, South Carolina, Tennessee.

TYPES: MCZ, No. 110, 3 specimens, North Carolina: Asheville, 8-5-1910.

♂♀*P. (P.) juventus* (Bollman, 1887)
— Bollman, 1887b: 254, 263, 266 [= 1893: 21, 30, 33] (*Lithobius*; subgenus *Neolithobius*; characters; list; notes; USA: Indiana). [**Note:** Bollman did not include this species in the key because the anal legs are missing from the specimen from which he described the species].
— Bollman, 1887c: 627 [= 1893: 44] (*Lithobius*; key; USA: Tennessee).
— Bollman, 1888c: 342 (*Lithobius*; note; USA: Tennessee).

— Bollman, 1888e:409 [= 1893: 111] (*Lithobius*; note; USA: Indiana).
— Bollman, 1893: 129 (*Lithobius*; subgenus *Neolithobius*; catalogue).
— Chamberlin, 1911c: 99, 101 (*Lithobius*; key to species; USA: Indiana).
— Chamberlin, 1912e: 175 (*Paitobius*; list).
— Chamberlin, 1922b: 285-286 (*Paitobius*; characters; USA: Indiana, Tennessee).
— Williams & Hefner, 1928: 143 (*Lithobius*; characters; USA: Ohio).
— Crabill, 1952b: 309, 423 (*Paitobius*; characters; plectrotaxy; USA: Indiana, Ohio).
— Summers, 1979: 696, 698 (*Paitobius*; checklist, key).
— Summers et al., 1980: 251; Fig. 27 (*Paitobius*; checklist; USA: Illinois).
— Summers et al., 1981: 62 (*Paitobius*; checklist; USA: Illinois).
— Watermolen, 1997: 6 (*Paitobius*; checklist, notes; USA: Wisconsin).

Distribution: USA: Illinois, Indiana, Ohio, Tennessee, Wisconsin.

HOLOTYPE: ?Unknown, Indiana: Bloomington (Bollman, coll.)

(Subgenus *Tunabius* Chamberlin, 1922)

Paitobius (*Tunabius*) Chamberlin, 1922: 182 (key). **Type Species:** *Paitobius* (*Tunabius*) *cantabrigensis zinus* (Chamberlin, 1911), fixed by original designation (see Eason, 1974b: 46).
— Eason, 1974b: 46-47 (synonym of *Paitobius exiguus*). [**Note:** Eason suggests " ... the identity of *exiguus* with *Paitobius zinus* is doubtful." Therefore the author has not recognized the synonymy until further taxonomic scrutiny].
— Jeekel, 2005: 33 (list).

♂♀ *P.* (*T.*) *atlantae* Chamberlin, 1922
— Chamberlin, 1922b: 298-299 (*Paitobius*; subgenus *Tunabius*; characters; USA: Georgia).
— Crabill, 1950c: 202 (*Paitobius*; USA: South Carolina).

Distribution: USA: Georgia, South Carolina.

TYPES: MCZ, No. 130, 4 specimens, Georgia: Atlanta, 7-30-1910. [**Note:** Needs holotype designation].

♂*P. (T.) watsuitus* (Chamberlin, 1911)
— Chamberlin, 1911b: 38 (*Lithobius*; characters; USA: Georgia, Virginia). [**Note:** According to Chamberlin, 1922b, "(ad part. Atlanta specimen)"].
— Chamberlin, 1922b: 297-298 (*Paitobius*; subgenus *Tunabius*; characters; USA: Georgia).

Distribution: USA: Georgia, Virginia. [**Note:** The author has left Virginia as a locality because he is not sure under which genus or species the Virginia specimens belong].

HOLOTYPE: ?MCZ; Unknown, Georgia: Atlanta, August, 1910.

♂♀*P. (T.) zinus* (Chamberlin, 1911)
— Chamberlin, 1911b: 41 (*Lithobius*; nominal taxon *cantabrigensis*; cited as variety *zinus*; characters; USA: Alabama, Georgia, North Carolina, Virginia).
— Chamberlin, 1922b: 301-302, pl. 5; Fig. 2 (*Paitobius*; subgenus *Tunabius*; characters; USA: Alabama, Georgia, North Carolina, Tennessee, Virginia).
— Brimley, 1938: 500 (*Lithobius*; nominal taxon *cantabrigensis*; cited as variety *zinus*; USA: North Carolina).
— Causey, 1940: 76 (*Paitobius*; characters; note).
— Chamberlin, 1940c: 56 (*Paitobius*; USA: North Carolina).
— Crabill, 1952b: 309-311, 424, pl. 6; Figs. 85-86 (*Paitobius*; characters; plectrotaxy; taxonomic notes; USA: Virginia).
— Crabill, 1960d: 157-160; Figs. 1-3 (*Paitobius*; report of sexual dimorphism; notes).
— Branson & Batch, 1967: 88-89 (*Paitobius*; brief description; habitat; USA: Kentucky).
— Wray, 1967: 155 (*Paitobius*; list).
— Hoffman, 1995a: 29 (*Paitobius*; list; notes).
— Jeekel, 2005: 33 (*Paitobius*; list).
?*cantabrigensis* Meinert, 1886a: 177 (*Lithobius*; characters; USA: Massachusetts). [**Note:** Meinert originally described *Lithobius cantabrigensis* from Cambridge, Massachusetts which was collected by Mr. H.H. James. It is not clear if this species is or is not the senior synonym of Chamberlin's *Paitobius cantabrigensis zinus*. The male of this species shows an interesting form of

sexual dimorphism in the prehensors which may help solve the enigma (see Crabill, 1960d)].
— Bollman, 1888a: 111 [= 1893: 85] (*Lithobius*; USA: Tennessee).
— Bollman, 1888c: 342 [=1893: 93] (*Lithobius*; note; USA: Tennessee).
— Chamberlin, 1911b: 41 (*Lithobius*; notes; USA: North Carolina, South Carolina, Tennessee, Virginia).
— Brimley, 1938: 500 (*Lithobius*; nominal taxon *cantabrigensis*; cited as variety *zinus*; USA: North Carolina).
— Wray, 1967: 155 (*Lithobius*; list).

Distribution: USA: Alabama, Georgia, Kentucky, North Carolina, South Carolina, Tennessee, Virginia.

SYNTYPES: MCZ, No. 131, 9 specimens, North Carolina: Brown's Summit, 8-14-1910.

PARATYPES: MCZ, No. 132-139, 175, Alabama, Georgia, Tennessee, Virginia.

♂♀*P.(T.) zygethus* Chamberlin & Wang, 1952
— Chamberlin & Wang, 1952a: 57 (*Paitobius*; subgenus *Tunabius*; characters; notes; USA: California).

Distribution: USA: California.

HOLOTYPE: NMNH, ♀, California.

Genus *Pampibius* Chamberlin, 1922
[1 species; Southeastern USA]

Pampibius Chamberlin, 1922b: 276; fig. 3 (characters; distribution). TYPE SPECIES: *Lithobius paitus* Chamberlin, 1911, fixed by subsequent designation and monotypy by Chamberlin, 1922b: 278).
— Attems, 1938a: 344 (synonym of *Lithobius* (*Alokobius*)).
— Crabill, 1952b: 354 (characters).
— Jeekel, 2005: 26 (list).

♂*P. paitus* (Chamberlin, 1911)
— Chamberlin, 1911b: 37-38, pl. 3; Fig. 6 (*Lithobius*; characters; USA: North Carolina, Tennessee).

— Chamberlin, 1922b: 278-279, pl. 3; Fig. 1-4 (*Pampibius*; characters; USA: North Carolina, Tennessee).
— Brimley, 1938: 501 (*Lithobius*; USA: North Carolina).
— Causey, 1940: 73-74 (*Pampibius*; characters; note).
— Crabill, 1952b: 354 (*Pampibius*; characters; note).
— Wray, 1967: 155 (*Pampibius*; list).
— Hoffman, 1995a: 29-30 (*Pampibius*; list; notes; USA: Virginia).
— Jeekel, 2005: 26 (*Pampibius*; list).

Distribution: USA: North Carolina, Tennessee, Virginia.

HOLOTYPE: NMNH, Tennessee: Unaka Springs.

Genus *Paobius* Chamberlin, 1916
[6 species; CANADA; Western & Northeastern USA]

Paobius Chamberlin, 1916: 160-162; Fig. 3 (characters; distribution; key to species; notes) TYPE SPECIES: *Paobius boreus* Chamberlin, 1916, fixed by original designation.
— Chamberlin, 1922a: 47-48 (key to species).
— Verhoeff, 1937: 242 (subgenus of *Lithobius*).
— Attems, 1938a: 344 (synonym of *Lithobius* (*Monotarsobius*)).
— Eason, 1974b: 30 (note).
— Eason, 1992: 3 (notes).
— Jeekel, 2005: 26 (list).

♂♀*P. albertanus* Chamberlin, 1922
— Chamberlin, 1922a: 47, 48 (*Paobius*; characters; CANADA: Alberta). [**Note:** It is highly likely that this species is a junior synonym to *P. columbiensis*].
— Kevan, 1983: 2949 (checklist).
— Behan-Pelletier, 1993: 26 (*Paobius*; list).

Distribution: CANADA: Alberta.

HOLOTYPE: NMNH, CANADA: Alberta, "the Spring Lakes Trip", 25-26 April 1918, by N. B. Sanson. [**Note:** MCZ has a vial (No. 2187) labeled as ♀ holotype with one specimen. The original locality label reads "Canada: Albert, Rauff[?], Alba, N.B. Sanson, Apr. 25/26, 1918"].

ALLOTYPE: MCZ, No. 2188, ♂, The original locality label reads "Canada: Alberta, Rauff[?], Alba, N.B. Sanson, 1921"].

♂♀*P. berkeleyensis* (Verhoeff, 1937)
— Verhoeff, 1937: 242-243 (*Lithobius*; subgenus *Paobius*; characters; USA: California).

Distribution: USA: California.

HOLOTYPE: ?Unknown; USA: California, Berkeley.

♂♀*P. boreus* Chamberlin, 1916
— Chamberlin, 1916: 163-166, pl. 3; Fig. 7-9, pl. 4; Fig. 1-3 (*Paobius*; characters (adult, stadia); notes; USA: Alaska).
— Chamberlin, 1919: 15H (*Paobius*; list).
— Chamberlin, 1946c: 186 (*Paobius*; list).
— Kevan, 1983: 2949 (checklist).
— Behan-Pelletier, 1993: 26 (*Paobius*; list).
— Jeekel, 2005: 26 (*Paobius*; list).

Distribution: USA: Alaska.

HOLOTYPE: MCZ, No. 345, Alaska: Forrester Island, by Harold & R. W. Heath.

PARATYPES: MCZ, No. 346-349, 418, Alaska: Forrester Island, by Harold & R. W. Heath.

♀*P. columbiensis* Chamberlin, 1916
— Chamberlin, 1916: 166-168, pl. 4; Fig. 4 (*Paobius*; characters; CANADA: British Columbia).
— Kevan, 1983: 2949 (checklist).
— Behan-Pelletier, 1993: 26 (*Paobius*; list).

Distribution: CANADA: British Columbia.

HOLOTYPE: MCZ, No. 350, CANADA: British Columbia: Kaslo.

PARATYPES: MCZ, No. 417, CANADA: British Columbia: Kaslo.

♀*P. orophilus* Chamberlin, 1916
— Chamberlin, 1916: 168-169 (*Paobius*; characters; notes; CANADA: British Columbia).
— Chamberlin, 1922a: 47 (*Paobius*; note).
— Kevan, 1983: 2949 (checklist).
— Behan-Pelletier, 1993: 26 (*Paobius*; list).

Distribution: CANADA: British Columbia.

HOLOTYPE: MCZ, No. 351, CANADA: British Columbia, Kaslo.

PARATYPES: MCZ, No. 416, CANADA: British Columbia, Kaslo.

♀*P. vagrans* Chamberlin, 1916
— Chamberlin, 1916: 169-171 (*Paobius*; characters; note; USA: Vermont).
— Chamberlin, 1951b: 32 (*Paobius*; USA: Vermont).
— Crabill, 1952b: 380 (*Paobius*; cited as new synonym of *Monotarsobius crassipes*).
— Kevan, 1983: 2949 (checklist).

Distribution: USA: Vermont.

HOLOTYPE: MCZ, No. 352; USA: Vermont, St. Johnsbury, by R.V. Chamberlin.

PARATYPES: MCZ, No. 412, USA: Vermont, St. Johnsbury; No. 414, Lake Carmi, by R.V. Chamberlin.

Genus *Pholobius* Chamberlin, 1940
[2 species; Southeastern and South Central USA]

Pholobius Chamberlin, 1940b: 49 (characters). **Type Species:** *Pholobius goffi* Chamberlin, 1940, fixed by original designation and monotypy.
— Jeekel, 2005: 27 (list).

♂*P. goffi* Chamberlin, 1940
— Chamberlin, 1940b: 49-50 (*Pholobius*; characters; USA: Florida). [**Note:** This species is apparently associated with the Florida Pocket Gopher (*Geomys floridanus*) making it the second lithobiomorph besides *Eulithobius hypogeus*].
— Jeekel, 2005: 27 (*Pholobius*; list).

Distribution: USA: Florida.

HOLOTYPE: NMNH, No. 418, ♂, Florida: Lake Co., Leesburg, 21 March 1938, by C. C. Goff.

♂♀*P. mundior* Chamberlin & Mulaik, 1940

— Chamberlin & Mulaik, 1940: 157-158 (*Pholobius*; characters; note; USA: Texas).

Distribution: USA: Texas.

HOLOTYPE: NMNH, Texas: Kerr County, Raven Ranch.

PARATYPE: NMNH, Texas.

Genus *Planobius* Chamberlin & Wang, 1952
[1 species; North America]

Planobius Chamberlin & Wang, 1952a: 57-58 (characters). TYPE SPECIES: *Planobius aletes* Chamberlin & Wang, 1952, fixed by original designation and monotypy. [**Note:** Chamberlin & Wang indicate that this genus is closely related with *Nampabius*.
— Jeekel, 2005: 28 (list).

♂*P. aletes* Chamberlin & Wang, 1952
— Chamberlin & Wang, 1952a: 58 (*Planobius*; North American).
— Jeekel, 2005: 28 (*Planobius*; list).

Distribution: North America: no country or state recorded.

HOLOTYPE: NMNH, ♂, North America; No other specific locality recorded.

Genus *Pokabius* Chamberlin, 1912
[29 species; CANADA & USA]

Pokabius Chamberlin, 1912: TYPE SPECIES: *Poabius verdescens* Chamberlin, 1912, fixed by direct substitution. [**Note:** Author suspects the type species of this genus is a junior synonym of *Pokabius bilabiatus* (Wood, 1867). In general, this genus appears to be a big mess and should be reassessed in its entirety].
— Chamberlin, 1912f: 316 (change of name from *Poabius* to *Pokabius*).
— Chamberlin, 1922b: 339-343 (characters; distribution; key to subgenera and species; notes).
— Attems, 1938a: 344 (key; list of synonyms).
— Crabill, 1952b: 349-351 (characters; key to subspecies of northeastern USA; nomenclatorial notes).
— Eason, 1974a: 66-67 (notes).

181

— Jeekel, 2005: 28 (list).

Poabius Chamberlin, 1912d: 153 (note). TYPE SPECIES: *Poabius verdescens* Chamberlin, 1912, fixed by original designation. [**Note: Preoccupied** by *Poabius* C.L. Koch, 1847]).

— Chamberlin, 1912f: 316 (change of name to *Pokabius*).

— Crabill, 1952b: 349, 350 (synonym of *Pokabius*; note).

— Jeekel, 2005: 28 (list; notes).

Physobius Chamberlin, 1945b: 197 (characters). TYPE SPECIES: *Physobius rappi* Chamberlin, 1945, fixed by original designation and monotypy. [**Note:** Synonymized by Crabill, 1981b: 359].

— Crabill, 1952b: 354-355 (synonym of *Pokabius*; characters; taxonomic note).

— Crabill, 1981b: 359 (synonym of *Pokabius*).

— Jeekel, 2005: 27 (list).

♂*P. aethes* Chamberlin, 1951

— Chamberlin, 1951c: 115 (*Pokabius*; characters; USA: California).

Distribution: USA: California.

HOLOTYPE: NMNH, ♂, California: Willow Creek, 14 July 1937, by R. V. Chamberlin.

♂♀*P. bilabiatus* (Wood, 1867)

— Wood, 1867b: 130 (*Lithobius*; characters; USA: Illinois).

— Stuxberg, 1875b: 18 (*Lithobius*; subgenus *Archilithobius*; list).

— Stuxberg, 1875c: 31[= 1876: 138] (*Lithobius*; subgenus *Archilithobius*; list).

— Bollman, 1887b: 254, 256 [=1893: 21, 23] (*Lithobius*; characters; key; USA: Michigan, Nebraska).

— Bollman, 1888e: 409 [= 1893: 110] (*Lithobius*; note; USA: Indiana).

— Bollman, 1893: 128, 185 (*Lithobius*; subgenus *Archilithobius*; catalogue; note; USA: Minnesota).

— Chamberlin, 1911b: 39 (*Lithobius*; range; note; USA: Mississippi).

— Chamberlin, 1911c: 100, 103 (*Lithobius*; key to species; USA: Illinois, Indiana, Iowa, Minnesota, Nebraska, Wisconsin).

— Gunthorp, 1913: 167 (*Poabius*; notes; USA: Kansas).

— Gunthorp, 1920: 113 (*Lithobius*; list).

— Chamberlin, 1922b: 347-352, pl. 8; Fig. 2 (*Pokabius*; characters (adult, stadia); notes; USA: Illinois, Indiana, Iowa, Kansas, Mississippi, Nebraska, Wisconsin).

— Holmquist, 1926: 411 (*Pokabius*; notes on hibernation such as: activity, date collected, developmental stage, ecological association/ niche, quantity found). [**Note:** Specific name misspelled *bilabiasus*].

— Filinger, 1928: 359 (*Poabius*; note).

— Holmquist, 1928: 83, Tab. 3 (*Pokabius*; co-inhabitant of *F. uleki* Emery ant nests; USA: Illinois).

— Williams & Hefner, 1928: 141-142 (*Lithobius*; characters; note; USA: Ohio).

— Matthews, 1935: 76, 77-79; Map: 80, pl. 6; Fig. 1-5 (*Pokabius*; characters; USA: Wisconsin).

— Chamberlin, 1942b: 16 (*Pokabius*; USA: Iowa).

— Chamberlin, 1942d: 187-188 (*Pokabius*; USA: Missouri).

— Chamberlin, 1943b: 105 (*Pokabius*; USA: Minnesota).

— Chamberlin, 1944d: 209 (*Pokabius*; USA: Illinois).

— Dowdy, 1944: 203, 214 (*Lithobius*; ecological note; habitat; USA: Ohio).

— Auerbach, 1946: 14 (*Pokabius*; list; USA: Illinois).

— Rapp, 1946: 667 (*Pokabius*; habitat; USA: Illinois).

— Auerbach, 1951a: 104 (*Pokabius*; desiccation; ecology; habitat; length to weight ratio; list; notes; population percentage; USA: Illinois).

— Auerbach, 1951b: 111, 113; Fig. 3 (*Pokabius*; key).

— Crabill, 1952b: 351-353, 439 (*Pokabius*; characters; plectrotaxy; range; USA: Illinois, Indiana, Wisconsin).

— Johnson, 1952: 176-180; Tab. 22; Map 15 (*Pokabius*; autecology; characters; distribution; natural history; USA: Michigan).

— Crabill, 1955f: 260 (*Pokabius*; checklist; USA: Kentucky).

Crabill, 1955h: 156 (*Pokabius*; note; USA: Missouri).

— Summers, 1979: 696, 698 (*Pokabius*; checklist, key).

— Summers & Uetz, 1979: 349, 350, 351; Tab. 1 & 2 (*Pokabius*; microhabitat; USA: Illinois).

— Summers et al., 1980: 251; Fig. 28 (*Pokabius*; checklist; USA: Illinois).

— Crabill, 1981b: 359 (*Pokabius*; taxonomic notes).

— Summers et al., 1981: 62 (*Pokabius*; checklist; USA: Illinois).

— Kevan, 1983: 2949 (*Pokabius*; subgenus *Pokabius*; checklist).

— Snider, 1991: 190 (*Pokabius*; list; notes).

— Watermolen, 1997: 6 (*Pokabius*; checklist, notes; USA: Wisconsin).

— Regier et al., 2005: 149; Tab. 1 (*Pokabius*; GenBank Accession No.; phylogeny).

— Mercurio, unpublished (*Pokabius*; USA: North Dakota). [**Note:** The author has determined specimens deposited at the AMNH as belonging to this species. To the author's knowledge this is the first report of a chilopod from North Dakota].

malterris Kenyon, 1893a: 18 (*Archilithobius*; characters; USA: Nebraska). [**Note:** Misspelled *Archilithabius*].

— Kenyon, 1893b: 161. (*Archilithobius*; characters; USA: Nebraska). [**Note:** Kenyon here reported *malterris* as a new species for the second time]. along with *A. nebrascensis* both of which appear to be described from an immature *P. bilabiatus*, hence, the placement here but if possible it should be verified].

— Chamberlin, 1911b: 103 (*Lithobius*; synonym of *bilabiatus*).

— Crabill, 1952: 351 (*Lithobius*; synonym of *bilabiatus*).

— Kevan, 1983: 2949 (*Lithobius*; synonym of *bilabiatus*).

?*nebrascensis* Kenyon, 1893a: 17-18 (*Archilithobius*; characters; USA: Nebraska).

— Kenyon, 1893b: 161 (*Archilithobius*; characters; USA: Nebraska).

rappi Chamberlin, 1945b: 197 (*Physobius*; characters; USA: Illinois). **HOLOTYPE:** NMNH.

— Rapp, 1946: 667 (*Physobius*; habitat; USA: Illinois).

— Crabill, 1952b: 355-356 (*Physobius*; characters; taxonomic notes). [**Note:** Crabill has suspicions of the validity of this species which he later synonomizes (see Crabill, 1981b)].

— Crabill, 1981b: 359 (*Physobius*; synonym of *bilabiatus*; taxonomic notes).

— Jeekel, 2005: 27 (*Physobius*; list).

— Regier et al., 2005: 149; Tab. 1 (*Pokabius*; table of taxon classification with GenBank accession numbers).

tuber Bollman, 1887b: 254, 256-257, 265 [=1893: 21, 23-24, 32] (*Lithobius*; subgenus *Archilithobius*; characters; key; list; notes; USA: Indiana). **HOLOTYPE:** NMNH.

— Bollman, 1888e: 409 [= 1893: 110] (*Lithobius*; synonym of *bilabiatus*).

— Bollman, 1893: 185 (*Lithobius*; note; USA: Minnesota).

— Chamberlin, 1911a: 39 (*Lithobius*; synonym of *bilabiatus*).

— Chamberlin, 1911b: 103 (*Lithobius*; synonym of *bilabiatus*).

— Crabill, 1952: 351 (*Lithobius*; synonym of *bilabiatus*).

— Kevan, 1983: 2949 (*Lithobius*; synonym of *bilabiatus*).

Distribution: USA: Illinois, Indiana, Iowa, Kansas, Kentucky, Michigan, Minnesota, Missouri, Mississippi, North Dakota, Nebraska, Ohio, Wisconsin.

HOLOTYPE: ?MCZ, Illinois: Rock Island, by Mr. Walsh. [**Note:** The MCZ has numerous specimens with the following catalog numbers: 191, 297, 298, 300-308, 2226-2228, 29179, 29180].

♂♀ *P. disantus* Chamberlin, 1922
   — Chamberlin, 1922b: 354-355, pl. 9; Fig. 3 (*Pokabius*; characters; note; USA: California).
   — Chamberlin, 1944d: 209 (*Pokabius*; USA: California).
   clavigerens Chamberlin, 1910: 373 (*Lithobius*; synonym to *disantus* in part; [**Note:** Preoccupied]; USA: California).

Distribution: USA: California.

TYPES: MCZ, No. 276 (Unique No. 14314), 2 specimens, California: Laurel Canyon, Los Angeles Co., June, 1909. [**Note:** Original locality/determined label reads "*Pokabius disantus* Ch., Cal: Laurel Canyon, June '09" an additional type vial (Unique No. 30006), also containing 2 specimens, has a locality label that reads "Laurel Canyon, Los Angeles Co., Cal., June, '09"].

♀ *P. eremus* Chamberlin, 1922
   — Chamberlin, 1922b: 343, 377-378 (*Pokabius*; characters; notes; key; CANADA: British Columbia).
   — Chamberlin, 1928a: 95 (*Lophobius*; key). [**Note:** transferred from *Pokabius* being placed in key to species of *Lophobius*? Or is this the first mention of *Juanobius eremus*?].
   — Kevan, 1983: 2949 (*Pokabius*; subgenus *Pokabius*; checklist).
   — Behan-Pelletier, 1993: 26 (*Pokabius*; list).

Distribution: CANADA: British Columbia.

HOLOTYPE: MCZ No. 187, ♀, CANADA: British Columbia, Kaslo.

♂ *P. hopianus* Chamberlin, 1938
   — Chamberlin, 1938b: 633 (*Pokabius*; characters; USA: Arizona).

Distribution: USA: Arizona.

HOLOTYPE: NMNH, ♂, Arizona: Wickensburg, 7 April 1938, by R.V. Chamberlin.

♂♀ *P. liber* Chamberlin, 1941
   — Chamberlin, 1941a: 11 (*Pokabius*; characters; USA: California).

Distribution: USA: California.

HOLOTYPE: NMNH, ♂, California: 2 mi. N of Independence, 17 March 1941, by S. & D. Mulaik.

ALLOTYPE: NMNH, ♀, California: 2 mi. N of Independence, 17 March 1941, by S. & D. Mulaik.

♂♀*P. linsdalei* Chamberlin, 1941
— Chamberlin, 1941a: 11-12 (*Pokabius*; characters; note; USA: California).

Distribution: USA: California.

HOLOTYPE: NMNH, California: Monterey Co., Hastings Reservation, by Dr. J.M. Linsdale.

♀*P. oreines* Chamberlin, 1941
— Chamberlin, 1941a: 12-13 (*Pokabius*; characters; USA: California).

Distribution: USA: California.

SYNTYPE: NMNH, California: 12 mi. NE of Hammond, between 3,000 - 4,000 ft., 21-22 March 1941, by S. & D. Mulaik.

♂*P. piedus* Chamberlin, 1930
— Chamberlin, 1930a: 111-112 (*Pokabius*; characters; USA: Utah).
— Chamberlin, 1940d: 51-52 (*Pokabius*; characters; note; USA: Utah). [**Notes:** Seemingly Chamberlin forgot that he already described this species ten years earlier from the same locality and there are now two male holotypes in existence from two different localities. The latter description was likely on a specimen of lesser maturity but due to his inconsistent descriptions no definitive answer can be made until the types are reexamined].

Distribution: USA: Utah.

HOLOTYPE: NMNH, ♂, Utah: Washington Co., St. George, 3 April 1929, by Lowell Woodbury.

♂♀*P. praefectus* Chamberlin, 1938

— Chamberlin, 1938b: 632-633 (*Pokabius*; characters; notes; USA: New Mexico).
— Chamberlin & Mulaik, 1940: 127 (*Pokabius*; USA: Texas).

Distribution: USA: New Mexico, Texas.

HOLOTYPE: NMNH, New Mexico: Bear Canyon, Camp Mary White, 15 SE of Cloudcroft, 8,000 ft., 20 August 1934, by S. Mulaik.

♂♀*P. simplex* Chamberlin, 1941
— Chamberlin, 1941a: 10 (*Pokabius*; characters; USA: California).

Distribution: USA: California.

HOLOTYPE: NMNH, ♂, 6 mi. W of Jackson, 27 March 1941, by S. & D. Mulaik.

♂*P. vaquero* Chamberlin, 1941
— Chamberlin, 1941a: 10-11 (*Pokabius*; characters; USA: Nevada).

Distribution: USA: Nevada.

HOLOTYPE: NMNH, ♂, Nevada: 29 mi. SW of Calloway Ranch, 15 March 1941, taken on "volcanic flow," by S. & D. Mulaik.

♂♀*P. verdescens* (Chamberlin, 1912)
— Chamberlin, 1912d: 154, 172, pl. 13; Fig. 1. (*Poabius*; cited as type species). [**Note:** Chamberlin failed to include an actual description of this new subspecies but has one figure. This subspecies is more than likely a synonym for the parent species].
— Chamberlin, 1922b: 353, pl. 8; Fig. 3 (*Pokabius*; subgenus *Pokabius*; nominal taxon *bilabiatus*; subspecies *verdescens*; characters; notes; USA: Illinois, Iowa, Mississippi).
— Matthews, 1935: 76, 77 (*Pokabius*; nominal taxon *bilabiatus*; taxonomic discussion; USA: Wisconsin).
— Crabill, 1952b: 349, 353-354, (*Pokabius*; characters; subgenus *Pokabius*; subspecies of *bilabiatus*; taxonomic notes; USA: Illinois, Kentucky).
— Crabill, 1955b: 40-41 (*Pokabius*; cited as subspecies of *bilabiatus*; taxonomic notes; USA: Missouri).
— Jeekel, 2005: 28 (*Poabius/Pokabius*; list).

Distribution: USA: Illinois, Iowa, Kentucky, Missouri, Mississippi.

SYNTYPES: MCZ, No. 312, 2 specimens, Illinois: East Peoria, 7-10-1910.

PARATYPES: MCZ, No. 310, 311, Iowa.

(Subgenus *Anobius* Chamberlin, 1922)

Pokabius (*Anobius*) Chamberlin, 1922b: 341, 343 (characters; key). TYPE SPECIES: *Lithobius centurio* Chamberlin, 1904, fixed by original designation.
— Jeekel, 2005: 11 (list; note).

♂♀*P.* (*A.*) *centurio* (Chamberlin, 1904)
— Chamberlin, 1904: 651 (*Lithobius*; characters; USA: New Mexico).
— Chamberlin, 1922b: 341, 343-346, pl. 10; Fig. 1, 2 (*Pokabius*; subgenus *Anobius*; characters; key; USA: New Mexico).
— Chamberlin, 1928d: 309 (*Anobius*; USA: Utah).
— Chamberlin, 1930a: 112 (*Anobius*; USA: Utah).
— Chamberlin, 1943b: 105 (*Pokabius*; USA: New Mexico).
— Jeekel, 2005: 11 (*Lithobius*; list).
nankus Chamberlin, 1912d: 153-154, pl. 12; Fig. 4 (*Poabius*; synonym of *centurio*; characters; note; USA: New Mexico). [**Note:** Holotype deposited at MCZ, No. 294].
— Chamberlin, 1922b: 343 (synonym of *centurio*).

Distribution: USA: New Mexico, Utah.

HOLOTYPE: NMNH, New Mexico: Las Vegas, Prof. T.D.A. Cockerell.

♂*P.* (*A.*) *gilae* Chamberlin, 1922
— Chamberlin, 1922b: 342, 346-347, pl. 11; Fig. 1 (*Pokabius*; subgenus *Anobius*; characters; habitat; key; USA: Arizona).

Distribution: USA: Arizona.

HOLOTYPE: MCZ, No. 184, ♂, Arizona: Thatcher, edge of Gila River, April, 1913.

(Subgenus *Lophobius* Chamberlin, 1922)

Pokabius (*Lophobius*) Chamberlin, 1922b: 343, 367 (characters; key). Tᴏᴘᴇ
Sᴘᴇᴄɪᴇs: *Lithobius socius* Chamberlin, 1901, fixed by subsequent
designation of Jeekel, 2005.
— Jeekel, 2005: 21 (designation of type).

♂*P.* (*L.*) *arizonae* (Chamberlin, 1922)
— Chamberlin, 1922b: 373-374 (*Pokabius*; subgenus *Lophobius*;
characters; note; USA: Arizona).
— Chamberlin, 1928a: 95 (*Lophobius*; key). [**Note:** transferred from
Pokabius being placed in key to species of *Lophobius*].
— Chamberlin, 1928d: 308 (*Lophobius*; USA: Utah).
— Chamberlin, 1958c: 133 (*Lophobius*; checklist; note).

Distribution: USA: Arizona, Utah.

HOLOTYPE: MCZ No. 285, ♂, Arizona: Tucson (W. M. Wheeler).

♂*P.* (*L.*) *castellopes* (Chamberlin, 1903)
— Chamberlin, 1903b: 158-159 (*Lithobius*; characters; key to species;
USA: California).
— Chamberlin, 1910: 373 (*Lithobius*; characters; key to species; USA:
California).
— Chamberlin, 1922b: 376-377, pl. 12; Fig. 1 (*Pokabius*; subgenus
Lophobius; characters; USA: California).
— Chamberlin, 1928a: 95 (*Lophobius*; key). [**Note:** transferred from
Pokabius being placed in key to species of *Lophobius*].
— Chamberlin, 1944c: 79-80 (*Lophobius*; USA: California).

Distribution: USA: California.

HOLOTYPE: ?MCZ, ♂, California: Shasta Springs, 1902.

♂♀*P.* (*L.*) *helenae* Chamberlin, 1922
— Chamberlin, 1922b: 374-375, pl. 12; Fig. 3 (*Pokabius*; subgenus
Lophobius; characters; USA: Montana).
— Chamberlin, 1928a: 95 (*Lophobius*; key). [**Note:** transferred from
Pokabius being placed in key to species of *Lophobius*?].
— Kevan, 1983: 2949 (*Pokabius*; subgenus *Lophobius*; checklist).

Distribution: USA: Montana.

HOLOTYPE: MCZ No. 185, Montana: Helena (W. M. Mann).

♂♀*P.* (*L.*) *loganus* (Chamberlin, 1925)
— Chamberlin, 1925e: 55-56 (*Lophobius*; characters; note; USA: Utah).
— Chamberlin, 1943b: 105 (*Lophobius*; USA: Idaho).
— Kevan, 1983: 2949 (*Pokabius*; subgenus *Lophobius*; list).

Distribution: USA: Idaho, Utah.

HOLOTYPE: NMNH, Utah: Cache Co., Logan Canyon.

♂*P.* (*L.*) *pungonius* (Chamberlin, 1912)
— Chamberlin, 1912d: 152-153 (*Sozibius*; characters; USA: Colorado).
— Chamberlin, 1922b: 372-373 (*Pokabius*; subgenus *Lophobius*; characters; USA: Colorado).
— Chamberlin, 1928a: 95 (*Lophobius*; key). [**Note:** transferred from *Pokabius* being placed in key to species of *Lophobius*?].

Distribution: USA: Colorado.

HOLOTYPE: NMNH, ♂, Colorado: Marshall, Prof. T.D.A. Cockerell.

♂♀*P.* (*L.*) *socius* (Chamberlin, 1901)
— Chamberlin, 1901: 23-24 (*Lithobius*; characters; key to species; USA: Utah)
— Chamberlin, 1922b: 367-370, pl. 12; Fig. 2 (*Pokabius*; subgenus *Lophobius*; characters; USA: Utah).
— Chamberlin, 1928a: 94 (*Lophobius*; key). [**Note:** transferred from *Pokabius* being placed in key to species of *Lophobius*?].
— Chamberlin, 1928d: 308 (*Lophobius*; USA: Utah).
— Chamberlin, 1958c: 133 (*Lophobius*; checklist; note).
— Jeekel, 2005: 21 (*Lithobius*; list). [**Note:** designated as type species of *Lophobius*].

Distribution: USA: Utah.

HOLOTYPE: NMNH, No. 784, Utah: Salt Lake City.

♂*P.* (*L.*) *stenenus* Chamberlin, 1938
— Chamberlin, 1938b: 632 (*Pokabius*; subgenus *Lophobius*; characters; USA: Oregon).
— Kevan, 1983: 2949 (*Pokabius*; subgenus *Lophobius*; checklist; misspelled *stenerus*).

Distribution: USA: Oregon.

HOLOTYPE: NMNH, ♂, Oregon: Grant's Pass, 12 June 1937.

(Subgenus *Pokabius* Chamberlin, 1922)

Pokabius (*Pokabius*) Chamberlin, 1922b: **Type Species:** *Pokabius* (*Pokabius*)
verdescens Chamberlin, 1921, fixed by original designation.
— Attems, 1938: 344 (synonym of *Lithobius*).

♂♀*P.* (*P.*) *clavigerens* (Chamberlin, 1903)
— Chamberlin, 1903b: 159-160 (*Lithobius*; characters; key to species;
USA: California).
— Chamberlin, 1909b: 187 (*Lithobius*; range; USA: California).
— Chamberlin(in part), 1910: 373 (*Lithobius*; characters; key to species;
USA: California).
— Anonymous, 1920: 35 (*Pokabius*; USA: California).
— Chamberlin, 1922b: 342, 360-363, pl. 9; Fig. 1, 2 (*Pokabius*; subgenus
Pokabius; characters (adult, stadia); habitat; key; USA: California).
manni Chamberlin, 1910: 373 (*Lithobius*; characters; key to species;
pseudomaturus variety of *clavigerens*; USA: California).
— Chamberlin, 1910: 374 (*Lithobius*; cited as variety *pia*; characters; key
to species; [Note: this must also be a variety of *clavigerens*]; USA:
California).
— Chamberlin, 1922b: 360 (synonym of *clavigerens*).

Distribution: USA: California.

HOLOTYPE: NMNH, California: Pacific Grove.

♂♀*P.* (*P.*) *iginus* (Chamberlin, 1912)
— Chamberlin, 1911e: 260 (*Lithobius*; USA: Washington).
— Chamberlin, 1912d: 154, pl. 13; Fig. 2 (*Poabius*; characters; note;
USA: Washington).
— Chamberlin, 1922b: 365-366 (*Pokabius*; subgenus *Pokabius*; characters
(adult, stadia); note; USA: Washington).
— Kevan, 1983: 2949 (*Pokabius*; subgenus *Pokabius*; checklist).

Distribution: USA: Washington.

TYPES: MCZ, No. 293, 4 specimens, Washington: Madison, Dr. E. Bergroth.

♂♀*P. (P.) iosemiteus* Chamberlin & Wang, 1952
— Chamberlin & Wang, 1952a: 57 (*Pokabius*; subgenus *Pokabius*; characters; USA: California).

Distribution: USA: California.

HOLOTYPE: NMNH; California: Yosemite National Park.

♂♀*P. (P.) pitophilus* (Chamberlin, 1903)
— Chamberlin, 1903b: 157 (*Lithobius*; characters; USA: California).
— Chamberlin, 1910: 373 (*Lithobius*; characters; USA: California).
— Chamberlin, 1922b: 363-364, pl. 9; Fig. 4, 5 (*Pokabius*; subgenus *Pokabius*; characters; habitat; USA: California).
— Chamberlin, 1944c: 79 (*Pokabius*; note; USA: California).

Distribution: USA: California.

HOLOTYPE: NMNH, California: Truckee, 1902.

♂♀*P. (P.) sokovus* (Chamberlin, 1909)
— Chamberlin, 1909b: 189 (*Lithobius*; habitat; USA: Nevada).
— Chamberlin, 1922b: 359-360, pl. 11; Fig. 2, 3 (*Pokabius*; subgenus *Pokabius*; characters; habitat; USA: Nevada).

Distribution: USA: Nevada.

HOLOTYPE: NMNH, Nevada, Las Vegas, 1909, in loose soil from six inches to a foot below the surface.

♂♀*P. (P.) utahensis* (Chamberlin, 1901)
— Chamberlin, 1901: 22-23 (*Lithobius*; characters; habitat; key to species; USA: Utah)
— Chamberlin, 1909b: 187 (*Lithobius*; USA: California, Oregon). [**Note:** some of this material reported belonged to *Oabius tiganus* Chamberlin].
— Chamberlin, 1912b: 153 (*Poabius*; list).
— Chamberlin, 1922b: 342, 355-359, pl. 10; Fig. 3, 4 (*Pokabius*; characters (adult, stadia); habitat; key; USA: Utah).
— Chamberlin, 1925e: 56 (*Pokabius*; habitat; USA: Utah).
— Chamberlin, 1928d: 308 (*Pokabius*; USA: Utah).

— Chamberlin, 1930a: 111 (*Pokabius*; habitat; USA: Utah).
— Chamberlin, 1943b: 105(*Pokabius*; USA: Idaho, Utah).
— Chamberlin, 1958c: 133 (*Lophobius*; checklist; note).
— Kevan, 1983: 2949 (*Pokabius*; subgenus *Pokabius*; nominal taxon *utahensis*; checklist).
obesus Bollman (not Stuxberg), 1888d: 347-348 [= 1893: 100] (*Lithobius*; characters; notes; USA: Utah).
— Chamberlin, 1922b: 355 (*Lithobius*; synonym of *utahensis*).

Distribution: USA: California, Idaho, Oregon, Utah.

SYNTYPE: NMNH, Utah: Neff's Canyon, Salt Lake Co.

♀*P. (P.) utahensis tidus* Chamberlin, 1962
— Chamberlin, 1962: 138 (*Pokabius*; nominal taxon *utahensis*; subspecies *tidus*; characters; USA: Oregon).
— Kevan, 1983: 2949 (*Pokabius*; subgenus *Pokabius*; checklist).

Distribution: USA: Oregon.

HOLOTYPE: ?Unknown, ♀, Oregon: Saddleback Mt.

Genus *Serrobius* Causey, 1942
[1 species; Southeastern USA]

Serrobius Causey, 1942: 79 (characters). **TYPE SPECIES:** *Serrobius pulchellus* Causey, 1942, fixed by original designation and monotypy.
— Jeekel, 2005: 30 (list).

♂♀*S. pulchellus* Causey, 1942
— Causey, 1942: 79-80; fig. 1 (*Serrobius*; characters; USA: North Carolina).
— Crabill, 1952b: 385 (*Serrobius*; note). [**Note:** According to Crabill "The affinity and identity of . . ." *Serrobius pulchellus* Causey " . . . is uncertain due to the inadequate original description[s] and the unavailablity of the type."]
— Wray, 1950: 155 (*Serrobius*; list).
— Wray, 1967: 155 (*Serrobius*; list).
— Hoffman, 1993: 18-20, Figs. 1-2, Tab. 1 (*Serrobius*; characters; habitat; taxonomic notes; USA: Virginia). [**Note:** Hoffman mentions that the types Causey was supposed to deposit at the ANSP were not found].

— Hoffman, 1995a: 30 (*Serrobius*; list; note).
— Jeekel, 2005: 30 (*Serrobius*; list).

Distribution: USA: North Carolina, Virginia.

HOLOTYPE: ?ANSP; North Carolina: Durham, Duke Forest, by Mr. James H. Starling. (see Hoffman, 1993).

Genus *Shosobius* Chamberlin & Wang, 1952
[1 species; Northwestern USA]

Shosobius Chamberlin & Wang, 1952a: 61 (characters). **TYPE SPECIES:** *Shosobius cordialis* Chamberlin & Wang, 1952, fixed by original designation and monotypy.
— Jeekel, 2005: 30 (list).

(imm. ♂)♀*S. cordialis* Chamberlin & Wang, 1952
— Chamberlin & Wang, 1952a: 61 (*Shosobius*; characters; USA: Idaho).
— Kevan, 1983: 2948 (*Shosobius*; checklist).
— Jeekel, 2005: 30 (*Shosobius*; list).

Distribution: USA: Idaho.

HOLOTYPE: NMNH, ♀, Idaho: Cour de Elaine, 4 September 1949, Mulaik.

PARATYPES: NMNH, 2 specimens, Idaho: Wallace, 3 September 1949, S.D. Mulaik.

Genus *Sigibius* Chamberlin, 1913
[5 species (1 Introduced); Eastern & Western USA]

Sigibius Chamberlin, 1913: **TYPE SPECIES:** *Sigibius puritanus* Chamberlin, 1913 [= *Lithobius* (*Sigibius*) *microps* Meinert, 1868], fixed by original designation.
— Chamberlin, 1913: 101-102 (characters; distribution; notes).
— Attems, 1938a: 344 (synonym of *Lithobius* (*Lithobius*)).
— Chamberlin, 1938b: 629 (key to known species).
— Crabill, 1952b: 335-339 (characters; key to species of northeastern USA; taxonomic notes).
— Matic, 1970: 12, 13 (synonym of *Monotarsobius*).
— Eason, 1974a: 71 (taxonomic notes).

— Eason, 1983: 142 (taxonomic notes).
— Watermolen, 1997: 7 (notes).
— Jeekel, 2005: 31 (list).
Haplolithobius Verhoeff, 1925: 153 (characters).
— Chamberlin, 1938b: 629 (synonym of *Sigibius*).

S. enans Chamberlin, 1938
— Chamberlin, 1938b: 630 (*Sigibius*; characters; USA: Maryland).
[**Note:** This species is recorded from a specimen at quarantine in
Baltimore and presumably originated in Honduras. It is probably
safe to say that the chances this species could establish itself in such
a different climate are highly unlikely].

Distribution: USA: ?Maryland.

HOLOTYPE: NMNH, Maryland: Baltimore, 25 May 1936 (Origin:
?Honduras).

♂♀*S. nidicolens* Chamberlin, 1938
— Chamberlin, 1938b: 630 (*Sigibius*; characters; habitat; USA: Virginia).
— Crabill, 1952b: 342 (*Sigibius*; characters).
— Hoffman, 1995a: 30 (*Sigibius*; list; note).

Distribution: USA: Virginia.

HOLOTYPE: NMNH, ♀, Virginia: Harrisburg, 22 September 1937, taken
from nest of Carolina Junco.

ALLOTYPE: NMNH, ♂, Virginia: Harrisburg, 22 September 1937, taken
from nest of Carolina Junco.

♂♀*S. siopius* Chamberlin & Wang, 1952
— Chamberlin & Wang, 1952a: 58 (*Sigibius*; characters; USA: Utah).

Distribution: USA: Utah.

SYNTYPE: NMNH, Utah: Provo, 22 May 1942.

PARATYPE: NMNH, Utah.

♂*S. starlingi* Causey, 1942

— Causey, 1942: 81-83 (*Sigibius*; characters; notes; USA: North Carolina).
— Crabill, 1952b: 385 (*Sigibius*; note). [**Note:** According to Crabill "The affinity and identity of ..." *Sigibius starlingi* Causey " ... is uncertain due to the inadequate original description[s] and the unavailablity of the type."].
— Wray, 1967: 155 (*Sigibius*; list).
— Hoffman, 1995a: 30 (*Sigibius*; list; note).

Distribution: USA: North Carolina.

HOLOTYPE: ANSP; ♂, North Carolina: Durham, Duke Forest, by Mr. James H. Starling.

♀*S. urbanus* Chamberlin, 1944
— Chamberlin, 1944d: 211 (*Sigibius*; characters; USA: Illinois).
— Crabill, 1952b: 341-342 (*Sigibius*; characters; taxonomic notes).
— Crabill, 1955h: 156 (*Sigibius*; characters; notes; USA: Missouri).
— Summers et al., 1980: 251 (*Sigibius*; checklist; USA: Illinois).
— Summers et al., 1981: 62 (*Sigibius*; checklist; USA: Illinois).

Distribution: USA: Illinois, Missouri.

HOLOTYPE: NMNH; ♀, Illinois, Chicago, 20 March 1942, by Henry S. Dybas.

Genus *Simobius* Chamberlin, 1922
[4 species; Western CANADA & USA]

Simobius Chamberlin, 1922b: 378-380; fig. 10 (characters; distribution; notes). **TYPE SPECIES:** *Lithobius ginampus* Chamberlin, 1909, fixed by original designation.
— Attems, 1938a: 344 (synonym of *Lithobius* (*Pokabius*)).
— Jeekel, 2005: 31 (list).

♂♀*S. gardneri* Auerbach, 1950
— Auerbach, 1950: 1-5; Tab. 1 (*Simobius*; characters (comparison of series); notes; USA: California).

Distribution: USA: California.

HOLOTYPE: SAC, ♂, California: Tehema County, Mill Creek, June 24, 1950, by George Gardner, Jr., berlesed from a sugar pine tree hole.

ALLOTYPE: SAC, ♀, California: Tehema County, Mill Creek, June 24, 1950, by George Gardner, Jr., berlesed from a sugar pine tree hole.

♂♀*S. ginampus* (Chamberlin, 1909)
— Chamberlin, 1909b: 187-188 (*Lithobius*; characters; USA: Washington).
— Chamberlin, 1922b: 380-382, pl. 12; Fig. 4, 5 (*Simobius*; characters; USA: Alaska, Washington).
— Kevan, 1983: 2949 (*Simobius*; checklist; CANADA: British Columbia).
— Behan-Pelletier, 1993: 27 (*Simobius*; list).
— Jeekel, 2005: 31 (*Simobius*; list).

Distribution: CANADA: British Columbia; USA: Alaska, Washington.

TYPES: MCZ, No. 215, 3 specimens, Washington: Pullman, 8-10-1910, by W. M. Mann. [**Note:** Chamberlin designated a ♂ as the holotype]

♂*S. lobophor* Chamberlin, 1941
— Chamberlin, 1941a: 4 (*Simobius*; characters; USA: Washington).

Distribution: USA: Washington.

HOLOTYPE: NMNH, ♂, Washington: Naches River, 15 mi. below summit, W. 121°7' N. 46°59', 5 July 1938, by W. Ivie.

♂(♀imm.)*S. opibius* Chamberlin & Wang, 1952
— Chamberlin & Wang, 1952a: 58 (*Simobius*; characters; USA: California).

Distribution: USA: California.

SYNTYPE: NMNH, California: Muir Woods, 5 September 1927.

Genus *Sonibius* Chamberlin, 1912
[3 species; Eastern CANADA & Northeastern & Central USA]

Sonibius Chamberlin, 1912e: 176-177 (characters; distribution; note). TYPE SPECIES: *Lithobius bius* Chamberlin, 1911, fixed by original designation.
— Chamberlin, 1922b: 312-314; fig. 6 (characters; distribution; key to species).

— Bailey, 1928b: 30 (characters; key to species of New York; species included: *parvus*, *numius*). [**Note:** Both *parvus* and *numius* are junior synonyms of *politus*].
— Attems, 1938a: 344 (synonym of *Lithobius* (*Monotarsobius*)).
— Crabill, 1952b: 317-318 (characters; key to species of northeastern USA; notes).
— Watermolen, 1997: 6 (notes).
— Jeekel, 2005: 31 (list).

♂*S. bius* (Chamberlin, 1911)
— Chamberlin, 1911c: 100, 102-103 (*Lithobius*; characters; key to species; USA: Michigan).
— Chamberlin, 1912e: 177 (*Sonibius*; cited as type species).
— Chamberlin, 1914b: 302 (*Sonibius*; habitat (p. 301); USA: Michigan).
— Chamberlin, 1922b: 314-315 (*Sonibius*; characters; USA: Michigan).
— Crabill, 1952b: 319-320, 428 (S*onibius*; characters; plectrotaxy; taxonomic notes: USA: Illinois).
— Johnson, 1952: 188-189 (*Sonibius*; autecology; characters; distribution; USA: Michigan).
— Summers, 1979: 696, 698 (*Sonibius*; checklist, key).
— Summers et al., 1980: 251; Fig. 29 (*Sonibius*; checklist; habitat (Table 1, p. 242-243); USA: Illinois).
— Summers et al., 1981: 62 (*Sonibius*; checklist; USA: Illinois).
— Kevan, 1983: 2949 (*Sonibius*; checklist; questioned synonym of *bius*).
— Snider, 1991: 190 (*Sonibius*; list; habitat; USA: Michigan).
— Jeekel, 2005: 31 (*Lithobius*; list).

Distribution: USA: Illinois, Michigan.

HOLOTYPE: NMNH, Michigan: Saunders.

♂♀*S. politus* (McNeill in Bollman, 1887)
— (McNeill in)Bollman, 1887b: 255, 261, 266 [=1893: 22, 28, 33] (*Lithobius*; subgenus *Lithobius*; characters; key; list; note; USA: Indiana, Michigan).
— Chamberlin, 1911c: 100, 103 (*Lithobius*; key; USA: Illinois).
— Chamberlin, 1912e: 177 (*Sonibius*; list).
— Chamberlin, 1920a: 95 (*Sonibius*; note; CANADA: Ontario, Quebec).
— Chamberlin, 1920b: 166 (*Sonibius*; CANADA: Ontario).

— Chamberlin, 1922b: 316-318 (*Sonibius*; characters; notes; CANADA: Ontario, Quebec; USA: Illinois, Indiana, Michigan).
— Verhoeff, 1925: 690 (*Lithobius*; list).
— Holmquist, 1926: 411 (*Sonibius*; notes on hibernation such as: activity, date collected, developmental stage, ecological association/niche, quantity found).
— Williams & Hefner, 1928: 142-143 (*Lithobius*; characters; USA: Ohio).
— Matthews, 1935: 47, 48-49; Map: 52, pl. 4; Figs. 1-6 (*Sonibius*; characters; USA: Wisconsin).
— Auerbach, 1951a: 101, 104, 109, 120, 121, 122; Tab. 1, 8, 9; Fig. 1, 4 (*Sonibius*; desiccation; habitat; length to weight ratio; list; notes; population percentage; USA: Illinois).
— Crabill, 1952b: 320-323, 429, pl. 6; Fig. 90 (*Sonibius*; characters; habitat; plectrotaxy; range; taxonomic notes; CANADA: Ontario, Quebec; USA: Illinois, Indiana, Kentucky, Michigan, New York, Ohio, Wisconsin). [**Note:** Crabill mentions Minnesota as part of the distribution of this species so it has been included as a questioned locality].
— Johnson, 1952: 184, 186, 188; Tab. 25; Map 17 (*Sonibius*; autecology; characters; distribution; USA: Michigan).
— Crabill, 1955f: 260 (*Sonibius*; checklist; USA: Kentucky).
— Crabill, 1955h: 156 (*Sonibius*; notes; USA: Missouri).
— Kevan, 1979: 298 (*Lithobius*; synopsis; CANADA: Ontario, Quebec).
— Kevan, 1983: 2949 (*Sonibius*; checklist).
— Summers, 1979: 696, 698 (*Sonibius*; checklist, key).
— Summers & Uetz, 1979: 347, 349, 350; Tab. 1 & 2 (*Sonibius*; notes; USA: Illinois).
— Snider, 1991: 190 (*Sonibius*; list; habitat; USA: Michigan).
— Behan-Pelletier, 1993: 27 (*Sonibius*; list).
— Watermolen, 1997: 6 (*Sonibius*; checklist, notes).
?*numius* Chamberlin, 1911c: 100, 102 (*Lithobius*; characters; key to species; synonym of *politus*; USA: Wisconsin). [**Note:** Holotype deposited at MCZ, Unique No. 14472, Wisconsin: Naugen, 6-30-10. The synonymy of *Lithobius numius* Chamberlin with *S. politus* (McNeill) is taken from Crabill (1952b: 320). Based on our knowledge of intra-specific variation of the Lithobiomorpha today it seems doubtless that *numius* and *parvus* are junior synonyms of *politus*. Furthermore, Crabill clearly discusses his reasoning for doing so in the "Systematic Notes" section under this species and is seemingly very acceptable. What is unclear is why Crabill (1955f) continued to report *numius* as a separate

species, although it may be because he didn't consider his thesis an official scientific publication].
— Chamberlin, 1912e: 177 (*Sonibius*; list).
— Chamberlin, 1922b: 318-319 (*Sonibius*; characters; USA: Wisconsin).
— Bailey, 1928a: 1082 (*Sonibius*; list).
— Bailey, 1928b: 19, 31 (*Sonibius*; characters; list; notes; USA: New York).
— Matthews, 1935: 47, 50-51; Map: 52, pl. 4; Fig. 1-6 (Sonibius; characters; USA: Wisconsin).
— Auerbach, 1946: 14 (*Sonibius*; list; USA: Illinois).
— Crabill, 1952b: 320 (*Lithobius*; cited as new synonym of *politus*).
— Crabill, 1955f: 260 (*Sonibius*; checklist; synonym of *politus*; USA: Kentucky).
— Lee, 1980: 2, 3, 4, 5; Tab. 1 (*Sonibius*; habitat; USA: Ohio).
— Watermolen, 1997: 6 (*Sonibius*; checklist, notes).

?*parvus* Chamberlin, 1922b: 315 (*Sonibius*; characters; USA: New York, Vermont). [**Note:** The synonomy of *Sonibius parvus* Chamberlin with *S. politus* (McNeill) is taken from Crabill (1952b: 320). Please see note under *numius* for further discussion].
— Bailey, 1928a: 1082 (*Sonibius*; list).
— Bailey, 1928b: 19, 30-31 (*Sonibius*; characters; list; notes; USA: New York).
— Crabill, 1952b: 320 (*Sonibius*; cited as new synonym of *politus*).
— Kevan, 1983: 2949 (*Sonibius*; questioned synonym of *politus*).

Distribution: CANADA: Ontario, Quebec; USA: Illinois, Indiana, Kentucky, Michigan, ?Minnesota, Missouri, New York, Ohio, Vermont, Wisconsin.

SYNTYPE: NMNH, Indiana: Dublin.

♀*S. scepticus* Chamberlin and Wang, 1952
— Chamberlin & Wang, 1952a: 59 (*Sonibius*; characters; notes; USA: New York).
— Crabill, 1952b: 323-324 (*Sonibius*; characters; taxonomic notes).
— Kevan, 1983: 2949 (*Sonibius*; checklist).

Distribution: USA: New York.

HOLOTYPE: NMNH; ♀, New York: Wilmington Notch, 26 August 1921.

Genus *Sozibius* Chamberlin, 1912

[7 species; Northwestern & Northeastern USA]

Sozibius Chamberlin, 1912d: 152 (note). TYPE SPECIES: *Lithobius tuobukus*
Chamberlin, 1911, fixed by original designation.
— Chamberlin, 1922b: 260-261; fig. 1 (characters; distribution; key to
species).
— Bailey, 1928b: 31 (characters; note).
— Attems, 1938a: 344 (synonym of *Lithobius* (*Alokobius*)).
— Crabill, 1952b: 327-331 (characters; key to species of northeastern
USA; taxonomic notes).
— Crabill, 1958b: 98 [**Note:** Crabill reports of finding a *Sozibius sp.*
in Wisconsin but was unable to identify species due to the male
specimen being in poor condition].
— Branson & Batch, 1967: 88 [**Note:** Branson & Batch report of a
Sozibius sp. from Kentucky but were not sure of it's status].
— Watermolen, 1997: 6 (notes).
— Jeekel, 2005: 31 (list).
Pearsobius Causey, 1942: 80 (characters; taxonomic notes). TYPE SPECIES:
Pearsobius carolinus Causey, 1942, fixed by original designation and
monotypy.
— Hoffman, 1995a: 30 (synonym of *Sozibius*). [**Note:** *Pearsobius* has
become a synonym of *Sozibius* because the only species of the genus
has been transferred].
— Jeekel, 2005: 27 (list).

♂*S. carolinus* (Causey, 1942)
— Causey, 1942: 80-81; fig. 2 (*Pearsobius*; characters; USA: North Carolina).
— Crabill, 1952b: 385 [**Note:** According to Crabill "The affinity and
identity of . . ." *Pearsobius carolinus* Causey " . . . is uncertain due to the
inadequate original description[s] and the unavailablity of the type."].
— Wray, 1967: 155 (*Pearsobius*; list).
— Hoffman, 1995a: 30 (*Sozibius*; list; notes; transferred to *Sozibius*;
USA: Virginia).
— Jeekel, 2005: 27 (*Pearsobius*; list).

Distribution: USA: North Carolina, Virginia.

HOLOTYPE: ANSP, North Carolina: Durham, Duke Forest, by Mr. James
H. Starling.

♂*S. mullanua* Chamberlin & Wang, 1952

— Chamberlin & Wang, 1952a: 59 (*Sozibius*; characters; notes; USA: Idaho).
— Kevan, 1983: 2949 (*Sozibius*; checklist).

Distribution: USA: Idaho.

HOLOTYPE: NMNH; ♂, Idaho: Mullan, 17 August 1929, by R. V. & E. S. Chamberlin.

♂♀*S. paurops* Chamberlin, 1944
— Chamberlin, 1944a: 33-34 (*Sozibius*; characters; note; USA: Georgia). [**Note:** Chamberlin states that this species essentially agrees with *proridens* and having minor differences such as fewer ocelli and spination of anterior legs].

Distribution: USA: Georgia.

HOLOTYPE: NMNH; ♀, Georgia: Dermorest, April 26, 1943.

ALLOTYPE: NMNH; ♂, Georgia: southeast of Pendergrass, April 23, 1943.

PARATYPE: NMNH; ♂, Georgia: Dermorest, April 26, 1943.

♂♀*S. pennsylvanicus* Chamberlin, 1922
— Chamberlin, 1922b: 266-268, pl. 1; Figs. 1-5 (*Sozibius*; characters (adult, stadia); USA: Pennsylvania).
— Brimley, 1938: 501 (*Lithobius*; USA: North Carolina).
— Causey, 1940: 74 (*Sozibius*; characters; USA: North Carolina).
— Crabill, 1952b: 331-332, 431, pl. 7; Figs. 97, 99, 101 (*Sozibius*; characters; habitat; plectrotaxy; USA: Connecticut, New York, Pennsylvania).
— Wray, 1967: 155 (*Sozibius*; list).
— Lee, 1980: 5; Tab. 1 (*Sozibius*; habitat; USA: Ohio).
— Kevan, 1983: 2949 (*Sozibius*; checklist).
— Hoffman, 1995a: 30 (*Sozibius*; list; note; USA: Virginia).

Distribution: USA: Connecticut, North Carolina, New York, Ohio, Pennsylvania, Virginia.

HOLOTYPE: MCZ, No. 94, Pennsylvania: Upsal, 20 October, 1912.

♂♀*S. proridens* (Bollman, 1887)

— Bollman, 1887a: 81 [= 1893: 19] (*Lithobius*; characters; USA: Indiana).
— Bollman, 1887b: 254, 257, 265 [= 1893: 21, 24, 32] (*Lithobius*; subgenus *Archilithobius*; characters (adult, stadium); habitat; key; list; USA: Indiana).
— Bollman, 1888a: 108, 111, 112 [= 1893: 79, 82, 85] (*Lithobius*; note; USA: Tennessee).
— Bollman, 1888b: 7 [= 1893: 79] (*Lithobius*; notes; USA: Arkansas).
— Bollman, 1888c: 341 [= 1893: 92] (*Lithobius*; note; USA: Tennessee).
— Bollman, 1888d: 347 [= 1893: 100] (*Lithobius*; note; USA: District of Columbia).
— Bollman, 1888e: 409 [= 1893: 110] (*Lithobius*; note; USA: Indiana). [**Note:** species name misspelled *protidens* in 1888 but corrected in 1893].
— Bollman, 1893: 129, 133-134 (*Lithobius*; subgenus *Archilithobius*; catalogue; characters; habitat; notes).
— Morse, 1902: 187 (*Lithobius*; USA: Ohio).
— Chamberlin, 1911b: 39 (*Lithobius*; USA: Arkansas, District of Columbia, Mississippi, Tennessee).
— Chamberlin, 1911c: 100, 103 (*Lithobius*; key to species; USA: Indiana).
— Chamberlin, 1918a: 24 (*Sozibius*; USA: Tennessee).
— Chamberlin, 1922b: 268-272, pl. 2; Fig. 3-6. (*Sozibius*; characters (adult, stadia); USA: Arkansas, Indiana, Tennessee). [**Note:** misspelling; Chamberlin sited the name as *providens* and the original name from Bollman (1887a) is *proridens*].
— Verhoeff, 1925: 690 (*Lithobius*; list).
— Bailey, 1928a: 1082 (*Sozibius*; list).
— Bailey, 1928b: 19, 31 (*Sozibius*; characters; list; USA: New York).
— Williams & Hefner, 1928: 141 (*Lithobius*; characters; USA: Ohio).
— Brimley, 1938: 501 (*Lithobius*; note; USA: North Carolina).
— Causey, 1940: 74 (*Sozibius*; USA: North Carolina).
— Chamberlin, 1944d: 211 (*Sozibius*; USA: Tennessee).
— Crabill,. 1950c: 202 (*Sozibius*; notes; USA: South Carolina).
— Crabill, 1952b: 332-333, 432 (*Sozibius*; characters; plectrotaxy; range; USA: District of Columbia, Indiana, Kentucky, New York, Ohio, Virginia).
— Johnson, 1952: 183-184; Tab. 24; Map 16 (*Sozibius*; characters; distribution; notes; USA: Michigan).

— Crabill, 1955b: 40 (*Sozibius*; note on spelling of specific name; range; USA: Missouri).
— Crabill, 1955f: 260 (*Sozibius*; checklist; USA: Kentucky).
— Crabill, 1955h: 156 (*Sozibius*; USA: Missouri).
— Wray, 1967: 155 (*Sozibius*; list).
— Summers, 1979: 696, 698 (*Sozibius*; checklist, key).
— Lee, 1980: 5; Tab. 1 (*Sozibius*; habitat; USA: Ohio).
— Summers et al., 1980: 251; Fig. 30 (*Sozibius*; checklist; habitat (Table 1, p. 242-243); USA: Illinois).
— Summers et al., 1981: 62 (*Sozibius*; checklist; USA: Illinois).
— Kevan, 1983: 2949 (*Sozibius*; checklist).
— Snider, 1991: 190 (*Sozibius*; list).
— Hoffman, 1995a: 30 (*Sozibius*; list; note; USA: Virginia).

Distribution: USA: Arkansas, District of Columbia, Illinois, Indiana, Kentucky, Michigan, Missouri, Mississippi, North Carolina, New York, Ohio, South Carolina, Tennessee, Virginia.

HOLOTYPE: NMNH, Indiana, Bloomington.

♂*S. texanus* Chamberlin, 1938
— Chamberlin, 1938b: 633-634 (*Sozibius*; characters; USA: Texas).
— Chamberlin & Mulaik, 1940: 127 (*Sozibius*; list).

Distribution: USA: Texas.

HOLOTYPE: NMNH, ♂, Texas: Bexar Co., 22 March 1937.

♂♀*S. tuobukus* (Chamberlin, 1911)
— Chamberlin, 1911b: 36-37, pl. 3; Fig. 7 (*Lithobius*; characters; notes; USA: Kentucky, North Carolina, South Carolina, Tennessee, Virginia, West Virginia).
— Chamberlin, 1922b: 261-266, pl. 1 fig. 6; pl. 2; Fig. 1,2 (*Sozibius*; characters (adult, stadia); note; USA: Kentucky, North Carolina, South Carolina, Tennessee, Virginia, West Virginia).
— Brimley, 1938: 501 (*Lithobius*; note; USA: North Carolina).
— Causey, 1940: 75 (*Sozibius*; characters; USA: North Carolina).
— Chamberlin, 1940c: 56 (*Sozibius*; USA: North Carolina).
— Crabill, 1952b: 333-334, 433, pl. 7; Fig. 98 characters; plectrotaxy; range; USA: Kentucky, Virginia, West Virginia).
— Crabill, 1955f: 260 (*Sozibius*; checklist; USA: Kentucky).
— Branson & Batch, 1967: 88 (*Sozibius*; characters; habitat; USA: Kentucky).

— Wray, 1967: 155 (*Sozibius*; list).
— Hoffman, 1995a: 30 (*Sozibius*; list; note).
— Jeekel, 2005: 31 (*Lithobius*; list).

Distribution: USA: Kentucky, North Carolina, South Carolina, Tennessee, Virginia, West Virginia.

HOLOTYPE: NMNH, North Carolina: Hot Springs.

Genus *Taiyubius* Chamberlin, 1912
[4 species; Eastern & Western United States]

Taiyubius Chamberlin, 1912e: 176 (characters; distribution; note). TYPE SPECIES: *Taiyubius angelus* (Chamberlin, 1903), fixed by original designation.
— Chamberlin, 1922b: 303-305; fig. 5 (characters; distribution; key to species).
— Attems, 1938a: 344 (synonym of *Lithobius* (*Lithobius*)).
— Watermolen, 1997: 7 (notes).
— Jeekel, 2005: 32 (list).

♀*T. angelus* (Chamberlin, 1903)
— Chamberlin, 1903b: 155 (*Lithobius*; characters; USA: California).
— Chamberlin, 1910: 372 (*Lithobius*; characters; USA: California).
— Chamberlin, 1922b: 305-307, pl. 5; Fig. 4, 5 (*Taiyubius*; characters; USA: California).
— Jeekel, 2005: 32 (*Lithobius*; list).

Distribution: USA: California.

TYPES: MCZ, No. 141, 3 specimens, California: Los Angeles Co., [?East ... Park], Los Angeles, 1902.

♂*T. dux* Chamberlin, 1940
— Chamberlin, 1940f: 76 (*Taiyubius*; characters; notes; USA: North Carolina).
— Causey, 1940: 75 (*Paitobius*; USA: North Carolina).
— Wray, 1967: 155 (*Llanobius*; list).
— Wray, 1967: 155 (*Paitobius*; list).
— Wray, 1967: 155 (*Taiyubius*; list).
— Hoffman, 1995a: 28 (*Llanobius*; list; note). [**Note:** Hoffman lists this species as belonging to *Llanobius* probably because he suggests it is a senior synonym of *Llanobius chamberlini* Causey,

1942. The author has left this species under *Taiyubius* until further systematic treatment].

Distribution: USA: North Carolina.

HOLOTYPE: NMNH, ♂, North Carolina: Linville, 14 October 1939, by Nelle B. Causey.

♀*T. purpureus* (Chamberlin, 1901)
— Chamberlin, 1901: 24-25 (*Lithobius*; characters, habitat; key to species; USA: Utah).
— Chamberlin, 1912e: 176 (*Taiyubius*; list).
— Chamberlin, 1922b: 311-312. (*Taiyubius*; characters; USA: Utah)

Distribution: USA: Utah.

HOLOTYPE: NMNH, No. 786, Utah: Salt Lake City, in and near growths of willows on the banks of the Jordan River.

♂*T. satanus* (Chamberlin, 1911)
— Chamberlin, 1911f: 380-381 (*Lithobius*; cited as subspecies *satanus* of *angelus*; characters; notes; USA: California).
— Chamberlin, 1912e: 176 (*Taiyubius*; list).
— Chamberlin, 1922b: 307-309 (*Taiyubius*; characters (adult, stadia); USA: California).

Distribution: USA: California.

HOLOTYPE: MCZ, No. 28929, ♂, California: Oakland.

PARATYPES: MCZ, No. 171, 173, California.

Genus *Texobius* Chamberlin & Mulaik, 1940
[1 species; Central Southern USA]

Texobius Chamberlin & Mulaik, 1940: 156 (characters). TYPE SPECIES: *Texobius unicus* Chamberlin & Mulaik, 1940, fixed by original designation and monotypy. [**Note:** Chamberlin compares this genus to *Sigibius*, which is currently recognized as a subgenus of *Lithobius*].
— Jeekel, 2005: 32 (list).

♂♀*T. unicus* Chamberlin & Mulaik, 1940

— Chamberlin & Mulaik, 1940: 157 (*Texobius*; characters; USA: Texas).

— Jeekel, 2005: 32 (*Texobius*; list).

Distribution: USA: Texas.

HOLOTYPE: NMNH, Texas: Kerr Co., Raven Ranch, December 1939.

Genus *Tidabius* Chamberlin, 1913
[15 species (1 Introduced); widespread USA]

Tidabius Chamberlin, 1913: 80-82 (characters; distribution; habitat; key to species; notes; range). **TYPE SPECIES:** *Tidabius tivius* (Chamberlin, 1909), fixed by original designation.

— Attems, 1938a: 344 (synonym of *Lithobius* (*Paitobius*)).

— Crabill, 1952b: 342-345 (characters; key to species of northeastern USA; taxonomic notes).

— Crabill & Lorenzo, 1957: 428-432 (taxonomic notes). [**Note:** Crabill & Lorenzo comment that "Those forms currently referable to *Tidabius* in North America actually may be reducible to a highly variable and genetically unstable pair of species . . ."].

— Lee, 1980: 5; Tab. 1 (*Tidabius*; habitat; USA: Ohio). [**Note:** Lee reports of a *Tidabius* sp. from Ohio but apparently was not able to identify the 3 specimens to species. It is probable that these specimens were referable to *tivius* as it is the only species known from Ohio at this time].

— Watermolen, 1997: 6 (notes).

— Jeekel, 2005: 32 (list).

♂*T. aberrans* Chamberlin, 1929

— Chamberlin, 1929: 37-38 (*Tidabius*; characters; USA: New York).

— Kevan, 1983: 2949 (*Tidabius*; checklist).

Distribution: USA: New York.

HOLOTYPE: NMNH, ♂, New York, near Ithaca, November 1927, by R. D. Harwood.

♂♀*T. anderis* Chamberlin, 1913

— Chamberlin, 1913: 90-92 (*Tidabius*; characters; USA: Washington).

207

— Crabill, 1952b: 346-347, 435 (*Tidabius*; characters; plectrotaxy; USA: Illinois). [**Note:** This report seems far removed from the type locality, therefore, the author has questioned this state].

— Summers, 1979: 696, 698 (*Tidabius*; checklist, key).

— Kevan, 1983: 2949 (*Tidabius*; checklist).

Distribution: USA: ?Illinois, Washington.

TYPES: MCZ, No. 509, 2 specimens, Washington: Pullman, [?] 1908.

♀*T. bonvillensis* (Chamberlin, 1909)

— Chamberlin, 1909b: 189-190 (*Lithobius*; characters; habitat; USA: Utah).

— Chamberlin, 1913: 101 (*Tidabius*; characters; habitat; USA: Utah).

— Crabill & Lorenzo, 1957: 430, see footnote 6 (*Tidabius*; notes). [**Note:** Crabill & Lorenzo comment that *bonvillensis* may be a stable species with its distinctive 3+3 basal spurs on the female gonopod].

Distribution: USA: Utah.

TYPES: MCZ, No. 511, 3 specimens, [**Note:** a ♀ was designated as holotype], Utah: Lake Point, May 30, 1909, under stones along foothills.

♀*T. emporus* Chamberlin, 1941

— Chamberlin, 1941a: 3-4 (*Tidabius*; characters; USA: Hawaii). [**Note:** A potentially invasive and introduced species from Japan].

— Zapparoli & Shelley, 2000: 47 (*Tidabius*; list; notes).

Distribution: USA: Hawaii.

HOLOTYPE: NMNH, ♀, Hawaii: Honolulu, 11 April 1938, in packing material about *Rhynchostylis retusa* from Japan.

♂♀*T. kansensis* (Gunthorp, 1913)

— Gunthorp, 1913: 166 (*Lithobius*; characters; habitat; USA: Kansas). [**Note:** Gunthorp suggests that this species is closely related to *Lithobius cantabrigensis* Meinert].

— Crabill, 1952b: 343-344 (*Tidabius*; taxonomic notes).

— Crabill & Lorenzo, 1957: 428-429 (*Tidabius*; characters (plectrotaxy); notes; USA: Kansas). [**Note:** According to Crabill & Lorenzo *T. pallidus* Chamberlin appears to be a junior synonym of *T. kansensis* Gunthorp based on their comparison of characters. Gunthorp's work of 1913 was in the month of January and that of Chamberlin's from November, therefore, *kansensis* is the senior synonym should they be synonymized. This work is a prime example of further studies that need to be undertaken for the entire family of Lithobiidae. Studies like this in conjunction with molecular work should make clear what species are conspecific. Due to the vagueness of Crabill & Lorenzo's conclusion regarding the synonomy of *kansensis* and *pallidus* the author has chosen to leave them as separate species for the time being. It is clear from their conclusions on plectrotaxic variability that the validity of many of the species in the genus *Tidabius* need further study].

Distribution: USA: Kansas.

LECTOTYPE: UKNHM; No. 159; ♀; Kansas: Douglas Co., Lawrence, April 1912, under the bark of old logs in damp weather, by Horace Gunthorp.

ALLOLECTOTYPE: UKNHM; ♂, Kansas: Douglas Co., Lawrence, April 1912, under the bark of old logs in damp weather, by Horace Gunthorp.

COTYPES: UKNHM; 7 specimens in poor condition; Kansas: Douglas Co., Lawrence, April 1912, under the bark of old logs in damp weather, by Horace Gunthorp.

♀ *T. nasintus* Chamberlin, 1913
— Chamberlin, 1913: 82-84 (*Tidabius*; characters; USA: Mississippi).

Distribution: USA: Mississippi.

HOLOTYPE: MCZ, No. 536, ♀, Mississippi: Jackson, 7/17/'10, (R.V.C.).

♀ *T. opiphilus* Chamberlin, 1913
— Chamberlin, 1913: 99-100, pl. 5; Fig. 9 (*Tidabius*; characters; USA: Wisconsin).
— Crabill, 1952b: 349, 438 (*Tidabius*; characters; plectrotaxy; USA: Michigan).
— Crabill, 1958b: 99 (*Tidabius*; a specimen in poor condition but is possibly referable to this species; USA: Wisconsin).

— Summers, 1979: 696, 698 (*Tidabius*; checklist, key).
— Kevan, 1983: 2949 (*Tidabius*; checklist).
— Watermolen, 1997: 6 (*Tidabius*; checklist, notes).

Distribution: USA: Michigan, Wisconsin.

HOLOTYPE: MCZ, Wisconsin: Beloit.

♀*T. pallidus* Chamberlin, 1913
— Chamberlin, 1913: 92-94 (*Tidabius*; characters; USA: Mississippi).
— Crabill & Lorenzo, 1957: 428 (*Tidabius*; characters (plectrotaxy); discussion of *kansensis* being conspecific with *pallidus*). [**Note:** see note under *kansensis*].

Distribution: USA: Mississippi.

HOLOTYPE: MCZ, No. 535, ♀, Mississippi: Jackson, July 17, 1910.

[**Note:** *Tidabius pallidus* may be a junior synonym of *T. kansensis*].

♀*T. pallidus alabamensis* Chamberlin, 1913
— Chamberlin, 1913: 94-95 (*Tidabius*; characters; nominal taxon *pallidus*; cited as subspecies *alabamensis*; notes; USA: Alabama).

Distribution: USA: Alabama.

HOLOTYPE: MCZ, No. 534, ♀, Alabama: Dallas: Selma, July 25, 1910.

♀*T. plesius* Chamberlin, 1945
— Chamberlin, 1945b: 198 (*Tidabius*; characters; USA: Illinois).
— Chamberlin - Rapp, 1946: 667 (*Tidabius*; habitat; USA: Illinois).
— Crabill, 1952b: 346 (*Tidabius*; characters; USA: Illinois).
— Summers et al., 1980: 254 (*Tidabius*; checklist; notes; USA: Illinois).
— Summers et al., 1981: 62 (*Tidabius*; checklist; notes; USA: Illinois).

Distribution: USA: Illinois.

HOLOTYPE: NMNH; ♀, Illinois: Urbana, 25 February 1945.

♂♀ *T. poaphilus* Chamberlin, 1913
— Chamberlin, 1913: 88-90 (*Tidabius*; characters; USA: Nebraska).
— Crabill, 1952b: 346 (*Tidabius*; comparison of *plesius* to *poaphilus*).

Distribution: USA: Nebraska.

HOLOTYPE: NMNH, ♂, Nebraska: Fremont.

♀ *T. suitus* (Chamberlin, 1911)
— Chamberlin, 1911b: 41 (*Lithobius*; nominal taxon *cantabrigensis*; cited as variety *suitus*; characters; notes; USA: Alabama, North Carolina).
— Chamberlin, 1913: 84-88 (*Tidabius*; characters (adult, immaturus, praematurus); USA: Alabama, New Jersey, North Carolina).
— Brimley, 1938: 500 (*Lithobius*; nominal taxon *cantabrigensis*; cited as variety *suitus*; note; USA: North Carolina).
— Causey, 1940: 73 (*Tidabius*; characters; note).
— Crabill, 1952b: 347-348, 436 (*Tidabius*; characters; plectrotaxy; range; USA: Illinois, New Jersey, New York, Virginia).
— Wray, 1967: 155 (*Tidabius*; list).
— Summers, 1979: 696, 698 (*Tidabius*; checklist, key).
— Summers et al., 1980: 254; Fig. 31 (*Tidabius*; checklist; habitat (Table 1, p. 242-243); USA: Illinois).
— Summers et al., 1981: 62 (*Tidabius*; checklist; USA: Illinois).
— Hoffman, 1995a: 30 (*Tidabius*; list; note). [**Note:** Hoffman expects that this species exists in Virginia].

Distribution: USA: Alabama, Illinois, North Carolina, New Jersey, New York, Virginia.

SYNTYPES: MCZ, No. 541, 2 specimens, North Carolina: Hot Springs, 6-8-1910, R.V. Chamberlin Coll

PARATYPE: MCZ, No. 543, 1 specimen, Alabama: Jefferson, Birmingham, July 27, 1910.

♂♀ *T. tivius* (Chamberlin, 1909)
— Chamberlin, 1909b: 188-189 (*Lithobius*; characters; habitat; notes; USA: Utah, Washington).
— Chamberlin, 1911a: 69-70 (*Lithobius*; characters; notes; USA: Colorado).

— Chamberlin, 1911c: 101, 104 (*Lithobius*; key to species; notes; USA: Nebraska).
— Chamberlin, 1913: 96-99, pl. 4; Fig. 5, pl. 5 fig. 2-7 (*Tidabius*; characters (adult, agenitalis, praematurus, pseudomaturus); USA: Alabama, Colorado, Illinois, Kentucky, New Jersey, Pennsylvania, Massachusetts, Mississippi, Tennessee, Utah, Virginia, Wisconsin).
— Chamberlin, 1928d: 308 (*Tidabius*; USA: Utah).
— Williams & Hefner, 1928: 142 (*Lithobius*; characters; USA: Ohio).
— Causey, 1940: 73 (*Tidabius*; characters; USA: North Carolina).
— Chamberlin, 1943b: 104 (*Tidabius*; USA: Utah).
— Crabill, 1952b: 348, 437 (*Tidabius*; characters; plectrotaxy; USA: Illinois, Kentucky, Massachusetts, Michigan, New York, Ohio, Pennsylvania, Virginia, Wisconsin).
— Johnson, 1952: 190-193; Tab. 26; Map 18 (*Tidabius*; autecology; characters; distribution; natural history; USA: Michigan). [**Note:** Crabill mentions that this species has been taken in Nevada but I am unaware of this publication, however, have included it as a questioned locality].
— Dearolf, 1953: 231 (*Tidabius*; cited as diplopod; list; USA: Kentucky). [**Note:** Cave record].
— Crabill, 1955f: 260 (*Tidabius*; checklist; USA: Kentucky).
— Chamberlin, 1958c: 133 (*Tidabius*; checklist; note).
— Crabill, 1958b: 99 (*Tidabius*; a specimen in poor condition but is possibly referable to this species; USA: Wisconsin).
— Wray, 1967: 155 (*Tidabius*; list).
— Summers, 1979: 696, 698 (*Tidabius*; checklist, key).
— Summers et al., 1980: 254; Fig. 31 (*Tidabius*; checklist; habitat (Table 1, p. 242-243); USA: Illinois).
— Summers et al., 1981: 62 (*Tidabius*; checklist; USA: Illinois).
— Kevan, 1983: 2949 (*Tidabius*; checklist).
— Snider, 1991: 190 (*Tidabius*; list).
— Hoffman, 1995a: 30 (*Tidabius*; list; notes).
— Watermolen, 1997: 6 (*Tidabius*; checklist, notes; USA: Wisconsin).
— Jeekel, 2005: 32 (*Lithobius*; list).

Distribution: USA: Alabama, Colorado, Illinois, Kentucky, Massachusetts, Michigan, Mississippi, Nebraska, New Jersey, North Carolina, New York, ?Nevada, Ohio, Pennsylvania, Tennessee, Utah, Virginia, Washington, Wisconsin.

HOLOTYPE: MCZ, No. 14575, Utah: Provo, R.V. Chamberlin.

PARATYPE: MCZ, No. 516, Utah: Provo, 4-1909, R.V. Chamberlin.

♂*T. vector* Chamberlin, 1931
— Chamberlin, 1931a: 190 (*Tidabius*; characters; USA: Hawaii).
— Zapparoli & Shelley, 2000: 47 (*Tidabius*; list; notes).

Distribution: USA: ?Hawaii.

HOLOTYPE: NMNH, ♂, Honolulu, Hawaii, in packing about orchids
from Mexico, 17 July 1930.

♀*T. zionicus* Chamberlin, 1925
— Chamberlin, 1925e: 55 (*Tidabius*; characters; USA: Utah).

Distribution: USA: Utah.

HOLOTYPE: NMNH, Utah: Zion National Park, May 1924.

Genus *Tigobius* Chamberlin, 1916
[1 species; Western USA]

Tigobius Chamberlin, 1916: 198-200; Fig. 6 (characters; distribution; notes).
Type Species: *Tigobius paralus* Chamberlin, 1916, fixed by original
designation and monotypy.
— Attems, 1938a: 344 (synonym of *Lithobius* (*Pokabius*)).
— Jeekel, 2005: 32 (list).

♂♀*T. paralus* Chamberlin, 1916
— Chamberlin, 1916: 200-201, pl. 10; Fig. 1-4 (*Tigobius*; characters;
USA: California).
— Jeekel, 2005: 32 (*Tigobius*; list).

Distribution: USA: California.

HOLOTYPE: MCZ, No. 366, California: Pacific Grove. R.V. Chamberlin.

PARATYPES: MCZ, No. 367, California: Pacific Grove; MCZ, No. 368,
452, Santa Barbara; MCZ, No. 451, 453, Santa Cruz Island. R.V. Chamberlin.

Genus *Typhlobius* Chamberlin, 1922

[2 species; Eastern & Western USA]

Typhlobius Chamberlin, 1922b: 272-273; fig. 2 (characters; distribution; key to species; notes). TYPE SPECIES: *Typhlobius kebus* Chamberlin, 1922, fixed by original designation. [**Note:** This is an eyeless genus of Lithobiomorphs].
— Crabill, 1952b: 335 (note).
— Jeekel, 2005: 33 (list).

♀*T. coecus* (Bollman, 1888)
— Bollman, 1888a: 111 [= 1893: 84-85] (*Lithobius*; subgenus *Archilithobius*; characters; USA: Tennessee).
— Bollman, 1893: 128 (*Lithobius*; subgenus *Archilithobius*; catalogue).
— Chamberlin, 1911b: 36 (*Lithobius*; USA: North Carolina).
— Chamberlin, 1922b: 274-275, pl. 3; Fig. 7; pl. 4; Fig. 1,2 (*Typhlobius*; characters; note; USA: North Carolina, Tennessee).
— Brimley, 1938: 500 (*Lithobius*; USA: North Carolina).
— Causey, 1940: 72 (*Typhlobius*; note).
— Crabill, 1952b: 335 (*Typhlobius*; note).
— Wray, 1967: 155 (*Typhlobius*; list). [**Note:** Both generic and specific name misspelled *Typhobius ocecus*].
— Hoffman, 1995a: 30 (*Typhlobius*; list; notes; USA: Virginia).

Distribution: USA: North Carolina, Tennessee, Virginia.

HOLOTYPE: ISM, Tennessee: Beaver Creek.

♀*T. kebus* Chamberlin, 1922
— Chamberlin, 1922b: 275-276, pl. 3; Fig. 5,6 (*Typhlobius*; characters; USA: California).
— Crabill, 1952b: 335 (*Typhlobius*; note).
— Jeekel, 2005: 33 (*Typhlobius*; list).

Distribution: USA: California.

HOLOTYPE: MCZ No. 151, ♀, California: Santa Barbara.

Genus ***Watobius*** Chamberlin, 1911
[1 species; Southeastern USA]

Watobius Chamberlin, 1911b: 35 (characters). TYPE SPECIES: *Watobius anderisus* Chamberlin, 1911, fixed by original designation and monotypy.

— Chamberlin, 1914a: 109-110 (characters).
— Attems, 1938a: 342 (key).
— Crabill, 1952b: 335 (characters).
— Jeekel, 2005: 34 (list).

♂♀ *W. anderisus* Chamberlin, 1911
— Chamberlin, 1911b: 35-36, pl. 3; Fig. 4,5; pl. 4, fig 3 (*Watobius*; characters; USA: Alabama, Georgia).
— Chamberlin, 1914a: 110-112, pl. 1; Fig. 1-6, pl. 2; Fig. 1-6 (*Watobius*; characters; habitat).
— Crabill, 1952b: 335 (*Watobius*; cited as type species).
— Jeekel, 2005: 34 (*Watobius*; list).

Distribution: USA: Alabama, Georgia.

HOLOTYPE: MCZ No. 83, Alabama: Anniston, 28 July, 1910, R.V. Chamberlin.

PARATYPES: MCZ No. 84, Alabama: Thomasville, 20 July, 1910; MCZ 85, Georgia: Tallulah Falls, 1 August, 1910; MCZ No. 86, Georgia: Bremen, 29 July, 1910, R.V. Chamberlin.

Genus *Zinapolys* Chamberlin, 1912
[2 species; Northwestern USA]

Zinapolys Chamberlin, 1912e: 174 (characters). **TYPE SPECIES:** *Zinapolys zipius* Chamberlin, 1912, fixed by original designation and monotypy.
— Chamberlin, 1925d: 441-443; fig. 1 (characters; distribution; notes).
— Attems, 1938a: 344 (synonym of *Lithobius* (*Australobius*)).
— Eason, 1992: 3 (Eason questions Attems' synonymy).
— Jeekel, 2005: 34 (list).

♂♀ *Z. zipius* Chamberlin, 1912
— Chamberlin, 1912e:174-175 (*Zinapolys*; characters; USA: Idaho).
— Chamberlin, 1925d: 443-445, pl. 1; Fig. 1-3 (*Zinapolys*; characters; USA: Idaho).
— Kevan, 1983: 2949 (*Zinapolys*; checklist).
— Jeekel, 2005: 34 (*Zinapolys*; list).

Distribution: USA: Idaho.

HOLOTYPE: ?MCZ, Idaho: Kooteno Co.

(Subgenus *Pygmobius* Chamberlin & Wang, 1952)

Pygmobius Chamberlin & Wang, 1952: 59 (characters). TYPE SPECIES: *Zinapolys uticola* Chamberlin & Wang, 1952, fixed by monotypy.
— Jeekel, 2005: 30 (list).

♂*Z. (P.) uticola* Chamberlin & Wang, 1952
— Chamberlin & Wang, 1952a: 59-60 (*Zinapolys*; subgenus *Pygmobius*; characters; notes; USA: Utah).
— Jeekel, 2005: 30 (*Zinapolys*; list).

Distribution: USA: Utah.

HOLOTYPE: NMNH; ♂, Utah: Daniel's Canyon, 15 October 1939, by S. Mulaik.

Subfamily **Pseudolithobiinae** (Matic, 1973)

Genus *Pseudolithobius* (Stuxberg, 1875)
[2 species; Southwestern USA]

Pseudolithobius Chamberlin, 1875a: 69 (species included: *megaloporus*). TYPE SPECIES: *Lithobius megaloporus* Stuxberg, 1875, fixed by subsequent designation by Chamberlin, 1910: 369.
— Chamberlin, 1917: 227-228; fig. 2 (characters, distribution).
— Attems, 1938a: 342 (key).
— Crabill, 1950a: 8-9 (characters; notes).
— Jeekel, 2005: 29 (list).

♀*P. festinatus* Crabill, 1950
— Crabill, 1950a: 10-12 (*Pseudolithobius*; characters; notes).

Distribution: USA: Arizona.

HOLOTYPE: ?Unknown, Crabill's collection No. C-483; ♀, Arizona: Navajo Co., 16 mi. SW of Show Low, on U.S. Highway 60, 10 August 1948, by George E. Ball & Howard E. Evans, under a rock.

♂♀*P. megaloporus* (Stuxberg, 1875)

— Stuxberg, 1875a: 69-70 [= 1875d: 190-191] (*Lithobius*; characters; notes; USA: California).

— Stuxberg, 1875b: 14-15 (*Lithobius*; subgenus *Pseudolithobius*; list).

— Stuxberg, 1875c: 29 [= 1877: 137] (*Lithobius*; subgenus *Pseudolithobius*; list).

— Bollman, 1887b: 266 [= 1893: 33] (*Lithobius*; subgenus *Pseudolithobius*; list).

— Bollman, 1893: 129 (*Lithobius*; subgenus *Pseudolithobius*; catalogue).

— Chamberlin, 1911f: 377, 381-382 (*Pseudolithobius*; characters; behavior; habitat; notes; USA: California).

— Chamberlin, 1911d: 470 (*Pseudolithobius*; characters; notes).

— Chamberlin, 1917: 229-233, pl. 3; Figs. 4-5; pl. 4; Figs. 1-2 (*Pseudolithobius*; characters (adult, stadia); habitat; notes; USA: California).

— Chamberlin, 1943b: 104 (*Pseudolithobius*; USA: California).

— Crabill, 1950a: 9-10 (*Pseudolithobius*; characters).

— Mundel, 1981: 160-163, 214; Tab. 21 (*Pseudolithobius*; characters; notes; USA: California). [**Note:** Mundel notes that "Eason (1973) reports that a number of Stuxberg's holotypes are extant, but he does not mention *P. megaloporus*."].

— Jeekel, 2005: 29 (*Lithobius*; list).

Distribution: USA: California.

HOLOTYPE: ?Unknown, California: near San Francisco, G. Eisen. [**Note:** The MCZ has numerous specimens with the following catalog numbers: 584 587, 618-621, 782, 830, 832, 29956, 29957. Additional locality information is as follows: California: Yoho Co., Putah Creek Canyon, 28 April 1967, by D.S. Horina Jr. (see Mundel, 1981)].

Subclass **Pleurostigmophora** (Pocock, 1902)
Superorder **Epimorpha** Haase, 1880
Order **Scolopendromorpha** Pocock, 1895
Family **Cryptopidae** Kohlrausch, 1881
Subfamily **Cryptopinae** Kohlrausch, 1881

Genus *Cryptops* Leach, 1815
[12 species (10 Introduced); CANADA & USA]

Cryptops Leach, 1815: 384 (characters). TYPE SPECIES: *Scolopendra hortensis* Donovan, 1810, fixed by monotypy.
— Bollman, 1893: 169 (etymology).
— Bailey, 1928b: 35 (note).
— Attems, 1930: 201-202 (characters; key to subgenera; subgenera included: *Chromatanops, Cryptops, Trigonocryptops*).
— Crabill, 1960a: 12 (notes).
— Crabill, 1969b: 202 (key to "Floridan" species).
— Shelley, 1978: 221 (note).
— Shelley, 2002: 88 (key to species of North America including introduced species; species included: *anomalans, floridanus, hortensis, leucopodus, parisi*).
— Lewis, 2003: 61-67 (problems of characterization). [**Note:** With particular reference to *Cryptops* Lewis discusses the following topics: sexual dimorphism, changes during post larval development, repair after damage; wear; variable spine position, longitudinal keels on tergites; and he also discusses the following problematical characters: size, color].
— Jeekel, 2005: 52 (list).
Mycotheres Rafinesque, 1820: 8 (characters; notes). TYPE SPECIES: *Mycotheres leucopoda* Rafinesque, 1820, fixed by subsequent designation of Crabill, 1953 (in Hoffman & Crabill, 1953).
— Hoffman & Crabill, 1953: 77, 79 (synonym of *Cryptops*).
— Jeekel, 2005: 56 (list).
Mycotheres (*Exocera*) Rafinesque, 1820: 8 (characters; notes). TYPE SPECIES: *Mycotheres oligopoda* Rafinesque, 1820, fixed by original designation.
— Jeekel, 2005: 16 (list).

(Subgenus *Cryptops* Leach, 1815)

♂♀*C.* (*C.*) *anomalans* Newport, 1844

— Newport, 1844b: 100 (*Cryptops*; characters).

— Newport, 1845b: 409; Tab. 33; Figs. 25, 26 (*Cryptops*; characters).

— Gervais, 1847: 293 (*Cryptops*; characters; note).

— Newport, 1856: 60 (*Cryptops*; characters; no locality).

— Attems, 1930: 221-223; Figs. 6, 12, 21, 26, 35, 281-286 (*Cryptops*; subgenus *Cryptops*; characters; distribution).

— Eason, 1964: 150-153; Figs. 241-251 (*Cryptops*; characters; distribution; habitat).

— Matic, 1972: 191–194; Fig. 73 (*Cryptops*; characters, ecology, Romania).

— Kevan, 1979: 298 (*Cryptops*; Eastern CANADA).

— Kevan, 1983: 2944 (*Cryptops*; subgenus *Cryptops*; checklist; note; CANADA: "eastern Canada, ?S Quebéc and S Ontario in greenhouses, unpublished . . .").

— Behan-Pelletier, 1993: 25 (*Cryptops*; subgenus *Cryptops*; list).

— Shelley, 2002: 96 (*Cryptops*; characters; notes). [**Note:** An introduced species from Europe, but the validity of its existence in the New England states still needs verification].

Distribution: CANADA: Ontario, Quebéc; USA: Northeast (see Kevan, 1983; Shelley, 2002).

HOLOTYPE: Unknown, (v. in Mus. Brit.) (see Newport, 1845b).

♀*C.* (*C.*) *floridanus* Chamberlin, 1925

— Chamberlin, 1925b: 36 (*Cryptops*; characters; USA: Florida).

— Shelley & Edwards, 1987: 2, 3, 4; Map 3 (*Cryptops*; key, note; USA: Florida).

— Shelley, 2002: 92-93; Figs. 160, 162 (*Cryptops*; characters; distribution; habitat; notes; USA: Florida).

denmarki Chamberlin, 1958a: 13-14 (*Cryptops*; characters; note; USA: Florida). **SYNTYPES:** NMNH.

— Shelley & Edwards, 1987: 2, 3, 4; Map 3 (*Cryptops*; key, note; USA: Florida).

— Shelley, 2002: 92 (synonym of *floridanus*).

parydrus Crabill, 1969b: 202-204; Figs. 2-3 (*Cryptops*; characters; habitat; USA: Florida). [**Note:** Holotype deposited at the MCZ].

— Shelley & Edwards, 1987: 2, 3, 4; Fig. 10; Map 3 (*Cryptops*; key; note; USA: Florida).

— Shelley, 2002: 92 (synonym of *floridanus*).

Distribution: USA: Florida.

HOLOTYPE: MCZ, Florida: Indian River Co., Sebastian, 11 February 1919, by R.V. Chamberlin.

PARATYPE: NMNH, Florida, Alachua Co., Paradise, 21 February 1919, by R.V. Chamberlin.

♂♀ *C.* (*C.*) *hortensis* (Donovan, 1810)
— Donovan, 1810: 23 (*Scolopendra*;). [**Note:** Reference not obtained].
— Leach, 1815: 408 (*Cryptops*; characters; habitat).
— Newport, 1845b: 408-409; Tab. 33, Figs. 23, 24 (*Cryptops*; characters).
— Newport, 1856: 59 (*Cryptops*; characters; Europe).
— Attems, 1930: 204, 207-209; figs. 254-258 (*Cryptops*; subgenus *Cryptops*; characters; distribution; key).
— Crabill, 1952b: 195-197 (Cryptops; characters; habitat; notes; USA: New York, Utah).
— Crabill, 1955b: 39 (*Cryptops*; notes; USA: Missouri).
— Shelley, 1978: 221 (*Cryptops*; note).
— Kevan, 1983: 2944 (*Cryptops*; subgenus *Cryptops*; checklist; note).
— Holsinger & Culver, 1988: 57 (*Cryptops*; USA: Virginia). [**Note:** Reported as an accidental occurrence in a cave].
— Shelley, 2000b: 41-42 (*Cryptops*; characters; notes; USA: Hawaii).
— Shelley, 2002: 93-96; Fig. 161, 163 (*Cryptops*; characters; distribution; notes; list of published records; CANADA: British Columbia; USA: California, Connecticut, Colorado, District of Columbia, Illinois, Louisiana, New Jersey, New York, Massachusetts, Maryland, Michigan, Minnesota, Missouri, Mississippi, Nevada, Ohio, Oregon, Pennsylvania, Rhode Island, South Carolina, Texas, Utah, Washington, Wisconsin, West Virginia).
— Jeekel, 2005: 52 (*Scolopendra*; list).
— Mercurio, Unpublished (*Cryptops*; USA: Vermont). [**Note:** During the 2008 Montpelier BioBlitz I collected one specimen in litter next to a rotten log].
hyalinus Bailey (not Say), 1928: 19, 35 (*Cryptops*; characters; list; USA: New York).
— Shelley, 2002: 93 (synonym of *hortensis*).

diego Chamberlin, 1944d: 175-176 (*Cryptops*; characters; USA: California).
   HOLOTYPE: FMNH; California: San Diego County, San Diego,
   Balboa Park, 25 August 1940, M. Moran.
   — Shelley, 2002: 93 (*Cryptops*; synonym of *hortensis*).
nana Attems, 1938b: 366, 369, 374-376; figs. 7-9 (*Cryptops*; characters;
   USA: Hawaii).
   — Nishida, 1994: 26 (*Cryptops*; checklist).
   — Nishida, 1997: 195 (*Cryptops*; checklist).
   — Shelley, 2000b: 41 (*Cryptops*; synonym of *hortensis*).

Distribution: CANADA: British Columbia; USA: California, Connecticut,
Colorado, District of Columbia, Hawaii, Illinois, Kansas, Louisiana, New
Jersey, New York, Massachusetts, Maryland, Michigan, Minnesota, Missouri,
Mississippi, Nevada, Ohio, Oregon, Pennsylvania, Rhode Island, South
Carolina, Texas, Utah, Vermont, Virginia, Washington, Wisconsin, West
Virginia. [**Notes:** This is an introduced species from Europe, but based on
its distribution there is no doubt it is well established in North America].

HOLOTYPE: Unknown, based on specimens from England, Dartmoor,
"in gardens at Exeter", by W.E. Leach. (see Shelley, 2002: 93).

C. (C.) leucopodus (Rafinesque, 1820)
   — Rafinesque, 1820: 8 (*Mycotheres*; characters; USA: Kentucky).
   — Hoffman & Crabill, 1953: 77-78 (*Mycotheres*; taxonomic discussion).
   — Crabill, 1955f: 259 (*Mycotheres*; checklist; USA: Kentucky).
   — Hoffman, 1995a: 25 (*Cryptops*; list; notes).
   — Shelley, 2000c: 15, 16 (*Mycotheres*; neotype designation).
   — Shelley, 2002: 89-92; Figs. 159, 162 (*Cryptops*; characters; habitat;
      notes; list of published records; notes, USA: Alabama, Arkansas;
      Florida, Georgia, Illinois, Indiana, Kentucky, Louisiana, Maryland,
      Michigan, Mississippi, Missouri, New York, North Carolina, Ohio,
      Oklahoma, Pennsylvania, South Carolina, Tennessee, Texas, Virginia,
      West Virginia).
   — McAllister et al., 2003: 113-114 (*Cryptops*; notes; USA: Arkansas, Texas).
   — Jeekel, 2005: 56 (*Mycotheres*; list).
asperipes Wood, 1867b: 129-130 (*Cryptops*; synonym of *hyalinus*; characters;
   USA: Virginia).
   — McNeill, 1887a: 326 (*Cryptops*; note; USA: Florida). [**Note:**
      Misspelled *asperites*].
   — Underwood, 1887: 65 (*Cryptops*; synonym of *hyalinus*).
   — Bollman, 1888b: 6 [= 1893: 78] (*Cryptops*; synonym of *hyalinus*).

— Bollman, 1893: 127, 147, 169 (*Cryptops*; synonym of *hyalinus*).

— Kraepelin, 1903: 47 (*Cryptops*; synonym of *hyalinus*).

— Gunthorp, 1920: 113 (*Cryptops*; list).

— Shelley, 2002: 89 (*Cryptops*; synonym of *leucopodus*).

centralis Chamberlin, 1943b: 98 (*Cryptops*; characters; USA: Texas).

— Shelley, 2002: 89 (synonym of *leucopodus*).

eques Chamberlin & Mulaik, 1940: 107-108 (*Cryptops*; characters; notes; USA: Texas). **PARATYPE:** NMNH.

— Shelley, 2002: 89 (*Cryptops*; synonym of *leucopodus*).

hyalinus Say, 1821: 111-112 (*Cryptops*; characters; habitat; USA: Florida).

— Lucas, 1840: 546-547 (*Cryptops*; characters; USA: Florida). [**Note:** Lucas cites the species as *hyalinus* as do the majority of the following authors, but they are all referring to the original species by Say who originally called it *hyalina*].

— Newport, 1845b: 409 (*Cryptops*; characters; USA: Florida, Georgia).

— Newport, 1856: 60 (*Cryptops*; characters; USA: Florida, Georgia).

— Wood, 1862: 34 (*Cryptops*; characters; list).

— Wood, 1865: 168 (*Cryptops*; characters).

— Gervais, 1847: 293 (*Cryptops*; characters; note).

— Koch, 1847: 175 (*Cryptops*; list).

— Kohlrausch, 1881: 129 (*Cryptops*; characters).

— Underwood, 1887:65 (*Cryptops*; USA: Florida, Georgia, Indiana).

— Bollman, 1888a: 107, 111, 112 [= 1893: 82, 84, 85] (*Cryptops*; USA: Tennessee).

— Bollman, 1888b: 5-6 [= 1893: 77-78] (*Cryptops*; taxonomic notes; USA: Arkansas, Florida, Indiana, Maryland, North Carolina, Pennsylvania, Tennessee).

— Bollman, 1888c: 341 [= 1893: 92] (*Cryptops*; note; USA: Tennessee).

— Bollman, 1888d: 347 [= 1893: 99] (*Cryptops*; note; USA: Georgia or Tennessee, Virginia). [**Note:** One of the localities is cited as Lookout Mountain which the majority of resides in Georgia, but a small portion of the mountain runs into Tennessee].

— Bollman, 1888e: 409 [= 1893: 110] (*Cryptops*; note; USA: Indiana).

— Bollman, 1893: 127, 147, 169 (*Cryptops*; catalogue; notes; range).

— Kraepelin, 1903: 47-48; Fig. 8 (*Cryptops*; characters; USA: Florida).

— Brölemann, 1904: 243 (*Cryptops*; USA: Louisiana).

— Chamberlin, 1911d: 476 (*Cryptops*; characters; USA: California).

— Chamberlin, 1918b: 375 (*Cryptops*; USA: Georgia, Louisiana).
— Chamberlin, 1921a: 230 (*Cryptops*; USA: Tennessee).
— Bailey, 1928a: 1082 (*Cryptops*; list).
— Williams & Hefner, 1928: 137 (*Cryptops*; characters; USA: Ohio).
— Attems, 1930: 224; Fig. 287 (*Cryptops*; subgenus *Cryptops*; characters; distribution).
— Cornwell, 1934: 291 (*Cryptops*; habitat; nesting behavior; notes; USA: North Carolina).
— Brimley, 1938: 501 (*Cryptops*; USA: North Carolina).
— Causey, 1940: 66 (*Cryptops*; characters; habitat; notes; USA: North Carolina).
— Chamberlin, 1944a: 32 (*Cryptops*; USA: Georgia).
— Chamberlin, 1944d: 176 (*Cryptops*; USA: Arkansas, Tennessee).
— Chamberlin, 1945c: 199 (*Cryptops*; USA: Utah).
— Chamberlin, 1946b: 194 (*Cryptops*; USA: Mississippi).
— Crabill, 1950c: 201 (*Cryptops*; notes; USA: South Carolina). [**Note:** misidentification].
— Wray, 1950: 156 (*Cryptops*; list).
— Chamberlin, 1951b: 33 (*Cryptops*; USA: Georgia).
— Crabill, 1952b: 193-195, pl. 5; Figs. 65-66, 70 (characters; habitat; range; USA: Illinois, Indiana, Kentucky, Maryland, New Jersey, New York, Virginia, West Virginia). [**Note:** The true identity of the species that Crabill cited as *hyalinus* may be a combination of what is *hortensis* and *leucopodus*. However, the descriptive characters are those of *leucopodus*].
— Johnson, 1952: 126-132; Tab. 15, 28; Map 9 (*Cryptops*; autecology; characters; distribution; natural history; USA: Michigan).
— Dearolf, 1953: 231 (*Cryptops*; cited as diplopod; list; USA: West Virginia). [**Note:** Cave record].
— Hoffman & Crabill, 1953: 78 (*Cryptops*; synonym of *leucopodus*).
— Crabill, 1955f: 259 (*Cryptops*; synonym of *leucopodus*; checklist; USA: Kentucky). [**Note:** Specific name spelled *hyaline*].
— Eason, 1964: 154-156, pl. 4 left; Figs. 252-269 (*Cryptops*; characters; distribution; habitat).
— Branson & Batch, 1967: 83 (*Cryptops*; characters; note; USA: Kentucky).
— Wray, 1967: 156 (*Cryptops*; list).
— Crabill, 1969b: 202, 203; Fig. 1 (*Cryptops*; key).
— Shelley, 1978: 221 (*Cryptops*; habitat, range).
— Summers, 1979: 696, 699 (*Cryptops*; checklist, key).

— Summers et al., 1980: 245; Fig 2 (*Cryptops*; checklist; habitat (Table 1, p. 242-243); USA: Illinois).

— Summers et al., 1981: 59 (*Cryptops*; checklist; USA: Illinois).

— Kevan, 1983: 2944 (*Cryptops*; subgenus *Cryptops*; specific name spelled *hyalina*; checklist; cited as questioned junior synonym of *leucopoda*).

— Gardner, 1986: 30 (*Cryptops*;). [**Note:** Reference not obtained].

— Shelley, 1987: 504-505; Fig. 12 (*Cryptops*; characters; habitat; notes; USA: North Carolina).

— Shelley & Edwards, 1987: 2, 3, 4; Fig. 9; Map 3 (*Cryptops*; key, note; USA: Florida).

— Holsinger & Culver, 1988: 57 (*Cryptops*; USA: Virginia). [**Note:** Reported as an accidental occurrence in a cave].

— Snider, 1991: 188 (*Cryptops*; list).

— Shelley, 2002: 89 (*Cryptops*; synonym of *leucopodus*).

— Regier et al., 2005: 149; Tab. 1 (*Cryptops*; table of taxon classification with GenBank accession numbers).

milberti Gervais, 1847: 592–593 (*Cryptops*; characters; note; USA: New Jersey).

— Wood, 1862: 34-35 (*Cryptops*; characters; specific name spelled *milbertii*).

— Wood, 1865: 168 (*Cryptops*; characters; specific name spelled *milbertii*).

— Saussure & Humbert, 1872: 199 (*Cryptops*; list).

— Underwood, 1887: 65 (*Cryptops*; key; specific name spelled *milbertii*).

— Bollman, 1888b: 6 [= 1893: 78] (*Cryptops*; questioned synonym of *hyalinus*).

— Bollman, 1893: 127, 147, 169 (*Cryptops*; questioned synonym of *hyalinus*).

— Kraepelin, 1903: 47 (*Cryptops*; questioned synonym of *hyalinus*).

— Attems, 1930: 224 (*Cryptops*; synonym of *hyalinus*).

— Shelley, 2002: 89 (*Cryptops*; synonym of *leucopodus*).

sulcatus Meinert, 1886a: 211-212 (*Cryptops*; characters; USA: Kentucky).

— Underwood, 1887:65 (*Cryptops*; synonym of *hyalinus*).

— Bollman, 1888b: 6 [= 1893: 78] (*Cryptops*; synonym of *hyalinus*).

— Bollman, 1893: 127, 147, 169 (*Cryptops*; synonym of *hyalinus*).

— Kraepelin, 1903: 47 (*Cryptops*; synonym of *hyalinus*).

— Attems, 1930: 224 (*Cryptops*; synonym of *hyalinus*).

— Shelley, 2002: 89 (*Cryptops*; synonym of *leucopodus*).

Distribution: USA: Alabama, Arkansas, Florida, Georgia, Illinois, Indiana, Kentucky, Louisiana, Maryland, Michigan, Missouri, Mississippi, North Carolina, New Jersey, New York, Ohio, Oklahoma, Pennsylvania, South Carolina, Tennessee, Texas, Virginia, West Virginia.

HOLOTYPE: Unknown. (see Shelley, 2002: 89).

NEOTYPE: NCSM; Kentucky: Estill Co., South Irvine, ca. 1 mi. (1.6 km) S Irvine, 10 August 2000, by R.M. Shelley.

PARANEOTYPES: NCSM; 14 specimens, Kentucky: Estill Co., South Irvine, ca. 1 mi. (1.6 km) S Irvine, 10 August 2000, by R.M. Shelley.

C. (C.) melanotypus Chamberlin, 1941
— Chamberlin, 1941b: 42 (*Cryptops*; characters; USA: Hawaii). [**Note:** A species with potential to become established, however, the extent of establishment in Hawaii is not currently known].

Distribution: USA: ?Hawaii.

HOLOTYPE: NMNH, Hawaii: Honolulu, in packing material about *Den. superbum* from the Philippine Islands, 27 December 1937.

C. (C.) nautiphilus Chamberlin, 1939
— Chamberlin, 1939: 63 (*Cryptops*; characters; notes; USA: Louisiana). [**Note:** Chamberlin indicates that this species is similar to *C. detectus* Silvestri].

Distribution: USA: ?Louisiana.

HOLOTYPE: NMNH, From Mexico, one specimen taken on banana leaf at quarantine in New Orleans, 13 August 1936.

C. (C.) navis Chamberlin, 1930
— Chamberlin, 1930c: 65 (*Cryptops*; characters; USA: Hawaii). [**Note:** One specimen was found in soil that originated from the Straits Settlement in Singapore, Malaysia].
— Lewis, 2003: 64 (*Cryptops*; taxonomic note). [**Note:** According to Lewis *navis* is apparently a senior synonym of *sinesicus*].
sinesicus Chamberlin, 1940d: 49 (*Cryptops*; characters; note; USA: Hawaii). [**Note:** This species described by Chamberlin was taken by a quarantine inspector in soil about *Litchi chinensis* from China. It is probably not established but remains listed here as a

potential invader. Chamberlin also notes the similarity of this species to *C. navis*. **HOLOTYPE:** ?Unknown, Hawaii: Honolulu, 28 April 1938.].

— Lewis, 2003: 64 (*Cryptops*; taxonomic note; synonym of *navis*). [**Note:** According to Lewis *sinesicus* is apparently a junior synonym of *navis*].

Distribution: USA: ?Hawaii. (Origin: Malaysia and China).

HOLOTYPE: NMNH, Hawaii: Honolulu, 10 March 1930.

C. (*C.*) *parisi* Brölemann, 1920
— Brölemann, 1920: 9; Figs. 1-5 (*Cryptops*). [**Note:** reference not obtained].
— Brölemann, 1930: 214-217; Figs. 351-354 (*Cryptops*; characters).
— Attems, 1930: 209-211; Figs. 259-262 (*Cryptops*; subgenus *Cryptops*; characters; distribution).
— Palmén, 1954: 138 (*Cryptops*; notes; CANADA: Newfoundland).
— Eason, 1964: 158, 160-161; Figs. 270-276 (*Cryptops*; characters; distribution; habitat).
— Kevan, 1979: 298 (*Cryptops*; CANADA: Newfoundland).
— Kevan, 1983: 2944 (*Cryptops*; subgenus *Cryptops*; checklist; CANADA: "Newfoundland, greenhouses; . . .").
— Behan-Pelletier, 1993: 25 (*Cryptops*; subgenus *Cryptops*; list).
— Shelley, 2002: 96 (*Cryptops*; characters; published records; notes). [**Note:** Another species that may be introduced, but the status of its existence needs further investigation].

Distribution: CANADA: Newfoundland.

HOLOTYPE: Unknown, Europe.

C. (*C.*) *positus* Chamberlin, 1939
— Chamberlin, 1939: 64-65 (*Cryptops*; characters; note; USA: Louisiana).

Distribution: USA: ?Louisiana.

HOLOTYPE: NMNH, Honduras, one specimen taken at New Orleans, 1 November 1937.

PARATYPE: NMNH, Nicaragua, two specimens taken at quarantine at New Orleans, 21 November 1936.

C. (C.) vector Chamberlin, 1931
— Chamberlin, 1931a: 189-190 (*Cryptops*; characters; notes; USA: Louisiana).

Distribution: USA: ?Louisiana.

HOLOTYPE: NMNH, taken in New Orleans, Louisiana, on banana debris, 1 July 1930, originating from Honduras.

C. (C.) venezuelae Chamberlin, 1939
— Chamberlin, 1939: 63-64 (*Cryptops*; characters; USA: District of Columbia).

Distribution: USA: District of Columbia.

HOLOTYPE: NMNH, From Venezuela, taken at quarantine in Washington D. C., 22 July 1936.

C. (C.) watsingus Chamberlin, 1939
— Chamberlin, 1939: 64 (*Cryptops*; characters; USA: Louisiana).

Distribution: USA: ?Louisiana.

HOLOTYPE: NMNH, From Guatemala, taken at New Orleans in debris on bananas, 23 July 1936.

Genus *Paracryptops* Pocock, 1891
[2 species (Introduced); Eastern USA & Hawaii]

Paracryptops Pocock, 1891a: 227 (characters). TYPE SPECIES: *Paracryptops weberi* Pocock, 1891, fixed by monotypy.
— Kraepelin, 1903: 60 (characters; key to species).
— Attems, 1930: 244 (characters; distribution; key to species; species included: *breviunguis, indicus, inexpectus, weberi*).
— Jeekel, 2005: 57 (list).

P. inexpectus Chamberlin, 1914
— Chamberlin, 1914c: 158-159. (*Paracryptops*; characters; USA: District of Columbia). [**Note:** This single specimen was taken from soil of plants imported from British Guiana. It is highly

unlikely that this species could establish itself in the Washington, D.C. area].
— Attems, 1930: 246 (*Paracryptops*; characters; distribution).

Distribution: USA: ?District of Columbia.

HOLOTYPE: MCZ, British Guiana.

P. weberi Pocock, 1891
— Pocock, 1891a: 227 (*Paracryptops*; characters; note; Indonesia).
— Pocock, 1894b: 316; Tab. 19; Fig. 8 (*Paracryptops*;). [**Note:** Reference not obtained].
— Kraepelin, 1903: 60; Fig. 18 (*Paracryptops*; characters).
— Chamberlin, 1930c: 65 (*Paracryptops*; USA: Hawaii). [**Note:** One specimen was found in soil which originated from the Straits Settlement in Singapore, Malaysia].
— Attems, 1930: 244-245; Fig. 320 (*Paracryptops*; characters; key (in German); distribution).
— Jeekel, 2005: 31 (*Paracryptops*; list).

Distribution: USA: ?Hawaii.

HOLOTYPE: ?Unknown, Hawaii: Honolulu, 10 March 1930. (Origin: Malaysia: Singapore, Straits Settlement). [**Note:** This genus does not normally occur in Hawaii (Honolulu); however, a specimen has been collected in soil from Singapore, Straits Settlement. It is also known to exist in Java].

Subfamily **Plutoniuminae** (Bollman, 1893)

Genus *Theatops* Newport, 1844
[4 species; USA]

Theatops Newport, 1844: 193 Type Species: *Cryptops postica* (recte -*cus*) Say, 1821, fixed by subsequent monotypy of Newport, 1845.
— Meinert, 1886a: 207 (questioned synonym of *Opisthemega*).
— Pocock, 1888a: 285 (characters; notes).
— Bollman, 1893: 142-143, 169 (notes on synonymy; etymology; key to species; species included: *posticus, spinicaudus*).
— Chamberlin, 1902a: 40-41 (key to American species).
— Attems, 1930: 250-251 (characters; distribution; key to species; species included: *erythrocephala, postica, spinicauda*).
— Chamberlin, 1951a: 100 (key to species).

— Crabill, 1957a: 345 (notes).
— Crabill, 1960a: 11 (notes).
— Shelley, 1978: 221 (habitat, note).
— Shelley, 1997: 71-74; Fig. 10 (characters; distribution; notes; key to species; species included: *californiensis, erythrocephalus, phanus, posticus, spinicaudus*).
— Shelley, 2002: 81-82 (key to species; notes).
— Jeekel, 2005: 60 (list).

Opisthemega Wood, 1862: 35 (characters). TYPE SPECIES: *Opisthemega postica* Wood, 1862, fixed by subsequent designation.
— Wood, 1865: 169 (characters).
— Saussure & Humbert, 1872: 200 (catalog). [**Note:** Misspelled *Opisthomega*].
— Latzel, 1880: 145-147
— Kohlrausch, 1881: 130 (characters).
— Meinert, 1886a: 207-208 (characters).
— Haase, 1887: 78 [**Note:** Misspelled *Opisthomega*].
— Underwood, 1887: 64-65 (distribution; key to species).
— Daday, 1889: 92 [**Note:** Reference not obtained].
— Chamberlin, 1920f: 203 (note).
— Shelley, 1997: 71 (synonym of *Theatops*).
— Jeekel, 2005: 56 (list; notes; designation of type). [**Note:** Jeekel claims " . . . *Opisthemega* cannot be treated as a junior objective synonym of *Theatops* Newport, 1844."].

♂♀ *T. californiensis* Chamberlin, 1902
— Chamberlin, 1902a: 41 (*Theatops*; characters; key to species; USA: California).
— Kraepelin, 1903: 66 (*Theatops*; synonym of *erythrocephalus*).
— Chamberlin, 1911d: 472 (*Theatops*; nominal taxon *erythrocephalus*; characters; USA: California).
— Kevan, 1983: 2945 (*Theatops*; checklist).
— Shelley, 1990: 2637, 2638, 2641, 2642, 2643; Fig. 4, 5 (*Theatops*; distribution; taxonomic discussion).
— Shelley, 1997: 83-87; Figs. 13-18 (*Theatops*; characters; distribution; habitat; USA: California, Oregon).
— Shelley, 2002: 86, 88; Figs. 153, 157 (*Theatops*; characters; distribution; notes; USA: California).

erythrocephalus Kraepelin (not Koch), 1903: 66-67; Fig. 26 (*Theatops*; characters; USA: California, Oregon).
— Chamberlin, 1911d: 472 (*Theatops*; note).
— Attems, 1930: 251-252; Figs. 331-335 (*Theatops*; characters; distribution).

— Kevan, 1983: 2945 (*Theatops*; checklist; cited as questioned synonym of *californiensis*).
— Shelley, 1997: 83 (*Theatops*; synonym of *californiensis*).
— Shelley, 2002: 86 (*Theatops*; synonym of *californiensis*).

Distribution: USA: California, Oregon.

SYNTYPES: NMNH, 3 specimens, California: Plumas Co., near Quincy, mining claim, 3,500 ft., summer 1901, by E. Garner.

♂♀*T. phanus* Chamberlin, 1951
— Chamberlin, 1951a: 101 (*Theatops*; characters; habitat; USA: Texas).
— Reddell, 1965: 166 (*Theatops*; checklist).
— Reddell, 1994: ? (*Theatops*; notes).
— Shelley, 1997: 94-98; Figs. 33-39 (*Theatops*; characters; distribution; habitat; USA: Texas).
— Shelley, 2002: 85; Figs. 152, 156 (*Theatops*; characters; distribution).
spinicauda Reddell (not Wood), 1965: 166 (*Theatops*; checklist).
— Shelley, 1990: 2638 (lists Reddell's *spinicaudus* as a misidentification of *phanus*).

Distribution: USA: Texas.

HOLOTYPE: NMNH, Texas: Sutton Co., near Sonora, from an unnamed cave on Stevenson's Ranch, 16.iv.1926, beneath stone on bottom of first drop in vertical cave, by G.G. Stephenson.

♂♀*T. posticus* (Say, 1821)
— Say, 1821: 112-113 (*Cryptops*; characters; USA: Florida, Georgia).
— Lucas, 1840: 547 (*Cryptops*; characters; USA: Florida, Georgia).
— Newport, 1844b: 100 (*Cryptops*; list).
— Newport, 1845b: 410 (*Theatops*; characters; USA: "Georgiâ Floridâque Orientali").
— Gervais, 1847: 294 (*Cryptops*; characters; notes).
— Newport, 1856: 61 (*Theatops*; characters; USA: "Georgiâ Floridâque Orientali").
— Wood, 1862: 36-37 (*Theatops*; characters).
— Wood, 1862: 35 (*Opisthemega*; characters; USA: North Carolina).
— Wood, 1865: 171 (*Theatops*; characters).

— Wood, 1865: 169-170; Figs. 8-9 (*Opisthemega*; characters; notes; USA: North Carolina).
— Cope, 1869: 179 (*Opisthemega*; habitat; notes; USA: Virginia).
— Saussure & Humbert, 1872: 200 (*Theatops*; catalog).
— Saussure & Humbert, 1872: 200 (*Opisthemega*; catalog).
— Underwood, 1887: 64, 65 (*Opisthemega*; distribution, key).
— Kohlrausch, 1881: 130 (*Opisthemega*; characters).
— Pocock, 1888a: 285 (*Cryptops*; characters; notes).
— Pocock, 1888a: 285-286, 289-290 (*Theatops*; characters; notes).
— Bollman, 1888c: 341 [= 1893: 92] (*Theatops*; note; USA: Tennessee).
— Bollman, 1888d: 347 [= 1893: 99] (*Theatops*; USA: Georgia, Virginia).
— Bollman, 1888e: 409 [= 1893: 110] (*Theatops*; note; USA: Indiana).
— Bollman, 1893: 147-148 (*Cryptops*; taxonomic notes).
— Bollman, 1893: 128, 147, 148, 170 (*Theatops*; catalogue; etymology; taxonomic notes).
— Brölemann, 1896: 50 (*Theatops*; list).
— Morse, 1902: 187 (*Theatops*; USA: Ohio).
— Chamberlin, 1902a: 41 (*Theatops*; key to species; range).
— Kraepelin, 1903: 65-66; Fig. 25 (*Theatops*; characters; USA: "Carolina", Florida).
— Brölemann, 1904: 244 (*Theatops*; USA: Louisiana, North Carolina).
— Chamberlin, 1918a: 23 (*Theatops*; USA: Tennessee).
— Chamberlin, 1918b: 375 (*Theatops*; USA: Georgia, Louisiana).
— Gunthorp, 1920: 113 (*Opisthemega*; list).
— Chamberlin, 1925e: 57 (*Theatops*; note; USA: Utah).
— Williams & Hefner, 1928: 137 (*Theatops*; characters; USA: Ohio).
— Attems, 1930: 251 (*Theatops*; characters; distribution).
— Brimley, 1938: 501 (*Theatops*; in part; USA: North Carolina).
— Causey, 1940: 66 (*Theatops*; characters; habitat; USA: North Carolina).
— Chamberlin, 1942d: 184-185 (*Theatops*; USA: Louisiana, Texas).
— Chamberlin, 1943b: 97 (*Theatops*; USA: Texas).
— Chamberlin, 1944a: 33 (*Theatops*; USA: Georgia).
— Chamberlin, 1944d: 178 (*Theatops*; USA: Arizona).
— Chamberlin, 1945d: 215 (*Theatops*; USA: Georgia).
— Chamberlin, 1946b: 194 (*Theatops*; USA: Mississippi).
— Crabill, 1950c: 201 (*Theatops*; notes; USA: South Carolina).
— Wray (in part), 1950: 156, (*Theatops*; list).

— Auerbach, 1951a: 100, 117 (*Theatops*; heart beat; list; notes).

— Chamberlin, 1951b: 33 (*Theatops*; USA: Florida).

— Crabill, 1952b: 188-190 (*Theatops*; characters; range; USA: Indiana, Kentucky, Ohio, Virginia, West Virginia).

— Johnson, 1952: 132-134; Fig. 21 (*Theatops*; autecology; characters; distribution; USA: Michigan).

— Crabill, 1955f: 259 (*Theatops*; checklist; USA: Kentucky).

— Branson & Batch, 1967: 83-84 (*Theatops*; characters; habitat; body measurements; USA: Kentucky).

— Wray (in part), 1967: 156, (*Theatops*; list).

— Summers, 1979: 696, 699 (*Theatops*; checklist; key).

— Summers et al., 1980: 245; Fig. 5 (*Theatops*; checklist; habitat (Table 1, p. 242-243); USA: Illinois).

— Summers et al., 1981: 59 (*Theatops*; checklist; USA: Illinois).

— Kevan, 1983: 2945 (*Theatops*; checklist).

— Shelley, 1987: 505; Figs. 3, 13 (*Theatops*; characters; habitat; note; USA: North Carolina).

— Shelley & Edwards, 1987: 3, 4; Fig. 8; Map 2 (*Theatops*; key, USA: Florida).

— Holsinger & Culver, 1988: 57 (*Theatops*; USA: Virginia). [**Note:** Reported as an accidental occurrence in a cave].

— Shelley, 1990: 2637-2644; Tab. 1; Figs. 1, 3, 4, 6-11 (*Theatops*; distribution; ecology; intraspecific variation; taxonomic discussion; USA: Arizona, California, New Mexico, Nevada, Utah).

— Snider, 1991: 188 (*Theatops*; list; note).

— Hoffman, 1995a: 25 (*Theatops*; list; notes).

— Shelley, 1997: 74-83; Figs. 1, 6, 11-12, 18, 32 (*Theatops*; characters; distribution; habitat; USA: Alabama, Arizona, Arkansas, California, Connecticut, Florida, Georgia, Illinois, Indiana, Kentucky, Louisiana, Maryland, Mississippi, New Jersey, New Mexico, New York, North Carolina, Oklahoma, Pennsylvania, South Carolina, Tennessee, Texas, Virginia, West Virginia).

— Shelley, 2002: 83-84; Figs. 149-150, 154 (*Theatops*; characters; distribution; habitat; USA: Alabama, Arizona, Florida, Georgia, Louisiana, Mississippi, North Carolina, Oklahoma, South Carolina, Tennessee, Texas).

— McAllister et al. 2003: 113 (*Theatops*; notes; USA: Arkansas, Oklahoma, Texas).

— Edgecombe G. & G. Giribet, 2004: 98, 102; figs. 3d, 6e, 6f (*Theatops*; electronmicrographs of mandible, first maxilla).

— McAllister et al. 2004: 73-74 (*Theatops*; notes; USA: Arkansas, Louisiana, Oklahoma, Texas).

— Jeekel, 2005: 56 (*Opisthemega* + *Cryptops*; list; notes).

— Regier et al., 2005: 149; Tab. 1 (*Theatops*; table of taxon classification with GenBank accession numbers).

— Giribet & Edgecombe, 2006: 533, 534, 536; Fig 1,2; Tab. 1 (*Theatops*; cladograms; GenBank accession code).

crassipes Meinert, 1886a: 209 (*Opisthemega*; characters; USA: Florida, Kentucky, Virginia).

— McNeill, 1887a: 326 (*Opisthemega*; notes; USA: Florida).

— Underwood, 1887: 64, 65 (*Opisthemega*; distribution, key).

— Bollman, 1888a: 110-111 [= 1893: 84] (*Theatops*; note; USA: Tennessee).

— McNeill, 1888: 16 (*Opisthemega*; characters; distribution; note).

— Bollman, 1893: 127, 147, 170 (*Opisthemega* + *Theatops*; catalogue; synonym of *posticus*).

— Kraepelin, 1903: 65 (*Opisthemega*; synonym of *posticus*).

— Shelley, 1997: 74 (*Opisthemega* + *Theatops*; synonym of *posticus*).

prolongé Gervais (not Say), 1847: 294 (*Cryptops*; characters; notes). [**Note:** Gervais listed *Cryptops postica* to the right of this species in parentheses].

— Shelley, 1997: 74 (*Cryptops*; synonym of *posticus*).

Distribution: USA: Alabama, Arkansas, Arizona, California, Connecticut, ?District of Columbia, ?Delaware, Florida, Georgia, Illinois, Indiana, Kentucky, Louisiana, Maryland, Michigan, ?Missouri, Mississippi, North Carolina, New Jersey, New Mexico, Nevada, New York, Ohio, Oklahoma, Pennsylvania, South Carolina, Tennessee, Texas, Utah, Virginia, West Virginia. [**Note:** Questioned states are according to Shelley's (2002) projected distribution].

SYNTYPE: BMNH; Florida: St. Johns Co., possibly near Picolata, winter 1818, by Say.

♂♀*T. spinicaudus* (Wood, 1862)

— Wood, 1862: 36 (*Opisthemega*; characters; USA: Illinois). [**Note:** Wood described this species from eleven specimens originally deposited in the NMNH as follows: NMNH, No. 264, 10 specimens, Illinois: "South Illinois" by R. Kennicott; and NMNH, No. 347, 1 specimen, Illinois: Cook Co., by R. Kennicott].

— Wood, 1865: 170-171; Figs. 8-11 (*Opisthemega*; characters; USA: Illinois, Pennsylvania).

— Saussure & Humbert, 1872: 200 (*Opisthemega*; catalog).
— Kohlrausch, 1881: 130 (*Opisthemega*; characters).
— Meinert. 1886: 208-209 (*Opisthemega*; characters; note).
— Underwood, 1887: 64, 65 (*Opisthemega*; distribution, key).
— Bollman, 1888b: 6 [= 1893: 78] (*Theatops*; note; USA: Arkansas).
— Bollman, 1888c: 341 [= 1893: 92] (*Theatops*; note; USA: Tennessee).
— Bollman, 1893: 128, 170 (*Theatops*; catalogue, distribution; etymology).
— Pocock, 1895: 28 (*Theatops*; USA: Illinois).
— Brölemann, 1896: 50-51 (*Theatops*; list).
— Chamberlin, 1902a: 41 (*Theatops*; key to species; range)
— Kraepelin, 1903: 65 (*Theatops*; characters; USA: Illinois, North Carolina).
— Brölemann, 1904: 244 (*Theatops*; USA: North Carolina).
— Chamberlin, 1920c: 10, 235 (*Theatops*; note; USA: Hawaii). [**Note:** Chamberlin (1920c) reported this species from the Hawaiian Islands and it is most likely it was found there because of a shipment originating from the southeastern U.S. Whether or not it is established there is unknown at this time].
— Gunthorp, 1920: 113 (*Opisthemega*; list).
— Chamberlin, 1928b: 153 (*Theatops*; USA: Missouri).
— Attems, 1930: 253 (*Theatops*; characters; distribution).
— Attems, 1938b: 366, 369 (*Theatops*; list; note).
— Brimley (not Say), 1938: 50, (*Theatops*; in part; USA: North Carolina).
— Bücherl, 1941: 326 (*Theatops*;). [**Note:** reference not obtained].
— Chamberlin, 1942d: 185 (*Theatops*; USA: Arkansas).
— Chamberlin, 1944a: 33 (*Theatops*; USA: Georgia).
— Chamberlin, 1944d: 177-178 (*Theatops*; USA: Arkansas, Illinois, Missouri, Tennessee).
— Crabill, 1950c: 201 (*Theatops*; range; USA: South Carolina).
— Wray (in part), 1950: 156, (*Theatops*; list).
— Crabill, 1952b: 187-188, pl. 5; Figs. 62-64 (*Theatops*; characters; range; Illinois, Pennsylvania).
— Crabill, 1955b: 39 (*Theatops*; notes; USA: Missouri).
— Crabill, 1955h: 157 (*Theatops*; note; USA: Missouri).
— Wray (in part), 1967: 156, (*Theatops*; list).
— Summers, 1979: 693, 696, 699 Figs. 7, 8 (*Theatops*; checklist, key).
— Summers et al., 1980: 245; Fig. 5 (*Theatops*; checklist; habitat (Table 1, p. 242-243); USA: Illinois).

— Summers et al., 1981: 59 (*Theatops*; checklist; USA: Illinois).
— Kevan, 1983: 2945 (*Theatops*; checklist).
— Gardner, 1986: 30 (*Theatops*;). [**Note:** reference not obtained].
— Shelley, 1987: 505-506; Figs. 4, 13 (*Theatops*; characters; habitat; notes; USA: North Carolina).
— Shelley, 1990: 2637, 2638 (*Theatops*; notes).
— Shelley, 1991: 192-184 (*Theatops*; notes; USA: Hawaii). [**Note:** Shelley deletes this species from the Hawaiian fauna].
— Hoffman, 1995a: 25 (*Theatops*; list; notes; USA: Virginia).
— Shelley, 1997: 87-94; Figs. 7-9, 19-32. (*Theatops*; characters; distribution; ecology; notes; USA: Alabama, Arkansas, Georgia, Illinois, Iowa, Kansas, Missouri, North Carolina, Oklahoma, South Carolina, Tennessee, Virginia). [**Note:** Shelley reports "A vial at the ANSP, supposedly containing a paratype taken by R. Kennicott in southern Illinois, is empty, and the holotype is not known to exist"].
— Shelley, 2002: 84-85; Figs. 151, 155 (*Theatops*; characters; distribution; habitat; USA: Alabama, Arkansas; Georgia; Missouri, Oklahoma; South Carolina; Tennessee).
— McAllister et al., 2003: 113 (*Theatops*; notes; USA: Arkansas, Oklahoma).
insulare Meinert, 1886a: 209-210 (*Opisthemega*; characters; USA: Hawaii).
— Haase, 1887: 79 (*Opisthemega*;). [**Note:** reference not obtained].
— Kraepelin, 1903: 65 (*Theatops*; cited as questioned synonym of *spinicaudus*; note).
— Attems, 1914: 48, 103 (*Theatops*; list).
— Shelley, 2002: 84 (*Opisthemega*; synonym of *spinicaudus*).

Distribution: USA: Alabama, Arkansas, Georgia, ?Hawaii, Illinois, Iowa, Kansas, Missouri, North Carolina, Oklahoma, Pennsylvania, South Carolina, Tennessee, Virginia.

HOLOTYPE: Unknown. (see Shelley, 1997: 90).

NEOTYPE: NMNH, Illinois: Cook Co., Chicago, unknown date and collector.

Family **Scolopendridae** Leach, 1815
Subfamily **Otostigminae** Kraepelin, 1903

Genus *Ethmostigmus* Pocock, 1844

[1 species (Introduced); Western USA]

Ethmostigmus Pocock, 1898: 327 (new name for *Heterostoma* Newport, 1844, **preoccupied** by *Heterostoma* Hartmann, 1843). TYPE SPECIES: *Scolopendra trigonopoda* Leach, 1817, fixed by direct substitution.
— Crabill, 1960a: 8, 9 (*Ethmostigmus*; notes (see footnote), key).
— Jeekel, 2005: 54 (list; notes).

E. californicus Chamberlin 1958
— Chamberlin, 1958b: 185-186 (*Ethmostigmus*; notes; USA: California).
— Shelley, 2002: 5-6. (*Ethmostigmus*; notes). [**Note:** Shelley claims that this species is based on a specimen that was subject to a labeling error because this is an Old World genus, and has deemed it an extra-limital species].

Distribution: USA: California.

HOLOTYPE: NMNH, California: El Dorado Co., Snowline Camp, 25 June 1948.

Genus ***Otostigmus*** Porat, 1876
[3 species (Introduced); USA: Hawaii]

Otostigmus Porat, 1876: 18-19 (characters). TYPE SPECIES: *Otostigmus carinatus* Porat, 1876, [=*Otostigmus scaber* Porat, 1876] fixed by subsequent designation by Pocock, 1891.
— Pocock, 1891a: 229 (key, type fixed).
— Attems, 1930: 128-139 (characters; distribution; key to species).
— Crabill, 1960a: 10-11 (notes).
— Kevan, 1983: 2944 (notes).
— Jeekel, 2005: 57 (list).

(Subgenus ***Otostigmus*** Porat, 1876)

Otostigmus (*Otostigmus*) Porat, 1876: 18 (characters). TYPE SPECIES: *Otostigmus carinatus* Porat, 1876, fixed by subsequent designation.

O. (O.) mians Chamberlin, 1930
— Chamberlin, 1930c: 67 (*Otostigmus*; characters; USA: Hawaii). [**Note:** "The holotype taken at Hawaii in 1928 on *Dioscorea* sp. from China. On this specimen one anal leg is in the initial stage of

regeneration. A paratype lacking both anal legs was taken, also on *Dioscorea*, in a different lot of specimens during the same year"].

Distribution: USA: ?Hawaii.

HOLOTYPE: NMNH, Hawaii: (Origin: China)

O. (*O.*) *scaber* Porat, 1876
— Porat, 1876: 20 (*Otostigmus*; characters; China).
— Kraepelin, 1903: 111 (*Otostigmus*; characters; distribution).
— Attems, 1930: 153 (*Otostigmus*; subgenus *Otostigmus*; characters; distribution).
— Nishida, 1994: 27 (*Otostigmus*; checklist).
— Nishida, 1997: 195 (*Otostigmus*; checklist). [**Note:** This species has most likely been introduced from Taiwan].

Distribution: USA: ?Hawaii.

HOLOTYPE: ?Unknown; China.

O. (*O.*) *sinicolens* Chamberlin, 1930
— Chamberlin, 1930c: 66-67. (*Otostigmus*; characters; USA: Hawaii). [**Note:** "The holotype was taken at Hawaii on specimens of *Eleocharis tuberosa*, imported from China in 1928"].

Distribution: USA: ?Hawaii.

HOLOTYPE: NMNH, Hawaii: (Origin: China)

Genus *Rhysida* Wood, 1862
[1 species (Introduced); CANADA; USA]

Rhysida Wood, 1862: 40 TYPE SPECIES: *Branchiostoma lithobioides* Newport, 1845, fixed by subsequent designation of Attems, 1930.
— Bollman, 1893: 143 (note on synonymy).
— Kraepelin, 1903: 139-143 (characters; key to species).
— Attems, 1930: 183-186 (characters; distribution; key to species).
— Bücherl, 1941: 315 [**Note:** reference not obtained].
— Crabill, 1960a: 11 (notes).
— Shelley, 2002: 49 (notes). [**Notes:** According to Shelley, "Neither *Rhysida* nor any other genus in the Otostigminae is native to the continental United States, but specimens have been found on the ground in Miami and Key West, Florida . . ." The localities below

for *longipes* indicate interceptions at quarantine except for Florida].
— Shelley & Edwards, 2004: 116 (notes).
— Jeekel, 2005: 59 (list).
Branchiostoma Newport (not Costa, 1834), 1845b: 411 (characters).
— Meinert, 1886a: 182 (characters).
— Kraepelin, 1903: 139 (synonym of *Rhysida*).
— Crabill, 1957a: 343 (synonym of *Rhysida*). [**Note:** *Branchiostoma* Costa, 1834 was for the reception of a cephalochordate and in 1893 Bollman proposed that *Rhysida* Wood, 1862 be used instead].

♂♀*R. longipes* (Newport, 1845)
— Newport, 1845b: 411 (*Branchiostoma*; characters).
— Gervais, 1847: 249 (*Branchiostoma*; characters).
— Wood, 1862: 26 (*Scolopendra*; characters; CANADA; USA: Florida). [**Note:** Wood suspected the specimen from Canada as incorrectly labeled, however, it is plausible that it arrived by ship].
— Wood, 1865: 164 (*Scolopendra*; characters; USA: Florida).
— Underwood, 1887: 63. (*Branchiostoma*; note).
— Bollman, 1893: 127 (*Rhysida*; catalogue).
— Pocock, 1896: 27, pl. 2, Fig. 11 (*Rhysida*; note; ventral view of anal somite; Mexico).
— Kraepelin, 1903: 148; Fig. 91 (*Rhysida*; characters; distribution).
— Gunthorp, 1920: 113 (*Scolopendra*; list).
— Attems, 1930: 193-194; Figs. 16, 24 (*Rhysida*; characters; distribution).
— Chamberlin, 1958a: 14 (*Rhysida*; note; USA: Florida).
— Crabill, 1960a: 11 (*Rhysida*; notes).
— Shelley & Edwards, 1987: 3, 4; Map 1 (*Rhysida*; key, USA: Florida).
— Shelley, 2002: 49; Fig. 49 (*Rhysida*; characters; distribution; notes; published records).
— Shelley & Edwards, 2004: 116-118 (*Rhysida*; nominal taxon *longipes*; discussion; USA: Florida).
— Shelley, 2006b: 18 (*Scolopendra*; catalog; distribution; notes).
?*celeris* Humbert & Saussure, 1870a: 155-156 (*Branchiostoma*; nominal taxon *celeris*; characters; USA: "Carolina"). [**Note:** Reference not obtained. After Shelley and Edwards (2004) verified populations of *longipes* in Florida, one has to wonder if what has been reported as *celeris* are actual specimens of *longipes*. Therefore, for convenience, I have included *celeris* as a questioned synonym

of *longipes*. The type locality is vague. The distribution of this species
is said to be Neotropical, including nearby places such as the Antillies,
which doesn't eliminate the possibility that specimens may find their
way to the Carolinas via natural means such as rafting with ocean
currents. During hurricane season it wouldn't seem difficult at all for
debris with specimens to wash up along the southeastern U.S. shores,
therefore, I have included both North and South Carolina as questioned
localities where it may seem possible to find this species].

— Humbert & Saussure, 1870b: 202-203 (*Branchiostoma*; characters;
USA: "Carolina").
— Kohlraush, 1881: 69-70 (*Branchiostoma*; characters; notes).
— Meinert, 1886a: 183 (*Branchiostoma*; characters; Jamaica,
Nicaragua).
— Pocock, 1896: 28 (*Rhysida*; transferred from *Branchiostoma*; characters;
distribution; notes).
— Kraepelin, 1903: 149-150 (*Rhysida*; characters; distribution; USA:
Georgia).
— Attems, 1930: 188-189 (*Rhysida*; characters; distribution; USA:
Georgia).
— Crabill, 1960a: 11 (*Rhysida*; notes).
— Bücherl, 1974: 119 (*Rhysida*; nominal taxon *celeris*; distribution).
— Shelley, 2002: 5 (*Rhysida*; notes).

Distribution: CANADA: Nova Scotia; USA: ?California, ?District of
Columbia, Florida, ?Georgia, ?North Carolina, ?New York, ?South Carolina.
[**Note:** State lists in Appendix III have *R. longipes* listed for either *longipes*
or *celeris* records].

HOLOTYPE: Unknown; USA: "Carolina". (see Shelley, 2002: 49).

Subfamily **Scolopendrinae** Leach, 1815

Tribe **Scolopendrini** Leach, 1815

Genus *Arthrorhabdus* Pocock, 1891
[1 species; Southwestern USA]

Arthrorhabdus Pocock, 1891a: 221-222 (characters; notes). TYPE SPECIES:
Arthrorhabdus formosus Pocock, 1891, fixed by original designation;
of *Arthrorhabdinus*, *Scolopendra pygmaea* Pocock, 1895, by original
designation.

— Attems, 1930: 58-59 (characters; key to species; species included: *formosus, mjöbergi, pygmaeus, spinifer*).
— Crabill, 1960a: 10 (notes).
— Lewis, 1986: 1086 (taxonomic note). [**Note:** Lewis notes that the distinction between *Trachycormocephalus* and *Scolopendra* cannot be maintained where *Trachycormocephalus* becomes a junior synonym of *Scolopendra*. Lewis adds that for the same reasons of the synonymy above the distinction between *Scolopendra, Cormocephalus* and *Arthrorhabdus* may also not be clear].
— Shelley, 2002: 46; fig. 72 (characters; distribution; map; notes).
— Shelley & Chagas, 2004: 532-536 (taxonomic notes).
— Jeekel, 2005: 50 (list).

A. pygmaeus (Pocock, 1895)
— Pocock, 1895: 15, pl. 2; Figs. 8, 8a-c (*Scolopendra*; characters; notes; Mexico).
— Kraepelin, 1903: 222-223; Figs. 147-148 (*Arthrorhabdus*; characters; USA: Texas).
— Verhoeff, 1907: 262 (*Arthrorhabdus*; subgenus *Arthrorhabdinus*; note).
— Attems, 1930: 60 (*Arthrorhabdus*; characters; distribution).
— Bücherl, 1941: 297 (*Arthrorhabdus*;). [**Note:** reference not obtained].
— Crabill, 1960a: 10 (*Arthrorhabdus*; note; USA: New Mexico).
— Shelley, 2002: 46-48; Figs. 66-68, 72 (*Arthrorhabdus*; characters; distribution; habitat; notes; published records including sites from Mexico).
— Shelley & Chagas, 2004: 532-536; Fig. 1 (*Arthrorhabdus*; characters; USA: Arizona, Texas).
— Shelley, 2006b: 25 (*Scolopendra*; catalog; distribution; notes).

Distribution: USA: Arizona, New Mexico, Texas.

HOLOTYPE: Unknown.

SYNTYPES: BMNH, Mexico: Guerrero, ca. 5.9 mi (9.5 km) NW Chilapa, Amula, ~ 6,000-7,000 feet, 17°38' N, 99°15' W (see Shelley, 2002).

Genus *Hemiscolopendra* Kraepelin, 1903
[1 species; Southeastern USA]

Hemiscolopendra Kraepelin, 1903: 212-214 (characters; distribution; key
to species). TYPE SPECIES: *Scolopendra punctiventris* Newport, 1844 [=
Hemiscolopendra marginata (Say, 1821)], fixed by subsequent designation
of Attems, 1930.
— Verhoeff, 1907: 261 (characters; notes).
— Chamberlin, 1911: 477 (notes).
— Crabill, 1960a: 10 (notes).
— Kevan, 1983: 2944 (list).
— Hoffman & Shelley, 1996: 41 (generic position).
— Shelley, 2002: 40-41 (characters; distribution).
— Jeekel, 2005: 54 (list).
— Shelley, 2008: 174 (genus revision; characters; distribution; notes).

♂♀*H. marginata* (Say, 1821)
— Say, 1821: 110 (*Scolopendra*; characters; USA: Florida, Georgia,
"Southern States").
— Brandt, 1840: 158 [= 1841: 68] (*Scolopendra*; list).
— Lucas, 1840: 546 (*Scolopendra*; characters).
— Gervais, 1847: 276 (*Scolopendra*; characters).
— Bollman, 1893: 147 (*Scolopendra*; note).
— Hoffman & Shelley, 1996: 37-41; Figs. 1-3 (*Hemiscolopendra*; review
of nomenclatorial history and distribution).
— Shelley, 2002: 41-46; Figs. 65, 71 (*Hemiscolopendra*; characters;
distribution; habitat; list of published records; notes; USA: Alabama,
Arkansas, Florida, Georgia, Illinois, Indiana, Kentucky, Louisiana,
Mississippi, Missouri, North Carolina, Ohio, Oklahoma, South
Carolina, Tennessee, Texas, Virginia, West Virginia). [**Note:**
Although this species has previously been recorded from Connecticut,
Massachusetts, New York, Nebraska and Pennsylvania, they have
been considered mislabelings and deleted from these states by
Shelley].
— McAllister et al., 2003: 112 (*Hemiscolopendra*; notes; USA: Arkansas,
Texas).
— Shelley, 2006b: 7 (*Scolopendra*; catalog; distribution).
— Shelley, 2008: 175-184, 199, Figs. 1-58 (*Hemiscolopendra*; characters;
distribution; habitat; map; notes; USA: Arkansas, Florida).
?*coerulescens* Cragin, 1885: 144 (*Scolopendra*; nominal taxon *morsitans*; USA:
Kansas).
— Gunthorp, 1913: 168 (*Scolopendra*; nominal taxon *morsitans*;
notes).

— Gunthorp, 1921: 89-90 (*Scolopendra*; nominal taxon *morsitans*; notes).

— Shelley, 2006b: 21 (*Scolopendra*; catalog; distribution; notes).

— Shelley, 2008: 172, 199, Fig. 1 (*Scolopendra*; nominal taxon *morsitans*; notes). [**Note:** Although this is still technically a valid species it is most likely a junior synonym of *marginata* and has been treated here as a questioned synonym].

inaequidens Gervais, 1847: 277-278 (*Scolopendra*; characters; USA: New York). [**Note:** Type deposited in Paris Museum].

— Wood, 1862: 24 (*Scolopendra*; characters; table showing characters variation; USA: Illinois).

— Wood, 1865: 162-163 (*Scolopendra*; characters; table showing character variation; USA: Illinois).

— Underwood, 1887: 64 (*Scolopendra*; cited as synonym of *woodii*, note). [**Note:** Underwood mentions " . . . the type is supposed to be in Paris."].

— Bollman, 1893: 127, 174 (*Scolopendra*; synonym of *woodii*).

— Kraepelin, 1903: 217 (*Scolopendra*; synonym of *punctiventris*).

— Attems, 1930: 50 (*Scolopendra*; listed as an unrecognized species).

— Hoffman & Shelley, 1996: 37, 39 (*Scolopendra*; synonym of *marginata*; notes).

— Shelley, 2006b: 11–12 (*Scolopendra*; catalog; distribution; notes).

— Shelley, 2008: 175 (*Scolopendra*; as subjective synonym of *marginata*).

morsitans Gunthorp (not Linnaeus), 1913: 168 (*Scolopendra*; notes; USA: Kansas). [**Note:** This record of *morsitans* is founded on Cragin's original report of this species from Kansas but it is actually *marginata*].

— Gunthorp, 1921: 89-90 (*Scolopendra*; notes). [**Note:** Again this listing is referring to Cragin's report, which is *marginata*].

— Brimley, 1938: 501 (*Scolopendra*; USA: North Carolina). [**Note:** This record according to Shelley (2002) is actually *marginata*).

— Chamberlin & Mulaik, 1940: 108 (*Scolopendra*; USA: Texas). [**Note:** Chamberlin referred a young specimen with some doubt to *morsitans*, but there is a possibility that this record could have been a specimen of *marginata*, which has been confused with *morsitans* before].

— Wray, 1950: 155 (*Scolopendra*; list). [**Note:** According to Shelley (2002) this is *Hemiscolopendra marginata*).

— Wray, 1967: 155 (*Scolopendra*; list). [**Note:** According to Shelley (2002) this is *Hemiscolopendra marginata*).

— Shelley, 2008: 175 (*Scolopendra*; as subjective synonyms of *marginata*).

parva Wood, 1861: 10 (*Scolopendra*; characters; USA: Georgia "Mountains of Georgia"). **Syntypes:** ANSP (see Shelley, 2008).

— Gunthorp, 1920: 113 (*Scolopendra*; list).

— Shelley, 2008: 175 (*Scolopendra*; synonym of *marginata*).

punctiventris Newport, 1844b: 100 (*Scolopendra*; characters; USA: Florida). **Holotype:** BMNH (see Shelley, 2008).

— Newport, 1845b: 386-387 (*Scolopendra*; characters; USA: Florida).

— Gervais, 1847: 277 (*Scolopendra*; characters; note).

— Newport, 1856: 33-34 (*Scolopendra*; characters; USA: Florida.

— Underwood, 1887: 63, 64 (*Scolopendra*; cited as synonym of *viridis*).

— Pocock, 1895: 17, pl. 2, Figs. 6, 6a-c (*Scolopendra*; characters; notes; USA: "Eastern States", Florida).

— Kraepelin, 1903: 217; Fig. 145 (*Hemiscolopendra*; characters; USA: "Carolina", Connecticut, Massachusetts, New York).

— Brölemann, 1904: 253 (*Hemiscolopendra*; USA: Louisiana).

— Chamberlin, 1918a: 24 (*Hemiscolopendra*; USA: Tennessee).

— Chamberlin, 1918b: 375 (*Hemiscolopendra*; USA: Georgia, Louisiana).

— Chamberlin, 1920d: 43 (*Hemiscolopendra*; USA: South Carolina).

— Attems, 1930: 111-112; figs. 134-135 (*Cormocephalus*; subgenus *Hemiscolopendra*; characters; distribution).

— Bücherl, 1939: 245 (*Scolopendra*; synonym of *Scolopendra viridis*).

— Chamberlin, 1944d: 182-183 (*Hemiscolopendra*; USA: Florida).

— Crabill, 1950c: 201 (*Cormocephalus*; subgenus *Hemiscolopendra*; synonym of *marginata*; USA: South Carolina).

— Crabill, 1952b: 214-216, pl. 6; Figs. 75-76 (*Cormocephalus*; subgenus *Hemiscolopendra*; synonym of *marginata*; characters; range; taxonomic notes; USA: Arizona, Connecticut, Illinois, Indiana, Kentucky, Massachusetts, New York, Virginia, West Virginia). [**Note:** Interestingly, Crabill states that "Like many species, its range seems to stop farther north at the twenty-inch isohyet." In other words, the range of this species may be restricted to those latitudes that receive about 20 inches of rainfall a year and not less].

— Chamberlin, 1955b: 43 (*Hemiscolopendra*; cited as generotype of *Hemiscolopendra*).

— Crabill, 1955b: 39 (*Cormocephalus*; subgenus *Hemiscolopendra*; range; synonym of *marginata*; USA: Missouri).

— Crabill, 1955f: 259 (*Cormocephalus*; subgenus *Hemiscolopendra*; synonym of *marginata*; USA: Kentucky).

— Crabill, 1960a: 10 (*Hemiscolopendra*; distribution notes).

— Bücherl, 1974: 110, Fig. 24e (*Hemiscolopendra*; dorsal view of tergite 1).

— Shelley, 1978: 221 (*Hemiscolopendra*; habitat, range).

— Summers, 1979: 696, 699 (*Hemiscolopendra*; checklist, key).

— Summers et al., 1980: 245; Fig. 6 (*Hemiscolopendra*; checklist; habitat (Table 1, p. 242-243); USA: Illinois).

— Summers et al., 1981: 59 (*Hemiscolopendra*; checklist; USA: Illinois).

— Kevan, 1983: 2944 (*Hemiscolopendra*; checklist).

— Shelley, 1987: 501-503; Figs. 1, 11 (*Hemiscolopendra*; characters; habitat; notes; USA: North Carolina).

— Shelley & Edwards, 1987: 3, 4; Map 1 (*Hemiscolopendra*; key, USA: Florida).

— Hoffman, 1994: 33-34, Fig. 1 (*Hemiscolopendra*; distribution; human envenomation; habitat; USA: Virginia).

— Hoffman, 1995a: 24-25 (*Hemiscolopendra*; list; notes).

— Hoffman & Shelley, 1996: 37 (*Scolopendra*; synonym of *marginata*).

— Jeekel, 2005: 54 (*Scolopendra*; list).

— Shelley, 2006b: 10 (*Scolopendra*; catalog; notes).

— Shelley, 2008: 175 (*Hemiscolopendra* & *Cormocephalus* (*Hemiscolopendra*); synonym of *marginata*).

viridis Cornwell, 1934: 290 (*Scolopendra*; habitat; nesting behavior; notes; USA: North Carolina). [**Note:** Based on the distribution of *viridis* by Shelley (2002), Cornwell's data is most likely referring to *H. marginata*].

— Brimley, 1938: 501 (*Scolopendra*; USA: North Carolina).

— Causey, 1940: 65 (*Scolopendra*; characters; USA: North Carolina).

— Wray, 1950: 155 (*Scolopendra*; list).

— Wray, 1967: 155 (*Scolopendra*; list).

— Shelley, 2008: 175 (*Scolopendra*; as subjective synonyms of *marginata*).

woodii Meinert, 1886a: 198-199 (*Scolopendra*; characters; notes; USA: Massachusetts, North Carolina, South Carolina, Virginia). **Holotype:** MCZ (see Shelley, 2008).

— McNeill, 1887a: 326 (*Scolopendra*; note; USA: Florida).

— Underwood, 1887: 63, 64 (*Scolopendra*; distribution; key).

— McNeill, 1888: 17 (*Scolopendra*; characters; distribution; note).
— Bollman, 1888b: 7 [= 1893: 78] (*Scolopendra*; note; USA: Arkansas).
— Bollman, 1888c: 341 [= 1893: 92] (*Scolopendra*; note; USA: Tennessee).
— Bollman, 1888d: 347 [= 1893: 100] (*Scolopendra*; USA: Georgia).
— Bollman, 1888e: 409 [= 1893: 110] (*Scolopendra*; note; USA: Indiana).
— Bollman, 1893: 127, 174 (*Scolopendra*; catalogue; etymology; range).
— Kenyon, 1893a: 16 (*Scolopendra*; USA: Nebraska).
— Kenyon, 1893b: 161 (*Scolopendra*; USA: Nebraska).
— Brölemann, 1896: 49-50 (*Scolopendra*; list).
— Morse, 1902: 187 (*Scolopendra*; USA: Ohio).
— Kraepelin, 1903: 217 (*Scolopendra*; synonym of *punctiventris*).
— Williams & Hefner, 1928: 137 (*Scolopendra*; characters; USA: Ohio).
— Brimley, 1938: 501 (*Scolopendra*; USA: North Carolina).
— Causey, 1940: 65 (*Scolopendra*; characters; habitat; notes; USA: North Carolina).
— Wray, 1950: 155 (*Scolopendra*; list).
— Wray, 1967: 155 (*Scolopendra*; list).
— Kevan, 1983: 2944 (*Scolopendra*; cited as synonym of *punctiventris*).
— Hoffman & Shelley, 1996: 37 (*Scolopendra*; synonym of *marginata*).
— Shelley, 2006b: 22 (*Scolopendra*; catalog; distribution; notes).
— Shelley, 2008: 175 (*Scolopendra*; synonym of *marginata*).

Distribution: USA: Alabama, Arizona, Arkansas, ?Connecticut, Florida, Georgia, Illinois, Indiana, ?Kansas, Kentucky, Louisiana, ?Massachusetts, Missouri, Mississippi, North Carolina, ?New York, Ohio, Oklahoma, South Carolina, Tennessee, Texas, Virginia, West Virginia. [**Note:** Questioned states are removed from the distribution by Shelley, 2002].

HOLOTYPE: Unknown. [**Note:** According to Bollman (1893: 147) the holotype of this species was sent to Dr. Leach and was said to be deposited in the British Museum of Natural History. Hoffman & Shelley (1996: 37) explain that the actual type has not been found after a thorough search and have designated a neotype].

NEOTYPE: VMNH, Florida: St. Johns Co., Picoloata, 21 February 1994, by R.L. Hoffman (see Hoffman & Shelley, 1996).

PARANEOTYPE: VMNH, Florida: St. Johns Co., Picoloata, 21 February 1994, by R.L. Hoffman (see Hoffman & Shelley, 1996).

Genus *Scolopendra* Linnaeus, 1758
[8 species; USA]

Scolopendra Linnaeus, 1758: 637 (characters). Type Species: *Scolopendra morsitans* Linnaeus, fixed by subsequent designation by Crabill (1955a).
— Pocock, 1895: 13-14 (key to species; species included: *copeana, gigas, heros, morsitans, pachygnatha, pomacea, punctiventris, pygmaea, subspinipes, sumichrasti, tenuitarsis, viridis*).
— Chamberlin, 1911d: 477 (key to species of California).
— Crabill, 1955a: 134-136 (designation of *S. morsitans* as type species).
— Opinion 454, 1957: 359-378 (designation of *S. morsitans* as the type species).
— Bücherl, 1971: 181-182 (characters).
— Haupt, 1979: 398 (note on vitreous body of eyes lacking).
— Jeekel, 2005: 59 (list).
Trachycormocephalus Kraepelin, 1903: 218-219 (characters, key to species; species included: *afer, mirabilis*). Type Species: *Cormocephalus mirabilis* Porat, 1876, fixed by original designation.
— Lewis, 1986: 1083-1087 (*Trachycormocephalus*; synonym of *Scolopendra*).

♂♀*S. alternans* Leach, 1813
— Leach, 1813: 383 (*Scolopendra*; characters). [**Note:** See note under Shelley, 2002: 35 for this species).
— Leach, 1815: 383 (*Scolopendra*; characters).
— Newport, 1844b: 98 (*Scolopendra*; characters).
— Newport, 1845b: 402-403 (*Scolopendra*; characters; notes; "Insulis Caribaeis").
— Gervais, 1847: 286-287 (*Scolopendra*; characters; notes).
— Meinert, 1886a: 193-194 (*Scolopendra*; characters; distribution; notes).
— Bollman, 1888f: 337 [= 1893: 88] (*Scolopendra*; note; Cuba).
— Pocock, 1888b: 472-473 (*Scolopendra*; distribution; notes; Dominica).

— Kraepelin, 1903: 244-245 (*Scolopendra*; characters; distribution; USA: Florida).
— Attems 1930: 37-38 (*Scolopendra*; characters; distribution).
— Bücherl, 1941: 286-287 (*Scolopendra*;.
— Chamberlin, 1944d: 184 (*Scolopendra*; USA: Florida).
— Chamberlin, 1950: 135 (*Scolopendra*;.
— Chamberlin, 1952b: 5 (*Scolopendra*; [**Note:** It seems appropriate to note that this species is common in the West Indies]).
— Chamberlin, 1958a: 14 (*Scolopendra*; USA: Florida).
— Crabill, 1960b: 170 (*Scolopendra*; note on being reported from southern peninsular Florida and the adjoining keys).
— Bücherl, 1974: 104 (*Scolopendra*;.
— Kevan, 1983: 2944 (*Scolopendra*; checklist).
— Shelley & Edwards, 1987: 3, 4; Fig. 1; Map 1 (*Scolopendra*; key, USA: Florida).
— Behan-Pelletier, 1993: 25 (*Scolopendra*; list).
— Sandefer, 1998, unnumbered page (*Scolopendra*). [**Note:** Reference not obtained].
— Shelley, 2002: 35-38; Figs. 43-48 (*Scolopendra*; characters; food habits; habitat; notes; published records). [**Note:** Shelley states the following: "I have been unable to obtain a copy of volume 7 of the 1st. 1813 edition, of the Edinburgh Encyclopedia, in which Leach addressed chilopods, to read the original account and see whether he mentioned a locality, The Library of Congress in Washington, D.C., has volumes 6 and 8 but not volume 7. Attempts to locate this volume in the United Kingdom have also proven fruitless, so the page number cannot be quoted"].
— Shelley, 2006b: 6 (*Scolopendra*; catalog; distribution; notes).
complanata Newport, 1844b: 98 (*Scolopendra*; characters; West Indies).
— Newport, 1845b: 404 (*Scolopendra*; characters).
— Kohlrausch, 1881: 118–119 (*Scolopendra*; characters).
— Daday, 1891b: 184-185 (*Scolopendra*; characters; Venezuela).
— Kraepelin, 1903: 244 (*Scolopendra*; synonym of *alternans*).
— Shelley, 2002: 35 (*Scolopendra*; synonym of *alternans*).
— Shelley, 2006b: 10 (*Scolopendra*; catalog; distribution; notes).
crudelis (not Koch): Meinert, 1886a: 194-195 (*Scolopendra*; characters; USA: Florida).
— Underwood, 1887: 63, 64 (*Scolopendra*; distribution; key).
— Daday, 1891b: 185 (synonym of *complanata*).
— Bollman, 1893: 127, 175, 199 (*Scolopendra*; catalogue; etymology; note; USA: Florida; West Indies).
— Kraepelin, 1903: 244 (*Scolopendra*; synonym of *alternans*).
— Kevan, 1983: 2944 (*Scolopendra*; checklist; see footnote 7).

— Shelley, 2002: 35 (*Scolopendra*; synonym of *alternans*).
— Shelley, 2006b: 13 (*Scolopendra*; catalog; distribution; notes).
cubensis Saussure, 1859: 387 [= Saussure, 1860: 129-130]; Fig. 47 (*Scolopendra*; characters; Cuba).
— Saussure & Humbert, 1872: 132, pl. 2; Fig. 132 (*Scolopendra*; characters).
— Kraepelin, 1903: 244 (*Scolopendra*; questioned synonym of *alternans*).
— Attems, 1930: 50 (*Scolopendra*; cited as an unrecognized species). [**Note:** Attems comments that perhaps this is conspecific with *alternans*].
— Shelley, 2002: 35 (*Scolopendra*; cited as a questioned synonym of *alternans*).
— Shelley, 2006b: 15–16 (*Scolopendra*; catalog; distribution; notes; synonym of *alternans*).
grayi Newport, 1844b: 98 (*Scolopendra*; characters).
— Newport, 1845b: 403-404 (*Scolopendra*; characters).
— Gervais, 1847: 289 (*Scolopendra*; characters).
— Kraepelin, 1903: 244 (*Scolopendra*; synonym of *alternans*).
— Shelley, 2002: 35 (*Scolopendra*; synonym of *alternans*).
inaequidens Kevan [not Gervais], 1983: 2944 (*Scolopendra*; synonym of *alternans*; CANADA: Quebec). [**Note:** Kevan reports *inaequidens* as a synonym of *alternans* and that the Canadian records are "casual imports only"].
incerta Newport, 1845b: 404-405 (*Scolopendra*; characters).
— Gervais, 1847: 289 (*Scolopendra*; characters; note).
— Kraepelin, 1903: 244 (*Scolopendra*; synonym of *alternans*).
— Shelley, 2002: 35 (*Scolopendra*; synonym of *alternans*).
hirsutipes Bollman, 1893: 198-199 (*Scolopendra*; characters; West Indies).
— Kraepelin, 1903: 244 (*Scolopendra*; characters).
— Attems, 1930: 27 (*Scolopendra*; characters; distribution).
— Shelley, 2002: 35 (*Scolopendra*; synonym of *alternans*).
— Shelley, 2006b: 24 (*Scolopendra*; catalog; distribution; notes).
longipes Wood, 1862: 26 (*Scolopendra*; characters; USA: Florida).
— Wood, 1865: 163 (*Scolopendra*; synonym of *alternans*).
— Bollman, 1893: 127, 175 (*Scolopendra*; synonym of *crudelis*).
— Kraepelin, 1903: 244 (*Scolopendra*; synonym of *alternans*).
— Shelley, 2002: 35 (*Scolopendra*; synonym of *alternans*).
multispinata Newport, 1844b: 98 (*Scolopendra*; characters; West Indies).
— Newport, 1845b: 405 (*Scolopendra*; characters).
— Daday, 1891b: 185 (synonym of *complanata*).
— Kraepelin, 1903: 244 (*Scolopendra*; synonym of *alternans*).

— Shelley, 2002: 35 (*Scolopendra*; synonym of *alternans*).
— Shelley, 2006b: 10 (*Scolopendra*; catalog; distribution; notes).
sagraea Gervais, 1837: 50 (*Scolopendra*; characters). [**Note:** Reference not obtained].
— Brandt, 1840: 157–158 (*Scolopendra*; characters; notes).
— Gervais, 1847: 281-282 (*Scolopendra*; characters; notes).
— Kraepelin, 1903: 244 (*Scolopendra*; synonym of *alternans*).
— Shelley, 2002: 35 (*Scolopendra*; synonym of *alternans*).
— Shelley, 2006b: 8 (*Scolopendra*; catalog; distribution; notes).

Distribution: CANADA: ?Quebec (see Kevan, 1983); USA: Florida.

HOLOTYPE: Unknown.

NEOTYPE: NMNH, British Virgin Islands, Fat Hog's Bay, Tortola, 12 March 1984, collected by A. Penn.

♂♀*S. heros* Girard, 1853
— Girard, 1853: 272-274, pl. 18; Figs. 1-5 (*Scolopendra*; USA: Texas). [**Note:** Reference not obtained].
— Wood, 1862: 18-20 (*Scolopendra*; characters; notes; table showing character variation; USA: Arizona, Arkansas, Louisiana, New Mexico, Texas).
— Wood, 1865: 155-156 (*Scolopendra*; characters).
— Porat, 1876: 8-9 (*Scolopendra*; characters; USA: Texas).
— Cragin, 1885: 143-144 (*Scolopendra*; notes; USA: Kansas).
— Meinert, 1886a: 195-196 (*Scolopendra*; characters; notes; USA: Alabama, California, Florida, Georgia, Kansas, Kentucky, Nebraska, New York, Texas).
— Underwood, 1887: 63, 64 (*Scolopendra*; distribution; key; USA: Utah). [**Note:** Underwood deleted this species from New York and it does not occur naturally in New York. However, it would be interesting to see if specimens could be taken in the deleted states by Shelley (2002) at the closest proximity to the known distribution to verify if this species indeed has a wider range].
— Bollman, 1888b: 6 [= 1893: 78] (*Scolopendra*; notes; USA: Arkansas).
— Bollman, 1888d: 347 [= 1893: 100] (*Scolopendra*; USA: Colorado, Florida).
— Bollman, 1893: 127, 175 (*Scolopendra*; catalogue; etymology; distribution; note).
— Kenyon, 1893a: 17 (*Scolopendra*; note; USA: Nebraska).

— Pocock, 1895: 18-19; Figs. 12, 12a-c (*Scolopendra*; notes; USA: Texas).

— Kraepelin, 1903: 237-238; Figs. 152-153 (*Scolopendra*; characters; distribution; USA: Alabama, Arizona, Georgia, Kansas, Louisiana, Texas).

— Brölemann, 1904: 317 (*Scolopendra*; USA: Texas).

— Chamberlin, 1911d: 478 (*Scolopendra*; characters; range; USA: California).

— Gunthorp, 1913: 167-168 (*Scolopendra*; notes; USA: Kansas). [**Note:** Gunthorp mentions that only two specimens of *heros* " . . . remain of Professor Cragin's Myriapod collection, as the remainder were destroyed by fire"].

— Chamberlin, 1918b: 375 (*Scolopendra*; USA: Louisiana).

— Gunthorp, 1921: 89 (*Scolopendra*; notes).

— Attems, 1930: 44-45; Fig. 60 (*Scolopendra*; characters; distribution).

— Chamberlin, 1931b: 97 (*Scolopendra*; USA: Oklahoma).

— Back, 1939: 2, Fig. 3 (*Scolopendra*; notes; USA: Texas).

— Chamberlin & Mulaik, 1940: 108 (*Scolopendra*; USA: Texas).

— Bücherl, 1941: 295 (*Scolopendra*;).[**Note:** Reference not obtained].

— Chamberlin, 1942d: 185 (*Scolopendra*; USA: Texas).

— Chamberlin, 1944d: 185 (*Scolopendra*; USA: Arkansas, Texas).

— Crabill, 1955f: 259 (*Scolopendra*; checklist; USA: Kentucky).

— Cloudsley-Thompson & Crawford, 1970b: 187 (*Scolopendra*; note).

— Easterla, 1975: 411 (*Scolopendra*; report of *heros* eating a long-nosed snake).

— Neck, 1985: 253-255 (*Scolopendra*; comparative behavior and color pattern of two sympatric species; USA: Texas).

— Sandefer, 1998, unnumbered page (*Scolopendra*). [**Note:** Reference not obtained].

— Shelley, 2002: 28-35; Figs. 1a, b, 27-42; Tab. 3 (*Scolopendra*; characters; distribution; habitat; notes; published records including Mexico). [**Note:** Shelley deleted Alabama, Colorado, Florida, Georgia, Kentucky, Nebraska, and Utah as states in which this species is found, hence their questionable presence in the distribution].

— McAllister et al., 2003: 112 (*Scolopendra*; note; USA: Oklahoma).

— Shelley, 2006b: 14 (*Scolopendra*; catalog; distribution; notes).

— Guarisco et al. 2007: 274-275; Fig. 1, 2 (*Scolopendra*; habitus photo, distribution map, notes; USA: Colorado).

arizonensis Kraepelin, 1903: 238 (*Scolopendra*; nominal taxon *heros*; cited as
   variety *arizonensis*).
   — Attems, 1930: 45 (*Scolopendra*; nominal taxon *heros*; cited as variety
      arizonensis; characters; distribution).
   — Sandefer, 1998: unnumbered page (*Scolopendra*; nominal taxon *heros*;
      cited as subspecies *arizonensis*). [**Note:** Reference not obtained].
   — Shelley, 2002: 28 (*Scolopendra*; nominal taxon *heros*; synonym of
      heros).
   — Shelley, 2006b: 25–26 (*Scolopendra*; catalog; distribution; notes).
castaneiceps Wood, 1861: 11 (*Scolopendra*; characters; USA: Texas).
   — Wood, 1865: 156-157 (*Scolopendra*; nominal taxon *heros*; cited as
      variety *castaneiceps*; characters; USA: Alabama, Georgia, Louisiana,
      New Mexico, Texas).
   — Cragin, 1885: 144 (*Scolopendra*; nominal taxon *heros*; cited as variety
      castaneiceps; notes; USA: Kansas).
   — Bollman, 1893: 127, 175 (*Scolopendra* + nominal taxon *heros* (cited
      as subspecies); synonym of *heros*).
   — Kraepelin, 1903: 237 (*Scolopendra*; synonym of *heros*).
   — Gunthorp, 1920: 113 (*Scolopendra*; list).
   — Attems, 1930: 45 (*Scolopendra*; nominal taxon *heros*; cited as variety
      castaneiceps; characters; distribution).
   — Bücherl, 1941: 295 (*Scolopendra*; nominal taxon *heros*; cited as variety
      castaneiceps). [**Note:** reference not obtained].
   — Sandefer, 1998: unnumbered page (*Scolopendra*; nominal taxon *heros*).
      [**Note:** Reference not obtained].
   — Shelley, 2002: 28 (*Scolopendra*; nominal taxon *heros*; synonym of
      heros).
   — Shelley, 2006b: 16 (*Scolopendra*; catalog; distribution; notes).
heros Attems, 1930: 45 (*Scolopendra*; nominal taxon *heros*; cited as variety
   heros; characters; USA: Southern United States; Mexico).
   — Bücherl, 1941: 295 (*Scolopendra*; nominal taxon *heros*; cited as variety
      heros). [**Note:** Reference not obtained].
   — Sandefer, 1998: unnumbered page (*Scolopendra*; nominal taxon *heros*).
      [**Note:** Reference not obtained].
   — Shelley, 2002: 28 (*Scolopendra*; nominal taxon *heros*; synonym of *heros*).
pernix Kohlrausch, 1881: 115-116 (*Scolopendra*; characters; USA: "America
   boreali").
   — Bollman, 1893: 127 (*Scolopendra*; catalogue).
   — Kraepelin, 1903: 237 (*Scolopendra*; synonym of *heros*).
   — Shelley, 2002: 28 (*Scolopendra*; synonym of *heros*).
   — Shelley, 2006b: 20 (*Scolopendra*; catalog; distribution; notes).

polymorpha—Bollman [not Wood], 1893: 127 (*Scolopendra*; synonym of *heros*).
prismatica Cragin, 1885: 144 (*Scolopendra*; nominal taxon *heros*; cited as variety *prismatica*; characters; USA: Kansas).
— Bollman, 1893: 175 (*Scolopendra*; nominal taxon *heros*; cited as variety *prismatica*; synonym of *heros*). [**Note:** Subspecies name misspelled *pusinatica*].
— Shelley, 2002: 28 (*Scolopendra*; nominal taxon *heros*; cited as variety *prismatica*; synonym of *heros*).
— Shelley, 2006b: 21 (*Scolopendra*; catalog; distribution; notes).

Distribution: USA: ?Alabama, Arkansas, Arizona, ?California,? Colorado,?Florida, ?Georgia, Kansas, ?Kentucky, Louisiana, ?Nebraska, New Mexico, New York, Oklahoma, Texas, ?Utah.

HOLOTYPE: Lost (see Shelley, 2002), Texas.

NEOTYPE: NMNH, USA: Oklahoma: Comanche County, Wichita Mountains Wildlife Refuge, ca. 22 mi. NW Lawton, 19 June 1928.

S. mima Chamberlin, 1942
— Chamberlin, 1942c: 18-19. (*Scolopendra*; characters; USA: New Jersey).
— Shelley, 2002: 5 (*Scolopendra*; note).
— Shelley, 2006b: 29 (*Scolopendra*; catalog; notes).

Distribution: USA: ?New Jersey. [**Note:** It is doubtful that this species could establish itself in the northeastern U.S. due to its tropical origin].

HOLOTYPE: NMNH, New Jersey: Hoboken, at quarantine, 31 May 1941, with Cattleya plants from Venezuela. [**Note:** Holotype is missing anal legs].

♂♀*S. morsitans* Linnaeus, 1758
— Linnaeus, 1758: 638 (*Scolopendra*; characters; distribution).
— Newport, 1845b: 378-379 (*Scolopendra*; characters; "Insulis Caribaeis").
— Wood, 1865: 161-162 (*Scolopendra*; characters; USA: Florida).
— Kohlrausch, 1881: 104 (*Scolopendra*; characters).
— Cragin, 1885: 144 (*Scolopendra*; nominal taxon *morsitans*; new variety *coerulescens*; characters; USA: Kansas). [**Note:** This new variety reported by Cragin is most likely a misidentification of *H. marginata*].
— Underwood, 1887: 64 (*Scolopendra*; distribution; USA: Florida).

— Bollman, 1893: 127 (*Scolopendra*; catalogue).
— Kraepelin, 1903: 250-253 (*Scolopendra*; characters; notes).
— Carlson, 1904: 270 (*Scolopendra*; ventral nerve cord physiology).
— Chamberlin, 1911d: 479 (*Scolopendra*; characters; range; USA: California, Georgia, Florida, Kansas, Utah).
— Attems, 1930: 23-25; figs. 29, 38-39 (*Scolopendra*; characters; distribution).
— Chamberlin & Mulaik, 1940: 108 (*Scolopendra*; USA: Texas). [**Note:** Chamberlin referred a young specimen with some doubt and Shelley (2002) has removed this record from the US fauna but has been included for future reference].
— Crabill, 1955a: 134, 135, 136 (*Scolopendra*; designation as type species).
— Bücherl, 1971: 179; fig. 9 (*Scolopendra*; note).
— Kevan, 1983: 2944 (*Scolopendra*; checklist; notes; USA: New York).
— Hoffman & Shelley, 1996: 35-38, (*Scolopendra*; discussion).
— Shelley, 2002: 39-40; Figs. 57-64 (*Scolopendra*; characters; distribution; notes; published records; USA: Pennsylvania). [**Note:** Although Shelley (2002) deleted all U. S. states for the distribution of this species based on apparent misidentifications, it has been included here because it has been intercepted at quarantine in Pennsylvania: Bucks Co., Philadelphia, in packing from the West Indies. The West Indies is also a place where *morsitans* has been introduced as well as Central American countries making it more likely that this species may become established in the U.S. Warmer climates are the more likely place for this species to establish itself (see also Hoffman & Shelley, 1996)].
— Jeekel, 2005: 57 (*Scolopendra*; list; notes).
— Shelley, Edwards & Chagas, 2005: 39-58; Fig. 1 (*Scolopendra*; notes; review of global occurrences; USA: Florida). [**Note:** Shelley et al., have reported that this species is now apparently established in Florida].
— Shelley, 2006b: 5 (*Scolopendra*; catalog; distribution).

Distribution: USA: ?California, Florida, ?Georgia, ?Kansas, ?New York, ?Pennsylvania, ?Texas, ?Utah.

HOLOTYPE: LSUK; locality, date and collector unknown (see Shelley, 2002; Shelley, 2006b).

S. pachygnatha Pocock, 1895

— Pocock, 1895: 23-24; Pl. 2, Figs. 3, 3a-b (*Scolopendra*; characters; Mexico).

— Kraepelin, 1903: 244 (*Scolopendra*; characters; Mexico).

— Attems, 1930: 47 (*Scolopendra*; characters; distribution).

— Chamberlin, 1951b: 33 (*Scolopendra*; USA: Florida). [**Note:** Chamberlin (1951b) reported two specimens of this species from the Archibold Biological Station, Lake Placid, Florida, which were collected 11 March 1950. It is not clear if this species has been misidentified by Chamberlin or if it is a valid species. Regardless, the original specimens from Pocock should be examined, if possible, and compared to those specimens from Florida in Chamberlin's report. In Selander & Vaurie's (1962) gazetteer they give the locality of Zacatecas as México, state of Zacatecas, city of Zacatecas, Capital of the state; 7377 feet; 22°47', 102°35'].

— Shelley, 2006b: 25 (*Scolopendra*; catalog; distribution; notes).

Distribution: USA: ?Florida.

HOLOTYPE: ?BMNH, México: Zacatecas, Mezquital del Oro (Buller coll.).

♂♀*S. polymorpha* Wood, 1861
— Wood, 1861: 11 (*Scolopendra*; characters; USA: Kansas).

— Wood, 1862: 20-21 (*Scolopendra*; characters; table showing character variation; USA: Arizona, Kansas, Texas).

— Wood, 1865: 158-159 (*Scolopendra*; characters; table showing character variation; USA: Kansas, Texas).

— Kohlrausch, 1881: 114-115 (*Scolopendra*; characters).

— Cragin, 1885: 144 (*Scolopendra*; habitat; notes; USA: Kansas).

— Kraepelin, 1903: 241-242; Fig. 155 (*Scolopendra*; characters; distribution; USA: Arizona, California, Kansas, Texas).

— Brölemann, 1904: 318 (*Scolopendra*; USA: Texas).

— Gunthorp, 1913: 168 (*Scolopendra*; note; USA: Kansas).

— Chamberlin, 1911d: 478; Fig. 156 c, d (*Scolopendra*; characters; range; USA: California).

— Chamberlin, 1914c: 193-194 (*Scolopendra*; note). [**Note:** a report from Brazil but may be a misidentification].

— Anonymous, 1920: 35 (*Scolopendra*; note; USA: California).

— Gunthorp, 1920: 113 (*Scolopendra*; list).

— Gunthorp, 1921: 90 (*Scolopendra*; notes).

— Chamberlin, 1923b: 390 (*Scolopendra*; USA: Arizona).

— Chamberlin, 1925e: 58 (*Scolopendra*; notes; USA: Utah).

— Chamberlin, 1928d: 307 (*Scolopendra*; note; USA: Utah).
— Chamberlin, 1930a: 116-117 (*Scolopendra*; USA: Utah). [**Note:**
According to Shelley (2002: 20) Chamberlin's report is only in part
a record of *polymorpha* while some specimens are *viridis*].
— Attems, 1930: 49 (*Scolopendra*; nominal taxon *viridis*; characters;
distribution).
— Chamberlin, 1931b: 97 (*Scolopendra*; USA: Oklahoma).
— Bücherl, 1939: 245 (*Scolopendra*; nominal taxon *viridis*; Brazil).
— Chamberlin & Mulaik, 1940: 108 (*Scolopendra*; USA: Arizona).
— Bücherl, 1941: 289 (*Scolopendra*; nominal taxon *viridis*;). [**Note:**
Reference not obtained].
— Chamberlin, 1943b: 98 (*Scolopendra*; USA: New Mexico, Texas).
— Chamberlin, 1944d: 183 (*Scolopendra*; USA: Arizona, California,
Kansas).
— Chamberlin, 1951b: 33 (*Scolopendra*; USA: Utah).
— Chamberlin, 1958c: 132 (*Scolopendra*; checklist; note).
— Cloudsley-Thompson & Crawford, 1970a: 26; tab. 1 (*Scolopendra*;
lethal temperatures; notes; supercooling; USA: New Mexico).
— Cloudsley-Thompson & Crawford, 1970b: 187-193; fig. 1; tab. 1-3
(*Scolopendra*; diurnal rhythms; notes; water and temperature relations;
USA: New Mexico).
— Crawford & Riddle, 1974: 86, 87, 88, 91; Tab. 1 & 2 (*Scolopendra*;
cold hardiness; USA: New Mexico).
— Shelley, 2000b: 44 (*Scolopendra*; characters; USA: Hawaii).
— Shelley, 2002: 20-28; Figs. 12-26; Tab. 2 (*Scolopendra*; characters;
distribution; habitat; list of published records; notes; USA: Arizona,
California, Colorado, Idaho, Kansas, Louisiana, Montana, Nebraska,
Nevada, New Mexico, Oklahoma, Oregon, South Dakota, Texas,
Utah, Washington, Wyoming).
— McAllister et al., 2003: 112 (*Scolopendra*; note; USA:
Oklahoma).
— Regier et al., 2005: 149; Tab. 1 (*Scolopendra*; table of taxon classification
with GenBank accession numbers).
— Shelley, 2006b: 16 (*Scolopendra*; catalog; distribution; notes).
californica Humbert & Saussure, 1870b: 203-204 (*Scolopendra*; characters;
USA: California).
— Humbert & Saussure, 1872: 127-128, pl. 5; Figs. 6d, 61 (*Scolopendra*;).
[**Note:** Reference not obtained].
— Shelley, 2002: 20 (*Scolopendra*; synonym of *polymorpha*).
copeiana Wood, 1862: 27-28 (*Scolopendra*; characters; USA: California).
TYPES: NMNH.

— Wood, 1865: 165-166 (*Scolopendra*; characters; distribution).
— Pocock, 1895: 19-20, pl. 2; Figs. 1, la-d (*Scolopendra*; characters; notes; USA: California, Texas).
— Kraepelin, 1903: 241 (*Scolopendra*; synonym of *polymorpha*).
— Chamberlin, 1914c: 193 (*Scolopendra*; synonym of *polymorpha*).
— Gunthorp, 1920: 113 (*Scolopendra*; list).
— Bücherl, 1939: 245 (synonym of *polymorpha*).
— Shelley, 2002: 20 (*Scolopendra*; synonym of *polymorpha*).
— Shelley, 2006b: 18 (*Scolopendra*; catalog; distribution; notes).

gaumeri Pocock, 1895: 20 (*Scolopendra*; nominal taxon *copeana*; notes; Honduras).
— Chamberlin, 1921b: 8 (*Scolopendra*; nominal taxon *polymorpha*; list).
— Bücherl, 1939: 245 (synonym of *polymorpha*).
— Shelley, 2002: 20 (*Scolopendra*; synonym of *polymorpha*).
— Shelley, 2006b: 25 (*Scolopendra*; catalog; distribution; notes).

leptodera Kohlrausch, 1879: 73 [?=1881: 116] (*Scolopendra*; characters; Brazil).
— Kraepelin, 1903: 241 (*Scolopendra*; synonym of *polymorpha*).
— Chamberlin, 1914c: 194 (*Scolopendra*; synonym of *polymorpha*).
— Bücherl, 1939: 245 (synonym of *polymorpha*).
— Shelley, 2002: 20 (*Scolopendra*; synonym of *polymorpha*).
— Shelley, 2006b: 20 (*Scolopendra*; catalog; distribution; notes).

michelbacheri Verhoeff, 1938: 282-283 (*Scolopendra*; characters; USA: California).
— Chamberlin, 1962: 134 (*Scolopendra*; USA: Nevada).
— Shelley, 2002: 20 (*Scolopendra*; synonym of *polymorpha*).
— Shelley, 2006b: 28 (*Scolopendra*; catalog).

mohavea Chamberlin, 1912d: 156-158 (*Scolopendra*; characters; USA: Arizona). **TYPE:** NMNH.
— Attems, 1930: 46 (*Scolopendra*; characters; distribution).
— Shelley, 2002: 20 (*Scolopendra*; synonym of *polymorpha*).
— Shelley, 2006b: 26 (*Scolopendra*; catalog).

mysteca Humbert & Saussure, 1869: 157 (*Scolopendra*; characters; Mexico).
— Kraepelin, 1903: 241 (*Scolopendra*; synonym of *polymorpha*).
— Chamberlin, 1914c: 193 (*Scolopendra*; synonym of *polymorpha*).
— Shelley, 2002: 20 (*Scolopendra*; synonym of *polymorpha*).
— Shelley, 2006b: 19 (*Scolopendra*; catalog; distribution; notes).

pachypus Kohlrausch, 1879: 73 [=1881: 113–114?] (*Scolopendra*; USA: California).
— Bollman, 1888d: 347 [= 1893: 100] (*Scolopendra*; note; USA: California).

— Bollman, 1893: 127, 174, (*Scolopendra*; catalogue; etymology; list of known localities).
— Pocock, 1895: 20 (Scolopendra; questioned synonym of *copeana*).
— Kraepelin, 1903: 241 (*Scolopendra*; synonym of *polymorpha*).
— Chamberlin, 1914c: 194 (*Scolopendra*; synonym of *polymorpha*).
— Bücherl, 1939: 245 (synonym of *polymorpha*).
— Shelley, 2002: 20 (*Scolopendra*; synonym of *polymorpha*).

pueblae Chamberlin 1915: 502 (*Scolopendra*; nominal taxon *polymorpha*; Mexico).
— Bücherl, 1939: 245 (synonym of *polymorpha*).
— Shelley, 2002: 20 (*Scolopendra*; synonym of *polymorpha*).
— Shelley, 2006b: 27 (*Scolopendra*; catalog; notes).

viridilimbata Daday, 1891a: 148 (*Scolopendra*; characters; "America borealis").
— Daday, 1891b: 186 (*Scolopendra*; characters; "Nord-Amerika").
— Kraepelin, 1903: 241 (*Scolopendra*; questioned synonym of *polymorpha*).
— Shelley, 2006b: 23 (*Scolopendra*; catalog; distribution; notes).

Distribution: USA: Arizona, California, Colorado, Hawaii, Idaho, Kansas, Louisiana, Montana, Nebraska, New Mexico, Nevada, Oklahoma, Oregon, South Dakota, Texas, Utah, Washington, Wyoming.

LECTOTYPE: ANSP, Kansas: Riley County, Ft. Riley, collected by Dr. Hammond.

PARALECTOTYPE: ANSP, 7 specimens, Kansas: Riley County, Ft. Riley, collected by Dr. Hammond.

TOPOTYPE: MCZ, Kansas: Riley County, Ft. Riley, but collected by H. Bravat.

♂♀*S. subspinipes* Leach, 1815
— Leach, 1815: 383 (*Scolopendra*; characters). [**Note:** Leach may have deposited a specimen at the British Museum].
— Newport, 1845b: 389-390 (*Scolopendra*; characters; notes).
— Kohlrausch, 1881: 96-100 (*Scolopendra*; characters; large synonym list).
— Underwood, 1887: 64 (*Scolopendra*; distribution).
— Bollman, 1893: 127, 172-173 (*Scolopendra*; catalogue; etymology; range).
— Kraepelin, 1903: 256-258 (*Scolopendra*; characters; notes).

— Chamberlin, 1911d: 478 (*Scolopendra*; USA: California, Florida).
— Chamberlin, 1920c: 30-31, 239 (*Scolopendra*; USA: Hawaii).
— Attems, 1930: 29-30; Fig. 1, 43 (*Scolopendra*; characters; distribution).
— Chamberlin, 1930c: 67 (*Scolopendra*; USA: Hawaii). [**Note:** "One specimen of this oriental species was taken on *Pachyrhizus* sp. from China in 1928"].
— Williams, 1931: fig. 148 (*Scolopendra*;). [**Note:** Reference not obtained].
— Chamberlin, 1944d: 184 (*Scolopendra*; USA: Hawaii).
— Remington, 1950: 453-455 (*Scolopendra*; bite and habits of this species in the Philippine Islands).
— Bücherl, 1971: 182, 193, 194; fig. 5, 7 (*Scolopendra*; characters; note; range; toxicity of venom).
— Lewis, 1980: 121-122 (*Scolopendra*; swimming behavior).
— Kevan, 1983: 2944 (*Scolopendra*; checklist; notes).
— Nishida, 1994: 27 (*Scolopendra*; checklist).
— Nishida, 1997: 195 (*Scolopendra*; checklist).
— Shelley, 2000b: 42-44 (*Scolopendra*; characters; notes; USA: Hawaii).
— McFee, et al., 2002: 573 (*Scolopendra*; envenomation).
— Shelley, 2002: 38-39; Figs. 50-56 (*Scolopendra*; characters; distribution; notes; published records; USA: California, District of Columbia, Florida, Maryland, New York, Pennsylvania, Texas). [**Note:** Shelley deleted the states of California, Florida and New York from the known distribution in the United States. Although there is unlikely that this species would establish itself in New York, there is a possibility for it to establish itself in the states of California and Florida, hence, their questionable listing in the distribution. Furthermore, it has been intercepted at quarantine in many states but appears to not have established itself anywhere in the contiguous states].
— Bouchard, et al., 2004: 312 (*Scolopendra*; envenomation).
— Shelley, 2006b: 7 (*Scolopendra*; catalog; distribution; notes).
bispinipes Wood, 1862: 28-29 (*Scolopendra*; characters; notes; USA: California).
— Bollman, 1893: 127 (*Scolopendra*; synonym of *dehaani*).
— Gunthorp, 1920: 113 (*Scolopendra*; list).
— Attems, 1930: 31 (*Scolopendra*; synonym of *Scolopendra subspinipes dehanni*).
— Shelley, 2002: 38 (*Scolopendra*; synonym of *subspinipes*).
— Shelley, 2006b: 18 (*Scolopendra*; catalog; distribution; notes).
byssina Wood, 1861: 10 (*Scolopendra*; characters; USA: ?Florida).

— Wood, 1862: 26-27 (*Scolopendra*; characters; USA: California).

— Wood, 1865: 164 (*Scolopendra*; characters; USA: possibly California or Florida).

— Bollman, 1893: 127, 173 (*Scolopendra*; synonym of *subspinipes*).

— Gunthorp, 1920: 113 (*Scolopendra*; list).

— Attems, 1930: 30 (*Scolopendra*; cited as synonym of *Scolopendra subspinipes subspinipes*).

— Shelley, 2002: 38 (*Scolopendra*; synonym of *subspinipes*).

— Shelley, 2006b: 16 (*Scolopendra*; catalog; distribution; notes).

?*dehaanii* Brandt, 1840: 152 [= 1841: 59] (*Scolopendra*; nominal taxon *subspinipes*; Indonesia, Java).

— Bollman, 1893: 127 (*Scolopendra*; catalogue).

— Kraepelin, 1903: 260 (*Scolopendra*; nominal taxon *subspinipes*; characters; notes).

— Chamberlin, 1911d: 477, 478. (*Scolopendra*; characters; key; USA: California.) [**Note:** Chamberlin states that this species is " . . . known only from California, where it has likely escaped from vessels."].

— Attems, 1930: 31 (*Scolopendra*; nominal taxon *subspinipes*; characters; distribution).

— Shelley, 2002: 38 (*Scolopendra*; nominal taxon subspinipes; see footnote). [**Note:** It is still not clear if this subspecies of *subspinipes* is valid and needs further scrutiny].

repens Wood, 1862: 31-32 (*Scolopendra*; characters; USA: Hawaii).

— Shelley, 200b: 42 (*Scolopendra*; synonym of *subspinipes*).

sandwichiana Gervais, 1847: 276 (*Scolopendra*; characters; USA: Hawaii). [**Note:** Type specimen deposited in the Paris Museum].

— Wood, 1862: 30-31 (*Scolopendra*; characters; notes).

— Shelley, 2000b: 42 (*Scolopendra*; synonym of *subspinipes*).

Distribution: USA: ?California, District of Columbia, ?Florida, Hawaii, ?Maryland, ?New York, ?Pennsylvania, ?Texas.

HOLOTYPE: Unknown, needs neotype designation (see Shelley, 2002: 38). [**Note:** This species has been introduced and is indigenous to countries such as China, New Guinea, New Zealand, and Thailand (see Attems, 1930; Chamberlin, 1944d)].

♂♀*S. viridis* Say, 1821

— Say, 1821: 110-111 (*Scolopendra*; characters; USA: Florida, Georgia).

— Lucas, 1840: 546 (*Scolopendra*; characters; USA: Florida, Georgia).
— Brandt, 1840: 158 [= 1841: 68] (*Scolopendra*; list, note).
— Gervais, 1847: 276-277 (*Scolopendra*; characters).
— Wood, 1862: 22-23; 1865: 159-160 (*Scolopendra*; characters; chart showing variation in characters; USA: Florida).
— Kohlrausch, 1881: 112 (*Scolopendra*; characters).
— Meinert, 1886a: 196-197 (*Scolopendra*; characters; USA: Georgia).
— McNeill, 1887a: 326 (*Scolopendra*; note; USA: Florida).
— Underwood, 1887: 63, 64 (*Scolopendra*; distribution; key; USA: Florida, Georgia, Tennessee).
— Bollman, 1893: 127, 147, 174-175 (*Scolopendra*; catalogue; etymology; notes; distribution).
— Kenyon, 1893a: 17 (*Scolopendra*; note; USA: Nebraska).
— Pocock, 1895: 21-23; Figs. 2, 2a-I (*Scolopendra*; notes; USA: Florida, Georgia, Texas).
— Kraepelin 1903: 242-243; Fig. 156 (*Scolopendra*; characters; distribution; notes).
— Brölemann, 1904: 318 (*Scolopendra*; USA: ?California). [**Note:** Locality cited as "Californie" and California is not impossible as a place for viridis to exist although according to Shelley, 2002, California is not part of the distribution).
— Chamberlin, 1918b: 375 (*Scolopendra*; USA: Louisiana).
— Chamberlin, 1921b: 8 (*Scolopendra*; list; Costa Rica, Guatemala, Honduras).
— Attems, 1930: 48; fig. 63 (*Scolopendra*; characters; distribution).
— Attems, 1930: 48 (*Scolopendra*; nominal taxon *viridis*; characters; distribution).
— Attems, 1930: 48 (*Scolopendra*; nominal taxon *viridis*; subspecies *viridis*; cited as variety *viridis*; characters; distribution).
— Bücherl, 1939: 245 (*Scolopendra*; characters; note).
— Chamberlin & Mulaik, 1940: 108 (*Scolopendra*; USA: Texas).
— Bücherl, 1941: 297 (*Scolopendra*;) [**Note:** Reference not obtained].
— Chamberlin, 1942d: 185 (*Scolopendra*; USA: Louisiana).
— Chamberlin, 1943b: 98 (*Scolopendra*; USA: New Mexico, Texas).
— Chamberlin, 1944a: 33 (*Scolopendra*; USA: Georgia).
— Chamberlin, 1944d: 183-184 (*Scolopendra*; USA: Arkansas, Florida, Georgia, Illinois, Louisiana, Texas).
— Chamberlin, 1945d: 215 (*Scolopendra*; USA: Georgia).
— Chamberlin, 1946b: 194 (*Scolopendra*; USA: Mississippi).

— Chamberlin, 1951a: 33 (*Scolopendra*; USA: Florida).
— Crabill, 1952b: 220-221, pl. 6; Fig. 74 (*Scolopendra*; cited as subspecies *viridis*; characters; notes; range; USA: Illinois).
— Chamberlin, 1958a: 14 (*Scolopendra*; USA: Florida).
— Crabill, 1960a: 10 (*Scolopendra*; distribution; note; USA: Illinois, Kentucky, Missouri, Virginia). [**Note:** Crabill mentions southern Illinois, Kentucky, Missouri & Virginia as part of the distribution for *viridis* but these states are not included as part of Shelley's (2002) distribution].
— Cloudsley-Thompson & Crawford, 1970b: 187 (*Scolopendra*; note).
— Reddell, 1970: 402 (*Scolopendra*; checklist; note).
— Shelley, 1978: 221 (*Scolopendra*; habitat, range).
— Summers, 1979: 693, 696, 699; Fig. 6 (*Scolopendra*; checklist; key).
— Summers et al., 1980: 245; Fig. 6 (*Scolopendra*; checklist; habitat (Table 1, p. 242-243); USA: Illinois).
— Summers et al., 1981: 59-60 (*Scolopendra*; checklist; USA: Illinois).
— Kevan, 1983: 2944 (*Scolopendra*; checklist; note; CANADA: Nova Scotia; USA: Oregon, Washington). [**Note:** All 3 reports are as imports].
— Neck, 1985: 253-255 (*Scolopendra*; comparative behavior and color pattern of two sympatric species; USA: Texas).
— Shelley, 1987: 503-504; Figs. 2, 11 (*Scolopendra*; characters; habitat; notes; USA: North Carolina).
— Shelley & Edwards, 1987: 3, 4; Fig. 5 (*Scolopendra*; key, USA: Florida).
  Bchan-Pelletier, 1993: 25 (*Scolopendra*; list).
— Hoffman & Shelley, 1996: 37-41; Fig. 4 (*Scolopendra*; taxonomic discussion).
— Sandefer, 1998: unnumbered page (*Scolopendra*). [**Note:** Reference not obtained].
— Shelley, 2002: 13-20; Figs. 5-11; Tab. 1 (*Scolopendra*; characters; distribution; habitat; notes; list of published records; USA: Alabama, Arizona, Colorado, Florida, Georgia, Kansas, Louisiana, Mississippi, Nevada, New Mexico, North Carolina, Oklahoma, South Carolina, Texas, Utah). [**Note:** Shelley lists eight or so species from Mexico he considers to conform to *viridis* and six of these are described by Verhoeff (1934) but he was unable to evaluate them. It would be interesting to further assess the biogeography of this species and its distribution into Mexico].

— Regier et al., 2005: 149; Tab. 1 (*Scolopendra*; table of taxon classification with GenBank accession numbers).
— Shelley, 2006b: 7–8 (*Scolopendra*; catalog; distribution; notes).
— Giribet & Edgecombe, 2006: 533, 534, 536; Fig 1,2; Tab. 1 (*Scolopendra*; cladograms; GenBank accession code).

azteca Saussure, 1859: 382-383 [= Saussure, 1860: 124-125]; Fig. 41 (*Scolopendra*; characters; Mexico).
— Saussure & Humbert, 1872: 128, pl. 5; Figs. 10, 14 (*Scolopendra*; characters; Mexico).
— Shelley, 2006b: 14 (*Scolopendra*; catalog; distribution; notes).

cuivis Pocock, 1891a: 62-63; Pl. 4, Fig. 7 (*Scolopendra*; characters). [**Note:** Pocock states that the locality is doubtful and that the specimen was taken from a bottle labeled "India and S. America". Based on Pocock's illustration and the probable localities this species is most likely not a synonym of *viridis*].
— Kraepelin, 1903: 242 (*Scolopendra*; questioned synonym of *viridis*).
— Shelley, 2002: 13 (*Scolopendra*; questioned synonym of *viridis*).

otomita Saussure, 1859: 383-384 [= Saussure, 1860: 125-126], Fig, 42 (*Scolopendra*; characters; Mexico).
— Porat, 1876: 9-10 (*Scolopendra*; characters; notes).
— Daday, 1893: 108, pl. 5, Fig. 8 (*Scolopendra*; characters; note).
— Kraepelin, 1903: 242 (synonym of *viridis*).
— Shelley, 2002: 13 (synonym of *viridis*).
— Shelley, 2006b: 14 (*Scolopendra*; catalog; distribution; notes).

parva Wood, 1861: 10 (*Scolopendra*; characters; USA: Georgia).
— Wood, 1862: 22 (*Scolopendra*; cited as synonym of *viridis*).
— Underwood, 1887: 64 (synonym of *viridis*).
— Bollman, 1893: 127, 147, 174 (*Scolopendra*; synonym of *viridis*).
— Bücherl, 1939: 245 (synonym of *viridis*). [**Note:** Bücherl also lists *punctiventris* as a synonym of *viridis* but this is a synonym of *Hemiscolopendra marginata*].
— Shelley, 2006b: 16 (*Scolopendra*; catalog; distribution; notes).

tenuitarsis Pocock, 1895: 18, pl. 2; Figs. 5, 5a-d (*Scolopendra*; characters; notes; Mexico).
— Shelley, 2002: 13 (synonym of *viridis*).
— Shelley, 2006b: 25 (*Scolopendra*; catalog; distribution; notes).

tolteca Saussure, 1859: 384-385 [= Saussure, 1860: 126]; Fig. 43 (*Scolopendra*; characters; Mexico).
— Saussure & Humbert, 1872: 129, pl. 5; Fig. 9 (*Scolopendra*; characters; Mexico).
— Shelley, 2002: 13 (synonym of *viridis*).
— Shelley, 2006b: 15 (*Scolopendra*; catalog; distribution; notes).

utahana Chamberlin, 1925e: 58 (*Scolopendra*; characters; notes; USA: Utah).
    HOLOTYPE: NMNH.
    — Shelley, 2002: 13 (synonym of *viridis*).
    — Shelley, 2006b: 27 (*Scolopendra*; catalog).

Distribution: CANADA: ?Nova Scotia; USA: Alabama, Arkansas, Arizona, ?California, Colorado, Florida, Georgia, Illinois, Kansas, ?Kentucky, Louisiana, Mississippi, North Carolina, Nebraska, New Mexico, Nevada, Oklahoma, ?Oregon, South Carolina, Tennessee, Texas, Utah, ?Virginia, Washington.

SYNTYPE: BMNH, Coastal region of Georgia or Florida, 1818 (see Shelley, 2002: 13).

Family **Scolopocryptopidae** Pocock, 1896
Subfamily **Kethopinae** Shelley, 2002

Genus *Kethops* Chamberlin, 1912
[2 species; Southwestern USA]

Kethops Chamberlin, 1912d: 154-155 (characters; notes). **TYPE SPECIES:** *Newportia utahensis* Chamberlin, 1909, fixed by original designation.
    — Attems, 1930: 265 (characters; distribution).
    — Crabill, 1958e: 238 (key to species).
    — Crabill, 1960a: 12 (notes).
    — Shelley, 2002: 76-77 (characters; distribution).
    — Jeekel, 2005: 55 (list).
    — Shelley, 2006a: 7 (list).

K. atypus Chamberlin, 1943
    — Chamberlin, 1943b: 97-98 (*Kethops*; characters; USA: Utah).
    — Shelley, 2002: 79; Fig. 141, 145 (*Kethops*; characters; notes; USA: Utah). [**Note:** Neither Chamberlin or Shelley report if the sex is known in this species].

Distribution: USA: Utah.

HOLOTYPE: NMNH, Utah: Salt Lake Co., Salt Lake City, April, 1942, by S. & D. Mulaik.

PARATYPE: NMNH, Utah: Salt Lake Co., Salt Lake City, April, 1942, by S. & D. Mulaik.

♂♀*K. utahensis* (Chamberlin, 1909)
— Chamberlin, 1909a: 29-30; Fig. 2 (*Newportia*; characters; notes; USA: Utah).
— Chamberlin, 1912d: 155-156, pl. 13; Figs. 3-6 (*Kethops*; characters; note; USA: New Mexico).
— Attems, 1930: 265-266; Figs. 353-356 (*Kethops*; characters; distribution).
— Shelley, 2002: 77-79; Figs. 142-145 (*Kethops*; characters; notes; published records with additional records as follows: USA: California, New Mexico).
— Jeekel, 2005: 55 (*Newportia*; list).
colomanus Chamberlin, 1941b: 41 (*Cryptops*; characters; USA: California). **TYPE:** NMNH.
— Shelley, 2002: 77 (*Cryptops*; synonym of *utahensis*).
euterpe Crabill, 1958e: 236-238; Fig. 1-3 (*Kethops*; characters; USA: New Mexico). **HOLOTYPE:** NMNH; No. 2454; ♂; USA: New Mexico; Otero County, Clouderoft, July 26, 1948, by G.E. Ball & H.E. Evans.
— Crabill, 1960a: 2, 5; Figs. 6, 8, 12 (*Kethops*; note).
— Shelley, 2002: 77 (synonym of *utahensis*).
glenvilleus Chamberlin, 1941b: 41-42 (*Cryptops*; characters; USA: California). **HOLOTYPE:** NMNH.
— Shelley, 2002: 77 (*Kethops*; synonym of *utahensis*).
leioceps Chamberlin, 1925e: 57 (*Kethops*; characters; USA: Utah). **HOLOTYPE:** NMNH.
— Shelley, 2002: 77 (synonym of *utahensis*).

Distribution: USA: California, New Mexico, Utah.

HOLOTYPE: Lost according to Shelley (2002), Utah: Salt Lake Co., Warm Springs. [**Note:** NMNH type database list this species as being a part of the collection].

Genus *Thalkethops* Crabill, 1960
[1 species; Southwestern USA]

Thalkethops Crabill, 1960a: 2-3, 12 (characters; note). **TYPE SPECIES:** *Thalkethops grallatrix* Crabill, 1960, fixed by original designation and monotypy.
— Shelley, 2002: 79 (characters; notes).

♀*T. grallatrix* Crabill, 1960
— Crabill, 1960a: 3-8; Figs. 1-5, 7, 9-11, 13-16 (*Thalkethops*; characters; USA: New Mexico).
— Barr & Reddell, 1967: 262 (*Thalkethops*; notes).
— Shelley, 2002: 79-80; Figs. 141, 146 (*Thalkethops*; characters; published records; USA: New Mexico).

Distribution: USA: New Mexico.

HOLOTYPE: NMNH, No. 2505, probably ♀, New Mexico: Eddy Co., Carlsbad Cave, 31 August 1957, by Dixon Freeland & Thomas Ela.

Subfamily **Scolopocryptopinae** Pocock, 1896

Genus *Dinocryptops* Crabill, 1953
[1 species (Introduced); Southeastern & Western USA]

Dinocryptops Crabill, 1953b: 96 (proposal of new genus; nomenclatorial notes).
TYPE SPECIES: *Scolopocryptops miersii* Newport, 1845, fixed by subsequent designation of Pocock, 1895, according to Crabill, 1953b and fixed by original designation according to Jeekel, 2005.
— Crabill, 1960a: 11 (notes).
— Shelley, 2002: 6 (notes).
— Jeekel, 2005: 53 (list).

D. miersii (Newport, 1845)
— Newport, 1845b: 405 (*Scolopocryptops*; characters; notes; Brazil).
— Gervais, 1847: 298 (*Scolopocryptops*; characters).
— Bollman, 1893: 128, 180 (*Scolopocryptops*; catalogue; list).
— Pocock, 1894a: 465 (*Scolopocryptops*; characters; South America).
— Kraepelin, 1903: 77-78 (*Scolopocryptops*; characters; distribution; USA: ?California, Georgia, Virginia).
— Chamberlin, 1911d: 475 (*Scolopocryptops*; characters; Chamberlin doubtfully records this species from California).
— Attems, 1930: 256 (*Scolopocryptops*; characters; distribution).
— Crabill, 1955g: 135-136 (*Dinocryptops*; notes).
— Crabill, 1960a: 11 (*Dinocryptops*; notes).
— Shelley, 2000a: 155, 156 (*Dinocryptops*; notes).
— Shelley, 2002: 6 (*Dinocryptops*; notes). [**Note:** Shelley briefly discusses the status of this species occurring in the US and it may

be possible that it actually does occur in the US but apparently specimens have not be procured yet to verify its existence. It does seem unlikely that this species exists in California but there is certainly a greater possibility that it can be found in the southeastern US].

— Jeekel, 2005: 53 (*Scolopocryptops*; list).

?*calcaratus* Bollman, 1893: 133 (*Scolopocryptops*; characters; USA: Indiana). [**Note:** Shelley (2002: 60) synonymized *calcaratus* with *nigridius* based on one character without seeing the presumed lost type specimen. Kraepelin (1903) suggested Bollman's *calcaratus* is a synonym of *miersii*, which deserves further investigation].

— Kraepelin, 1903: 77 (*Scolopocryptops*; synonym of *miersii*).

Distribution: USA: ?California, ?Georgia, ?Indiana, ?Virginia.

HOLOTYPE: ?Unknown.

Genus *Scolopocryptops* Newport, 1844
[6 species; Canada, USA]

Scolopocryptops Newport, 1844: 275 **Type Species:** *Scolopocryptops melanostoma* Newport, 1845, fixed by subsequent designation of Lucas, 1849.

— Cope, 1869: 179 (note). [**Note:** According to Bollman (1893: 179) the undescribed *Scolopocryptops* species from Virginia that Cope was referring to was *nigridius*].

— Meinert, 1886a: 179 (characters).

— Attems, 1930: 255-256 (characters; distribution; key to species; species included: *brolemanni, miersii*).

— Crabill, 1953b: 96 (nomenclatorial notes).

— Crabill, 1957a: 345 (notes).

— Crabill, 1960a: 12 (notes).

— Shelley, 1978: 221 (habitat, notes).

— Shelley, 2002: 51-53 (key to species; notes).

— Jeekel, 2005: 60 (list).

Anethops Chamberlin, 1902: 39 (characters; key). **Type Species:** *Anethops occidentalis* Chamberlin, 1902, fixed by monotypy.

— Attems, 1930: 246-247 (characters; distribution).

— Crabill, 1960a: 12 (synonym of *Scolopocryptops*; note).

— Shelley, 2002: 51 (synonym of *Scolopocryptops*).

— Jeekel, 2005: 50 (list).

Otocryptops Haase, 1887: 96. (characters). [**Note:** reference not obtained].
TYPE SPECIES: *Otocryptops rubiginosus* Koch, 1878, fixed by
monotypy.
— Bailey, 1928b: 36 (characters; note).
— Crabill, 1952a: 129 (synonym of *Scolopocryptops*; key to species east
of the Rocky Mountains).
— Crabill, 1953b: 96 (synonym of *Scolopocryptops*; nomenclatorial
notes).
— Shelley, 2002: 51 (synonym of *Scolopocryptops*).
— Jeekel, 2005: 57 (list).

♂♀*S. gracilis* Wood, 1862
— Wood, 1862: 38-39 (*Scolopocryptops*; characters; USA: California).
— Wood, 1865: 173-174; Fig. 14 (*Scolopocryptops*; characters; USA:
California).
— Wood, 1867b: 128 (*Scolopocryptops*; characters; USA: California).
— Bollman, 1893: 128, 179, 180 (*Scolopocryptops*; catalogue; etymology;
list; range).
— Kraepelin, 1903: 70-71 (*Otocryptops*; characters; notes; USA:
California).
— Brölemann, 1904: 244 (*Otocryptops*; USA: California).
— Chamberlin, 1911d: 473 (*Otocryptops*; characters; notes; USA:
California).
— Anonymous, 1920: 35 (*Otocryptops*; note; USA: California).
— Gunthorp, 1920: 113 (*Scolopocryptops*; list).
— Attems, 1930: 264 (*Otocryptops*; characters).
— Crabill, 1952a: 123, 124, 125, figs, 2, 6, 11 (*Otocryptops*; notes; cited
as subspecies *gracilis*).
— Shelley, 2002: 70-72; Figs. 109-113 (*Scolopocryptops*; characters;
habitat; published records; USA: Arizona; California; Idaho; Nevada;
Oregon; Utah; Washington). [**Note:** Although Chamberlin (1943b)
reported this species from Texas, Shelley, 2002 has removed it from
this state as it seems far removed from the rest of the population.
Hence, Texas is left off the distribution].
— Shelley & Six, 2004: 257-258 (*Scolopocryptops*; habitat; map; notes;
range; USA: Montana).
berkeleyensis Verhoeff, 1938: 384-385 (*Otocryptops*; characters; USA:
California).
— Shelley, 2003: 57 (*Scolopocryptops*; nominal taxon *gracilis*; synonym
of *gracilis*).

californica Humbert & Saussure, 1870b: 204 (*Scolopendra*; characters; USA: California).
— Saussure & Humbert, 1872: 134-135, pl. 6; Figs. 19, 19d, 19v (*Scolopocryptops*; characters; Mexico).
— Bollman, 1893: 128, 179 (*Scolopocryptops*; synonym of *gracilis*).
— Kraepelin, 1903: 70 (*Otocryptops*; synonym of *gracilis*).
— Shelley, 2006b: 19 (*Scolopendra*; catalog; distribution; notes).
lanatipes Wood, 1862: 39 (*Scolopocryptops*; synonym of *gracilis*).
— Wood, 1865: 175; Figs. 16-17 (*Scolopocryptops*; synonym of *gracilis*).
— Kohlrausch, 1881: 56 (*Scolopocryptops*; characters).
— Underwood, 1887: 62, 63 (*Scolopocryptops*; key; cited as synonym of *gracilis*).
— Bollman, 1893: 128, 179 (*Scolopocryptops*; synonym of *gracilis*).
— Kraepelin, 1903: 70 (*Otocryptops*; synonym of *gracilis*).
— Gunthorp, 1920: 113 (*Scolopocryptops*; list).
mundus Chamberlin, 1911d: 473; Fig. 156a, b, e, f (*Otocryptops*; characters; USA: Idaho). **TYPE:** NMNH, Idaho: Latah County, Kendrick.
— Kevan, 1983: 2945 (*Scolopocryptops*; checklist; questioned form of *gracilis*).
— Shelley, 2002: 70 (*Otocryptops*; synonym of *gracilis*).
occidentalis Chamberlin, 1902a: 40 (*Anethops*; Type locality: USA: California; Los Angeles county, San Gabriel Canyon, 25.V.1901, by Mr. Charles E. Hutchinson). **TYPE:** ?NMNH.
— Kraepelin, 1903: 61-62; Fig. 20 (*Anethops*; characters; USA: California).
— Chamberlin, 1911d: 475 (*Anethops*; characters; USA: California).
— Attems, 1930: 247; Fig. 324 (*Anethops*; characters; distribution).
— Shelley, 2002: 70 (synonym of *gracilis*).
— Jeekel, 2005: 50 (*Anethops*; list).
sexspinosus Chamberlin (not Say), 1925e: 57 (*Otocryptops*; USA: Utah).
— Shelley, 2002: 70 (cited record from Chamberlin as a synonym of *gracilis*).

Distribution: USA: Arizona, California, Idaho, Montana, Nevada, Oregon, Utah, Washington.

HOLOTYPE: Lost.

PARATYPES: ANSP, California: Kern Co., Ft. Tejon, unknown date & collector.

♂♀*S. nigridius* McNeill, 1887

— McNeill, 1887b: 333 (*Scolopocryptops*; characters; USA: Indiana).

— Bollman, 1888a: 107, 110, 112 [= 1893: 82, 84, 85] (*Scolopocryptops*; note; USA: Tennessee).

— Bollman, 1888c: 341 [= 1893: 92] (*Scolopocryptops*; note; USA: Tennessee).

— Bollman, 1888e: 409 [= 1893: 110] (*Scolopocryptops*; note; USA: Indiana).

— Bollman, 1893: 128, 179, 180 (*Scolopocryptops*; catalogue; characters; etymology; list, note; USA: Indiana, Pennsylvania, Tennessee).

— Brölemann, 1896: 50 (*Scolopocryptops*; list).

— Attems, 1930: 264 (*Otocryptops*; cited as inadequate description).

— Brimley, 1938: 501 (*Otocryptops*; USA: North Carolina).

— Causey, 1940: 66 (*Scolopocryptops*; characters; habitat; notes; USA: North Carolina).

— Crabill, 1950c: 201 (*Otocryptops*; characters; range; USA: South Carolina).

— Wray, 1950: 156 (*Otocryptops*; list).

— Crabill, 1952a: 126-128; Figs. 3, 7, 9-10 (*Otocryptops*; characters; USA: Alabama, Kentucky, South Carolina, Virginia).

— Crabill, 1952b: 207-209, pl. 5; Figs. 68, 73 (*Scolopocryptops*; characters; range; USA: Indiana, Kentucky, Pennsylvania, Virginia).

— Crabill, 1955f: 259 (*Scolopocryptops*; checklist; USA: Kentucky).

— Branson & Batch, 1967: 85 (*Scolopocryptops*; characters; habitat; body measurements; USA: Kentucky).

— Wray, 1967: 156 (*Otocryptops*; list).

— Shelley, 1978: 221 (*Scolopocryptops*; habitat, range).

— Summers, 1979: 694, 696, 699; Fig. 10 (*Scolopocryptops*; checklist; key).

— Shelley, 1987: 506; Figs. 5, 14 (*Scolopocryptops*; characters; habitat; note; USA: North Carolina).

— Summers et al., 1980: 245; Fig. 3 (*Scolopocryptops*; checklist; habitat (Table 1, p. 242-243); USA: Illinois).

— Summers et al., 1981: 59 (*Scolopocryptops*; checklist; USA: Illinois).

— Shelley & Edwards, 1987: 3, 4; Fig. 6; Map 2 (*Scolopocryptops*; key; USA: Florida).

— Hoffman, 1995a: 25 (*Scolopocryptops*; list; notes).

— Shelley, 2002: 60-63; Figs. 84-92 (*Scolopocryptops*; characters; distribution; habitat; list of published records; USA: Alabama,

District of Columbia, Florida, Georgia, Illinois, Indiana, Kentucky, Maryland, Mississippi, North Carolina, Ohio, Pennsylvania, South Carolina, Tennessee, Virginia, West Virginia).

calcaratus Bollman, 1893: 133 (*Scolopocryptops*; characters; USA: Indiana).

— Shelley, 2002: 60 (*Scolopocryptops*; synonym of *nigridius*).

Distribution: USA: Alabama, District of Columbia, Florida, Georgia, Illinois, Indiana, Kentucky, Maryland, Mississippi, North Carolina, Ohio, Pennsylvania, South Carolina, Tennessee, Virginia, West Virginia.

SYNTYPE: NMNH, Indiana: Monroe Co., near Bloomington.

NEOTYPE: NMNH, Indiana: Brown Co., Brown County State Park, 22 August 1960, by W.L. Brown, Jr.

♂♀*S. peregrinator* (Crabill, 1952)
— Crabill, 1952a: 124-126; Figs. 4-5 (*Otocryptops*; characters; cited as subspecies of *gracilis*; USA: Maryland, Virginia).
— Crabill, 1952b: 210-212 (*Otocryptops*; cited as subspecies of *gracilis*; characters; USA: Maryland, Virginia).
— Branson & Batch, 1967: 85 (*Scolopocryptops*; nominal taxon *gracilis*; brief description; habitat; note; USA: Kentucky).
— Shelley, 1987: 507-510; Figs. 8-9, 15-16 (*Scolopocryptops*; characters; distribution; habitat; notes; USA: District of Columbia, North Carolina, Pennsylvania, Virginia, West Virginia).
— Hoffman, 1995a: 25-26 (*Scolopocryptops*; list; notes; USA: Pennsylvania).
— Shelley, 2002: 64-66; Figs. 93-95 (*Scolopocryptops*; characters; distribution; habitat; list of published records; USA: District of Columbia, Kentucky, Maryland, New York, North Carolina, Pennsylvania, Virginia, West Virginia).

Distribution: USA: District of Columbia, Kentucky, Maryland, New York, North Carolina, Pennsylvania, Virginia, West Virginia.

HOLOTYPE: NMNH, Virginia: Albemarle, Co., Charlottesville, 9 March 1949, by R.L. Hoffman.

PARATYPE: MCZ, Maryland: Montgomery Co., Woodside, unknown date, by J.E. Benedict.

♂♀ *S. rubiginosus* Koch, 1878
  — Koch, 1878: 792-793 (*Scolopocryptops*; characters).
  — Haase, 1887: 97 (*Otocryptops*). [**Note:** reference not obtained].
  — Kraepelin, 1903: 71-72; Figs. 28-30 (*Otocryptops*; characters; distribution; USA: Indiana, Minnesota).
  — Attems, 1914: 103 (*Otocryptops*; distribution; list).
  — Attems, 1930: 259-260; Figs. 23, 30, 339-344 (*Otocryptops*; characters; distribution).
  — Bücherl, 1941: 331-332 (*Otocryptops*;). [**Note:** Reference not obtained].
  — Crabill, 1952a: 129 (*Otocryptops*; notes; USA: Illinois, Iowa, Kansas, Nebraska).
  — Crabill, 1952b: 209-210 (*Scolopocryptops*; characters; range; USA: Illinois, Indiana). [**Note:** Crabill states that "It is interesting to note that its distribution describes a great arc beginning in the Great Plains of North America, extending through Alaska, and passing down into eastern Asia."].
  — Crabill, 1953b: 96 (nomenclatorial notes).
  — Crabill, 1955b: 39-40 (*Scolopocryptops*; character discussion; notes; USA: Missouri).
  — Crabill, 1955h: 157, 158 (*Scolopocryptops*; notes; USA: Missouri).
  — Crabill, 1958b: 96-97 (*Scolopocryptops*; notes; USA: Wisconsin).
  — Crabill, 1960a: 12 (*Scolopocryptops*; range).
  — Summers, 1979: 693, 694, 696, 699; Figs. 9 (*Scolopocryptops*; checklist; key).
  — Summers et al., 1980: 245; Fig. 3 (*Scolopocryptops*; checklist; habitat (Table 1, p. 242-243); USA: Illinois).
  — Summers et al., 1981: 55 (*Scolopocryptops*; checklist; USA: Illinois).
  — Kevan, 1979: 298 (*Scolopocryptops*; CANADA). [**Note:** Shelley (2002:68) has deleted Kevan's probable report of this species from southern Ontario to Southern Manitoba].
  — Kevan, 1983: 2944-2945 (*Scolopocryptops*; checklist; notes; CANADA: "? W Ontario and S Manitoba, unconfirmed, S. Alaska"; USA: Minnesota, Ohio, Wisconsin). [**Note:** These are not authenticated localities].
  — Shelley, 1992: 23, 24, 25 (*Scolopocryptops*; notes).
  — Behan-Pelletier, 1993: 25 (*Scolopocryptops*; list).
  — Watermolen, 1997: 2 (*Scolopocryptops*; checklist, notes).

— Shelley, 2002: 66-70; Figs. 96-108 (*Scolopocryptops*; characters; distribution; habitat; list of published records; notes; USA: Arkansas, Illinois, Iowa, Kansas, Minnesota, Missouri, Nebraska, Oklahoma, Texas, Wisconsin).

— McAllister et al., 2003: 113 (*Scolopocryptops*; notes; USA: Texas).

— Jeekel, 2005: 57 (*Scolopocryptops*; list).

confucii Karsch, 1884: 65 (*Scolopocryptops*; characters; China).

— Kraepelin, 1903: 71 (*Otocryptops*; synonym of *rubiginosus*).

— Attems, 1930: 259 (*Scolopocryptops*; synonym of *rubiginosus*).

— Kevan, 1983: 2944 (synonym of *rubiginosus*).

Distribution: CANADA: ?Manitoba, ?Ontario; USA: ?Alaska, Arkansas, Illinois, Indiana, Iowa, Kansas, Missouri, Minnesota, Nebraska, ?Ohio, Oklahoma, Texas, Wisconsin.

HOLOTYPE: Unknown.

SYNTYPES: BMNH, 5 specimens, Japan: unknown dates. (see Shelley, 2002: 66)

♂♀*S. sexspinosus* (Say, 1821)

— Say, 1821: 112 (*Cryptops*; characters; habitat; USA).

— Lucas, 1840: 547 (*Cryptops*; characters). [**Note:** Specific name cited as *sex-spinosus*].

— Newport, 1844b: 100 (*Cryptops*; list).

— Newport, 1845a: 302; Tab. 33, Figs. 20-24 (*Cryptops*; illustrations).

— Newport, 1845b: 407 (*Scolopocryptops*; characters; USA: Florida, Georgia). [**Note:** Specific name was cited as 6-*spinosa*].

— Gervais, 1847: 297-298 (*Scolopocryptops*; characters; note).

— Newport, 1856: 57 (*Scolopocryptops*; characters; notes; USA: Florida, Georgia). [**Note:** Specific name cited as 6-*spinosa*].

— Wood, 1862: 37-38 (*Scolopocryptops*; characters; notes; USA: Illinois, Mississippi, Missouri, North Carolina, Pennsylvania, South Carolina).

— Wood, 1865: 172-173; Figs. 12-13 (*Scolopocryptops*; characters; notes; USA: "The Atlantic United States").

— Porath, 1876: 26-27 (*Scolopocryptops*; characters; notes).

— Kohlrausch, 1881: 54-55 (*Scolopocryptops*; characters).

— Brodie & White, 1883: 67 (*Scolopocryptops*; CANADA).

— Meinert, 1886a: 179-180 (*Scolopocryptops*; characters; notes; USA: California, Georgia, Iowa, Maryland, Massachusetts, New York, Ohio, Pennsylvania, Virginia, West Virginia).

— McNeill, 1887a: 326 (*Scolopocryptops*; note; USA: Florida).

— McNeill, 1888: 16 (*Scolopocryptops*; characters; note; USA: Indiana).

— Bollman, 1888a: 110 [= 1893: 84] (*Scolopocryptops*; notes; USA: Tennessee).

— Bollman, 1888b: 7 [= 1893: 78] (*Scolopocryptops*; note; USA: Arkansas).

— Bollman, 1888c: 341 [= 1893: 92] (*Scolopocryptops*; note; USA: Tennessee).

— Bollman, 1888d: 347 [= 1893: 99] (*Scolopocryptops*; notes; USA: Georgia, Virginia).

— Bollman, 1888e: 408 [= 1893: 110] (*Scolopocryptops*; note; USA: Indiana).

— Bollman, 1893: 128, 147, 177-178, 180, 184 (*Cryptops* + *Scolopocryptops*; catalogue; list, notes; range; USA: Minnesota).

— Kenyon, 1893a: 16 (*Scolopocryptops*; USA: Nebraska).

— Kenyon, 1893b: 161 (*Scolopocryptops*; USA: Nebraska).

— Brölemann, 1896: 50 (*Scolopocryptops*;). [**Note:** Reference not obtained].

— Kraepelin, 1903: 72 (*Otocryptops*; characters; notes; USA: California, "Carolina", Florida, Georgia).

— Brölemann, 1904: 244 (*Otocryptops*; USA: Indiana, Louisiana, North Carolina, Washington). [**Note:** The record from Washington is most likely of *spinicaudus*].

— Carlson, 1904: 270 (*Scolopocryptops*; ventral nerve cord physiology).

— Cook, 1904: 73 (*Otocryptops*; list).

— Chamberlin, 1911d: 473, 475 (*Otocryptops*; characters; USA: California).

— Gunthorp, 1913: 167 (*Otocryptops*; characters; notes; USA: Kansas).

— Attems, 1914: 103 (*Otocryptops*; distribution; list).

— Chamberlin, 1918a: 24 (*Otocryptops*; note; USA: Tennessee).

— Chamberlin, 1918b: 375 (*Otocryptops*; USA: Louisiana).

— Chamberlin, 1919: 15H (*Otocryptops*; list; USA: Alaska).

— Chamberlin, 1921a: 230 (*Otocryptops*; USA: Tennessee).

— Gunthorp, 1921: 89 (*Otocryptops*; notes).

— Holmquist, 1926: 411 (*Otocryptops*; notes on hibernation such as: activity, date collected, developmental stage, ecological association/niche, quantity found).

— Bailey, 1928a: 1082 (*Otocryptops*; list).
— Bailey, 1928b: 19, 36 (*Otocryptops*; characters; list; note; USA: New York).
— Chamberlin, 1928b: 153 (*Otocryptops*; USA: Missouri).
— Williams & Hefner, 1928: 137-138 (*Otocryptops*; characters; USA: Ohio).
— Attems, 1930: 260 (*Otocryptops*; characters; distribution).
— Chamberlin, 1931b: 97 (*Otocryptops*; USA: Oklahoma).
— Matthews, 1935: 28-30; Map: 31, plate 2, a, c-e (*Scolopocryptops*; characters; habitat; USA: Wisconsin).
— Brimley, 1938: 501 (*Otocryptops*; USA: North Carolina).
— Bücherl, 1939: 295 (*Otocryptops*; characters; note; Colombia).
— Causey, 1940: 67, 83, 87, 88, 89, 93, 96, 97; Tab. 1, 7 (*Otocryptops*; characters; desiccation; habitat; notes; USA: North Carolina).
— Bücherl, 1941: 332 (*Otocryptops*;.
— Chamberlin, 1942d: 184 (*Otocryptops*; USA: Arkansas, Louisiana, Missouri).
— Chamberlin, 1942b: 15 (*Otocryptops*; USA: Iowa).
— Chamberlin, 1944d: 176-177 (*Otocryptops*; USA: Arkansas, Georgia, Illinois, Indiana, Missouri, Tennessee). [**Note:** Although Chamberlin reports of *sexspinosus* from California and Washington, according to Shelley (2002) the report from Washington and California are *spinicaudus* and are excluded from the distribution listing].
— Chamberlin, 1944c: 79 (*Otocryptops*; note; USA: California).
— Chamberlin, 1945d: 215 (*Otocryptops*; USA: Georgia).
— Auerbach, 1946: 6, 10-11, 14, 15, 18, 21, 22, 34, 35; chart 3: 48-49 (*Otocryptops*; habitat; hygrotaxis; list; notes; USA: Illinois).
— Chamberlin, 1946b: 194 (*Otocryptops*; USA: Mississippi).
— Auerbach, 1949: 222, 225, 226; Fig. 5 (*Otocryptops*; ecology; hygrotaxis; notes; USA: Illinois).
— Crabill, 1950c: 201 (*Otocryptops*; notes; USA: South Carolina).
— Wray, 1950: 156 (*Otocryptops*; list).
— Auerbach, 1951a: 101, 104, 105, 106, 107, 108, 120; Tab. 1, 3, 5, 7-9, 14, 15; Fig. 1, 3 (*Otocryptops*; desiccation; ecology; habitat; heart beat; hygrotaxis; length to weight ratio; list; notes; oviposition period; population percentage; USA: Illinois).
— Chamberlin, 1951b: 33 (*Otocryptops*; USA: Florida, New York, North Carolina). [**Note:** According to Shelley (2002: 75) the report of *sexspinosus* from California is *spinicaudus*, hence the state is removed from the locality information].
— Crabill, 1952a: 128-129; Figs. 1, 8 (*Otocryptops*; notes).

— Crabill, 1952b: 204-207, pl. 5; Figs. 67, 69, 71-72 (*Scolopocryptops*;
behavioral note; characters; habitat; range; USA: Connecticut,
District of Columbia, Illinois, Indiana, Kentucky, Maryland,
Massachusetts, Michigan, New York, Ohio, Pennsylvania, Virginia,
West Virginia).

— Johnson, 1952: 134-138; Tab. 16, 28; Map 10 (*Otocryptops*; autecology,
characters; distribution; natural history; USA: Michigan).

— Dearolf, 1953: 231 (*Otocryptops*; cited as diplopod; list; USA:
Tennessee). [**Note:** Cave records].

— Crabill, 1953b: 96 (nomenclatorial notes).

— Hoffman & Hubricht, 1954: 193 (*Scolopocryptops*; reported from
stomach contents of the salamander *Plethodon*).

— Crabill, 1955f: 259 (*Scolopocryptops*; checklist; USA: Kentucky).

— Crabill, 1955g: 134 (*Scolopocryptops*; character notes).

— Crabill, 1955h: 157 (*Scolopocryptops*; notes; USA: Missouri).

— Branson & Batch, 1967: 84 (*Scolopocryptops*; characters; habitat; body
measurements; USA: Kentucky).

— Wray, 1967: 156 (*Otocryptops*; list).

— Holsinger & Peck, 1971: 32 (*Scolopocryptops*; USA: Georgia). [**Note:**
reference not obtained].

— Shelley, 1978: 221 (*Scolopocryptops*; habitat, range).

— Summers, 1979: 694, 696, 699; Fig. 11 (*Scolopocryptops*; checklist;
key).

— Summers & Uetz, 1979; Tab. 1, 2: 347-348, 349, 350, 351
(*Scolopocryptops*; microhabitat; notes; USA: Illinois).

— Lee, 1980: 2, 3, 5; Tab. 1 (*Scolopocryptops*; habitat; USA: Ohio).

— Summers et al., 1980: 245; Fig. 4 (*Scolopocryptops*; checklist; habitat
(Table 1, p. 242-243); USA: Illinois).

— Summers et al., 1981: 59 (*Scolopocryptops*; checklist; USA:
Illinois).

— Kevan, 1983: 2945 (*Scolopocryptops*; checklist; CANADA: British
Columbia). [**Note:** According to Shelley, 2002, this would be *S.*
spinicaudus].

— Gardner, 1986: 30 (*Scolopocryptops*). [**Note:** reference not
obtained].

— Shelley, 1987: 506-507; Figs. 6-7, 15 (*Scolopocryptops*; characters;
habitat; notes; USA: North Carolina).

— Shelley & Edwards, 1987: 3, 4; Fig. 7; Map 2 (*Scolopocryptops*; key;
USA: Florida).

— Holsinger & Culver, 1988: 57 (*Scolopocryptops*; USA: Tennessee).
[**Note:** Reported as an accidental occurrence in a cave].

— Snider, 1991: 188, 189 (*Scolopocryptops*; list; note).
— Shelley, 1992: 23, 25; Figs. 1-5 (*Scolopocryptops*; characters; notes; CANADA; USA: Alaska).
— Behan-Pelletier, 1993: 25 (*Scolopocryptops*; list).
— Hoffman, 1995a: 26 (*Scolopocryptops*; list; notes).
— Schileyko, 1995: 75; Fig. 1 (*Scolopocryptops*; characters; distribution; Vietnam).
— Schileyko & Minelli, 1995: 75 (*Scolopocryptops*; characters; distribution).
— Watermolen, 1997: 2 (*Scolopocryptops*; checklist, notes; USA: Wisconsin).
— Shelley, 1999: 5; Fig. 5 (*Scolopocryptops*; note).
— Shelley, 2002: 53-60; Figs. 76-83 (*Scolopocryptops*; characters; distribution; habitat; list of published records; notes; CANADA: Ontario; USA: Alabama, Arkansas, Connecticut, District of Columbia, Florida, Georgia, Illinois, Kansas, Kentucky, Louisiana, Maryland, Massachusetts, Michigan, Mississippi, Missouri, New Jersey, New York, North Carolina, Ohio, Oklahoma, Pennsylvania, South Carolina, Tennessee, Texas, Virginia, West Virginia, Wisconsin).
— McAllister et al., 2003: 112-113 (*Scolopocryptops*; notes; USA: Arkansas, Oklahoma, Texas).
— Edgecombe G. and G. Giribet, 2004: 91, 98, 104; figs. 1c, 3c, 3h, 8h (*Scolopocryptops*; electron micrographs of second maxilla, mandibles, and collared sensilla on antenna).
— Hickerson, et al., 2004: 679-686 (*Scolopocryptops*; behavioral interactions with salamanders; USA: Ohio).
— Regier et al., 2005: 149; Tab. 1 (*Scolopocryptops*; table of taxon classification with GenBank accession numbers).
— Giribet & Edgecombe, 2006: 533, 534, 536; Fig 1,2; Tab. 1 (*Scolopocryptops*; cladograms; GenBank accession code).
georgicus Meinert, 1886a: 180 (*Scolopocryptops*; characters; USA: Georgia).
— Underwood, 1887: 63 (*Scolopocryptops*; key).
— Bollman, 1888d: 347 [= 1893: 99] (*Scolopocryptops*; synonym of *sexspinosus*).
— Bollman, 1893: 128, 147, 177 (*Scolopocryptops*; synonym of *sexspinosus*).
— Attems, 1930: 260 (*Scolopocryptops*; synonym of *sexspinosus*).
— Shelley, 2002: 53 (*Scolopocryptops*; synonym of *sexspinosus*).
gracilis Chamberlin (not Wood), 1943b: 97 (*Otocryptops*; USA: Texas).
  [**Note:** According to Shelley (2002: 53) this report is *sexspinosus*].

helvola C. L. Koch, 1863: 34-35; Tab. 76, Fig. 156 (*Scolopendropsis*; characters; North America).
— Bollman, 1893: 128, 147, 152, 177 (*Scolopendropsis*; notes; synonym of *sexspinosus*).
— Attems, 1930: 260 (*Scolopendropsis*; synonym of *sexspinosus*).

nigridius Kraepelin [not McNeill], 1903: 72 (*Otocryptops*; cited as questioned synonym of *sexspinosus*).
— Attems [not McNeill], 1930: 260 (*Scolopocryptops*; cited as synonym of *sexspinosus*). [**Note:** This was an improper synonymy listing that was most likely based on Kraepelin's questioned synonymy listing above].

punctatus Pocock, 1891: 159 (*Otocryptops*; characters; note; Southeast Korea). [**Note:** Type specimen deposited at the BMNH]. [**Note:** Pocock (1891) described the species *Otocryptops punctatus* from southeast Korea which is a junior synonym of *S. sexspinosus*. Furthermore, this species was reported from Vietnam for the first time by Schileyko (1992) and interestingly appears to be part of the Vietnamese fauna].
— Cook, 1904: 73 (*Otocryptops*; synonym of *sexspinosus*).
— Attems, 1930: 260 (*Otocryptops*; synonym of *sexspinosus*).

rubiginosus Chamberlin (not Koch), 1962: 134 (*Scolopocryptops*; USA: Oregon). [**Note:** According to Shelley (2002: 53) this report by Chamberlin of *rubiginosus* from Oregon is actually *sexspinosus*].

spinicaudus Bollman (not Wood), 1893: 128 (*Scolopocryptops*; cited as synonym of *sexspinosus*).

Distribution: CANADA: ?British Columbia, Ontario; USA: Alabama, ?Alaska, Arkansas, ?California, Connecticut, District of Columbia, Florida, Georgia, Illinois, Indiana, Iowa, Kansas, Kentucky, Louisiana, Massachusetts, Maryland, Michigan, Minnesota, Missouri, Mississippi, North Carolina, Nebraska, New Jersey, New York, Ohio, Oklahoma, Pennsylvania, South Carolina, Tennessee, Texas, Virginia, ?Washington, Wisconsin, West Virginia.

HOLOTYPE: Unknown: ?Pennsylvania: Philadelphia.

LECTOTYPE: BMNH, unknown locality, date, & collector, possibly from Georgia or Florida (see Shelley, 2002: 53).

♂♀ *S. spinicaudus* Wood, 1862
— Wood, 1862: 39 (*Scolopocryptops*; characters; USA: Oregon, Washington).

— Wood, 1865: 174; Fig. 15 (*Scolopocryptops*; characters; USA: Oregon).

— Wood, 1867b: 128 (*Scolopocryptops*; USA: California).

— Bollman, 1893: 178-179, 180 (*Scolopocryptops*; nominal taxon *sexspinosus*; etymology; list, notes; USA: California, Oregon, Washington).

— Gunthorp, 1920: 113 (*Scolopocryptops*; list).

— Chamberlin & Wang, 1952b: 179 (*Otocryptops*; Japan).

— Shelley, 2002: 72-76; Figs. 114-140 (*Scolopocryptops*; characters; distribution; habitat; list of published records; notes; CANADA: British Columbia; USA: Alaska, California, Colorado, Oregon, Washington).

nipponicus Shinohara, 1990: 63-64; Figs. 1-4 (*Scolopocryptops*; characters).

— Shelley, 2002: 73 (*Scolopocryptops*; synonym of *spinicaudus*).

rubiginosis Chamberlin (not Koch), 1919: 15H (*Otocryptops*; list). [**Note:** Shelley (1992: 23) considered this report as *sexspinosus* but based on distribution maps from Shelley (2002) *sexspinosus* is not a western species].

sexspinosus Brölemann (not Say), 1904: 244 (*Otocryptops*; USA: Washington). [**Note:** The record from Washington state is *spinicaudus* according to Shelley (2002) while the remaining localities are true *sexspinosus*].

— Chamberlin (not Say), 1944d: 176-177 (*Otocryptops*; USA: California and Washington). [**Note:** Although Chamberlin reports of *sexspinosus* from California and Washington, the reports from these localities must be *spinicaudus* according to Shelley (2002)].

— Chamberlin (not Say), 1951b: 33 (*Otocryptops*; USA: California). [**Note:** Although Chamberlin reports *sexspinosus* from California, Shelley (2002) has determined that these reports are of *spinicaudus*].

— Kevan (not Say), 1979: 298 (*Scolopocryptops*; distribution; list). [**Note:** Kevan reporting of *sexspinosus* was actually *spinicaudus* (see Shelley, 2002: 75)].

— Kevan (not Say), 1983: 2945 (*Scolopocryptops*; checklist; note).

— Shelley (not Say), 1992: 23-27 (*Scolopocryptops*; characters; distribution; CANADA; USA: Alaska).

Distribution: CANADA: British Columbia; USA: Alaska, California, Colorado, Oregon, Washington.

HOLOTYPE: Lost (see Shelley, 2002: 73).

NEOTYPE: NMNH, Oregon: Curry Co., Loeb State Park, 23 May 1957, by B. Malkin.

PARANEOTYPES: NMNH, 4 specimens; Oregon: Curry Co., Loeb State Park, 23 May 1957, by B. Malkin.

Subclass **Pleurostigmophora** (Pocock, 1902)
Superorder **Epimorpha** Haase, 1880
Order **Geophilomorpha** Pocock, 1895

Superfamily **Dignathodontoidea** Crabill, 1970

Family **Dignathodontidae** Cook, 1895

[**Note:** Crabill (1953d) considered *Scolioplanes* a synonym of *Linotaenia* Koch. Crabill (1960b) synonymized *Linotaenia* with *Strigamia*. Crabill (1970a) also considered the family "Scolioplanidae" equivalent to Dignathodontidae. Therefore, Linotaeniidae and Scolioplanidae are combined here under Dignathodontidae. It seems Linotaeniidae is still published as an accepted family but I am not certain of how the ICZN works for this situation. To my knowledge this issue has not been clearly addressed in a publication but I will leave the determination to chilopodologists more familiar with the literature and ICZN].

Subfamily **Agathothinae** Chamberlin, 1966

Genus *Agathothus* Bollman, 1893
[1 species; Southeastern USA]

Agathothus Bollman, 1893: 166 (characters; etymology). TYPE SPECIES: *Scolioplanes gracilis* Bollman, 1888, fixed by original designation.
— Cook, 1895b: 73 (distribution; list; notes).
— Chamberlin, 1912c: 428-429 (characters; notes).
— Attems, 1929: 231 (characters; distribution).
— Attems, 1947: 130 (key) [**Note:** Misspelled *Agathotus*].
— Crabill, 1952b: 85 (characters; note) [**Note:** Crabill states in his footnote "Brimley, (Insects of North Carolina, 1938), reports the presence of *A. carolinae* Chamberlin in the Duke Forest, North Carolina; however, Chamberlin has admitted (in litt.) that this name must be considered a nomen nudum." The author has not found this literature Crabill speaks of but *carolinae* is considered a nomen nudum at this time].
— Chamberlin, 1912c: 428-429 (characters).
— Chamberlin, 1966: 215, 216, 218 (key; list; notes). [**Note:** Chamberlin erected the subfamily Agathothinae].
— Jeekel, 2005: 68 (list).

♂♀*A. gracilis* (Bollman, 1888)

— Bollman, 1888a: 110 [= 1893: 84] (*Scolioplanes*; characters; notes; USA: Tennessee).

— Bollman, 1888c: 341 [= 1893: 92] (*Scolioplanes*; note; USA: Tennessee).

— Bollman, 1893: 126, 166 (*Linotaenia* + *Scolioplanes*; catalogue; cited as type species of the *Agathothus*).

— Cook, 1895b: 73 (*Agathothus*; list; notes).

— Attems, 1903b, 269 (*Scolioplanes*; list).

— Chamberlin, 1912c: 429-430, pl. 2; Figs. 7-11 (*Agathothus*; characters; USA: Tennessee). [**Note:** Chamberlin mentions that his description " ... is chiefly that of partly grown female"].

— Attems, 1929: 231 (*Agathothus*; characters; distribution). [**Note:** In German].

— Causey, 1940: 60-61; Figs. 31, 33a-b (*Agathothus*; characters; note; USA: Tennessee). [**Note:** Causey misspelled the genus as *Agothothus* and cited the author incorrectly as being Harger, 1872 when the actual author of this species was Bollman, 1888. It is clear that she confused *Geophilus gracilis* Harger, 1872, now a junior synonym of *Schendyla nemorensis* Koch, with *A. gracilis*].

— Crabill, 1952b: 85-86 (*Agathothus*; cited as type species; note; USA: South Carolina). [**Note:** I question the South Carolina locality by Crabill as I have not seen a published locality from this state].

— Chamberlin, 1966: 217; fig. 1 (figure of labrum).

— Hoffman, 1995a: 22 (*Agathothus* [misspelled *Agathotus*]; list; notes).

— Jeekel, 2005: 68 (*Agathothus*; list; note).

Distribution: USA: ?South Carolina, Tennessee.

HOLOTYPE: ?ISM; Tennessee: Jefferson Co., Beaver Creek, open cedar thickets May 21-26, 1887.

Genus *Horonia* Chamberlin, 1966
[1 species; Southwestern USA]

Horonia Chamberlin, 1966: 216 (characters). TYPE SPECIES: *Horonia bella* Chamberlin, 1966, fixed by original designation and monotypy.
— Shelley, 2006a: 11 (list).

H. bella Chamberlin, 1966

— Chamberlin, 1966: 216-218; Fig. 2-4 (*Horonia*; characters; USA: New Mexico). [**Note:** Known from two specimens and sex not mentioned in description].

Distribution: USA: New Mexico.

HOLOTYPE: ?Unknown, New Mexico, 13 mi. NE Bernabello, 17 September 1946, Stanley Mulaik.

PARATYPE: ?Unknown, New Mexico, 13 mi. NE Bernabello, 17 September 1946, Stanley Mulaik.

Genus *Malochora* Chamberlin, 1941
[1 species; Western USA]

Malochora Chamberlin, 1941d: 773 (characters). TYPE SPECIES: *Malochora linsdalei* Chamberlin, 1941, fixed by original designation and monotypy.
— Chamberlin, 1966: 219 (list).
— Jeekel, 2005: 88 (list).

♂*M. linsdalei* Chamberlin, 1941
— Chamberlin, 1941d: 773-774 (*Malochora*; characters; USA: California).
— Jeekel, 2005: 88 (*Malochora*; list).

Distribution: USA: California.

HOLOTYPE: NMNH, ♂, California: Monterey Co., Hastings Reservation, 1 March 1941, by Dr. J.M. Linsdale.

Genus *Zantaenia* Chamberlin, 1960
[1 species; Northwestern USA]

Zantaenia Chamberlin, 1960b: 240 (characters). TYPE SPECIES: *Zantaenia idahona* Chamberlin, 1960, fixed by original designation and monotypy.
— Chamberlin, 1966: 218-219 (list).
— Shelley, 2006a: 15 (list).

♂♀*Z. idahona* Chamberlin, 1960
— Chamberlin, 1960b: 240-241 (*Zantaenia*; characters; USA: Idaho).

Distribution: USA: Idaho.

HOLOTYPE: NMNH, Idaho: Wallace, 3 September 1949, by S. Mulaik.
[**Note:** Vial label may only have the genus noted].

Subfamily **Dignathodontinae** Attems, 1926

Genus *Chaetechelyne* Meinert, 1870
[1 species (Introduced); Northeastern USA & ?CANADA]

Chaetechelyne Meinert, 1870: 44-46; Tab. 3, Figs. 20-26 (characters; notes).
TYPE SPECIES: *Geophilus vesuvianus* Newport, 1845, fixed by subsequent
designation of Cook, 1896.
— Latzel, 1880: 200-201 (characters; distribution).
— Bollman, 1893: 166 (in key to subfamily Geophilinae).
— Attems, 1929: 236-238 (characters; distribution; key to species).
— Jeekel, 2005: 73 (list).

♂♀*C. vesuviana* (Newport, 1845)
— Newport, 1845b: 435 (*Geophilus*; characters; "Prope Neapolin").
— Newport, 1856: 87 (*Geophilus*; characters; "Prope Neapolin").
— Meinert, 1870: 46-47 (*Chaetechelyne*; characters; notes).
— Latzel, 1880: 201 (*Chaetechelyne*; characters; distribution).
— Attems, 1929: 237, 238-239; Fig. 200 (*Chaetechelyne*; characters;
distribution).
— Crabill, 1955e: 248, 249 (*Chaetechelyne*; habitat; notes; USA: New
York). [**Note:** This is most likely an introduced species from Europe.
In North America it was found for the first time in Rochester, New
York. The extent of its current distribution is unknown but it does
appear to have some establishment].
— Eason, 1964: 82-86; Figs. 82-90 (*Chaetechelyne*; characters;
distribution). [**Note:** Eason notes that this species also occurs in
Iberia, the Mediterranean region and northern Africa apart from
Europe].
— Matic, 1972: 135-139; Figs. 52, 53 (*Chaetechelyne*; characters;
ecological note; Romania).
— Kevan, 1979: 298 (*Chaetechelyne*; ?CANADA). [**Note:** Kevan
suspected that this species may occur in Canada].
— Kevan, 1983: 2945 (*Chaetechelyne*; notes; unconfirmed report from
CANADA: Ontario).
— Jeekel, 2005: 73 (*Chaetechelyne*; list).

Distribution: CANADA: ?Ontario; USA: New York.

HOLOTYPE: "*v. in* Mus. D. Hope"(see Newport, 1845b), Europe: Prope Neapolin.

Genus *Damothus* Chamberlin, 1960
[2 species; Western USA]

Damothus Chamberlin, 1960b: 239 (characters; note). **Type Species:** *Damothus montis* Chamberlin, 1960, fixed by original designation and monotypy.
— Shelley, 2006a: 10 (list).

♂*D. alastus* Crabill, 1962
— Crabill, 1962a: 82-84, pl. 5; Figs. a-d (*Damothus*; characters; USA: Utah).

Distribution: USA: Utah.

HOLOTYPE: MCZ (Unique No. 14227; TC-664) ♂, Utah: Tooele Co., Ophir, Oquirrh Mountains, 2000 m, 25 April 1961, by Herbert W. Levi, in cottonwood, sage.

♂*D. montis* Chamberlin, 1960
— Chamberlin, 1960b: 239-240 (*Damothus*; characters; USA: Utah).
— Chamberlin, 1961: 99 (*Damothus*; note; USA: Utah).

Distribution: USA: Utah.

HOLOTYPE: NMNH, Utah: Wahsatch Mountains, Brighton, 19 November 1939.

Genus *Korynia* Chamberlin, 1941
[5 species; South Central and Western USA]

Korynia Chamberlin, 1941d: 774 (characters). **Type Species:** *Korynia carmela* Chamberlin, 1941, fixed by original designation.
— Chamberlin, 1954: 122 (notes).
— Jeekel, 2005: 86 (list).

♂*K. auxa* Chamberlin, 1954
— Chamberlin, 1954: 122 (*Korynia*; characters; USA: California).

Distribution: USA: California.

HOLOTYPE: NMNH, ♂, California: Squaw Valley, 23 March 1941, S. & D. Mulaik.

♂*K. carmela* Chamberlin, 1941
— Chamberlin, 1941d: 774-775 (*Korynia*; characters; note; USA: California).
— Jeekel, 2005: 86 (*Korynia*; list).

Distribution: USA: California.

HOLOTYPE: NMNH, ♂, California: Monterey Co., Hastings Reservation, 1 February 1941, by Dr. J.M. Linsdale. [**Note:** Holotype label reads *Linotaenia* as the genus].

♀*K. texensis* Chamberlin, 1941
— Chamberlin, 1941d: 775-776 (*Korynia*; characters; note; USA: Texas).

Distribution: USA: Texas.

HOLOTYPE: ?Unknown, ♀, Texas: Bangs, 10 February 1938, by Christensen, Jones & Clancy. [**Note:** This type my be at the NMNH with the specific name as "*amplipes*"?].

♀*K. tripora* Chamberlin, 1941
— Chamberlin, 1941d: 775 (*Korynia*; characters; USA: California).

Distribution: USA: California.

HOLOTYPE: NMNH, ♀, California: Tulare Co., 12 mi. NE of Hammond, 21 March 1941, by S. & D. Mulaik.

♂♀*K. urania* (Crabill, 1954)
— Crabill, 1954c: 418; Figs. 1-3 (*Tomotaenia*; cited as subgenus *Korynia*; characters; key to subspecies; USA: Missouri).
— Crabill, 1955h: 159 (*Tomotaenia*; notes; USA: Missouri). [**Note:** *Tomotaenia* is a junior synonym of *Strigamia* so it has been transferred to *Korynia*].

Distribution: USA: Missouri.

HOLOTYPE: NMNH, Crabill collection No. 1764, with slides 677-678; ♀, Missouri: St. Louis Co., ~ 23 mi. W of St. Louis, 5 mi. W of Valley Park and 2 mi. E of Eureka, Beaumont Boy Scout Reservation, 18 October 1936, by Dr. E.P. Meiners.

Genus *Steneurytion* Attems, 1909
[1 species; USA: Hawaii]

Steneurytion Attems, 1909: 28 (*Eurytion*; subgenus *Steneurytion*; characters). TYPE SPECIES: *Geophilus sitocola* Attems, 1903, fixed by original designation.
— Attems, 1926: ? (raised to genus level from subgenus of *Eurytion*). [**Note:** Reference not obtained].
— Bonato, Pereira & Minelli, 2007: 8-9 (characters; species list; taxonomic & nomenclatorial notes).
Zelanion Chamberlin, 1920c: 39 (characters; notes). TYPE SPECIES: *Zelanion dux* Chamberlin, 1920, fixed by original designation. [**Note:** The type species *dux* is described from New Zealand; holotype at MCZ (No. 1906); paratype specimen is deposited in the NMNH].
— Archey, 1936: 53 (synonym of *Steneurytion*). [**Note:** Reference not obtained].
— Jeekel, 2005: 110 (list).
— Bonato, Pereira & Minelli, 2007: 8, 9 (characters; synonym of *Steneurytion*).

S. hawaiiensis (Chamberlin, 1953)
— Chamberlin, 1953b: 82-83 (*Zelanion*; characters; note; USA: Hawaii).
— Archey, 1936: 53 (transferred to *Steneurytion*). [**Note:** Reference not obtained].
— Nishida, 1994 26 (*Zelanion*; checklist).
— Nishida, 1997, 195 (*Zelanion*; checklist).
— Nishida, 2002, 235 (*Zelanion*; checklist).
— Bonato, Foddai, Minelli & Shelley, 2004: 26, 27, 28 (*Zelanion*; characters; key; notes).
— Bonato, Pereira & Minelli, 2007: 8 (*Steneurytion*; list; note; transferred to *Steneurytion*).

Distribution: USA: Hawaii.

HOLOTYPE: NMNH: ♂, Hawaii.

Genus *Strigamia* Gray, 1842
[15 species; CANADA and USA]

Strigamia Gray, 1842: 547 (characters). **TYPE SPECIES:** *Strigamia fulva* Sager, 1856, fixed by subsequent monotypy (see Crabill, 1953d: 171). [**Note:** There has long been confusion regarding this genus with respect to three other genera *Linotaenia* Koch, 1847, *Tomotaenia* Cook, 1895, and *Scolioplanes* Bergsöe & Meinert, 1866. Crabill (1953d) clearly explains the situation and the reader is referred to this publication for the full discussion of the matter].

— Wood, 1862: 46; fig. 11 (characters).
— Wood, 1865: 181; fig. 20 (characters).
— Bollman, 1893: 126 (*Strigamia*; questioned synonym of *Linotaenia*).
— Cook, 1895c: 864-866 (taxonomic discussion; cited as synonym of *Geophilus*; note). [**Note:** Cook considers *Strigamia* a synonym of *Linotaenia* and of *Tomotaenia* as well].
— Chamberlin, 1920f: 203 (note).
— Crabill, 1952b: 86-99 (characters; key to species; taxonomic notes). [**Note:** Crabill's taxonomic discussion is lengthy and has important insight into the long standing confusion of this genus].
— Crabill, 1953d: 169-172 (nomenclatorial discussion). [**Note:** Crabill cites *Strigamia fulva* Sager, as the type species of *Strigamia*].
— Weaver, 1982: unnumbered page (key to species of eastern United States; note).
— Chipman, et. al., 2004: 1250-1255 (segment generation in development). [**Note:** *Strigamia maritima* (Leach) is used as a model organism to investigate double segment periodicity].
— Jeekel, 2005: 106 (list; notes).

Diplochora Attems, 1903b: 281 (characters). **TYPE SPECIES:** *Diplochora fusata* Attems, 1903 [=*Tomotaenia fusata* (Attems)], fixed by monotypy.
— Crabill, 1962a: 85, 86 (notes; synonym of *Tomotaenia*). [**Note:** see footnote 2 pg. 85-86].
— Crabill, 1962c: 181 (synonym of *Tomotaenia*).
— Chamberlin, 1963a: 90 (synonym of *Tomotaenia*).
— Jeekel, 2005: 77 (list).

Linotaenia C. Koch, 1847: 86 (characters). **TYPE SPECIES:** *Linotaenia rosulans* Koch, 1847, fixed by subsequent designation of Pocock, 1890 (see, Jeekel, 2005).

— Bollman, 1893: 142 (notes on synonymy).
— Cook, 1895c: 865-866 (distribution; taxonomic discussion).
— Chamberlin, 1912a: 658-659 (characters).
— Chamberlin, 1912c: 423 (key to species).
— Holmquist, 1926: 411 (notes on hibernation such as: activity, date collected, developmental stage, ecological association/niche, quantity found).
— Bailey, 1928b: 41-42 (characters; key to species of New York; species included: *bidens, chionophila, fulva*).
— Crabill, 1953d: 169-172 (nomenclatorial discussion).
— Crabill, 1960b: 190 (synonym of *Strigamia*).
— Jeekel, 2005: 87 (list; notes).

Paraplanes Verhoeff, 1933: 22-23 (characters). **TYPE SPECIES:** *Paraplanes svenhedini* Verhoeff, 1933, fixed by monotypy.
— Chamberlin, 1941d: 775 (synonym of *Tomotaenia*). [**Note:** see footnote at bottom of page 775].
— Chamberlin, 1954: 118, 119 (synonym of *Tomotaenia*).
— Crabill, 1954c: 416-417 (subjective synonym of *Tomotaenia*).
— Crabill, 1962c: 181 (synonym of *Tomotaenia*).
— Chamberlin, 1963a: 90 (synonym of *Tomotaenia*).
— Jeekel, 2005: 96 (list).

Stenotaenia Koch, 1847: 85-86 (characters). **TYPE SPECIES:** *Geophilus linearis* Koch, 1835, fixed by subsequent designation.
— Bollman, 1893: 126 (*Stenotaenia*; in part, synonym of *Linotaenia*). [**Note:** misspelled *Stenonia*].
— Jeekel, 2005: 105 (list).
— Bonato & Minelli, 2008: 253-286; Figs. 1-8 (characters; notes).

Scolioplanes Bergsöe & Meinert, 1866: (characters). **TYPE SPECIES:** *Geophilus maritimus* Leach, 1817 [= *Strigamia* (*Strigamia*) *maritima* (Leach)] by subsequent designation by Cook, 1895b: 75 and Cook, 1895c: 866.
— Crabill, 1953d: 169-172 (nomenclatorial discussion; synonym of *Linotaenia*).
— Cook, 1895c: 865, 866 (taxonomic discussion; cited as synonym of *Linotaenia*).
— Bollman, 1893: 126 (*Scolioplanes*; synonym of *Linotaenia*).
— Crabill, 1953d: 170-172 (nomenclatorial discussion).
— Crabill, 1960b: 190 (synonym of *Strigamia*).
— Jeekel, 2005: 103 (list).

Tomotaenia Cook, 1895c: 866 (notes). **TYPE SPECIES:** *Strigamia parviceps* Wood, 1862, fixed by original designation.
— Cook, 1895b: 73 (distribution; list).
— Crabill, 1953d: 171, 172 (taxonomic notes; list).

— Chamberlin, 1954: 118–119 (taxonomic notes).
— Crabill, 1962c: 179, 180, 181 (discussion of characters).
— Chamberlin, 1963a: 88-90 (taxonomic notes).
— Matic, 1972: 139 (synonym of *Strigamia*).
— Jeekel, 2005: 108 (list).

♂♀*S. acuminata* (Leach, 1815)
— Leach, 1815: 386 (*Geophilus*; characters; habitat; England: Battersea fields, Plymouth).
— Newport, 1845b: 434 (*Geophilus*; characters; notes; England).
— Koch, 1847: 188 (*Stenotaenia*; list).
— Bergsöe & Meinert, 1866: 101-102 (*Scolioplanes*; characters; notes; Denmark).
— Cook, 1895c: 866 (*Linotaenia*; distribution note).
— Attems, 1909a: 9, 25–26 (*Scolioplanes*; characters; list).
— Attems, 1929: 222; fig. 194 (*Scolioplanes*; characters; distribution; USA: Alaska).
— Verhoeff, 1935: 19 (*Scolioplanes*; subgenus *Scolioplanes*; within key to species in German).
— Crabill, 1952b: 99-102 (*Strigamia*; subgenus *Linotaenia*; characters; taxonomic notes; USA: District of Columbia, New York). [**Note:** Crabill mentions that this widespread European species is not known to have established itself in the United States, but has been intercepted at Washington D.C. quarantine in soil from England, and among apple cuttings from Germany in New York City].
— Crabill, 1953d:172 (*Strigamia*; subgenus *Linotaenia*; cited as type of *Linotaenia*).
— Crabill, 1954a: 45; Fig. 1 (*Strigamia*; labrum compared to *bothriopa*).
— Eason, 1964: 92-95; Figs. 111-123. (*Strigamia*; characters (adult and stadia); distribution; habitat). [**Note:** Eason suspects that *Strigamia chionophila* Wood is a junior synonym of *acuminata*].
— Behan-Pelletier, 1993: 25 (*Strigamia*; list).
— Watermolen, 1997: 3 (*Strigamia*; checklist, notes).
crassipes Bollman [not Koch], 1893: 142 (*Linotaenia*; notes; synonym of *acuminata*).
— Brölemann [not Koch], 1932: 128 (*Scolioplanes*; synonym of *acuminata*). [**Note:** Reference not obtained].
— Crabill, 1952b: 100 (synonym of *acuminata*).
?*hirsutipes* Attems, 1927: 293-294; fig. 1, 3 (*Scolioplanes*; characters; Japan). [**Note:** Because Attems compares this species to *acuminatus* it is listed here for reference].

rosulans C. L. Koch, 1847: 188-189 (*Linotaenia*; characters; "Baiern" Germany).
— Plateau, 1872: 417 (synonym of *acuminata*).
— Attems, 1929: 222 (*Linotaenia*; synonym of *acuminata*).
— Crabill, 1953d: 170, 172 (nomenclatorial discussion; synonym of *acuminata*).
— Jeekel, 2005: 87 (*Linotaenia*; list).
sanguineus Gervais, 1847: 316-317 (*Geophilus*; characters; notes).
— Attems, 1929: 222 (*Geophilus*; synonym of *acuminata*).
variabilis Verhoeff, 1895: 354 (*Scolioplanes*;). [**Note:** Reference not obtained].
— Attems, 1929: 222 (*Scolioplanes*; synonym of *acuminata*).

Distribution: USA: Alaska, ?District of Columbia, ?New York.

HOLOTYPE: ?Unknown, (v. in Mus. Brit.) (see Newport, 1845)

♂♀*S. bidens* Wood, 1862
— Wood, 1862: 47 (*Strigamia*; characters; notes; USA: "near Philadelphia").
— Wood, 1865: 183 (*Strigamia*; characters).
— Bollman, 1888c: 340 [= 1893: 92] (*Linotaenia*; note; USA: Tennessee).
— Bollman, 1893: 126 (*Linotaenia*; catalogue).
— Brölemann, 1896: 58–60, pl. 6; Figs. 10-13 (*Scolioplanes*; characters).
— Chamberlin, 1909b: 186 (*Linotaenia*; USA: North Carolina).
— Chamberlin, 1912c: 409, 426-428, pl. 2; Figs. 1-3 (*Linotenia*; characters; notes; USA: Georgia, Kentucky, Mississippi, North Carolina, Tennessee, Virginia).
— Gunthorp, 1920: 113 (*Strigamia*; list).
— Chamberlin, 1928b: 153 (*Linotaenia*; USA: Missouri).
— Williams & Hefner, 1928: 134 (*Linotenia*; characters; USA: Ohio).
— Attems, 1929: 226 (*Scolioplanes*; characters; distribution).
— Brimley, 1938: 502 (*Linotenia*; USA: North Carolina).
— Causey, 1940: 61 (*Linotaenia*; characters; USA: North Carolina).
— Chamberlin, 1940c: 56 (*Linotaenia*; note; USA: North Carolina).
— Chamberlin, 1942d: 185 (*Linotaenia*; habitat; USA: Missouri).
— Crabill, 1952b: 102-104, pl. 2; Figs. 19, 25, 28 (*Strigamia*; subgenus *Linotaenia*; characters; range; taxonomic notes; USA: Indiana, Kentucky, Ohio, Pennsylvania, Virginia). [**Note:** Crabill synonymized *Strigamia walkeri* with *bidens*].

— Crabill, 1955b: 38 (*Strigamia*; range; USA: Missouri). [**Note:** Crabill
states that this ". . . species is now believed to range from eastern
Pennsylvania south to northern Georgia and Louisiana, and west to
eastern Missouri"].

— Crabill, 1955f: 258 (*Strigamia*; checklist; USA: Kentucky).

— Crabill, 1955h: 158 (*Strigamia*; note; USA: Missouri).

— Branson & Batch, 1967: 82 (*Strigamia*; characters; habitat; note;
USA: Kentucky).

— Wray, 1967: 156 (*Linotaenia*; list).

— Summers, 1979: 697, 699 (*Strigamia*; checklist; key).

— Lee, 1980: 2, 5; Tab. 1 (*Strigamia*; habitat; USA: Ohio).

— Summers et al., 1980: 247; Fig. 8 (*Strigamia*; checklist; habitat (Table
1, p. 242-243); USA: Illinois).

— Summers et al., 1981: 60 (*Strigamia*; checklist; USA: Illinois).

— Hoffman, 1995a: 22 (*Strigamia*; list; notes).

?*laevipes* Wood, 1862: 48 (*Strigamia*; characters; USA: Georgia). [**Note:**
Wood (1862) described the species *Strigamia laevipes* from Georgia.
It is tentatively considered a junior synonym of *S. bidens* here for
reasons explained in the notes under *Tomotaenia parviceps* (see notes
for *T. parviceps*)].

— Wood, 1865: 184-185 (*Strigamia*; synonym of *bidens*).

— Wood, 1867b: 128 (*Strigamia*; synonym of *bidens*).

— Gunthorp, 1920: 113 (*Strigamia*; synonym of *bidens;* list).

ruber Bollman, 1887a: 82 [= 1893: 20] (*Scolioplanes*; characters; USA: Indiana).

— Bollman, 1888a: 107, 110 [= 1893: 82, 84] (*Scolioplanes*; note; USA:
Tennessee).

— Bollman, 1888c: 340 [= 1893: 91] (*Linotaenia*; note; USA:
Tennessee).

— Bollman, 1888e: 408 [= 1893: 109] (*Linotaenia*; note; USA:
Indiana).

— Bollman, 1893: 126, 132-133 (*Linotaenia* + *Scolioplanes*; catalogue;
characters; note).

— Cook, 1895c: 866 (*Tomotaenia*; note).

— Chamberlin, 1912c: 409 (synonym of *bidens*).

?*walkeri* Wood, 1865: 184 (*Strigamia*; characters; USA: Pennsylvania).

— Bollman, 1893: 126 (*Linotaenia*; catalogue).

— Gunthorp, 1920: 113 (*Strigamia*; list).

— Crabill, 1952b: 102 (*Strigamia*; synonym of *bidens*). [**Note:**
Crabill synonymized *walkeri* with *bidens* in his thesis but it
was never published and he never explains why. If you simply
look at Wood's description he claims *walkeri* has 64 pairs

of legs, which is an even number. All centipedes exhibit an odd number of pairs of legs. Although this is technically still a valid species the author has included it here as a questioned synonym of *bidens* which has 65-81 pairs of legs. According to Wood the type specimen should be deposited at the NMNH, however, it does not appear in their type database].

Distribution: USA: Georgia, Illinois, Indiana, Kentucky, Louisiana, Missouri, Mississippi, North Carolina, Ohio, Pennsylvania, Tennessee, Virginia.

HOLOTYPE: ANSP, 11227, Pennsylvania, near Philadelphia.

♂♀*S. bothriopa* Wood, 1862
— Wood, 1862: 46 (*Strigamia*; characters; USA: Pennsylvania).
— Meinert, 1886a: 222-223 (*Scolioplanes*; characters; USA: Massachusetts).
— Bollman, 1888a: 110 [= 1893: 84] (*Scolioplanes*; note; USA: Tennessee).
— Bollman, 1888b: 4 [= 1893: 76] (*Linotaenia*; note; USA: Arkansas).
— Kenyon, 1893a: 16 (*Scolioplanes*; USA: Nebraska).
— Kenyon, 1893b: 161 (*Scolioplanes*; USA: Nebraska). [**Note:** Nebraska seems far away from the general distribution of this species so I have questioned this state as a locality].
— Cook, 1895: 866 (*Tomotaenia*; note).
— Attems, 1903b: 268 (*Scolioplanes*; list).
— Gunthorp, 1920: 113 (*Strigamia*; list).
— Crabill, 1952b: 104-106, pl. 2; Figs. 18, 24, 26-27; pl. 3; Fig. 36 (*Strigamia*; subgenus *Linotaenia*; characters; taxonomic note; USA: Indiana, Kentucky, Massachusetts, New York, Pennsylvania, Virginia).
— Crabill, 1954a: 43-44; Figs. 3, 6, 8-9 (*Strigamia*; characters; comparative discussion; USA: Indiana, Kentucky, Massachusetts, New York, Pennsylvania, Virginia).
— Crabill, 1955b: 38 (*Strigamia*; notes; USA: Missouri).
— Crabill, 1955f: 258 (*Strigamia*; checklist; USA: Kentucky).
— Crabill, 1955h: 158 (*Strigamia*; note; USA: Missouri).
— Crabill, 1958b: 96 (*Strigamia*; habitat; notes; USA: Wisconsin).
— Branson & Batch, 1967: 82. (*Strigamia*; note; USA: Kentucky).
— Summers, 1979: 697, 699; Fig. 12 (*Strigamia*; checklist; key).
— Summers & Uetz, 1979; Tab. 1, 2: 348, 349, 350 (*Strigamia*; microhabitat; USA: Illinois).

— Summers et al., 1980: 247; Fig. 9 (*Strigamia*; checklist; habitat (Table 1, p. 242-243); USA: Illinois).

— Summers et al., 1981: 60 (*Strigamia*; checklist; USA: Illinois).

— Hoffman, 1995a: 22 (*Strigamia*; list; notes).

— Watermolen, 1997: 4 (*Strigamia*; checklist, notes; USA: Wisconsin).

— Regier et al., 2005: 149; Tab. 1 (*Strigamia*; table of taxon classification with GenBank accession numbers).

— Giribet & Edgecombe, 2006: 533, 534, 536; Fig 1,2; Tab. 1 (*Strigamia*; cladograms; GenBank accession code).

?*fulva* Sager, 1856: 109 (*Strigamia*; characters).

— Meinert, 1886a: 222 (*Strigamia*; questioned synonym of *bothriopa*). [**Note:** species name misspelled *flava* and author is misspelled Sayer].

Distribution: USA: Arkansas, Illinois, Indiana, Kentucky, Massachusetts, Missouri, ?Nebraska, New York, Pennsylvania, Tennessee, Virginia, Wisconsin.

HOLOTYPE: ANSP, 11228, Pennsylvania: W Pennsylvania.

♂♀*S. branneri* (Bollman, 1888)

— Bollman, 1888b: 4-5 [= 1893: 76-77] (*Linotaenia*; characters; USA: Arkansas).

— Bollman, 1888c: 342 [= 1893: 93] (*Linotaenia*; note; USA: Tennessee).

— Cook, 1895: 866 (*Tomotaenia*; note).

— Bollman, 1893: 126 (*Linotaenia*; catalogue).

— Attems, 1903b: 268 (*Scolioplanes*; list).

— Chamberlin, 1912c: 425-426, pl. 2; Figs. 4-6 (*Linotenia*; characters; USA: Georgia, North Carolina, South Carolina, Tennessee, Virginia).

— Chamberlin, 1928b: 153 (*Linotaenia*; USA: Missouri).

— Attems, 1929: 229 (*Scolioplanes*; characters; distribution).

— Crabill, 1950c: 199-200 (*Linotaenia*; notes; range; USA: South Carolina).

— Crabill, 1952b: 110-113, pl. 2; Figs. 22, 30, 32; pl. 3; Fig. 37 (*Strigamia*; subgenus *Strigamia*; characters; taxonomic notes; USA: New York, Virginia).

— Crabill, 1955f: 258 (*Strigamia*; checklist; USA: Kentucky).

— Branson & Batch, 1967: 82 (*Strigamia*; brief description; habitat; USA: Kentucky).

— Wray, 1967: 156 (*Linotaenia*; list).

— Summers, 1979: 697, 699 (*Strigamia*; checklist; key).
— Summers et al., 1980: 247; Fig. 10 (*Strigamia*; checklist; habitat (Table 1, p. 242-243); USA: Illinois).
— Summers et al., 1981: 60 (*Strigamia*; checklist; USA: Illinois).
— Snider, 1991: 187 (*Strigamia*; list; notes; USA: Michigan).
— Hoffman, 1995a: 22 (*Strigamia*; list; notes).
— Watermolen, 1997: 7 (*Strigamia*; notes).

Distribution: USA: Arkansas, Georgia, Illinois, Kentucky, Michigan, Missouri, North Carolina, New York, South Carolina, Tennessee, Virginia.

HOLOTYPE: NMNH, ♀, Arkansas: Little Rock. [**Note:** Vial at NMNH has no locality data. ISM may have a type specimen].

S. branneri **var.** *miura* Chamberlin, 1912
— Chamberlin, 1912c: 426 (*Strigamia*; characters; USA: Georgia, Kentucky, North Carolina, Tennessee).

Distribution: USA: Georgia, Kentucky, North Carolina, Tennessee.

SYNTYPES: MCZ, No. 978, 2 specimens, Tennessee: Russellville, 8-10-10. [**Note:** Chamberlin did not designate a state or specimen in the literature to represent this variety. Based on his differentiating characters of shorter body and antennal length it may be safe to refer this species as *branneri* but it shall be left for future study].

S. cephalica Wood, 1862
— Wood, 1862: 49 (*Strigamia*; characters; USA: California).
— Bollman, 1893: 125 (*Himantarium*; catalogue).
— Gunthorp, 1920: 113 (*Strigamia*; list).

Distribution: USA: California.

HOLOTYPE: ?Unknown; California.

♂♀*S. chionophila* Wood, 1862
— Wood, 1862: 50 (*Strigamia*; characters; undeterminable locality but probably Canada).
— Wood, 1865: 189 (*Strigamia*; characters; undeterminable locality but probably Canada).
— Meinert, 1886a: 223-224 (*Scolioplanes*; characters; USA: Massachusetts).

— Bollman, 1888d: 346 [= 1893: 98] (*Linotaenia*; note; USA: District of Columbia).
— Bollman, 1888e: 408 [= 1893: 109] (*Linotaenia*; note; USA: Indiana).
— Bollman, 1893: 126 (*Linotaenia*; catalogue).
— Cook, 1895: 866 (*Tomotaenia*; note).
— Brölemann, 1896: 60 (*Scolioplanes*; list; note).
— Attems, 1903b: 268 (*Scolioplanes*; list).
— Cook, 1904: 73-74 (*Tomotaenia*; notes; USA: Alaska).
— Attems, 1909a: 9 (*Tomotaenia*; list; USA: Alaska).
— Chamberlin, 1909b: 186 (*Linotenia*; USA: Kentucky, New York, Texas).
— Chamberlin, 1911e: 260-262 (*Linotenis* (misspelling); taxonomic notes; USA: Alaska).
— Chamberlin, 1912b: 68 (*Linotenia*; habitat; taxonomic notes; USA: Wisconsin). [**Notes:** Chamberlin states "It is very close to *Linotaenia acuminata* (Leach) of Europe, and may have to be merged with it"].
— Chamberlin, 1912c: 424-425 (*Linotenia*; notes; USA: Kentucky).
— Chamberlin, 1914b: 301 (*Linotenia*; habitat (p. 301); USA: Michigan).
— Chamberlin, 1919: 15H (*Linotaenia*; list).
— Chamberlin, 1920a: 95 (*Linotaenia*; notes; CANADA: Ontario).
— Gunthorp, 1920: 113 (*Strigamia*; list).
— Chamberlin, 1923a: 242 (*Linotaenia*; taxonomic notes; USA: Alaska, New York). [**Note:** Chamberlin discusses the possibility of *chionophila* being conspecific with the European *acuminatus* of Leach (It appears that somehow in error the specific name "*attenuatus*" was subsequently and repeatedly used instead of *acuminatus*). He also compares the pairs of legs in males and females from various localities as well as coxal pores].
— Chamberlin, 1925e: 59 (*Linotaenia*; notes; USA: Utah).
— Holmquist, 1926: 411 (*Linotaenia*; notes on hibernation such as: activity, date collected, developmental stage, ecological association/niche, quantity found).
— Bailey, 1928a: 1083 (*Linotaenia*; list).
— Bailey, 1928b: 19, 42 (*Linotaenia*; list; notes; USA: New York).
— Williams & Hefner, 1928: 134 (*Linotaenia*; characters; USA: Ohio).
— Chamberlin, 1928d: 310 (*Linotaenia*; USA: Utah).
— Attems, 1929: 228 (*Scolioplanes*; characters; distribution).
— Chamberlin, 1930a: 114 (*Linotaenia*; note; USA: Utah).

— Matthews, 1935: 24; Map: 25 (*Linotaenia*; characters; USA: Wisconsin).
— Chamberlin, 1942d: 185 (*Linotaenia*; USA: Missouri).
— Chamberlin, 1943b: 100 (*Linotaenia*; USA: Colorado, Idaho, Montana, Wyoming).
— Chamberlin, 1944d: 194 (*Linotaenia*; USA: Indiana).
— Auerbach, 1946: 14, (*Linotaenia*; list; USA: Illinois).
— Chamberlin, 1946c: 183 (*Linotaenia*; USA: Alaska).
— Chamberlin, 1951b: 35 (*Linotaenia*; USA: California).
— Crabill, 1952b: 113-116, pl. 2; Figs. 21, 31 (*Strigamia*; subgenus *Strigamia*; characters; distributional notes; taxonomic notes; CANADA: Alberta, Ontario; USA: District of Columbia, Indiana, Kentucky, New York, Virginia).
— Johnson, 1952: 114, 116-122; Tab. 13, 28; Map 7 (*Linotenia*; autecology; characters; distribution; USA: Michigan).
— Crabill, 1955f: 258 (*Strigamia*; checklist; USA: Kentucky).
— Crabill, 1958b: 95 (*Strigamia*; habitat; notes; USA: Wisconsin).
— Chamberlin, 1961: 99 (*Linotaenia*; USA: Utah).
— Eason, 1964: 94 (*Strigamia*; note). [**Note:** Eason suggests that *chionophila* may be synonymous with *S. acuminata*].
— Kevan, 1979: 298 (*Strigamia*; CANADA: Northwest Territories). [**Note:** Kevan considers this species transcontinental and suspects that it may be a synonym of *acuminata* (Leach)].
— Summers, 1979: 697, 699 (*Strigamia*; checklist; key).
— Summers & Uetz, 1979: 347; Tab. 1 (*Strigamia*; USA: Illinois).
— Summers et al., 1980: 247 (*Strigamia*; checklist; habitat (Table 1, p. 242-243); USA: Illinois).
— Summers et al., 1981: 60 (*Strigamia*; checklist; USA: Illinois).
— Snider, 1991: 187 (*Strigamia*; list).
— Behan-Pelletier, 1993: 25 (*Strigamia*; list).
— Hoffman, 1995a: 22-23 (*Strigamia*; list; notes; USA: Virginia).
— Watermolen, 1997: 4 (*Strigamia*; checklist, notes; USA: Wisconsin).
acuminatus Leach, 1815: 386 (*Geophilus*; characters; habitat; England).
— Chamberlin, 1920a: 95 (*Linotaenia*; note on suspecting this as the senior synonym of *chionophila*).
— Chamberlin, 1923a: 242 (*Geophilus*; cited as questioned synonym of *chionophila*).
— Chamberlin, 1923a: 242 (*Scolioplanes*; synonym of *chionophila*).
— Chamberlin, 1925e: 59 (*Geophilus*; cited as questioned synonym of *chionophila*).
— Chamberlin, 1925e: 59 (*Scolioplanes*; synonym of *chionophila*).

albus Cook, 1904: 77 (*Escaryus*; characters; notes; USA: Alaska). [**Note:** **SYNTYPE:** NMNH, No. 794, Alaska: Pribilof Islands, St. Paul Island, 1897, by T. Kincaid.

— Attems, 1909a: 9 (*Escaryus*; list).
— Chamberlin, 1919: 15H (*Escaryus*; list).
— Chamberlin, 1923a: 244 (*Escaryus*; USA: Alaska).
— Attems, 1929: 97 (*Escaryus*; characters; distribution).
— Chamberlin, 1946c: 177 (*Escaryus*; list).
— Crabill, 1961: 68-69 see footnote (*Strigamia*; notes). [**Note:** Crabill claims that the cotypes were juveniles and undoubtedly referable to some species of the genus Strigamia].
— Pereira & Hoffman, 1993: 7-8 (*Escaryus*; taxonomic discussion). [**Note:** Based on the conclusions of Pereira & Hoffman the author has decided to assign *albus* as a tentative synonym of *chionophila*, however, in agreement with them the author feels this should be a case held open to further investigation and that the species should be considered a *species inquirenda*].

chionophilus Meinert, 1886a: 223-224 (*Scolioplanes*; characters; USA: Massachusetts).
— Cook, 1904: 73 (synonym of *chionophila*).

miuropus Chamberlin, 1902c: 478 - 479 (*Linotaenia*; characters; habitat). Type locality: USA: Utah, 25.v.1901. [**Note:** the exact type locality is not known but is most likely Logan Canyon in Cache county].
— Chamberlin, 1925e: 59 (*Linotaenia*; synonym of *chionophila*).
— Chamberlin, 1961: 99 (*Linotaenia*; synonym of *chionophila*).

Distribution: CANADA: Alberta, Northwest Territories, Ontario; USA: Alaska, California, Colorado, District of Columbia, Idaho, Illinois, Indiana, Kentucky, Massachusetts, Michigan, Missouri, Montana, New York, Ohio, Texas, Utah, Virginia, Wisconsin, Wyoming.

HOLOTYPE: ?Unknown, CANADA: Fort Simpson. [**Note:** MCZ has numerous specimens with the following catalog numbers: 975, 977, 979, 1826, 2192, 2232, 32109-32124, 73197].

♂♀ *S. fulva* Sager, 1856
— Sager, 1856: 109 (*Strigamia*; characters; note). [**Note:** Sager does not give a locality and states that this species "Approximates closely to the *Geophilus rubens* of Say"].
— Bollman, 1888c: 341 [= 1893: 92] (*Linotaenia*; note; USA: Tennessee).

— Bollman, 1888d: 346 [= 1893: 98] (*Linotaenia*; note; USA: Georgia).
— Bollman, 1888e: 408 [= 1893: 109] (Linotaenia; note; USA: Indiana).
— Bollman, 1893: 126, 184 (*Linotaenia*; catalogue; note; USA: Minnesota).
— Cook, 1895: 866 (*Tomotaenia*; note).
— Chamberlin, 1909b: 186 (*Linotenia* (misspelled); habitat; USA: Michigan, New York).
— Chamberlin, 1912b: 68 (*Linotenia*; notes; USA: Illinois; Iowa).
— Chamberlin, 1912c: 423-424, pl. 1; Figs. 8-10 (*Linotaenia*; notes; USA: Georgia, Kentucky, North Carolina, South Carolina, Tennessee, Virginia, West Virginia).
— Gunthorp, 1913: 171 (*Linotaenia*; note; USA: Kansas).
— Chamberlin, 1918b: 376 (*Linotenia*; USA: Louisiana).
— Chamberlin, 1921a: 230 (*Linotaenia*; USA: Tennessee).
— Holmquist, 1926: 411 (*Linotaenia*; notes on hibernation such as: activity, date collected, developmental stage, ecological association/ niche, quantity found).
— Bailey, 1928a: 1083 (*Linotaenia*; list).
— Bailey, 1928b: 19, 42 (*Linotaenia*; list; note; USA: New York).
— Williams & Hefner, 1928: 134-135 (*Linotaenia*; characters; USA: Ohio).
— Attems, 1929: 228 (*Scolioplanes*; characters; distribution).
— Matthews, 1935: 22; Map: 23 (*Linotaenia*; characters; USA: Wisconsin).
— Brimley, 1938: 502 (*Linotaenia*; USA: North Carolina).
— Causey, 1940: 61 (*Linotaenia*; characters; USA: North Carolina).
— Chamberlin, 1942d: 185 (*Linotaenia*; USA: Missouri).
— Chamberlin, 1943b: 100 (*Linotaenia*; USA: Colorado).
— Chamberlin, 1944d: 194 (*Linotaenia*; USA: Illinois, Louisiana).
— Dowdy, 1944: 203 (*Linotenia*; ecological note; USA: Ohio).
— Auerbach, 1946: 14 (*Linotaenia*; list; USA: Illinois).
— Rapp, 1946: 666 (*Linotaenia*; habitat; USA: Illinois).
— Auerbach, 1951a: 100, 106; Tab. 1, 14, 15; Fig. 1 (*Linotaenia*; food habits; habitat; heart beat; list; notes; population percentage; USA: Illinois).
— Crabill, 1952b: 106-110, pl. 2; Figs. 20, 23, 33; pl. 3; Fig. 38 (*Strigamia*; subgenus *Strigamia*; characters; distribution notes; taxonomic notes; USA: New York).
— Johnson, 1952: 122-126; Tab. 14; Map 8; Figs. 22-23 (*Linotenia*; autecology; characters; distribution; USA: Michigan).

— Crabill, 1954a: 41-43; Figs. 2, 4-5, 7, 10 (*Strigamia*; characters;
   distribution notes; USA: Connecticut, New York).
— Crabill, 1955f: 258 (*Strigamia*; checklist; USA: Kentucky).
— Chamberlin, 1961: 99 (*Linotenia*; USA: Utah).
— Wray, 1967: 156 (*Linotaenia*; list).
— Snider, 1991: 186-187 (*Strigamia*; list; notes).
— Jeekel, 2005: 106 (*Strigamia*; list).

bothriopa Wood, 1862: 46 (*Strigamia*; characters; USA: Pennsylvania).
— Wood, 1865: 182 (*Strigamia*; synonym of *fulva*).
— Bollman, 1888c: 341 [= 1893: 92] (*Linotaenia*; synonym of *fulva*).
— Bollman, 1893: 126 (*Strigamia* + *Scolioplanes*; synonym of *fulva*).
?— Kenyon, 1893a: 16 (*Scolioplanes*; USA: Nebraska).
?— Kenyon, 1893b: 161 (*Scolioplanes*; USA: Nebraska).
— Attems, 1929: 228 (*Scolioplanes*; synonym of *fulvus*).

?*laevipes* Wood, 1862: 48 (*Strigamia*; characters; USA: Georgia)
— Bollman, 1893: 126 (*Linotaenia*; catalogue).
— Chamberlin, 1909b: 186 (*Linotenia*; habitat; possible synonym of
   fulva; USA: California).
— Chamberlin, 1912a: 659, 664; fig. 217c; fig. 218c (*Linotenia*;
   subjective synonym of *fulva*; characters; range; USA: California).
— Anonymous, 1920: 35 (*Linotaenia*; note; USA: California).
— Gunthorp, 1920: 113 (*Strigamia*; list).
— Chamberlin, 1951b: 35 (*Linotenia*; USA: California).

micropus Chamberlin, 1902c: 479-480 (*Linotaenia*; characters (adult,
   stadia); habitat; key to species; range). Type locality: USA: Utah,
   Pole Canyon, San Pete County, 15.vii.1901.
— Chamberlin, 1961: 99 (*Linotenia*; synonym of *fulva*).

robustus Meinert, 1886a: 224 (*Scolioplanes*; characters; North America).
   [**Note:** Holotype deposited at the MCZ, No. 987; a label in the vial
   from Crabill reads "The spcm. is almost surely = Wood's *bothriopa*,
   1862, & nearly certainly came from the U.S. east of the Rocky Mtns.
   R. Crabill, VII.1959"].
— Bollman, 1888b: 4 [= 1893: 76] (*Linotenia*; characters; note; USA:
   Arkansas).
— Bollman, 1888c: 341 [= 1893: 92] (*Linotenia*; note; USA:
   Tennessee).
— Bollman, 1893: 126 (*Linotaenia*; catalogue).
— Cook, 1896: 240 (*Scolioplanes*; notes; USA: New York,
   Pennsylvania). [**Note:** Cook comments "I have collected
   what is evidently the same in central New York and southern
   Pennsylvania, and am unable to separate this species from Sager's

Strigamia fulva, the probable type of which I have seen in the Museum of the Academy of Natural Sciences at Philadelphia"].
— Attems, 1903b: 269 (*Scolioplanes*; list).
— Chamberlin, 1912c: 424 (synonym of *fulva*).

Distribution: USA: Arkansas, California, Colorado, Georgia, Illinois, Indiana, Iowa, Kansas, Kentucky, Louisiana, Michigan, Minnesota, Missouri, North Carolina, ?Nebraska, New York, Ohio, Pennsylvania, South Carolina, Tennessee, Utah, Virginia, West Virginia.

HOLOTYPE: ANSP, 11229, no locality info available.

S. gracilis Wood, 1867
— Wood, 1867b: 128 (*Strigamia*; characters; USA: California).
— Bollman, 1893: 125 (*Himantarium*; catalogue).
— Gunthorp, 1920: 113 (*Strigamia*; list).

Distribution: USA: California.

HOLOTYPE: Academy of Natural Sciences, PA?; California: San Jose.

S. inermis Wood, 1867
— Wood, 1867b: 129 (*Strigamia*; characters; USA: California).
— Bollman, 1893: 125 (*Himantarium*; catalogue).
— Chamberlin, 1912a: 671 (*Nothobius*; notes). [**Note:** Misspelled *Notobius*].
— Gunthorp, 1920: 113 (*Strigamia*; list).
— Attems, 1929: 329 (*Strigamia*; incertae sedis).

Distribution: USA: California.

HOLOTYPE: ANSP, 11231, California: Santa Cruz Mountains. [**Note:** MCZ has a vial (Unique No. 35232) with a label that seems to read "*S. inermis*, Cotype, Santa Cruz Mountains"].

♀*S. kerrana* (Chamberlin & Mulaik, 1940)
— Chamberlin & Mulaik, 1940: 110 (*Linotaenia*; characters; USA: Texas). [**Note:** This species has been included under the genus *Strigamia* and the subgenus *Linotaenia* following the reasoning of Crabill (see Crabill, 1953d)].

Distribution: USA: Texas.

HOLOTYPE: NMNH, Texas, Kerr Co., Raven Ranch, December 1939.

S. maculaticeps Wood, 1862
— Wood, 1862: 48 (*Strigamia*; characters; USA: ?Colorado "Upper Colorado").
— Wood, 1865: 186 (*Strigamia*; characters; USA: Upper Colorado).
— Bollman, 1893: 126 (*Linotaenia*; catalogue; USA: Texas). [**Note:** Although the first two localities for this species only specify "Upper Colorado", Texas is the state reported in the catalogue].
— Gunthorp, 1920: 113 (*Strigamia*; list).

Distribution: USA: ?Colorado, Texas. [**Note:** According to a map on the United States Bureau of Reclamation website (*http://www.usbr.gov/uc/index.html*) the Upper Colorado Region includes the following states: Colorado, New Mexico, Utah, Arizona, Texas, Wyoming, Idaho and Nevada. Based on the map, an estimate of the majority of the land area includes almost the entire states of Utah and New Mexico with Colorado following in relative percentage of land area per state. For the time being the locality of this enigmatic species will be considered as Colorado until further studies are taken to confirm, if even possible, the true identity of this species].

HOLOTYPE: ?; ?Colorado "Upper Colorado"; Lieut. J.C. Ives, U.S.A., H.B. Mollhausen.

♂♀ *S. parviceps* Wood, 1862
— Wood, 1862: 49 (*Strigamia*; characters; USA: California).
— Wood, 1865: 187-188 (*Strigamia*; characters; USA: California). [**Note:** 71 pairs of legs, length 2¼ inches].
— Bollman, 1888d: 346 [= 1893: 98] (*Linotaenia*; note; USA: California). [**Note:** Misspelled *parriceps*; specimen examined as follows: NMNH, Acc. 17414, California: ♂ with 79 pairs of legs, Shasta Co., Baird, L.M. Green].
— Cook, 1895b: 73 (*Tomotaenia*; cited as type species of *Tomotaenia*).
— Cook, 1895c: 866 (*Tomotaenia*; transferred to *Tomotaenia*; list).
— Bollman, 1893: 126 (*Linotaenia*; catalogue).
— Cook, 1896: 240 (*Scolioplanes*; notes). [**Note:** Cook apparently discusses a type specimen deposited at the MCZ that Meinert had once studied. He also mentions having collected "it", presumably *parviceps* and not *bidens*, " . . . in the vicinity of Washington", and not the District of Columbia, Washington].

— Attems, 1903b: 269 (*Scolioplanes*; list).
— Gunthorp, 1920: 113 (*Strigamia*; list).
— Chamberlin, 1941d: 775 (*Tomotaenia*; comparison of *Korynia tripora* with *parviceps*).
— Chamberlin, 1954: 119-121 (*Tomotaenia*; characters; notes).
— Chamberlin, 1963a: 90 (*Tomotaenia*; list of references and synonyms).
— Behan-Pelletier, 1993: 25 (*Tomotaenia*; list). [**Note:** Misspelled *paviceps*].
— Jeekel, 2005: 108 (*Strigamia*; list).

californicus Verhoeff, 1938a: 283-284 (*Paraplanes*; synonym of *parviceps*; characters; USA: California).
— Verhoeff, 1938b: 372-373 (*Paraplanes*; synonym of *parviceps*; characters; USA: California).
— Chamberlin, 1954: 119 (*Paraplanes*; synonym of *parviceps*).
— Crabill, 1962a: 85 (*Paraplanes*; synonym of *fusata*).
— Chamberlin, 1963a: 90 (*Paraplanes*; synonym of *parviceps*).

epileptica Wood, 1862: 49-50 (*Strigamia*; characters; notes; USA: Washington). **Type locality:** Puget's Sound [**Note:** A type specimen is located at the Utah Museum of Natural History in a vial with a locality of Oregon: Puget Sound].
— Wood, 1865: 188-189; Figs. 21, 22. (*Strigamia*; USA: Oregon). [**Note:** 81 pairs of legs, 5½ in].
— Bollman, 1893: 126 (*Linotaenia*; catalogue).
— Gunthorp, 1920: 113 (*Strigamia*; list).
— Chamberlin, 1954: 118, 119; Fig. 1-4 (*Strigamia*; synonym of *parviceps*). [**Note:** figures are of *Strigamia epileptica* Wood, holotype].
— Crabill, 1962c: 181 (*Tomotaenia*; synonym of *parviceps*).
— Chamberlin, 1963a: 90 (*Strigamia*; synonym of *parviceps*).

imperialis Brölemann, 1896: 60-63 (*Scolioplanes*; USA "Washington Territory").
— Attems, 1903b: 269 (*Scolioplanes*; list).
— Chamberlin, 1912a: 659 (*Scolioplanes*; synonym of *laevipes*).
— Chamberlin, 1954: 119 (*Scolioplanes*; synonym of *parviceps*).
— Chamberlin, 1963a: 90 (*Scolioplanes*; synonym of *parviceps*).

?*laevipes* Chamberlin, 1909b: 186 (*Linotenia*; habitat; synonym of *parviceps*; USA: California).
— Chamberlin, 1912a: 659, 664; Fig. 217c; fig. 218c (*Linotenia*; synonym of *parviceps*; characters; range; USA: California).
— Anonymous, 1920: 35 (*Linotaenia*; note; USA: California).
— Gunthorp, 1920: 113 (*Strigamia*; list).
— Chamberlin, 1944c: 79 (*Tomotaenia*; note; USA: California).

— Chamberlin, 1951b: 35 (*Linotenia*; USA: California).

— Chamberlin, 1954: 119 (*Linotaenia*; synonym of *parviceps*).

fusata Attems, 1903b: 281-282, pl. 13; Figs. 24, 25 (*Diplochora*; characters; Mexico).

— Crabill, 1962c: 182; Figs. 1-5 (*Tomotaenia*; composite description; synonym of *parviceps*; USA: California). [**Note:** Holotype in the Attems Collection of the Naturhistorisches Museum, Vienna; mouthparts and head on microscopic slide, remainder of specimen in alcohol. Collected at Ventanas in western Mexico by Rorrer, March 3, 1883].

— Chamberlin, 1963a: 90 (*Diplochora*; synonym of *parviceps*).

— Jeekel, 2005: 77 (*Diplochora*; list).

rubelliana Chamberlin, 1904: 656-657 (*Linotaenia*; characters; notes; USA: California) Type locality: USA: California: Palo Alto. [**Note:** Cotype specimen located at the University of Utah in a vial with a second determined label as *Tomotaenia parviceps*].

— Chamberlin, 1909b: 187 (*Linotaenia*; USA: California).

— Chamberlin, 1954: 119 (*Linotaenia*; synonym of *parviceps*).

— Chamberlin, 1963a: 90 (*Linotaenia*; synonym of *parviceps*).

Distribution: CANADA: British Columbia (see Chamberlin, 1954 & 1963); USA: California, Oregon, ?Washington.

HOLOTYPE: ANSP, No. 1080, Oregon, vicinity of Puget Sound, by Dr. Kennedy. [**Note:** When Wood (1862) originally described *Strigamia parviceps* he said the type depository was the NMNH and the data is as follows: No. 311, 1 specimen, California. Wood (1862) also described *Strigamia epileptica* and it too was said to be deposited in the NMNH with the following data: No. 297, 1 specimen, Puget's Sound, A. Campbell, Com. N. W. B. S., Dr. C.B. Kennerly. Chamberlin (1912a) considered *Strigamia parviceps* as a junior synonym to *Linotaenia* [sic] *laevipes* and regards this as the only species of *Linotaenia* known in California at that time. The enigmatic thing is that Wood (1862, 1865) originally states the habitat of *laevipes* as being in Georgia not California! The genus *Linotaenia* is a junior synonym to *Strigamia* and the only known species of *Strigamia* found in Georgia with 69 pairs of legs is *Strigamia bidens* Wood. Considering all references to localities following Wood for *Linotaenia laevipes* are from California by Chamberlin, (one Anonymous (1920) but specimens identified by Chamberlin) it will be tentatively proposed in this catalog as a synonym to *Strigamia bidens*. All records from Chamberlin of *Linotaenia laevipes* are tentatively considered records for *Tomotaenia parviceps*. Verhoeff (1938a, 1938b) described *Paraplanes californicus*, which was from Berkeley, California].

Superfamily **Geophiloidea** ?Cook, 1896

Family **Ballophilidae** Cook, 1895

Genus *Ityphilus* Cook, 1899
[2 species; Southeastern USA]

Ityphilus Cook, 1899: 306 (characters). Type Species: *Ityphilus lilanicus* Cook, 1899, fixed by original designation.
— Attems, 1929: 106 (characters; distribution; key to species; species included: *ceibanus, guianensis, lilacinus*).
— Jeekel, 2005: 85 (list).
— Bonato, Pereira & Minelli, 2007: 3 (characters; taxonomic & nomenclatorial notes).

I. lilacinus Cook, 1899
— Cook, 1899: 306-307, pl. 5; Figs. 1a-1e (*Ityphilus*; characters; USA: Florida).
— Attems, 1903b: 184 (*Ityphilus*; list).
— Attems, 1929: 106 (*Ityphilus*; characters; distribution).
— Chamberlin, 1950: 157 (*Ityphilus*; Puerto Rico).
— Chamberlin, 1952b: 4 (*Ityphilus*; [**Note:** It seems appropriate to report that this species is also found in the West Indies]).
— Foddai, Pereira & Minelli, 2000: 154 (*Ityphilus*; catalogue).
— Jeekel, 2005: 85 (*Ityphilus*; list).
— Bonato, Pereira & Minelli, 2007: 3 (*Ityphilus*; list).

Distribution: USA: Florida.

SYNTYPE: NMNH; No. 777; Florida: N side of Sugar Loaf Key, near shore.

♂*I. nemoides* Chamberlin, 1943
— Chamberlin, 1943b: 99 (*Ityphilus*; characters; note; USA: Texas).
— Bonato, Pereira & Minelli, 2007: 4 (*Ityphilus*; list).

Distribution: USA: Texas.

HOLOTYPE: NMNH, ♂, Texas: Jim Wells Co., 12 mi. N of Alice, 6 June 1941.

Family **Chilenophilidae** Attems, 1909

Subfamily **Chilenophilinae** Attems, 1909

Genus *Arctogeophilus* Attems, 1909
[8 species; CANADA & USA]

Arctogeophilus Attems, 1909a: 23 (characters). TYPE SPECIES: *Geophilus glacialis* Attems, 1909, fixed by monotypy.
— Chamberlin, 1912a: 660 (characters; notes).
— Crabill, 1952b: 160-161 (characters, notes).
— Jeekel, 2005: 69 (list).
Cryophilus Chamberlin, 1919: 15H, 17H-18H (characters; list; notes). TYPE SPECIES: *Cryophilus alaskanus* Chamberlin, 1919, fixed by original designation.
— Kevan, 1983: 2946 (synonym of *Arctogeophilus*).
— Jeekel, 2005: 75 (list).
Gnathomerium Ribaut, 1910 (synonym of *Arctogeophilus*). TYPE SPECIES: *Gnathomerium inopinatus* Ribaut, 1912, fixed by subsequent designation.
— Attems, 1914: 128 (synonym of *Arctogeophilus*).
— Jeekel, 2005: 81 (list).
Idiona Chamberlin, 1946e: 69 (characters). TYPE SPECIES: *Idiona shelfordi* Chamberlin, 1946, fixed by original designation and monotypy.
— Crabill, 1968b: 331 (catalogue; synonym of *Arctogeophilus*).
— Jeekel, 2005: 84 (list).

♂♀*A. atopus* (Chamberlin, 1902)
— Chamberlin, 1902c: 176-177 (*Geophilus*; characters; habitat; key to species; USA: Utah).
— Attems, 1929: 180-181 (*Geophilus*; characters; distribution).
— Chamberlin, 1961: 96 (*Arctogeophilus*; characters; USA: Arizona, Utah, Wyoming).

Distribution: USA: Arizona, Utah, Wyoming.

HOLOTYPE: MCZ, No. 1915, ♂, Utah: Salt Lake City, 3 June 1900. [**Note:** Crabill note in vial reads "Head and mouthparts missing when rest found in 1956"].

A. corvallis Chamberlin, 1941
— Chamberlin, 1941d: 783 (*Arctogeophilus*; characters; notes; USA: Oregon).
— Crabill, 1952b: 165 (note). [**Note:** Crabill suspects *corvallis* might be a junior synonym of *A. fulvus* (Wood)].

Distribution: USA: Oregon.

HOLOTYPE: NMNH, Oregon: 6 mi. W of Corvallis on Oak Creek, 9 May 1936, by J.C. Chamberlin.

♀*A. fulvus* (Wood, 1862)
— Wood, 1862: 41 (*Mecistocephalus*; characters; USA: "around Philadelphia").
— Wood, 1865: 176-177 (*Mecistocephalus*; characters; habitat; notes).
— Cook, 1895a: 60-61 (*Mecistocephalus*; cited as synonym of *attenuatus*).
— Bollman, 1893: 125 (*Mecistocephalus*; catalogue).
— Attems, 1929: 155 (*Mecistocephalus*; *incertae sedis*).
— Crabill, 1950d: 253, 254-256; Figs. 1-5 (*Arctogeophilus*; redescription; taxonomic notes; USA: Virginia). [**Note:** Crabill reports the type as missing and designated a female "digm", "and two "paradigms" from Virginia (see Crabill, 1950d). Needs neotype designation].
— Crabill, 1952b: 164-165 (*Arctogeophilus*; characters; taxonomic notes; USA: Pennsylvania, Virginia).
— Crabill, 1955b: 38 (*Arctogeophilus*; notes; USA: Missouri).
— Crabill, 1955h: 158 (*Arctogeophilus*; note; USA: Missouri).
— Hoffman, 1995a: 23 (*Arctogeophilus*; list; notes).
attenuatus Say, 1821, see: *bipuncticeps* Wood, 1862.

Distribution: USA: Missouri, Pennsylvania, Virginia.

HOLOTYPE: Missing (see Crabill, 1950d), Pennsylvania: Philadelphia.

♂♀*A. glacialis* (Attems, 1909)
— Attems, 1909a: 9, 23-25, t. 1; Figs. 1-4 (*Geophilus*; subgenus *Arctogeophilus*; characters; list; USA: Alaska).
— Chamberlin, 1919: 15H (*Arctogeophilus*; list).
— Attems, 1929: 297; Figs. 12, 32, 261, 262 (*Arctogeophilus*; characters; distribution).

— Chamberlin, 1946c: 182 (*Arctogeophilus*; notes; USA: Alaska).
— Chamberlin, 1952c: 209 (*Arctogeophilus*; USA: Alaska).
— Behan-Pelletier, 1993: 25 (*Arctogeophilus*; list).
— Jeekel, 2005: 69 (*Geophilus*; list).

alaskanus Chamberlin, 1919: 18H-19H; Figs. 1-5 (*Cryophilus*; characters; habitat; notes; USA: Alaska). **HOLOTYPE:** MCZ #2085, Type Locality: Nome, Alaska.

— Chamberlin, 1946: 179 (synonym of *glacialis*).
— Jeekel, 2005: 75 (*Cryophilus*; list).

Distribution: USA: Alaska.

COTYPES: MCZ, (Unique No. 14351), 2 specimens, Nunamo. [**Note:** Label in vial from Crabill reads "Original holotype label in error. Specimens can only be Cotypes. New label prepared IX.1959 by Crabill"].

A. insularis Attems, 1947
— Attems, 1947: 68-69; Fig. 51, Taf. VII (*Arctogeophilus*; characters; CANADA: British Columbia).

Distribution: CANADA: British Columbia.

HOLOTYPE: ?NMW; "Vancouver, Manaimo" [**Note:** Manaimo = Nanaimo].

♂♀*A. melanonotus* (H. C. Wood, 1862)
— Wood, 1862: 41-42 (*Mecistocephalus*; characters; USA: Georgia).
— Wood, 1865: 177 (*Mecistocephalus*; characters; USA: Georgia).
— Bollman, 1893: 125 (*Mecistocephalus*; catalogue).
— Attems, 1903b: 210 (*Mecistocephalus*; list).
— Chamberlin, 1912a: 661; Figs. 218d-e, 219a-b, 220b (*Gnathomerium*; characters; range; USA: California).
— Chamberlin, 1919: 15H (*Gnathomerium*; list).
— Gunthorp, 1920: 113 (*Mecistocephalus*; list).
— Attems, 1929: 298 (*Arctogeophilus*; characters; distribution).
— Chamberlin, 1946c: 182 (*Arctogeophilus*; notes; USA: Alaska).
— Crabill, 1950d, p 253 (*Mecistocephalus*; taxonomic notes). [**Note:** Crabill mentions that *melanonotus* should not belong to the genus *Arctogeophilus* due to the fact that the " . . . type lacks superficially opening coxopleural pores . . ." and suggests that this is possibly *Geophilus rubens* (= *Geophilus vittatus*). Crabill also indicates that the head is detached and missing from the type. Could this

mean that the head was possibly mounted on a slide and has been overlooked?].

— Kevan, 1979: 298 (*Arctogeophilus*; CANADA: British Columbia).

— Kevan, 1983: 2946 (*Geophilus*; questioned synonym of *vittatus*).

glaber Bollman, 1887d: 229 [= 1893: 66] (*Geophilus*; characters; USA: California). [**Note:** According to Underwood the actual description of this species was omitted and did not appear in Bollman's (1887d) manuscript. The actual description appears in Bollman's 1893 manuscript under Omissions and Corrections. Two types are located at the NMNH, one as *Arctogeophilus* and the other as *Geophilus*, which is the Syntype].

— Bollman, 1893: 124, 201 (*Geophilus*; catalogue; characters; USA: California).

— Attems, 1903b: 263 (*Geophilus*; list).

— Chamberlin, 1909b: 185 (*Geophilus*; habitat; notes; USA: California, Oregon, Washington).

— Chamberlin, 1911e: 260 (*Geophilus*; USA: Washington).

— Chamberlin, 1912a: 661 (*Geophilus*; synonym of *melanonotus*).

— Attems, 1929: 328 (*Geophilus*; *incertae sedis*).

quadratus Wood, 1867b: 128 (*Mecistocephalus*; synonym of *melanonotus*; characters; USA: California).

— Bollman, 1893: 125 (*Mecistocephalus*; synonym of *melanonotus*; catalogue).

— Chamberlin, 1912a: 661 (*Mecistocephalus*; cited as synonym of *Gnathomerium melanonotum*).

— Gunthorp, 1920: 113 (*Mecistocephalus*; list).

— Attems, 1929: 298 (*Mecistocephalus*; synonym of *melanonotus*).

— Behan-Pelletier, 1993: 25 (*Arctogeophilus*; list).

Distribution: CANADA: ?British Columbia; USA: ?Alaska, ?California, Georgia, ?Oregon, ?Washington. [**Note:** Georgia is the type locality of this species which is far removed form all other reports of this species, therefore, I have questioned all of the west coast reports. Could all of the other localities from the west coast represent another species or are all of these records an introduced species? Furthermore, Wood's *melanonotus* could be a synonym of a southeastern *Geophilus* sp. and/or *melanonotus* records form the western US and Canada could be synonymous with a western *Geophilus* sp.].

HOLOTYPE: ANSP; Georgia, by Dr. John Le Conte.

♀*A. shelfordi* (Chamberlin, 1946)

— Chamberlin, 1946e: 69-71; Figs. 1-4 (*Idiona*; characters; USA: California).
— Crabill, 1968b: 331 (*Arctogeophilus*; catalogue).
— Jeekel, 2005: 84 (*Idiona*; list).

Distribution: USA: California.

HOLOTYPE: NMNH; ♀; California: Angeles National Forest, Glendora, taken in oak chaparral, July 15, 1944.

♂♀*A. umbraticus* (McNeill, 1887)
— McNeill, 1887b: 332 (*Mecistocephalus*; characters; note; USA: Indiana).
— Bollman, 1888a: 107, 109, 112 [= 1893: 82, 84, 85] (*Geophilus*; notes; USA: Tennessee).
— Bollman, 1888c: 341 [= 1893: 92] (*Geophilus*; note; USA: Tennessee).
— Bollman, 1888d: 346 [= 1893: 98] (*Geophilus*; note; USA: Colorado).
— Bollman, 1888e: 408 [= 1893: 109] (*Geophilus*; note; USA: Indiana).
— Bollman, 1893: 125 (*Geophilus*; catalogue).
— Cockerell, 1893: 370 (*Geophilus*; USA: Colorado).
— Attems, 1903b: 214 (*Mecistocephalus*; list).
— Chamberlin, 1912b: 68 (*Gnathomerium*; USA: Colorado).
— Chamberlin, 1912c: 422-423 (*Gnathomerium*; notes; USA: Alabama, Georgia, Kentucky, Mississippi, North Carolina, South Carolina, Tennessee, Virginia, West Virginia).
— Chamberlin, 1918a: 23 (*Gnathomerium*; USA: Tennessee).
— Verhoeff, 1925: 695 (*Arctogeophilus*; subgenus *Gnathomerium*; list).
— Chamberlin, 1925e: 59 (*Gnathomerium*; habitat; notes; USA: Utah).
— Chamberlin, 1928a: 96 (*Gnathomerium*; habitat; note; USA: Utah).
— Chamberlin, 1928b: 153 (*Gnathomerium*; USA: Missouri).
— Chamberlin, 1928d: 310 (*Gnathomerium*; USA: Utah).
— Attems, 1929: 155 (*Mecistocephalus*; distribution; *incertae sedis*).
— Brimley, 1938: 501 (*Gnathomerium*; note; USA: North Carolina).
— Causey, 1940: 62 (*Gnathomerium*; characters; USA: North Carolina).
— Chamberlin, 1943b: 101 (*Arctogeophilus*; USA: Texas).
— Chamberlin, 1944d: 195 (*Arctogeophilus*; USA: Tennessee).

— Chamberlin, 1951b: 34 (*Arctogeophilus*; USA: North Carolina).
— Crabill, 1952b: 162-164 (*Arctogeophilus*; characters; range; taxonomic notes; USA: Indiana, Kentucky, Ohio, Virginia, West Virginia).
— Crabill, 1955f: 258 (*Arctogeophilus*; checklist; USA: Kentucky).
— Chamberlin, 1961: 97 (*Arctogeophilus*; habitat; note; USA: Utah).
— Branson & Batch, 1967: 81-82 (*Arctogeophilus*; characters; habitat; note; USA: Kentucky).
— Wray, 1967: 156 (*Arctogeophilus*; list).
— Summers, 1979: 697, 699 (*Arctogeophilus*; checklist; key).
— Lee, 1980: 2, 3, 4, 5; Tab. 1 (*Arctogeophilus*; habitat; USA: Ohio).
— Summers et al., 1980: 247; Fig. 7 (*Arctogeophilus*; checklist; habitat (Table 1, p. 242-243); USA: Illinois).
— Summers et al., 1981: 60 (*Arctogeophilus*; checklist; USA: Illinois).
— Holsinger & Culver, 1988: 57 (*Arctogeophilus*; USA: Virginia). [**Note:** Reported as an accidental occurrence in a cave].
— Hoffman, 1995a: 23 (*Arctogeophilus*; list; notes).
americanum Ribaut, 1912: 120; Figs. 12-17 (*Gnathomerium*; characters; USA: North Carolina).
— Chamberlin, 1912c: 423 (*Gnathomerium*; synonym of *umbraticum*).
— Attems, 1929: 298 (*Arctogeophilus*; characters; distribution).
— Chamberlin, 1944d: 195 (*Gnathomerium*; synonym of *umbraticus*).
— Chamberlin, 1961: 97 (*Gnathomerium*; synonym of *umbraticus*).
xenoporus Chamberlin, 1902c: 475-476 (*Geophilus*; characters (adult, stadia); habitat; key to species; Type locality: USA: Utah, City Creek Canyon, 1.iv.1901.
— Chamberlin, 1925a: 54 (*Gnathomerium*; note; USA: New Mexico).
— Attems, 1929: 327 (*Geophilus*; characters; distribution; *species inquirenda*).
— Chamberlin, 1930a: 114-115 (*Gnathomerium*; notes; USA: Utah).
— Chamberlin, 1943b: 102 (*Arctogeophilus*; USA: Colorado, Idaho, Utah, Wyoming).
— Chamberlin, 1951b: 34 (*Arctogeophilus*; USA: California).
— Chamberlin, 1958c: 133 (*Arctogeophilus*; checklist; note).
— Chamberlin, 1961: 97 (*Geophilus*; synonym of *umbraticus*).

Distribution: USA: Alabama, California, Colorado, Georgia, Idaho, Illinois, Indiana, Kentucky, Missouri, Mississippi, North Carolina, New Mexico,

Ohio, South Carolina, Tennessee, Texas, Utah, Virginia, West Virginia, Wyoming.

HOLOTYPE: ?Unknown, Indiana: near Bloomington. [**Note:** MCZ has what is said to be a holotype according to their database; however, the type locality does not match. Catalog number 31903, 1 specimen, Tennessee, Knox, Knoxville, March 17, 1921, Geo. G. Ainslee].

Genus *Cheiletha* Chamberlin, 1946
[3 species; Western USA & Alaska]

Cheiletha Chamberlin, 1946c: 182 (characters). TYPE SPECIES: *Cheiletha alaska* Chamberlin, 1946, fixed by original designation.
— Jeekel, 2005: 73 (list).

C. alaska Chamberlin, 1946
— Chamberlin, 1946c: 182-183, pl. 2; Figs. 8-9 (*Cheiletha*; characters; note; USA: Alaska).
— Behan-Pelletier, 1993: 25 (*Cheiletha*; list).
— Jeekel, 2005: 73 (*Cheiletha*; list).

Distribution: USA: Alaska.

HOLOTYPE: NMNH, ♀, Alaska: Juneau, 28-29 April 1945, by J.C. Chamberlin. [**Note:** 1 vial + slide].

PARATYPES: NMNH, 3 vials + slides, numerous specimens.

C. kincaidi Chamberlin, 1955
— Chamberlin, 1955a: 180 (*Cheiletha*; characters; USA: Washington).

Distribution: USA: Washington.

HOLOTYPE: ?NMNH, Washington: Ocean Park, 22 July 1954, by Professor Trevor Kincaid. [**Note:** NMNH has two vials, one labeled as *vasingtona* (type) and the other as *washingtonia* (syntype). It may be that these specimens are the types of *kincaidi*. Chamberlin may have changed the specific name in the literature but not on the holotype vials.]

♀*C. phoenix* Chamberlin, 1955
— Chamberlin, 1955a: 179-180 (*Cheiletha*; characters; note; USA: Arizona).

311

Distribution: USA: Arizona.

HOLOTYPE: NMNH; ♀, Arizona: Phoenix, South Mountains, 10 April 1953.

Genus *Navajona* Chamberlin, 1930
[1 species; Southwestern USA]

Navajona Chamberlin, 1930a: 115 (characters). TYPE SPECIES: *Navajona miuropus* Chamberlin, 1930, fixed by original designation and monotypy.
— Jeekel, 2005: 91 (list).

♂*N. miuropus* Chamberlin, 1930
— Chamberlin, 1930a: 116 (*Navajona*; characters; USA: Arizona).
— Attems, 1947: 147 (*Navajona*; list). [**Note:** Specific name misspelled *micropus*].
— Jeekel, 2005: 91 (*Navajona*; list).

Distribution: USA: Arizona.

HOLOTYPE: NMNH, ♂, Arizona, Flagstaff, May 1929.

Genus *Nesidiphilus* Chamberlin, 1915
[1 species; Southeastern USA]

Nesidiphilus Chamberlin, 1915: 511-512 (characters; notes). TYPE SPECIES: *Nesidiphilus latus* Chamberlin, 1915, fixed by original designation.
— Attems, 1929: 283 (characters; distribution; key to species; species included: *juvenis, latus, montis, nicaraguae*). [**Note:** Apparently Attems placed *marginalis* into the genus *Pachymerium* based on the fact that Chamberlin (1912) transferred *marginalis* from *Geophilus* to *Polycricus*, and that *Polycricus* was determined to be a junior synonym to *Pachymerium*. However, Attems never referenced Chamberlin from 1912 or 1915 when he transferred the species to *Nesidiphilus*, therefore, the author has decided to leave *marginalis* in *Nesidiphilus* until further study can be made upon this species. Furthermore, Attems disregarded the fact that Chamberlin (1915) synonymized *floridanus* with *marginalis*, and kept *floridanus* as a distinct species. Apparently *Nesidiphilus* is synonymous with *Telocricus* Chamberlin, 1915. I am unclear as

to the proper placement of this genus and *marginalis*, however, Matic, et al. 1977 may clarify the taxonomy, a paper I was unable to obtain].

♂♀*N. marginalis* (Meinert, 1886)
— Meinert, 1886a: 218 (*Geophilus*; characters; USA: Florida).
— Bollman, 1893: 125 (*Geophilus*; catalogue).
— Attems, 1903b: 261 (*Geophilus*; list).
— Chamberlin, 1909b: 183 (*Geophilus*; USA: Florida).
— Chamberlin, 1912c: 415 (*Polycricus*; note; USA: Florida).
— Chamberlin, 1915: 512 (*Nesidiphilus*; transferred from *Polycricus*).
— Attems, 1929: 253 (*Pachymerium*; characters; distribution).
floridanus Cook, 1899: 307-308, pl. 4; Figs. 2a-2f (*Polycricus*; characters; notes; USA: Florida). [**Note:** The syntype of Cook's *P. floridanus* is deposited at the NMNH, No. 778, and was found on Sugar Loaf Key].
— Attems, 1903b: 260 (*Geophilus*; subgenus *Pachymerium*; list).
— Chamberlin, 1915: 512 (*Polycricus*; synonym of *marginalis*).
— Attems, 1929: 252 (*Pachymerium*; characters; distribution).
— Matic, et al., 1977: 287 [**Note:** Reference not obtained].

Distribution: USA: Florida.

HOLOTYPE: MCZ, No. 970, Florida: Key West, 12 February 1878. [**Note:** The original determined label reads "*Geophilus marginalis* n. sp. No. 159" and there is also what appears to be an original piece of paper written on by Meinert with a doodle. A label from Crabill reads "N.B. See small vials: one with head, antenna, maxillae; one with posterior segments and prosternum with prehensors; one with bulk of segments. R. Crabill, Jan. 1959].

Genus *Queenslandophilus* Verhoeff, 1925
[1 species; Western USA]

Queenslanophilus Verhoeff, 1925: 49 (cited as subgenus of *Arctogeophilus*; Type Species: *Arctogeophilus sjöstedti* Verhoeff, 1925, fixed by monotypy.
— Attems, 1929: 299 (characters; distribution; key to species; species included: *sjöstedti, viridicans*).
— Jeekel, 2005: 101 (list; notes).

♀*Q. elongatus* Verhoeff, 1938

— Verhoeff, 1938b: 375-376, t. 7; Figs. 40-42 (*Queenslanophilus*; characters; USA: California).
— Attems, 1947: 145 (*Queenslanophilus*; list).

Distribution: USA: California.

HOLOTYPE: ?Unknown, ♀, California: Berkeley or Calistoga, April or May, 1935, by Herr Michelbacher.

Genus *Taiyuna* Chamberlin, 1912
[6 species (1 Introduced); USA]

Taiyuna Chamberlin, 1912a: 661 (characters). TYPE SPECIES: *Geophilus occidentalis* Meinert, 1886, fixed by original designation.
— Attems, 1929: 304-305 (characters; distribution).
— Crabill, 1952b: 165-166 (characters; note).
— Jeekel, 2005: 107 (list).

T. australis Chamberlin, 1914
— Chamberlin, 1914c: 218-220 (*Taiyuna*; characters; USA: District of Columbia). [**Note:** This species was found in pots of imported plants from British Guiana. The chances of it becoming established in the Washington, D.C. area is highly unlikely].
— Attems, 1929: 305 (characters; distribution).

Distribution: USA: ?District of Columbia.

HOLOTYPE: MCZ, British Guiana.

T. claremontus Chamberlin, 1912
— Chamberlin, 1912a: 662 (*Taiyuna*; notes; USA: California). [**Note:** Chamberlin did not even indicate this as a new species when he published it for the first time and does not give a description of this species other than that he mentions it differs from *occidentalis* only by having the anal legs with a distinct claw].
— Anonymous, 1920: 35 (*Taiyuna*; notes; USA: California).
— Attems, 1929: 306 (*Taiyuna*; characters; distribution).

Distribution: USA: California.

HOLOTYPE: MCZ, No. 14283, 1 specimen (slide), California: Claremont.

♂*T. idahoana* Chamberlin, 1941
— Chamberlin, 1941d: 783-784 (*Taiyuna*; characters; note; USA: Idaho).

Distribution: USA: Idaho.

HOLOTYPE: NMNH, ♂, Idaho: Idaho Co., "From next "No. 120" of burrowing owl, Speotyto," 14 June 1939.

T. isantus (Chamberlin, 1909)
— Chamberlin, 1909b: 185-186, pl. 26; Figs. 7-9 (*Geophilus*; characters; USA: California).
— Chamberlin, 1912a: 661 (*Taiyuna*; note; transferred to *Taiyuna*).
— Attems, 1929: 192 (*Brachygeophilus*; characters; distribution). [**Note:** Apparently Attems felt that this species belonged to *Brachygeophilus* instead of *Geophilus*, however, *Brachygeophilus* is now considered a junior synonym of *Geophilus* and *Taiyuna* is a valid genus so the species remains here until the genus is reassessed].

Distribution: USA: California.

HOLOTYPE: NMNH, California: Los Angeles.

♀*T. moderata* Chamberlin, 1941
— Chamberlin, 1941d: 784 (*Taiyuna*; characters; note; USA: California).
— Chamberlin, 1943: 101-102 (*Taiyuna*; USA: New Mexico). [**Note:** The genus was misspelled *Taiyna*].

Distribution: USA: California, New Mexico.

HOLOTYPE: NMNH, ♀, California: Monterey Co., Hastings Reservation, 30 January 1941, Dr. J.M. Linsdale.

PARATYPE: NMNH, ♀, California: Monterey Co., Hastings Reservation, 22 February 1941, Dr. J.M. Linsdale.

♀*T. occidentalis* (Meinert, 1886)
— Meinert, 1886a: 220(*Geophilus*; characters; USA: California).
— Bollman, 1893: 125 (*Geophilus*; catalogue).
— Attems, 1903b: 264 (*Geophilus*; list).
— Chamberlin, 1909b: 185 (*Geophilus*; USA: California)

— Chamberlin, 1912a: 661-662 (*Taiyuna*; characters; notes; USA California).
— Anonymous, 1920: 35 (*Taiyuna*; USA: California).
— Attems, 1929: 305 (*Taiyuna*; characters; distribution).
— Chamberlin, 1944d: 195 (*Taiyuna*; USA: California).
— Behan-Pelletier, 1993: 25 (*Taiyuna*; list).
— Jeekel, 2005: 107 (*Geophilus*; list).
californiensis Bollman, 1887c: 624-625 [= 1893: 41-42] (*Geophilus*; characters; note; USA: California). [**Note:** Type located at the NMNH].
— Bollman, 1893: 124 (*Geophilus*; catalogue).
— Attems, 1903b: 263 (*Geophilus*; list).
— Chamberlin, 1912a: 661 (*Geophilus*; synonym of *occidentalis*).
— Attems, 1929: 328 (*Geophilus*; *incertae sedis*).

Distribution: USA: California.

HOLOTYPE: MCZ, No. 974, California: Ukiah, by Mr. J.H. Burke. [**Note:** Specimen at MCZ is said to be the type presumably by Crabill, however, the typed locality label as follows: "San Francisco, Cal. Presented by T. G. Cary Jr.", disagrees with the original published locality as above. There also seems to be a pencil label agreeing with the handwriting of Meinert, which reads "*Geophilus occidentalis* n. sp., No. 162"].

♀*T. opita* Chamberlin, 1912
— Chamberlin, 1912b: 67-68 (*Taiyuna*; characters; USA: Michigan).
— Attems, 1929: 305 (*Taiyuna*; characters; distribution).
— Crabill, 1952b: 166 (*Taiyuna*; characters).
— Crabill, 1958b: 96 (*Taiyuna*; habitat; notes; USA: Wisconsin).
— Kevan, 1979: 298 (*Taiyuna*; habitat; CANADA: British Columbia).
— Summers, 1979: 697, 699 (*Taiyuna*; checklist; key).
— Summers et al., 1980: 247; Fig. 8 (*Taiyuna*; checklist; habitat (Table 1, p. 242-243); USA: Illinois).
— Summers et al., 1981: 60 (*Taiyuna*; checklist; USA: Illinois).
— Snider, 1991: 188 (*Taiyuna*; list; notes; USA: Michigan).
— Watermolen, 1997: 3 (*Taiyuna*; checklist, notes; USA: Wisconsin).

Distribution: CANADA: British Columbia; USA: Illinois, Michigan, Wisconsin.

HOLOTYPE: ?Unknown, ♀, Michigan, Posers & Kimball's.

Genus *Watophilus* Chamberlin, 1912
[6 species; Central, Southeastern & Western USA]

Watophilus Chamberlin, 1912a: 662 (characters; key to species of California).
TYPE SPECIES: *Watophilus alabamae* Chamberlin, by subsequent
designation (see Crabill, 1976: 397-398, footnote 2). [**Note:** *Watophilus*
was not presented as a new genus until Chamberlin, 1912c: 420 (see
Crabill, 1976: 397-398, footnote 2)].
— Chamberlin, 1912a: 662 (characters; key to species of California).
— Chamberlin, 1912c: 420-421 (characters).
— Attems, 1929: 311 (characters; distribution).
— Crabill, 1976: 395-398 (notes; list of known species).
— Jeekel, 2005: 110 (list).

(Subgenus *Parawatophilus* Crabill, 1955)

Watophilus (*Parawatophilus*) Crabill, 1955d: 221-222 (characters). **TYPE SPECIES:**
Geophilus dolicocephalus Gunthorp, 1913 [=*Watophilus* (*Parawatophilus*)
dolicocephalus (Gunthorp), by original designation and monotypy.
— Jeekel, 2005: 96 (list).

♂♀ *W.* (*P.*) *dolicocephalus* (Gunthorp, 1913)
— Gunthorp, 1913: 169-170, pl. 20; Figs. 4-6 (*Geophilus*; characters;
notes; USA: Kansas). [**Note:** Gunthorp compared *Geophilus strigosus*
(McNeill) to this species].
— Crabill, 1955d: 222-223, figs, 1, 3-4 (*Watophilus*; cited as subgenus
Parawatophilus; characters; notes; USA: Kansas).
— Crabill, 1976: 398 (*Watophilus*; list).
— Jeekel, 2005: 96 (*Geophilus*; list).

Distribution: USA: Kansas.

LECTOTYPE: UKNHM; no. 162; ♀, Kansas: Cowley Co., March 1911,
by Horace Gunthorp.

(Subgenus *Watophilus* Chamberlin, 1912)

Watophilus (*Watophilus*) Chamberlin, 1912a: 662 (characters; key to species
of California). **TYPE SPECIES:** *Watophilus* (*Watophilus*) *alabamae*

Chamberlin, 1912 by subsequent designation (see Crabill, 1976: 397-398, footnote 2).
— Chamberlin, 1912c: 420-421 (characters).

♂♀*W. (W.) alabamae* Chamberlin, 1912
— Chamberlin, 1912c: 421-422, pl. 1; Figs. 5-7 (*Watophilus*; characters; USA: Alabama, Georgia).
— Attems, 1929: 311-312 (*Watophilus*; characters; distribution).
— Crabill, 1950c: 200 (*Watophilus*; notes; USA: South Carolina).
— Crabill, 1952b: 160 (*Watophilus*; cited as type species, note).
— Crabill, 1955d: 223; Fig. 2 (*Watophilus*; cited as subgenus *Watophilus*).
— Crabill, 1976: 397 (*Watophilus*; list).
— Jeekel, 2005: 110 (*Watophilus*; list).

Distribution: USA: Alabama, Georgia, South Carolina.

HOLOTYPE: MCZ; Alabama: Anniston or Maplesville; possibly Tallulah Falls, Georgia.

♂♀*W. (W.) errans* Chamberlin, 1912
— Chamberlin, 1912a: 663 (*Watophilus*; characters; USA: California).
— Attems, 1929: 312 (*Watophilus*; characters; distribution).
— Crabill, 1976: 398 (*Watophilus*; list).

Distribution: USA: California.

HOLOTYPE: ?MCZ; Unknown, California: Berkeley, April, 1911, by R.V. Chamberlin.

♀*W. (W.) knowltoni* Crabill, 1976
— Crabill, 1976: 396-397 (*Watophilus*; characters; USA: Utah).

Distribution: USA: Utah.

HOLOTYPE: NMNH, Utah: Box Elder Co., Kelton Pass, in juniper duff, April 28, 1969, George F. Knowltoni.

♂*W. (W.) laetus* Chamberlin, 1912
— Chamberlin, 1912a: 663 (*Watophilus*; characters; USA: California).
— Attems, 1929: 312 (*Watophilus*; characters; distribution).

— Crabill, 1976: 398 (*Watophilus*; list).

Distribution: USA: California.

HOLOTYPE: ?MCZ; Unknown, California: Berkeley, R.V. Chamberlin.

♂♀ *W.* (*W.*) *utus* Chamberlin, 1928
 — Chamberlin, 1928a: 95-96 (*Watophilus*; characters; notes; USA:
   Utah).
 — Chamberlin, 1928d: 310 (*Watophilus*; notes; USA: Utah).
 — Crabill, 1955d: 221 (*Watophilus*; notes).
 — Chamberlin, 1958c: 133 (*Watophilus*; checklist; note).
 — Chamberlin, 1961: 97 (*Watophilus*; notes; USA: Utah).
 — Crabill, 1976: 398 (*Watophilus*; list).

Distribution: USA: Utah.

HOLOTYPE: NMNH, ♀, Utah: La Sal Mountains, July 1927, by Dr.
V.M. Tanner.

Subfamily Pachymeriinae Verhoeff, 1925

Genus Eremerium Chamberlin, 1941
[1 species; Southwestern USA]

Eremerium Chamberlin, 1941d: 776 (characters). Type Species: *Eremerium
apachum* Chamberlin, 1941, fixed by original designation.
 — Jeekel, 2005: 78 (list).

E. apachum Chamberlin, 1941
 — Chamberlin, 1941d: 776 (*Eremerium*; characters; USA: Arizona).
 — Jeekel, 2005: 73 (*Eremerium*; list).

Distribution: USA: Arizona.

SYNTYPE: NMNH, Arizona: Phoenix, 29 January 1939, by Christensen,
Jones & Clancy.

Genus Pachymerium Koch, 1847
[2 species (1 Introduced); USA]

Pachymerium Koch, 1847: 85, 187 (characters). TYPE SPECIES: *Geophilus ferrugineus* C. L. Koch, 1835, fixed by monotypy.
— Cook, 1895a: 60 (cited as synonym of *Mecistocephalus*).
— Bailey, 1928b: 40 (characters; note).
— Jeekel, 2005: 96 (list; notes; junior objective synonym of *Mecistocephalus*).

♂♀*P. ferrugineum* (Koch, 1835)
— Koch, 1835: 2 (*Geophilus*;). [**Note:** reference not obtained].
— Koch, 1847: 187 (*Pachymerium*; list).
— Cook, 1895a: 60, 61 (*Geophilus* + *Pachymerium*; cited as synonym of *Mecistocephalus attenuatus*; habitat; notes).
— Attems, 1903b: 255 (*Geophilus*; subgenus *Pachymerium*; characters; distribution; North America).
— Cook, 1904: 74 (*Geophilus* + *Pachymerium* + *Mecistocephalus*; synonym of *Mecistocephalus attenuatus*).
— Attems, 1909a: 9 (*Pachymerium*; list).
— Chamberlin, 1912b: 67 (*Pachymerium*; habitat; notes; range; USA: Illinois, Iowa, Wisconsin).
— Chamberlin, 1912c: 415-416 (*Pachymerium*; habitat; notes).
— Chamberlin, 1919: 15H (*Pachymerium*; list).
— Chamberlin, 1923a: 244 (*Pachymerium*; notes; USA: Alaska).
— Bailey, 1928a: 1083 (*Pachymerium*; list).
— Bailey, 1928b: 19, 41 (*Pachymerium*; habitat; list; notes; USA: New York).
— Williams & Hefner, 1928: 134 (*Pachymerium*; characters; USA: Ohio).
— Attems, 1929: 245-246; Fig. 205 (*Pachymerium*; characters; distribution).
— Attems, 1938b: 365, 369 (*Pachymerium*; list; note; USA: Hawaii).
— Brimley, 1938: 501 (*Pachymerium*; USA: North Carolina).
— Causey, 1940: 64 (*Pachymerium*; characters; USA: North Carolina).
— Chamberlin & Mulaik, 1940: 110 (*Pachymerium*; USA: Texas).
— Chamberlin, 1944a: 33 (*Pachymerium*; USA: Georgia).
— Chamberlin, 1946c: 179 (*Pachymerium*; USA: Alaska).
— Chamberlin, 1951b: 34 (*Pachymerium*; USA: Nebraska).
— Crabill, 1952b: 169-170, pl. 4; Figs. 46, 50 (*Pachymerium*; characters; range; USA: Illinois, Indiana, Maryland, Michigan, New York, Ohio, Virginia, Wisconsin).

— Johnson, 1952: 103-108, tabs. 10-11, 28; Map 5 (*Pachymerium*; autecology; characters; distribution; natural history; USA: Michigan).

— Crabill, 1955b: 37-38 (*Pachymerium*; habitat; notes; USA: Missouri). [**Note:** Crabill thought this species was introduced. Need to determine if this is introduced or just a widespread holarctic species of Asia, Europe and North America (see Chamberlin, 1923a: 244)].

— Crabill, 1955h: 158 (*Pachymerium*; notes; USA: Missouri).

— Eason, 1964: 102-106; Figs. 137-146 (*Pachymerium*; characters; distribution; habitat).

— Wray, 1967: 156 (*Pachymerium*; list).

— Matic, 1972: 160–164; Fig. 62–63 (*Pachymerium*; characters; ecology; Romania).

— Kevan, 1979: 298 (*Pachymerium*; CANADA: "Subarctic").

— Summers, 1979: 697, 699 (*Pachymerium*; checklist; key).

— Summers et al., 1980: 247; Fig. 7 (*Pachymerium*; checklist; habitat (Table 1, p. 242-243); USA: Illinois).

— Summers et al., 1981: 60 (*Pachymerium*; checklist; USA: Illinois).

— Snider, 1991: 188 (*Pachymerium*; list; USA: Michigan).

— Behan-Pelletier, 1993: 25 (*Pachymerium*; list).

— Nishida, 1994: 26 (*Pachymerium*; checklist).

— Hoffman, 1995a: 24 (*Pachymerium*; list; notes).

— Nishida, 1997: 195 (*Pachymerium*; checklist).

— Watermolen, 1997: 3 (*Pachymerium*; checklist, notes).

— Bonato, Foddai, Minelli & Shelley, 2004: 27, 28; Figs. 4, 12, 14; Tab. 1 (*Pachymerium*; characters; key; notes; USA: Hawaii).

— Jeekel, 2005: 96 (*Geophilus*; list).

attenuatus Say, 1821, see: *bipuncticeps* Wood, 1862.

foveatus McNeill, 1887a: 325 (*Mecistocephalus*; note; USA: Florida). [**Note:** Syntype deposited at the NMNH].

— McNeill, 1887b: 333 (*Mecistocephalus*; characters; note; USA: Indiana).

— Bollman, 1888d: 346 [= 1893: 98] (*Geophilus*; note; USA: Georgia or Tennessee). [**Note:** The locality was cited as Lookout Mountain which the majority of resides in Georgia but some of the mountain runs into Tennessee].

— Bollman, 1888e: 408 [= 1893: 109] (*Geophilus*; note; USA: Indiana). [**Note:** Species name misspelled *foreatus* in 1888, but corrected in 1893].

— Bollman, 1893: 124, 184 (*Geophilus*; catalogue; USA: Minnesota).

— Attems, 1903b: 213 (*Mecistocephalus*; list). [**Note:** Misspelled *forcatus*].

— Chamberlin, 1912b: 67 (*Geophilus*; synonym of *ferrugineum*).

— Chamberlin, 1923a: 244 (*Mecistocephalus*; synonym of *ferrugineum*)

Distribution: CANADA: "Subarctic" (see Kevan, 1979); USA: Alaska, Florida, Georgia, ?Hawaii, Illinois, Indiana, Iowa, Maryland, Michigan, Minnesota, Missouri, Mississippi, North Carolina, Nebraska, New York, Ohio, ?Tennessee, Texas, Virginia, Wisconsin.

HOLOTYPE: Unknown?; Europe.

P. idium Chamberlin, 1960
— Chamberlin, 1960b: 241 (*Pachymerium*; characters; habitat; USA: California). [**Note:** Known from a single specimen and sex of type is not mentioned in description].

Distribution: USA: California.

HOLOTYPE: NMNH, California: Marin Co., 1½ mi. N of Dillon Beach, in the upper zone in a fissure in rock kept moist by spray but not covered at high tide, 26 July 1960.

Family **Geophilidae** Leach, 1815

[**Note:** Crabill (1970a) synonymized Chamberlin's Sogonidae and Soniphilidae under Geophilidae].

Subfamily **Geophilinae** Attems, 1926

Genus *Arenophilus* Chamberlin, 1912
[6 species; CANADA, USA]

Arenophilus Chamberlin, 1912c: 416-417 (characters; key to species; species included: *bipuncticeps, unaster, watsingus*). **TYPE SPECIES:** *Geophilus unaster* Chamberlin, 1909, [= *Arenophilus unaster* (Chamberlin)], fixed by original designation.
— Chamberlin, 1912a: 657-658 (characters; notes).
— Chamberlin, 1912c: 416-417 (characters; key to species of southeastern United States).

— Holmquist, 1926: 411 (notes on hibernation such as: activity, date collected, developmental stage, ecological association/niche, quantity found).
— Bailey, 1928b: 40 (characters; note).
— Holmquist, 1928: 83, Tab. 3 (co-inhabitant of *F. uleki* ant nests).
— Attems, 1929: 217 (characters; distribution; key to species; species included: *bipuncticeps, unaster, watsingus*).
— Crabill, 1952b: 122-125 (characters; key to species; taxonomic notes).
— Crabill, 1969a: 7-9, 10 (characters; key to species; systematic notes).
— Jeekel, 2005: 69 (list).

Zygomerium Chamberlin, 1943b: 100 (characters). **TYPE SPECIES:** *Zygomerium euphanum* Chamberlin, 1943, fixed by original designation.

— Crabill, 1952b: 122, 124, 129 (junior subjective synonym of *Arenophilus*; taxonomic notes).
— Crabill, 1969a: 7 (junior subjective synonym of *Arenophilus*). [**Note:** Crabill (see footnote 4) claims that the type species is conspecific with *Arenophilus watsingus* Chamberlin].
— Jeekel, 2005: 110 (list).

♂♀*A. bipuncticeps* (H. C. Wood, 1862)
— Wood, 1862: 45 (*Geophilus*; characters; USA: Illinois, South Carolina).
— Wood, 1865: 180-181 (*Geophilus*; characters; USA: Illinois).
— Brodie & White, 1883: 67 (*Geophilus*; CANADA). [**Note:** Although Brodie & White listed the species name as *bipunctatus* they must have meant *bipuncticeps*].
— Cragin, 1885: 145 (*Geophilus*; note; USA: Kansas).
— Bollman, 1888d: 347 [= 1893: 99] (*Geophilus*; note; USA: Georgia).
— Bollman, 1893: 124, 184 (*Geophilus*; catalogue; note; USA: Minnesota).
— Kenyon, 1893a: 16 (*Geophilus*; USA: Nebraska).
— Kenyon, 1893b: 161 (*Geophilus*; USA: Nebraska).
— Chamberlin, 1912a: 658 (*Arenophilus*; characters; notes).
— Chamberlin, 1912b: 66-67 (*Arenophilus*; range; notes; USA: Illinois, Iowa, Nebraska, Wisconsin).
— Chamberlin, 1912c: 417-418 (*Arenophilus*; notes; USA: Alabama, Georgia, Mississippi, Tennessee).
— Gunthorp, 1913: 168 (*Arenophilus*; habitat; notes; USA: Kansas).
— Chamberlin, 1918a: 23 (*Arenophilus*; USA: Tennessee).

— Chamberlin, 1918b: 376 (*Arenophilus*; USA: Louisiana).
— Anonymous, 1920: 35 (*Arenophilus*; USA: California).
— Gunthorp, 1920: 113 (*Geophilus*; list).
— Bailey, 1928a: 1083 (*Arenophilus*; list).
— Bailey, 1928b: 19, 40 (*Arenophilus*; characters; list; notes; USA: New York).
— Williams & Hefner, 1928: 131-132 (*Arenophilus*; characters; USA: Ohio).
— Attems, 1929: 218 (*Arenophilus*; characters; distribution).
— Chamberlin, 1931b: 97 (*Arenophilus*; USA: Oklahoma).
— Matthews, 1935: 18; Map: 19 (*Geophilus*; characters; USA: Wisconsin).
— Causey, 1940: 64 (*Arenophilus*; characters; USA: North Carolina).
— Chamberlin, 1942d: 185 (*Arenophilus*; USA: Louisiana, Missouri, Oklahoma).
— Chamberlin, 1942b: 15 (*Arenophilus*; USA: Iowa).
— Chamberlin, 1943b: 100 (*Arenophilus*; USA: Wyoming).
— Chamberlin, 1944a: 33 (*Arenophilus*; USA: Georgia).
— Chamberlin, 1944d: 192 (*Arenophilus*; USA: Arkansas, Florida, Illinois).
— Auerbach, 1946: 14, (*Arenophilus*; list; USA: Illinois).
— Auerbach, 1951a: 101, 106, 109, 112, 115, 118, 119, 120, 121, 122; Tab. 1, 8; Fig. 1 (*Arenophilus*; desiccation; habitat; list; notes; USA: Illinois).
— Chamberlin, 1951b: 34 (*Arenophilus*; USA: Georgia, Louisiana, Nebraska, Virginia).
— Crabill, 1952b: 125-128, pl. 4; Fig. 39-41, 43-44, 48 (*Arenophilus*; characters; range; taxonomic notes; CANADA: Nova Scotia; USA: District of Columbia, Illinois, Indiana, Kentucky, Massachusetts, Michigan, New York, Ohio, Virginia, Wisconsin).
— Johnson, 1952: 108-114; Tab. 12; Map 6 (*Arenophilus*; autecology; characters; distribution; natural history; USA: Michigan).
— Peters, 1954: 135 (*Arenophilus*; CANADA: Ontario).
— Crabill, 1955f: 258 (*Arenophilus*; checklist; USA: Kentucky).
— Crabill, 1955h: 158 (*Arenophilus*; notes; USA: Missouri).
— Crabill, 1957a: 344 (*Arenophilus*; notes).
— Crabill, 1958b: 95 (*Arenophilus*; behavior; notes; range; USA: Wisconsin).
— Wray, 1967: 156 (*Arenophilus*; list).
— Crabill, 1969a: 10 (comparative characters with *psednus*; notes).

— Summers, 1979: 697, 699 (*Arenophilus*; checklist; key).

— Summers et al., 1980: 247; Fig. 11 (*Arenophilus*; checklist; habitat (Table 1, p. 242-243); USA: Illinois).

— Summers et al., 1981: 60 (*Arenophilus*; checklist; USA: Illinois).

— Snider, 1991: 188 (*Arenophilus*; list).

— Behan-Pelletier, 1993: 25 (*Arenophilus*; list).

— Hoffman, 1995a: 22 (*Arenophilus*; list; notes).

— Watermolen, 1997: 3 (*Arenophilus*; checklist, notes).

attenuatus Say, 1821: 114 (*Geophilus*; characters; habitat; USA).

— Wood, 1865: 176, 177 (*Geophilus*; cited as possible synonym of *fulvus*).

— Bollman, 1888e: 408 [= 1893: 110] (*Geophilus*; note; USA: Indiana).

— Bollman, 1893: 124, 148 (*Geophilus*; catalogue; taxonomic notes). [**Note:** Bollman regards *bipuncticeps* as conspecific with *attenuatus* which would make *attenuatus* the senior synonym, but due to the poor description of *attenuatus* by Say it seems most authors have retained *bipuncticeps* as the valid species. Some authors, such as Wood (1865: 176, 177), have regarded *attenuatus* as conspecific with *Arctogeophilus fulvus*. Yet other authors such as Chamberlin (1912c: 415; 1923a: 244) have cited it as a synonym of *Pachymerium ferrugineum*. While Attems (1929: 348) designated *attenuatus* as *incertae sedis*. Under the circumstances, the author agrees that *attenuatus* should remain designated as *incertae sedis*].

— Cook, 1895a: 59, 60, 61 (*Geophilus* + *Mecistocephalus*; cited as type species of *Mecistocephalus*; notes).

— Attems, 1903b: 255 (*Mecistocephalus*; synonym of *ferrugineum*).

— Cook, 1904: 74-75 (*Mecistocephalus*; characters; notes; USA: Alaska).

— Attems, 1909a: 9 (*Mecistocephalus*; synonym of *Pachymerium ferrugineum*).

— Chamberlin, 1912b: 66 (*Geophilus*; from Bollman cited as synonym of *bipuncticeps*; Chamberlin was not certain if *attenuatus* of Say was a synonym of *bipuncticeps*).

— Chamberlin, 1912c: 415, 417 (*Geophilus*; cited as synonym of *ferrugineum* and also a questionable synonym of *bipuncticeps*).

— Chamberlin, 1923a: 244 (*Geophilus*; synonym of *ferrugineum*).

— Attems, 1929: 348 (*Geophilus*; incertae sedis).

— Attems, 1929: 245 (*Mecistocephalus*; synonym of *ferrugineum*).

— Chamberlin, 1946c: 179 (*Mecistocephalus*; synonym of *ferrugineum*).

— Crabill, 1954b: 187 (cited as *species inquirenda*; notes).

— Crabill, 1957a: 344 (*Geophilus*; notes).

georgianus Meinert, 1886a: 219-220 (*Geophilus*; characters; USA: Georgia). [**Note:** Holotype deposited at the MCZ, No. 958, and the locality label simply reads Georgia].

— Bollman, 1893: 124, 148 (*Geophilus*; catalogue; synonym of *attenuatus*).

— Cook, 1896: 239 (*Geophilus*; notes). [**Note:** Cook mentions that the coxal pores are close to the structure and location of *G. rubens*].

— Attems, 1903b: 236 (*Geophilus*; list).

— Chamberlin, 1912b: 66 (*Geophilus*; cited as synonym of *bipuncticeps*).

— Chamberlin, 1912c: 417 (*Geophilus*; cited as synonym of *bipuncticeps*).

huronicus Meinert, 1886a: 220-221 (*Geophilus*; characters; note; USA: Massachusetts). [**Note:** Two type specimens deposited at the MCZ, No. 929. Interestingly, Meinert comments that two specimens of the four specimens he examined were labeled "N. Engl."].

— Bollman, 1893: 125 (*Geophilus*; catalogue).

— Crabill, 1951: 314 (*Geophilus*; taxonomic notes; synonym of *bipuncticeps* and *Geophilus longicornis*).

— Attems, 1903b: 236 (*Geophilus*; list).

— Chamberlin, 1912c: 414 (*Geophilus*; notes; USA: Tennessee).

latro Meinert, 1871: 79 (*Geophilus*; characters). [**Note:** Four type specimens deposited at the MCZ, No. 955; Meinert's original locality label reads "Ship "Monsoon", J. M. Barnard"; original determined label reads "*Geophilus latro* Mein. No. 165"].

— Bollman, 1893: 125 (*Geophilus*; catalogue).

— Chamberlin, 1912b: 66 (*Geophilus*; cited as synonym of *bipuncticeps*).

— Attems, 1903b: 236 (*Geophilus*; list).

— Chamberlin, 1912c: 417 (*Geophilus*; cited as synonym of *bipuncticeps*).

perforata McNeill, 1887a: 325-326, pl. 11; Figs. 6-7 (*Schendyla*; characters; USA: Florida).

— Bollman, 1888a: 109 [= 1893: 84] (*Geophilus*; USA: Tennessee).

— Bollman, 1888b: 5 [= 1893: 77] (*Geophilus*; note; USA: Arkansas).

— Bollman, 1893: 125 (*Geophilus*; catalogue).

— Chamberlin, 1912b: 66 (*Schendyla*; cited as synonym of *bipuncticeps*).

— Chamberlin, 1912c: 417 (*Schendyla*; cited as questionable synonym of *bipuncticeps*).

— Attems, 1929: 349 (*Schendyla*; questioned subgenus; *incertae sedis*).

Distribution: CANADA: Ontario, Nova Scotia; USA: Alabama, Alaska, Arkansas, California, District of Columbia, Florida, Georgia, Illinois, Indiana, Iowa, Kansas, Kentucky, Louisiana, Massachusetts, Michigan, Minnesota, Missouri, Mississippi, North Carolina, Nebraska, New York, Ohio, Oklahoma, South Carolina, Tennessee, Virginia, Wisconsin, Wyoming.

SYNTYPES: NMNH, No. 266, 3 specimens, Illinois: "South Illinois", by R. Kennicott; NMNH, No. 285, 1 specimen, Illinois: Charleston; NMNH, No. 279, 5 specimens, ?Sonora (presumably somewhere in Illinois and not Mexico), by T.D. Graham. [**Note:** The author has designated these specimens as syntypes because at this time it is unclear if Wood (1862) designated any one specific specimen from these as a holotype].

A. iugans Chamberlin, 1944
— Chamberlin, 1944d: 192-193, pl. 14; Fig. 10 (*Arenophilus*; characters; USA: California).

Distribution: USA: California.

HOLOTYPE: FMNH, California: Los Angeles, 22 December 1936, by Gordon Grant.

PARATYPE: NMNH, California: Los Angeles, 1 March 1936, by Gordon Grant.

♂♀*A. osborni* Gunthorp, 1913
— Gunthorp, 1913: 169, pl. 16; Figs. 1-3 (*Arenophilus*; characters; notes; USA: Kansas).

Distribution: USA: Kansas.

HOLOTYPE: ?UKNHM, ♀, Kansas: Cowley Co.

♀*A. psednus* Crabill, 1969
— Crabill, 1969a: 10-11; Figs. 3-5 (*Arenophilus*; characters; USA: Kentucky). [**Note:** According to Andrew Weaver, from unpublished notes made on a reprint of this paper, on April 16, 1969:

"I looked at all my specimens of *Arenophilus* I had identified as *A. bipuncticeps* (Wood). Some of those from Nebraska lack the densely setose ventral areas on the anterior legs. These seem to be the smaller specimens and seem to have the second antennal segment ectally setose while the larger specimens with typical setose legs have few setae on the second antennal segment. All have long telopodite lappets and lack any dorsal, dark geminate band. Very careful examination of one specimen revealed stubby setae ventrally on the anterior legs. Had these been normal length the legs would have been typically *bipuncticeps*. Another specimen, however, gave no such appearance and looked more like *psednus*. I wonder whether Crabill has examined large series of *bipuncticeps*, esp. immature ones, to see how much variation there is in the armature of the ventral surfaces of the legs?" Apparently on a latter date he remarks: "I have now seen specimens of *psednus* from Georgia. They fit the description in every way except for length of telopodite lappets. They are about the same as in *bipuncticeps*. Differences not mentioned by Crabill are: lateral margins of cephalic plate densely pilose in *psednus*, not at all pilose in *bipuncticeps* and 1ˢᵗ antennal segment pilose in *psednus*, not so in *bipuncticeps*. Question: is there a lot of intraspecific variability in setae in this species and are we dealing with 2 species or one highly variable one?" Then on June 7, 1971 he adds: "3 specimens of *psednus* from Florida show 1ˢᵗ, 2 and half of 3ʳᵈ antennal seg pilose, lateral margins of cephalic plate not pilose!"].

Distribution: USA: Kentucky.

HOLOTYPE: NMNH; ♀, Kentucky: Boyle Co., Lawrence Cave at the SW edge of Perryville, 23 June 1967, by T. C. Barr, Jr.

A. unaster (Chamberlin, 1909)
— Chamberlin, 1909b: 179, pl. 25, figs, 4-6 (*Geophilus*; nominal taxon *attenuatus*; characters; USA: Louisiana, Texas).
— Chamberlin, 1912c: 417 (*Arenophilus*; notes; USA: Texas). [**Note:** Chamberlin discusses the similarity of *unaster* to *bipuncticeps* and *watsingus*].
— Attems, 1929: 217-218 (*Arenophilus*; characters; distribution).
— Chamberlin & Mulaik, 1940: 110 (*Arenophilus*; USA: Texas).
— Chamberlin, 1944d: 192 (*Arenophilus*; USA: Texas).
— Jeekel, 2005: 69 (*Geophilus*; list).

— Bonato, Zapparoli & Minelli, 2008: 348; Tab. 1 (*Geophilus*; nominal taxon *attenuatus*; "Main diagnostic characters differentiating *Diphyonyx* gen. n. from all other genera tentatively recognized in the subgroup of Geophilidae to which *Diphyonyx* belongs.").

Distribution: USA: Louisiana, Texas.

HOLOTYPE: ?Unknown, Texas: Austin, by Prof. J.H. Comstock.

♂♀*A. watsingus* Chamberlin, 1912
— Chamberlin, 1912c: 418-420, pl. 1; Figs. 1-4 (*Arenophilus*; characters; notes; USA: Alabama, Georgia, Kentucky, Mississippi, North Carolina, South Carolina, Virginia).
— Chamberlin, 1918b: 376 (*Arenophilus*; USA: Louisiana).
— Attems, 1929: 218 (*Arenophilus*; characters; distribution).
— Brimley, 1938: 501 (*Arenophilus*; USA: North Carolina).
— Causey, 1940: 64; Figs. 32, 36 (*Arenophilus*; characters; USA: North Carolina).
— Chamberlin, 1944d: 192 (*Arenophilus*; USA: Arkansas, Missouri).
— Crabill, 1950c: 200-201 (*Arenophilus*; range; USA: South Carolina).
— Crabill, 1952b: 128-129, pl. 4; Fig. 42 (*Arenophilus*; characters; range; taxonomic notes; USA: Kentucky, Virginia).
— Crabill, 1955b: 37 (*Arenophilus*; notes; USA: Missouri).
— Crabill, 1955f: 258 (*Arenophilus*; checklist; USA: Kentucky).
— Wray, 1967: 156 (*Arenophilus*; list).
— Hoffman, 1995a: 23 (*Arenophilus*; list; notes).
euphanum Chamberlin, 1943b: 100-101 (*Zygomerium*; characters; USA: New Mexico). [**Note:** See note under *Z. rotarium* for an explanation of the synonymy of *Z. euphanum*].
— Crabill, 1952b: 128 (*Zygomerium*; synonym of *watsingus*).
— Crabill, 1969a: 7 (*Zygomerium*; synonym of *watsingus*).
— Jeekel, 2005: 110 (*Zygomerium*; list).

Distribution: USA: Alabama, Arkansas, Georgia, Kentucky, Louisiana, Missouri, Mississippi, New Mexico, North Carolina, South Carolina, Virginia.

SYNTYPES: MCZ, No. 953, 5 specimens, ♂ [**Note:** apparently Chamberlin designated a male specimen as the holotype], Virginia: Chatham, 14 August 1910 (8-14-10).

Genus *Brachygonarea* Ribaut, 1911
[1 species; Northwestern USA]

Brachygonarea Ribaut, 1911: 123, 126 (characters; key). TYPE SPECIES: *Polygonarea apora* Attems, 1909, fixed by subsequent designation.
— Attems, 1929: 306 (characters; distribution; species included: *aporus*).
— Attems, 1947: 146 (key to species; species included: *aporus*, *borealis*).

B. borealis Attems, 1934
— Attems, 1934: 314-315; fig. 7 (*Brachygonarea*; characters; notes; USA: Washington).
— Attems, 1947: 146 (*Brachygonarea*; list).

Distribution: USA: Washington.

HOLOTYPE: Hamburg Museum; Washington: Puget Sound, Port Ludlow.

Genus *Californiphilus* Verhoeff, 1938
[1 species; Western USA]

Californiphilus Verhoeff, 1938: 371 TYPE SPECIES: *Californiphilus michelbacheri* Verhoeff, fixed by monotypy. [**Note:** It is not clear if *Californiphilus* Verhoeff is a synonym of *Gosiphilus* Chamberlin which was synonymized to *Chomatobius* Humbert & Saussure by Crabill in 1968 and requires further investigation (see, Crabill, 1953e: 88; Crabill, 1968a)].
— Crabill, 1953e: 88 (note). [**Note:** Crabill states "I am relatively certain that Verhoeff's *Californiphilus*, described from California, is actually referable to Chamberlin's *Gosiphilus*, which is itself represented by southwestern, Californian, and Mexican forms, one of which Verhoeff might well have had before him when he described *C. michaelbacheri*, the type (by monotypy) of his genus." Should *Californiphilus* become synonymized under *Chomatobius* it would cause secondary homonymy of *Californiphilus mexicanus* (Attems, 1947) with *Chomatobius mexicanus* (Saussure, 1858)]
— Crabill, 1959a: 155, 158-159 (suspected synonym of *Gosiphilus*; notes).
— Jeekel, 2005: 73 (list).

♂*C. michelbacheri* Verhoeff, 1938
— Verhoeff, 1938b: 371-372, taf. 7; Figs. 31-36 (*Californiphilus*; characters; USA: California).
— Attems, 1947: 80 (*Californiphilus*; list).
— Crabill, 1959a: 155, 156 (*Californiphilus*; cited as type-species; list).
— Jeekel, 2005: 73 (*Californiphilus*; list).

Distribution: USA: California.

HOLOTYPE: ?Unknown, ♂, California: San Lucas, 18.viii.1934.

Genus *Caliphilus* Chamberlin, 1941
[1 species; Western USA]

Caliphilus Chamberlin, 1941d: 782 (characters). Type Species: *Caliphilus alamedanus* Chamberlin, 1941, fixed by original designation and monotypy.
— Jeekel, 2005: 73 (list).

♀*C. alamedanus* Chamberlin, 1941
— Chamberlin, 1941d: 782 (*Caliphilus*; characters; habitat; USA: California).
— Jeekel, 2005: 73 (*Caliphilus*; list).
— Bonato, Zapparoli & Minelli, 2008: 348; Tab. 1 (*Caliphilus*; "Main diagnostic characters differentiating *Diphyonyx* gen. n. from all other genera tentatively recognized in the subgroup of Geophilidae to which *Diphyonyx* belongs.").

Distribution: USA: California.

HOLOTYPE: ?Unknown, ♀, California: Alameda Co., Castro Valley, 9 March 1941, from sifting material of wood rat nest, by W.M. Pearce.

Genus *Condylona* Chamberlin, 1941
[2 species; Western USA]

Condylona Chamberlin, 1941d: 780 (characters). Type Species: *Condylona sontipes* Chamberlin, 1941, fixed by original designation.
— Jeekel, 2005: 74 (list).

♀*C. isabella* Chamberlin, 1941

— Chamberlin, 1941d: 780-781 (*Condylona*; characters; USA: California).

Distribution: USA: California.

HOLOTYPE: NMNH, ♀, California: Kern Co., Isabella, 2, 525 ft., 18 March 1941, by S. & D. Mulaik.

♀*C. sontipes* Chamberlin, 1941
— Chamberlin, 1941d: 780 (*Condylona*; characters; USA: California).
— Jeekel, 2005: 74 (*Condylona*; list).
— Bonato, Zapparoli & Minelli, 2008: 348; Tab. 1 (*Condylona*; "Main diagnostic characters differentiating *Diphyonyx* gen. n. from all other genera tentatively recognized in the subgroup of Geophilidae to which *Diphyonyx* belongs.").

Distribution: USA: California.

HOLOTYPE: NMNH, California: Inyo Co., 16 March 1941. [**Note:** Genus labeled as *Tomotaenia*].

PARATYPE: NMNH, California: Inyo Co., 16 March 1941.

Genus ***Erithophilus*** Cook, 1899
[1 species; Southeastern USA]

Erithophilus Cook, 1899: 308, 310 (characters; notes). **Type Species:** *Erithophilus neopus* Cook, 1899, fixed by monotypy.
— Attems, 1903b: 271 (list).
— Jeekel, 2005: 78 (list).

♂♀*E. neopus* Cook, 1899
— Cook, 1899: 310, pl. 4; Figs. 1a-1d (*Erithophilus*; characters; habitat; note; USA: Florida).
— Attems, 1903b: 271 (*Erithophilus*; list).
— Jeekel, 2005: 78 (*Erithophilus*; list).
— Bonato, Zapparoli & Minelli, 2008: 348; Tab. 1 (*Erithophilus*; "Main diagnostic characters differentiating *Diphyonyx* gen. n. from all other genera tentatively recognized in the subgroup of Geophilidae to which *Diphyonyx* belongs.").

Distribution: USA: Florida.

SYNTYPE: NMNH; No. 779; Florida: Dade Co.?: Sugar Loaf Key, under stones in the ground.

Genus *Garrina* Chamberlin, 1915
[4 species; Southern Central USA]

Garrina Chamberlin, 1915: 506-507 (characters; note). **TYPE SPECIES:** *Garrina ochra* Chamberlin, 1915, fixed by original designation and monotypy.
— Crabill, 1968b: 331 (catalogue).
— Jeekel, 2005: 80 (list).
Pycnona Chamberlin, 1943a: 18 (characters). **TYPE SPECIES:** *Pycnona pujola* Chamberlin, 1943, fixed by original designation.
— Crabill, 1968b: 331 (catalogue; synonym of *Garrina*).
— Jeekel, 2005: 101 (list).
Simoleptus Chamberlin, 1941d: 776 (characters). **TYPE SPECIES:** *Simoleptus parapodus* Chamberlin, 1941, [= *Garrina parapodus*] fixed by original designation and monotypy.
— Chamberlin, 1944d: 193 [**Note:** Chamberlin transferred the type species *Simoleptus parapodus* to *Garrina* making *Simoleptus* a junior synonym of *Garrina*. Following this the author has tentatively transferred the remaining species *Simoleptus cruzanus* to *Garrina*).
— Jeekel, 2005: 104 (list).

♀*G. alicea* Chamberlin, 1943
— Chamberlin, 1943b: 102 (*Garrina*; characters; USA: Texas).

Distribution: USA: Texas.

HOLOTYPE: NMNH, ♀, Texas: Jim Wells Co., 12 mi. N of Alice, 6 June 1941. [**Note:** The author could not locate the holotype at the NMNH].

PARATYPE: NMNH; ♀, Texas: Edinburg, 4 June 1941, Mulaik.

G. cruzanus (Chamberlin, 1942)
— Chamberlin, 1942c: 19-20 (*Simoleptus*; characters; USA: Texas). [**Note:** This species was originally described as belonging to the genus *Simoleptus*, however, Chamberlin transferred the type species of *Simoleptus* to *Garrina* therefore this species is tentatively, for convenience, placed in *Garrina* until further study].

333

Distribution: USA: ?Texas.

HOLOTYPE: NMNH, Texas: Laredo, 5 June 1941, with orchids in baggage from Vera Cruz, Mexico.

G. ochra Chamberlin, 1915
— Chamberlin, 1915: 507, pl. 3; Figs. 1-3 (characters; Mexico).
— Attems, 1929: 345 (*Garrina*; characters; distribution).
— Chamberlin & Mulaik, 1940: 125-126 (*Garrina*; note; USA: Texas).
— Crabill, 1968b: 331 (*Garrina*; catalogue; cited as type species).
— Jeekel, 2005: 80 (*Garrina*; list).

Distribution: USA: Texas.

HOLOTYPE: MCZ, No. 1723, Mexico: Pachuca, by W.M. Mann. [**Note:** This species was established for a specimen originally described from Mexico but was later found to be in Texas].

♂ *G. parapodus* (Chamberlin, 1941)
— Chamberlin, 1941d: 777 (*Simoleptus*; characters; USA: Texas).
— Chamberlin, 1944d: 193 (*Garrina*; USA: Missouri).
— Crabill, 1955h: 159 (*Garrina*; note; USA: Missouri).
— Jeekel, 2005: 104 (*Simoleptus*; list).

Distribution: USA: Missouri, Texas.

HOLOTYPE: NMNH, ♂, Texas: Bangs, 10 February 1938, by L.D. Christensen, L.S. Jones & D.W. Clancy.

PARATYPE: NMNH.

Genus *Geophilus* Leach, 1815
[50 species(4-5 Introduced); CANADA and USA]

Geophilus Leach, 1815: 384 (characters). **Type Species:** *Scolopendra electricus* Linnaeus, 1758, fixed by monotypy. [**Note:** see Crabill (1952b: 140) for his discussion on the type species of the genus *Geophilus*).
— Bollman, 1887c: 624 (key to species; species included: *bipuncticeps, cephalicus, georgianus; latro, mordax, okolonae, oweni, perforatus*).

— Chamberlin, 1902c: 473 (key to species of western United States).
— Chamberlin, 1912a: 656 (characters; key to species of California).
— Chamberlin, 1912c: 411-412 (key to species of southeastern United States).
— Bailey, 1928b: 39 (characters; key to species of New York; species included: *deducens, mordax, rubens*).
— Crabill, 1952b: 136-142 (characters; key to species of northeastern United States; notes; taxonomic notes).
— Crabill, 1958a: 15, 16 (key to species of eastern North America).
— Haupt, 1979: 394; fig. 2b (chemotactile sensilla).
— Jeekel, 2005: 81 (list; notes).

Arthronomalus Newport, 1845b: 430 (characters). TYPE SPECIES: *Arthronomalus longicornis* Leach, 1815, fixed by subsequent designation.

— Gervais, 1847: 308 (synonym of *Geophilus*).
— Crabill, 1957a: 343 (subjective synonym of *Geophilus*, notes).
— Jeekel, 2005: 70 (list; junior objective synonym of *Necrophloeophagus*).

Brachygeophilus Brölemann, 1909: 338 (characters). TYPE SPECIES: *Geophilus truncorum* Bergsöe & Meinert, 1866: 94-95, by original designation and monotypy. [**Note:** see *Brachygeophilus truncorum* (Bergsöe & Meinert) in Eason, 1964: 139-143; Figs. 230-240 (characters (adolescens); distribution)].

— Attems, 1929: 189-190 (characters; distribution; key to species; species included: *claremontus, conjungens, isantus, richardi, strictus, truncorum, truncorum ribauti*).
— Foddai, et al., 1995: 9 (*Brachygeophilus* transferred to *Geophilus*).
— Jeekel, 2005: 72 (list).

Dysmesus Chamberlin, 1944d: 193-194 (characters). TYPE SPECIES: *Dysmesus orites* Chamberlin, 1944, fixed by original designation and monotypy.

— Crabill, 1952b: 129-130 (synonym of *Brachygeophilus*; characters; notes).
— Crabill, 1981a: 174 (synonym of *Brachygeophilus*).
— Jeekel, 2005: 77 (list).

Necrophloeophagus Newport, 1843: 180 (cited as a subgenus). TYPE SPECIES: *Necrophloeophagus longicornis* Leach, 1815, fixed by monotypy.

— Gervais, 1847: 308 (synonym of *Geophilus* sensu strictu).
— Jeekel, 2005: 92 (list).

Soniphilus Chamberlin, 1912b: 68-69 (characters; synonym of *Brachygeophilus*)
   TYPE SPECIES: *Soniphilus embius* Chamberlin, 1912, fixed by original
   designation.
   — Attems, 1929: 331 (species included: *embius*, *geronimmo* and *secundus*;
   characters; distribution; key to species).
   — Crabill, 1952b: 130-133 (characters; cited as synonym of
   Brachygeophilus; notes; taxonomic notes).
   — Jeekel, 2005: 105 (list).

♂♀*G. admarinus* (Chamberlin, 1952)
   — Chamberlin, 1952a: 83-84 (*Brachygeophilus*; characters; habitat; notes;
   USA: Alaska). [**Note:** Chamberlin reports "Numerous specimens
   taken under stones near the low tide mark . . .".].
   — Behan-Pelletier, 1993: 25 (*Brachygeophilus*; list).
   — Foddai, et al., 1995: 9 (*Brachygeophilus* transferred to *Geophilus*).

Distribution: USA: Alaska.

SYNTYPES: NMNH, Alaska: Redd Bay, Kuiu Id., 23-25 July 1951, by Borys
Malkin. [**Note:** Literature records the collection month as July, however, the
locality label in the vial has August as the month collected. Peter Mundel left
a hand written note in the vial "This is apparently the type series. P. Mundel
Nov. 1984". There are numerous specimens in the vial].

♂*G. alaskanus* Cook, 1904
   — Cook, 1904: 75 (*Geophilus*; characters; note; USA: Alaska).
   — Attems, 1909a: 9 (*Geophilus*; list).
   — Chamberlin, 1919: 15H (*Geophilus*; list).
   — Attems, 1929: 321 (*Geophilus*; characters; distribution; *species
   inquirenda*).
   — Chamberlin, 1946c: 179 (*Geophilus*; USA: Alaska).
   — Behan-Pelletier, 1993: 25 (*Geophilus*; list).

Distribution: ?CANADA; USA: Alaska.

HOLOTYPE: NMNH, No. 793, ♂, Alaska: Sitka, June 1899, by A.H. Twitchell.

G. ampyx Crabill, 1954
   — Crabill, 1954b: 183-184; Figs. 9-11 (*Geophilus*; characters; systematic
   notes; USA: Alabama, South Carolina).
   — Crabill, 1955f: 258 (*Geophilus*; checklist; USA: Kentucky).

— Branson & Batch, 1967: 81 (*Geophilus*; brief description; habitat; notes; USA: Kentucky).

— Summers et al., 1980: 247; Fig. 13 (*Geophilus*; checklist; habitat (Table 1, p. 242-243); USA: Illinois).

— Summers et al., 1981: 60 (*Geophilus*; checklist; USA: Illinois).

— Hoffman, 1995a: 23 (*Geophilus*; list; notes: USA: Virginia).

Distribution: USA: Alabama, Illinois, Kentucky, South Carolina, Virginia.

HOLOTYPE: NMNH, Crabill collection; No. C-1590; ♀, South Carolina: Oconee Co., Clemson, 28 March 1950, by E.C. Turner, Jr. & W.R. Mason. [**Note:** 2 slides, 368 & 369].

ALLOTYPE: NMNH, Crabill collection; No. C-1591; ♂, South Carolina: Oconee Co., Clemson, 25 March 1952, by E.C. Turner, Jr.

PARATYPES: NMNH, Crabill collection, (1♂, 2♀), South Carolina: Oconee Co., Clemsson, 27 March 1950, by E. C. Turner & W. R. Mason. (6♀, 4♂), South Carolina: Oconee Co., Clemsson, 24-27 March 1951, by G. E. Ball, J. C. Martin & W. R. Mason. [**Note:** These are just a couple, please see Crabill, 1954b for the full paratype series].

♀*G. anonyx* (Chamberlin, 1941)
— Chamberlin, 1941d: 777 (*Brachygeophilus*; characters; note; USA: Oregon).
— Foddai, et al., 1995: 9 (*Brachygeophilus* transferred to *Geophilus*).

Distribution: USA: Oregon.

HOLOTYPE: NMNH, ♀, Oregon: Grant's Pass, 7 December 1937, by L.D. Christensen, L.S. Jones & D.W. Clancy.

♀*G. atopodon* Chamberlin, 1903
— Chamberlin, 1903a: 37-38 (*Geophilus*; characters; USA: New Mexico).
— Attems, 1929: 180 (*Geophilus*; characters; distribution).

Distribution: USA: New Mexico.

HOLOTYPE: NMNH, ♀, New Mexico: Beulah, ca. 8,000 ft.

♀*G. becki* Chamberlin, 1951

— Chamberlin, 1951b: 34-35 (*Geophilus*; characters; habitat; note; USA: California).

Distribution: USA: California.

SYNTYPES: NMNH, 5 specimens, California: Cabbrillo Beach near San Pedro, 22 August 1950, found under rocks and kelp at ocean water's edge. [**Note:** Although Chamberlin designated a female as the holotype, the type vial at the NMNH contains 5 specimens. Chamberlin also states that there are 2 female paratypes. Three specimens appear mature which are referable to the holotype and paratype specimens, the remaining 2 specimens are immature].

♂ *G. brevicornis* Wood, 1862
— Wood, 1862: 45 (*Geophilus*; characters; USA: Illinois, Texas). [**Note:** The locality for Texas said "In route from New Orleans to Galveston" but Wood (1865) clarifies that the specimen was collected in Texas and not Louisiana].
— Wood, 1865: 179-180 (*Geophilus*; characters; USA: Illinois, Texas).
— Bollman, 1893: 124 (*Geophilus*; catalogue).
— Gunthorp, 1920: 113 (*Geophilus*; list).
— Attems, 1929: 328 (*Geophilus*; distribution; *incertae sedis*).
— Williams & Hefner, 1928: 132 (*Geophilus*; characters; USA: Ohio).
— Crabill, 1954b: 187 (cited as *species inquirenda*; notes).

Distribution: USA: Illinois, Ohio, Texas.

HOLOTYPE: Unknown?

♂♀ *G. brunneus* McNeill, 1887
— McNeill, 1887b: 331 (*Geophilus*; characters; note; USA: Indiana).
— Bollman, 1888e: 408 [= 1893: 109] (*Geophilus*; note; USA: Indiana).
— Bollman, 1893: 124 (*Geophilus*; catalogue).
— Attems, 1903b: 263 (*Geophilus*; list).
— Williams & Hefner, 1928: 132 (*Geophilus*; characters; USA: Ohio).
— Attems, 1929: 328 (*Geophilus*; distribution; *incertae sedis*).
— Crabill, 1954b: 187 (cited as *species inquirenda*; notes).

Distribution: USA: Indiana, Ohio.

HOLOTYPE: NMNH, Indiana: Monroe Co., Bloomington Township.

♂♀ *G. carpophagus* Leach, 1815
— Leach, 1815: 384-385 (*Geophilus*; characters; habitat).
— Brodie & White, 1883: 67 (*Geophilus*; CANADA).
— Attems, 1929: 163; fig. 169 (*Geophilus*; characters; distribution).

Distribution: CANADA.

HOLOTYPE: ?Unknown. [**Note:** MCZ has some specimens as follows: 945, 947, 31038 & 31041 from Ireland, Germany, Italy & England respectively].

♂♀ *G. cayugae* Chamberlin, 1904
— Chamberlin, 1904: 655 (*Geophilus*; characters; notes; USA: New York).
— Attems, 1929: 322 (*Geophilus*; characters; distribution; *species inquirenda*).
— Crabill, 1952b: 149-150, pl. 5; Fig. 61 (*Geophilus*; characters; notes; USA: New York, Virginia).
— Crabill, 1954b: 181-182; Figs. 2, 8, 13 (*Geophilus*; characters; habitat; notes; USA: New York, North Carolina, Virginia).
— Hoffman, 1995a: 23 (*Geophilus*; list; notes; range extension; USA: Tennessee, Virginia).

Distribution: USA: North Carolina, New York, Tennessee, Virginia.

HOLOTYPE: NMNH; New York, Ithaca.

G. claremontus Chamberlin, 1909
— Chamberlin, 1909b: 184, pl. 25; Figs. 7-9 (*Geophilus*; characters; notes; USA: California).
— Attems, 1929: 192 (*Brachygeophilus*; characters; distribution). [**Note:** Apparently Attems felt that this species belonged to *Brachygeophilus* instead of *Geophilus*, however, *Brachygeophilus* is now considered a junior synonym of *Geophilus*].

Distribution: USA: California.

HOLOTYPE: ?Unknown; California: Claremont, one specimen.

G. compactus (Attems, 1934)
— Attems, 1934: 313-314; fig. 6 (*Brachygeophilus*; characters; notes; USA: Washington).
— Attems, 1947: 123 (*Brachygeophilus*; list).
— Foddai, et al., 1995: 9 (*Brachygeophilus* transferred to *Geophilus*).

Distribution: USA: Washington.

HOLOTYPE: HM; Washington: Puget Sound: Tacoma Puget Sound, Port Blackley.

G. delotus (Chamberlin, 1941)
— Chamberlin, 1941d: 779 (*Brachygeophilus*; characters; USA: California). [**Note:** Although multiple localities are listed; the actual holotype locality is not distinguishable based on data. It is assumed that the first of these is the holotype locality until verified].
— Foddai, et al., 1995: 9 (*Brachygeophilus* transferred to *Geophilus*).

Distribution: USA: California.

HOLOTYPE: NMNH, California, Kern Co., Isabella, elevation 2,525 feet, March 18, 1941, by S. & D. Mulaik.

♂♀**G. electricus** (Linnaeus, 1758)
— Linnaeus, 1758: 638 (*Scolopendra*; characters; "Europae terra").
— Bergsöe & Meinert, 1866: 90-91 (*Geophilus*; characters; notes).
— Meinert, 1870: 84-85 (*Geophilus*; characters (adult, stadia); notes).
— Latzel, 1880: 187-189 (*Geophilus*; characters (adult, stadia); distribution).
— Attems, 1903b: 224-225, taf. 15; Figs. 42-44 (*Geophilus*; characters; distribution).
— Attems, 1929: 167-169; Fig. 172 (*Geophilus*; characters; distribution).
— Palmén, 1954: 134; Fig. 1(2) (*Geophilus*; notes; CANADA: Newfoundland).
— Eason, 1964: 116-120; Figs. 167-176 (*Geophilus*; characters (adult, adolescens); distribution; habitat).
— Kevan, 1979: 298 (*Geophilus*; CANADA: Newfoundland).
— Behan-Pelletier, 1993: 25 (*Geophilus*; list).

— Jeekel, 2005: 81 (*Scolopendra*; list).

Distribution: CANADA: Newfoundland. [**Note:** Introduced from Europe].

HOLOTYPE: Unknown by author. Europe.

♀*G. embius* (Chamberlin, 1912)
— Chamberlin, 1912b: 69, pl. 1; Figs. 1-5 (*Soniphilus*; characters; notes; USA: Iowa).
— Holmquist, 1926: 411 (*Brachygeophilus*; notes on hibernation such as: activity, date collected, developmental stage, ecological association/ niche, quantity found).
— Holmquist, 1928: 83, Tab. 3 (*Brachygeophilus*; co-inhabitant of *F. uleki* ant nests).
— Attems, 1929: 331 (*Soniphilus*; characters; distribution).
— Auerbach, 1951a: 101 (*Soniphilus*; notes).
— Auerbach, 1951b: 114 (*Soniphilus*; key).
— Crabill, 1952b: 133-134 (*Brachygeophilus*; characters; key to species of northeastern US).
— Summers, 1979: 697, 699 (*Brachygeophilus*; checklist; key).
— Foddai, et al., 1995: 9 (*Brachygeophilus* transferred to *Geophilus*).
— Watermolen, 1997: 3 (*Soniphilus*; checklist, notes).
— Jeekel, 2005: 105 (*Soniphilus*; list).

Distribution: USA: Illinois, Iowa.

HOLOTYPE: Unknown by author, ♀, Iowa: DeWitt.

♂♀*G. flavus* (De Geer, 1778)
— DeGeer, 1778: 561 (*Scolopendra*; no locality stated).
— Newport, 1856: 85 (*Geophilus*; characters; Europe).
— Attems, 1929: 173 (*Geophilus*; synonym of *longicornis*).
— Crabill, 1952b: 157 (*Scolopendra*; cited as a possible synonym of *longicornis*).
— Crabill, 1952b: 157 (*Geophilus*; cited as a possible synonym of *longicornis*).
huronicus Meinert, 1886a: 220-221 (*Geophilus*; characters; USA: Massachusetts).
— Chamberlin, 1912c: 414 (*Geophilus*; notes; USA: Tennessee).
— Attems, 1929: 323-324 (*Geophilus*; characters; distribution).
— Crabill, 1951: 314, 315 (*Geophilus*; taxonomic notes; cited as synonym of *longicornis* and *Arenophilus bipuncticeps*). [**Note:**

Crabill mentions that of the four specimens Meinert used to describe this species two of them are referable to *A. bipuncticeps* while the other two are synonyms of *longicornis*].

— Crabill, 1952b: 158 (*Geophilus*; cited as synonym of *longicornis*; notes).

longicornis Leach, 1815: 386 (*Geophilus*; characters; habitat).

— Newport, 1842: 180 (*Geophilus*; subgenus *Necrophloeophagus*). [**Note:** Reference not obtained].

— (Leach, 1844): 430 (*Geophilus* (*Necrophloeophagus*)). [**Note:** Reference not obtained].

— Newport, 1845a: 292; Figs. 3, 15 (*Arthronomalus*; subgenus *Geophilus*; note on comparative anatomy of head).

— Attems, 1929: 173-174 (*Geophilus*; characters; distribution).

— Brölemann, 1930: 151 (*Necrophloeophagus*; characters).

— Crabill, 1951: 314, 315 (*Geophilus*; notes).

— Crabill, 1952b: 157-158 (*Necrophloeophagus*; characters; notes; CANADA: Ontario; USA: Massachusetts). [**Note:** Crabill suspects that this species has probably been introduced from Europe before 1886].

— Palmén, 1954: 133-134; Fig. 1(2,4,5,7) (*Necrophloeophagus*; notes; CANADA: Newfoundland).

— Eason, 1964: 134-139, pl. 1 & 3; Figs. 217-229 (*Necrophloeophagus*; characters (adult, adolescens); distribution; habitat).

— Kevan, 1979: 298 (*Necrophloeophagus*; CANADA: Newfoundland, Ontario).

— Behan-Pelletier, 1993: 25 (*Necrophloeophagus*; list).

— Jeekel, 2005: 92 (*Geophilus*; list).

similis Newport, 1845b: 432 (*Arthronomalus*; subgenus *Necrophloeophagus*; characters; notes).

— Crabill, 1952b: 157 (*Arthronomalus*; subgenus *Necrophloeophagus*; cited as synonym of *longicornis*).

?*tristanicum* Attems, 1928: 157-160, text-figs. 58-62, pl. 10, figs 489, 490 (*Pachymerium*; characters; notes). **Type Locality:** Tristan da Cunha, a group of Islands off the southwestern coast of Africa. [**Note:** Description in English. Attems comments on how remarkable the distribution of this species is because he claimed to have found the same species in Europe. Being that these islands are a territory of the U.K. it would seem more plausible that this species has been introduced. This further exemplifies the reason why this is more likely a synonym of *flavus* and why it is found in the USA as well].

— Attems, 1929: 247-248; figs. 206-208 (*Pachymerium*; characters; distribution).

— Crabill, 1950b: 101 (*Pachymerium*; notes; CANADA: Ontario). [**Note:** misidentification].

— Crabill, 1951: 315 (*Pachymerium*; Crabill's report of this species in 1950b is cited as a synonym of *Geophilus longicornis*).

— Crabill, 1952b: 157 (*Pachymerium*; cited as a possible synonym of *longicornis*).

— Kevan, 1979: 298 (*Pachymerium*; CANADA: Ontario).

Distribution: CANADA: Newfoundland, Ontario; USA: Massachusetts, Tennessee.

HOLOTYPE: ?Unknown.

♀ *G. fruitanus* Chamberlin, 1928
— Chamberlin, 1928d: 310 (*Geophilus*; characters; note; USA: Utah).
— Chamberlin, 1930a: 114 (*Geophilus*; USA: Utah).
— Attems, 1947: 117 (*Geophilus*; list).
— Chamberlin, 1961: 98 (*Geophilus*; note).

Distribution: USA: Utah.

HOLOTYPE: NMNH; ♀, Utah: Wayne Co., Fruita.

G. geronimo (Chamberlin, 1912)
— Chamberlin, 1912d: 158-159, pl. 12; Figs. 1-3 (*Soniphilus*; characters; USA: New Mexico).
— Attems, 1929: 332 (*Soniphilus*; characters; distribution).
— Foddai, et al., 1995: 9 (*Brachygeophilus* transferred to *Geophilus*).

Distribution: USA: New Mexico.

HOLOTYPE: ?Unknown, New Mexico: San Geronimo, Mrs. W.P. Cockerell & Miss Mary Cooper.

♂♀ *G. glyptus* (Chamberlin, 1902)
— Chamberlin, 1902c: 477-478 (*Geophilus*; characters (adult, stadia); habitat; key to species; USA: Utah).
— Chamberlin, 1911e: 260 (*Geophilus*; USA: Washington).
— Attems, 1929: 182 (*Geophilus*; characters; distribution).

— Chamberlin, 1930a: 114 (*Geophilus*; note; USA: Utah).
— Chamberlin, 1943b: 100 (*Geophilus*; USA: Utah).
— Chamberlin, 1961: 97-98 (*Brachygeophilus*; notes; USA: ?Idaho). [**Note:** Chamberlin comments "While it appears to range into Idaho, . . .".].
— Foddai, et al., 1995: 9 (*Brachygeophilus* transferred to *Geophilus*).

Distribution: USA: ?Idaho, Utah, Washington.

HOLOTYPE: ?Unknown, Utah: Salt Lake Co., Neff's Canyon, 30.vi.1900, by S.C. Chamberlin.

G. indianae McNeill, 1887
— McNeill, 1887a: 331-332 (*Geophilus*; characters; note; USA: Indiana).
— Bollman, 1888e: 408 [= 1893: 109] (*Geophilus*; note; USA: Indiana).
— Bollman, 1893: 125 (*Geophilus*; catalogue).
— Attems, 1903b: 263 (*Geophilus*; list).
— Attems, 1929: 328 (*Geophilus*; distribution; *incertae sedis*).
— Crabill, 1954b: 187 (cited as *species inquirenda*; notes).

Distribution: USA: Indiana.

HOLOTYPE: NMNH, Indiana: near La Fayette, by Prof. Henry L. Osborne.

♀*G. leionyx* (Chamberlin, 1938)
— Chamberlin, 1938a: 255 (*Brachygeophilus*; characters; USA: Oregon).
— Foddai, et al., 1995: 9 (*Brachygeophilus* transferred to *Geophilus*).

Distribution: USA: Oregon.

HOLOTYPE: NMNH, ♀, Oregon: Boyer, by J.A. Macnab.

PARATYPE: NMNH, ♀, Oregon: Boyer, by J.A. Macnab.

♂♀*G. mordax* Meinert, 1886
— Meinert, 1886a: 217 (*Geophilus*; characters; USA).
— Bollman, 1893: 125 (*Geophilus*; catalogue).
— Attems, 1903b: 236 (*Geophilus*; list).

— Chamberlin, 1912c: 412-414 (*Geophilus*; characters; notes; USA: Alabama, Louisiana, Mississippi, North Carolina, Tennessee, Virginia).
— Gunthorp, 1913: 171 (*Geophilus*; habitat; note; USA: Kansas).
— Chamberlin, 1918a: 23 (*Geophilus*; USA: Tennessee).
— Chamberlin, 1918b: 376 (*Geophilus*; USA: Louisiana).
— Bailey, 1928a: 1083 (*Geophilus*; list).
— Bailey, 1928b: 19, 40 (*Geophilus*; list; notes; USA: New York).
— Williams & Hefner, 1928: 132 (*Geophilus*; characters; USA: Ohio).
— Attems, 1929: 324 (*Geophilus*; characters; distribution; *species inquirenda*).
— Brimley, 1938: 501 (*Geophilus*; USA: North Carolina).
— Causey, 1940: 62-63 (*Geophilus*; characters; notes; USA: North Carolina).
— Chamberlin, 1944a: 33 (*Geophilus*; USA: Georgia).
— Chamberlin, 1944d: 191 (*Geophilus*; USA: Tennessee).
— Chamberlin, 1945d: 216 (*Geophilus*; USA: Georgia).
— Chamberlin, 1946b: 194 (*Geophilus*; USA: Mississippi).
— Crabill, 1950c: 200 (*Geophilus*; taxonomic notes; USA: South Carolina).
— Chamberlin, 1951b: 34 (*Geophilus*; USA: Florida).
— Crabill, 1952b: 151-152, pl. 5; Fig. 60 (*Geophilus*; characters; notes; distribution; USA: District of Columbia, Indiana, Kentucky, New York, Ohio, Virginia, West Virginia).
— Crabill, 1954b: 182-183; Figs. 3, 12 (*Geophilus*; characters; notes; USA: Arkansas, Indiana). [**Note:** Crabill mentions that he was unable to find the type of *mordax* at the MCZ].
— Crabill, 1955b: 37 (*Geophilus*; character notes; USA: Missouri). [sensu strictu]
— Crabill, 1955h: 158 (*Geophilus*; notes; USA: Missouri).
— Chamberlin, 1958a: 14 (*Geophilus*; USA: Florida).
— Wray, 1967: 156 (*Geophilus*; list).
— Summers, 1979: 697, 699; Fig. 13 (*Geophilus*; checklist; key).
— Summers et al., 1980: 247; Fig. 13 (*Geophilus*; checklist; habitat (Table 1, p. 242-243); USA: Illinois).
— Summers et al., 1981: 60 (*Geophilus*; checklist; USA: Illinois).
— Hoffman, 1995a: 23-24 (*Geophilus*; list; notes).
a*topleurus* Chamberlin, 1909b: 181, pl. 24; Figs. 1-3 (*Geophilus*; notes; USA: North Carolina).
— Attems, 1929: 170 (*Geophilus*; characters; distribution).
— Brimley, 1938: 501 (*Geophilus*; USA: North Carolina).
— Causey, 1940: 62 (*Geophilus*; synonym of *mordax*).

— Wray, 1967: 156 (*Geophilus*; list).

louisianae Brölemann, 1896: 55, t. 5, f. 4-6 (*Geophilus*). [**Note:** Reference not obtained].

— Attems, 1903b: 236 (*Geophilus*; synonym of *mordax*).

— Attems, 1929: 170-171 (*Geophilus*; characters; distribution).

salemensis Bollman, 1887e: 82 [= 1893: 71] (*Geophilus*; characters; notes; USA: Indiana). [**Note:** Syntype is at the NMNH].

— Bollman, 1888b: 5 [= 1893: 77] (*Geophilus*; note; USA: Arkansas).

— Bollman, 1888e: 408 [= 1893: 109] (*Geophilus*; note; USA: Indiana).

— Bollman, 1893: 125 (*Geophilus*; catalogue).

— Attems, 1903b: 264 (*Geophilus*; list).

— Attems, 1929: 329 (*Geophilus*; distribution; *incertae sedis*).

— Crabill, 1952b, 145-147 (*Geophilus*; synonym of *mordax*).

?*virginiensis* Bollman, 1888d: 346-347 [= 1893: 98-99] (*Geophilus*; characters; notes; USA: Virginia).

— Bollman, 1893: 125 (*Geophilus*; catalogue).

— Brölemann, 1896: 54–55 (*Geophilus*; characters).

— Attems, 1903b: 237 (*Geophilus*; list).

— Causey, 1940: 62 (*Geophilus*; synonym of *mordax*).

— Crabill, 1952b, 152-153, pl. 5; Fig. 59 (*Geophilus*; synonym of *mordax*).

— Hoffman, 1995a: 24 (*Geophilus*; list; notes; USA: Virginia). [**Note:** Although Causey (1940) and Crabill (1952b) considered this to be a synonym of *Geophilus mordax* Hoffman believes this to be a valid species].

Distribution: USA: Alabama, Arkansas, District of Columbia, Florida, Georgia, Illinois, Indiana, Kansas, Kentucky, Louisiana, Missouri, Mississippi, North Carolina, Nebraska, New York, Ohio, South Carolina, Tennessee, Virginia, West Virginia.

HOLOTYPE: Unknown; MCZ.

G. nasintus Chamberlin, 1909

— Chamberlin, 1909b: 183-184, pl. 24; Figs. 7-9 (*Geophilus*; locality not certain, USA: California). [**Note:** Apparently Chamberlin was not certain if this specimen even came from California, but he suggests that if it did it was from the southern part of the state].

— Chamberlin, 1912a: 657 (*Geophilus*; characters; notes; USA: California).

— Attems, 1929: 177-178 (*Geophilus*; characters; distribution).

Distribution: USA: California.

HOLOTYPE: University of PA?, California.

♂♀ *G. nealotus* Chamberlin, 1902
- Chamberlin, 1902c: 474-475 (*Geophilus*; characters (adult, stadia); habitat; key to species; USA: Utah).
- Attems, 1929: 177 (*Geophilus*; characters; distribution).

Distribution: USA: Utah.

HOLOTYPE: ?Unknown, Utah: north of Salt Lake City, 28 March 1901.

♀ *G. nicolanus* Chamberlin, 1940
- Chamberlin, 1940a: 4 (*Geophilus*; characters; USA: California).

Distribution: USA: California.

HOLOTYPE: NMNH, ♀, California: San Nicholas Island, by Harry Allen.

♀ *G. oregonus* (Chamberlin, 1941)
- Chamberlin, 1941d: 777-778 (*Brachygeophilus*; characters; USA: Oregon).
- Chamberlin, 1961: 97 (*Brachygeophilus*; notes).
- Foddai, et al., 1995: 9 (*Brachygeophilus* transferred to *Geophilus*).

Distribution: USA: Oregon.

HOLOTYPE: NMNH, ♀, Oregon: Medford, 4 December 1937, by Christensen, Jones & Clancy.

PARATYPE: NMNH, 3 immatures, Oregon: Medford, 4 December 1937, by Christensen, Jones & Clancy.

G. orites (Chamberlin, 1944)
- Chamberlin, 1944d: 194 (*Dysmesus*; characters; USA: Tennessee).
- Crabill, 1952b: 129-130 (*Dysmesus*; cited as type species of *Dysmesus*).
- Crabill, 1981a: 174 (*Brachygeophilus*; notes).

— Foddai, et al., 1995: 9 (*Brachygeophilus* transferred to *Geophilus*).
— Jeekel, 2005: 73 (*Dysmesus*; list).
— Bonato, Zapparoli & Minelli, 2008: 348; Tab. 1 (*Dysmesus*; "Main diagnostic characters differentiating *Diphyonyx* gen. n. from all other genera tentatively recognized in the subgroup of Geophilidae to which *Diphyonyx* belongs.").

Distribution: USA: Tennessee.

HOLOTYPE: NMNH; Tennessee: Great Smoky Mountains National Park, Greenbrier Cove, 14-19 June 1942, by H.S. Dybas.

PARATYPE: FMNH; Tennessee: Great Smoky Mountains National Park, Greenbrier Cove, 14-19 June 1942, by H.S. Dybas.

♂♀*G. oweni* Bollman, 1887
— Bollman, 1887c: 623, 624 [= 1893: 40, 41] (*Geophilus*; characters; key, notes; USA: Indiana).
— Bollman, 1888e: 408 [= 1893: 109] (*Geophilus*; note).
— Bollman, 1893: 125 (*Geophilus*; catalogue).
— Attems, 1903b: 264 (*Geophilus*; list).
— Williams & Hefner, 1928: 132 (*Geophilus*; characters; USA: Ohio).
— Attems, 1929: 329 (*Geophilus*; distribution; *incertae sedis*).
— Matthews, 1935: 16; Map: 17 (*Geophilus*; characters; USA: Wisconsin).
— Crabill, 1952b: 147-148, pl. 4; Figs. 49, 52-57 (*Geophilus*; characters; distribution).
— Crabill, 1954b: 180; Fig. 1, 7 (*Geophilus*; characters; USA: Missouri).
— Summers, 1979: 697, 700 (*Geophilus*; checklist; key).
— Watermolen, 1997: 3 (*Geophilus*; checklist, notes).
missouriensis Chamberlin, 1928b: 153-154 (*Geophilus*; characters; USA: Missouri). [**Note:** Holotype located at the NMNH].
— Chamberlin & Mulaik, 1940: 108 (*Geophilus*; USA: Texas).
— Attems, 1947: 117 (*Geophilus*; list).
— Crabill, 1952b: 147 (*Geophilus*; cited as synonym of *oweni*).

Distribution: USA: Indiana, Missouri, Ohio, Texas, Wisconsin.

HOLOTYPE: NMNH; Indiana: Posey Co., New Harmony, by Dr. Richard D. Owen.

♂♀ *G. parki*, (Auerbach, 1954)
— Auerbach, 1954: 1-5; Tab. 1; Fig. 1 (*Brachygeophilus*; characters; notes; USA: Illinois).
— Crabill, 1955f: 258 (*Brachygeophilus*; checklist; USA: Kentucky).
— Summers et al., 1980: 247; Fig. 12 (*Brachygeophilus*; checklist; habitat (Table 1, p. 242-243); USA: Illinois).
— Summers et al., 1981: 60 (*Brachygeophilus*; checklist; USA: Illinois).
— Foddai, et al., 1995: 9 (*Brachygeophilus* transferred to *Geophilus*).

Distribution: USA: Illinois, Kentucky.

HOLOTYPE: ?FMNH; Kentucky, Louisville, by Theodore J. Spilman.

♀ *G. phanus* Chamberlin, 1943
— Chamberlin, 1943b: 99-100 (*Geophilus*; characters; notes; USA: Montana).

Distribution: USA: Montana.

HOLOTYPE: NMNH, ♀, Montana: Brown, 5060 feet, 19 September 1941.

♂ *G. piedus* Chamberlin, 1930
— Chamberlin, 1930a: 114 (*Geophilus*; characters; USA: Utah).
— Attems, 1947: 117 (*Geophilus*; list).
— Chamberlin, 1961: 98 (*Geophilus*; note).

Distribution: USA: Utah.

HOLOTYPE: NMNH; ♂, Utah: Washington Co., St. George, 3 April 1929, by Lowell Woodbury.

♂♀ *G. proximus* C. L. Koch, 1847
— Koch, 1847: 186 (*Geophilus*; characters; "Baiern").
— Latzel, 1880: 184-186 (*Geophilus*; characters).
— Attems, 1929: 164; Figs. 170, 171 (*Geophilus*; characters; distribution).
— Crabill, 1958a: 15, 16 (*Geophilus*; key to species of eastern North America; notes; CANADA: Ontario; USA: Pennsylvania). [**Note:** The specimen from Pennsylvania was taken at quarantine in

Philadelphia " . . . from soil about the roots of *Amaryllis* plants imported from Germany." The specimen from Canada was presumably found outdoors in Cobden].
— Kevan, 1979: 298 (*Geophilus*; CANADA: Ontario).
— Behan-Pelletier, 1993: 25 (*Geophilus*; list).

Distribution: CANADA: Ontario; USA: Pennsylvania.

HOLOTYPE: Europe.

G. regnans Chamberlin, 1904
— Chamberlin, 1904: 654-655 (*Geophilus*; characters; notes; USA: California).
— Chamberlin, 1909b: 185 (*Geophilus*; USA: California).
— Chamberlin, 1912a: 657 (*Geophilus*; characters; notes; USA: California).
— Anonymous, 1920: 35 (*Geophilus*; note; USA: California).
— Attems, 1929: 326 (*Geophilus*; characters; distribution; *species inquirenda*).
— Chamberlin, 1944d: 191 (*Geophilus*; USA: California).

Distribution: USA: California.

HOLOTYPE: ?UPA, California: Southern California (Los Angeles, etc.)

♂♀*G. rupestris* (Crabill, 1949)
— Crabill, 1949b: 210-213 (*Brachygeophilus*; characters; notes; USA: New York).
— Crabill, 1952b: 134-135. (*Brachygeophilus*; characters; taxonomic notes; USA: Kentucky, New York, Virginia). [**Note:** Crabill suspects that *rupestris* may be a junior synonym to *G. setiger* Bollman, but the lack of the type makes it difficult to determine. Bollman (1893: 184) reports of a male specimen from Minnesota " . . . that agrees perfectly well with the types of this species" but the whereabouts of the material which was said to be in his collection is not known at this time. Crabill also compares *rupestris* to *truncorum* and these species should be compared in further detail with the knowledge of geophilomorph variation today. The reader is referred to Crabill's thesis (1952b) for details of the discussion. Those states of the distribution with questionmarks refer only to localities of *G. setiger* by the listed authors. A reference to *truncorum* is given under the TYPE SPECIES of this genus].

— Summers et al., 1980: 247; Fig. 12 (*Brachygeophilus*; checklist; habitat (Table 1, p. 242-243); USA: Illinois).

— Summers et al., 1981: 60 (*Brachygeophilus*; checklist; USA: Illinois).

— Hoffman, 1995a: 23 (*Brachygeophilus*; list; notes; USA: Virginia).

— Foddai, et al., 1995: 9 (*Brachygeophilus* transferred to *Geophilus*).

?*setiger* Bollman, 1887e: 82-83 [= 1893: 71-72]. (*Geophilus*: characters; USA: Indiana). [**Note:** According to Crabill (1952b) a possible synonym of *rupestris*).

— Crabill, 1952b: 134 (*Geophilus*; cited as a questionable synonym of *rupestris*).

— Bollman, 1888e: 408 [= 1893: 109] (*Geophilus*; note; USA: Indiana).

— Bollman, 1893: 125, 184 (*Geophilus*; catalogue; note; USA: Minnesota).

— Attems, 1903b: 264 (*Geophilus*; list).

— Williams & Hefner, 1928: 133 (*Geophilus*; characters; USA: Ohio).

— Attems, 1929: 329 (*Geophilus*; *incertae sedis*).

— Johnson, 1952: 74-78; Tab. 5; Map 6 (*Geophilus*; autecology; characters; distribution; USA: Michigan).

Distribution: USA: Illinois, ?Indiana, Kentucky, ?Michigan, ?Minnesota, New York, ?Ohio, Virginia.

HOLOTYPE: Crabill collection; No. C-393, ♀, New York: Cattaraugas Co., Olean, 30 March 1949, by Walter Kempf, O.F.M.; under a rock along bank of Alleghany River.

ALLOTYPE: Crabill's collection, No. C-394, ♂, Virginia: Alleghany Co., Clifton Forge, 6 March 1949, by R.L. Hoffman.

PARATYPES: Crabill's collection, No. C-395, ♀, Virginia: Alleghany Co., Clifton Forge, 6 March 1949, by R.L. Hoffman.

♂♀ **G. secundus** (Chamberlin, 1912)

— Chamberlin, 1912a: 665; Fig. 218 a-b (*Soniphilus*; characters; notes; USA: California).

— Attems, 1929: 332 (*Soniphilus*; characters; distribution).

— Chamberlin & Mulaik, 1940: 125 (*Soniphilus*; note; USA: Texas).

— Foddai, et al., 1995: 9 (*Brachygeophilus* transferred to *Geophilus*).

Distribution: USA: California, Texas.

HOLOTYPE: ?Unknown, California: Sausalito, April, 1911, or Pacific Grove, July, 1909.

♂*G. sequoia* Chamberlin, 1941
— Chamberlin, 1941d: 781-782 (*Geophilus*; characters; USA: California).

Distribution: USA: California.

SYNTYPE: NMNH; ♂, California: 1 mi. E of Hammond, 20 March 1941, by S. & D. Mulaik.

♂♀*G. setiger* Bollman, 1887
— Bollman, 1887e: 82-83 [= 1893: 71-72] (*Geophilus*; characters; notes; USA: Indiana).
— Bollman, 1888e: 408 [= 1893: 109] (*Geophilus*; note; USA: Indiana).
— Bollman, 1893: 125, 184 (*Geophilus*; catalogue; note; USA: Minnesota).
— Williams & Hefner, 1928: 133 (*Geophilus*; characters; USA: Ohio).
— Attems, 1929: 329 (*Geophilus*; distribution; *incertae sedis*).
— Johnson, 1952: 74-78; Tab. 5; Map 6 (*Geophilus*; autecology; characters; distribution; USA: Michigan).
— Crabill, 1954b: 187 (cited as *species inquirenda*; notes).
— Snider, 1991: 187 (*Geophilus*; list; notes; *species inquirenda*).
?*rupestris* Crabill, 1949b: 210-213 (*Brachygeophilus*; characters; notes; USA: New York).
— Crabill, 1952b: 134-135 (*Brachygeophilus*; possible synonym of *setiger*). [**Note:** Crabill suspects that *B. rupestris* may be a junior synonym to *G. setiger* Bollman, but the lack of the type makes it difficult to determine. Crabill also compares *B. rupestris* to *B. truncorum* and these species should be compared in further detail with the knowledge of geophilomorph variation today. The reader is referred to Crabill's thesis (1952b) for details of the discussion. Those distribution states with a questionmark in front of them refer only to reports of *B. rupestris* by the listed authors].

— Summers et al., 1980: 247; Fig. 12 (*Brachygeophilus*; habitat (Table 1, p. 242-243); possible synonym of *setiger*).
— Summers et al., 1981: 60 (*Brachygeophilus*; possible synonym of *setiger*).

Distribution: USA: ?Illinois, Indiana, ?Kentucky, Michigan, Minnesota, ?New York, Ohio, ?Virginia. [**Note:** States with a ? are from the distribution of *rupestris*].

HOLOTYPE: ?ISM, Indiana, Salem.

♂♀ *G. seurati* Brölemann, 1924
— Brölemann, 1924: 17 (*Geophilus*; nominal taxon *fucorum*; Algeria). [**Note:** Reference not obtained].
— Attems, 1929: 172 (*Geophilus*; nominal taxon *fucorum*; characters).
gracilis Meinert, 1870: 82 (*Geophilus*; characters; "Bona"). [**Note: Preoccupied**; junior homonym of *Geophilus gracilis* Gervais, 1849].
— Harger, 1872: 118 (*Geophilus*; characters; habitat; locality of "New Haven" USA: ?Connecticut). [**Note:** This European species has probably been introduced and it was recorded in N. America by Harger as *Geophilus gracilis* as early as 1872.].
— Bollman, 1893: 184 (*Geophilus*; cited as synonym of *Escaryus urbicus*).
— Eason, 1961: 390 (*Geophilus*; synonym of *Geophilus seurati* Brölemann, 1924). [**Note:** Synonymy taken from Chilobase, 7/12/2009; Original reference not obtained].

Distribution: USA: ?Connecticut.

HOLOTYPE: ?Unknown. USA: ?Connecticut, New Haven. [**Note:** The type specimen is possibly located at the Yale Peabody Museum].

G. shoshoneus Chamberlin, 1925
— Chamberlin, 1925e: 59 (*Geophilus*; characters; USA: Utah).
— Attems, 1929: 183 (*Geophilus*; distribution).
— Chamberlin, 1961: 98 (*Geophilus*; USA: Utah).

Distribution: USA: Utah.

HOLOTYPE: NMNH, Utah: Logan Canyon.

♀*G. smithi* Bollman, 1888
— Bollman, 1888d: 347 [= 1893: 99] (*Geophilus*; characters; notes; USA: District of Columbia).
— Bollman, 1888e: 408 [= 1893: 109] (*Geophilus*; note; USA: Indiana).
— Bollman, 1893: 125 (*Geophilus*; catalogue).
— Attems, 1903b: 264 (*Geophilus*; list). [**Note:** Spelled *smithii*].
— Chamberlin, 1909b: 182 (*Geophilus*; USA: District of Columbia).
— Attems, 1929: 329 (*Geophilus*; distribution; *incertae sedis*). [**Note:** Spelled *smithii*].
— Crabill, 1954b: 187 (cited as *species inquirenda*; notes).

Distribution: USA: District of Columbia, Indiana.

HOLOTYPE: NMNH, Washington D.C., by J.B. Smith.

G. strigosus (McNeill, 1887)
— McNeill, 1887b: 332-333 (*Mecistocephalus*; characters; note; USA: Indiana).
— Bollman, 1888e: 408 [= 1893: 109] (*Geophilus*; note; USA: Indiana).
— Bollman, 1893: 125 (*Geophilus*; catalogue).
— Attems, 1903b: 214 (*Mecistocephalus*; list).
— Williams & Hefner, 1928: 133 (*Geophilus*; characters; USA: Ohio).
— Attems, 1929: 155 (*Mecistocephalus*; cited as *incertae sedis*; distribution).
— Dowdy, 1944: 203, 208, 211, 214, 217; Tab. 1-5 (*Geophilus*; ecological note; habitat; USA: Ohio). [**Note:** Is this *Pachymerium ferrugineum*?].

Distribution: USA: Indiana, Ohio.

HOLOTYPE: NMNH, Indiana: near Bloomington.

G. tampophor (Chamberlin, 1953)
— Chamberlin, 1953a: 39 (*Brachygeophilus*; characters; notes; USA: Oregon). [**Note:** There were seven specimens obtained, but there was no mention of the sex of these].
— Foddai, et al., 1995: 9 (*Brachygeophilus* transferred to *Geophilus*).

Distribution: USA: Oregon.

HOLOTYPE: ?Unknown or Oregon State College, Oregon: Tombstone Prairie, Santiam Pass, 13 August 1949, by V. Roth.

♂♀ *G. terrae-novae* Palmén, 1954
— Palmén, 1954: 134-138; Figs. 5-15 (*Geophilus*; characters; CANADA: Newfoundland). [**Note:** Palmén suggests that this species has been overlooked in the past and may have been identified as *Linotaenia chionophila* (=*Strigamia chionophila*)].
— Kevan, 1979: 298 (*Geophilus*; note; CANADA: Newfoundland).
— Behan-Pelletier, 1993: 25 (*Geophilus*; list).

Distribution: CANADA: Newfoundland.

HOLOTYPE: CMN; ♂, CANADA: Newfoundland: Grand Bruit, 13 June 1949.

PARATYPES: CMN; 8 ♂, 7♀, CANADA: Newfoundland: Grand Bruit, 13 June 1949.

G. transitus (Chamberlin, 1941)
— Chamberlin, 1941d: 778 (*Brachygeophilus*; characters; USA: Oregon).
— Foddai, et al., 1995: 9 (*Brachygeophilus* transferred to *Geophilus*).

Distribution: USA: Oregon.

SYNTYPE: NMNH, Oregon: Grant's Pass, 7 December 1937, by L.D. Christensen, L.S. Jones & D.W. Clancy.

♂♀ *G. varians* McNeill, 1887
— McNeill, 1887b: 332 (*Geophilus*; characters; note; USA: Indiana).
— Bollman, 1888c: 341 [= 1893: 92] (*Geophilus*; note; USA: Tennessee).
— Bollman, 1888e: 408 [= 1893: 109] (*Geophilus*; note; USA: Indiana).
— Bollman, 1893: 125 (*Geophilus*; catalogue).
— Attems, 1903b: 265 (*Geophilus*; list).
— Chamberlin, 1912c: 414 (*Geophilus*; USA: Indiana, North Carolina).
— Williams & Hefner, 1928: 133 (*Geophilus*; characters; USA: Ohio).

— Attems, 1929: 329 (*Geophilus*; distribution; *incertae sedis*).
— Brimley, 1938: 501 (*Geophilus*; USA: North Carolina).
— Causey, 1940: 63 (*Geophilus*; characters; USA: North Carolina).
— Chamberlin, 1940c: 56 (*Pleurogeophilus*; note; USA: North Carolina).
— Crabill, 1952b: 148-149 (*Geophilus*; characters; range; USA: District of Columbia, Indiana, Ohio, Pennsylvania, Virginia). [**Note:** Crabill mentions the distribution of this species as ranging "From South Carolina north to Indiana, Ohio and Pennsylvania." South Carolina is not known to be in any publication as a locality for this species but I have included it as a questioned locality].
— Johnson, 1952: 78-79 (*Geophilus*; characters; distribution; habitat; USA: Michigan).
— Crabill, 1954b: 180-181, figs, 5, 7 (*Geophilus*; characters; USA: District of Columbia, Indiana, Kentucky, Ohio, Pennsylvania, Virginia).
— Crabill, 1955f: 258 (*Geophilus*; checklist; USA: Kentucky).
— Branson & Batch, 1967: 81 (*Geophilus*; characters; habitat; notes; USA: Kentucky).
— Wray, 1967: 156 (*Pleurogeophilus*; list).
— Summers, 1979: 697, 699 (*Geophilus*; checklist; key).
— Snider, 1991: 188 (*Geophilus*; list).
— Hoffman, 1995a: 24 (*Geophilus*; list; notes).
lanius Brölemann, 1896: 51 (*Geophilus*; synonym of *varians*).
— Attems, 1929: 199 (*Pleurogeophilus*; characters; distribution).
— Crabill, 1952b: 148 (*Geophilus* + *Pleurogeophilus*; cited as synonym of *varians*).
— Chamberlin, 1912c: 414 (*Geophilus*; cited as synonym of *varians*).
legiferens Chamberlin, 1909b: 182-183, pl. 24; Figs. 4-6 (*Geophilus*; characters; synonym of *varians*; Type locality: USA: District of Columbia).
— Chamberlin, 1912c: 414 (*Geophilus*; synonym of *varians*).
— Attems, 1929: 178 (*Geophilus*; characters; distribution).
— Crabill, 1952b: 148 (*Geophilus*; cited as new synonym of *varians*).
— Wray, 1967: 156 (*Geophilus*; list; synonym of *varians*).

Distribution: USA: District of Columbia, Indiana, Kentucky, Michigan, North Carolina, Ohio, Pennsylvania, ?South Carolina, Tennessee, Virginia.

SYNTYPE: NMNH; Indiana: near Bloomington.

♂***G. virginiensis*** Bollman, 1888
- Bollman 1888d: 346-347 [= 1893: 98-99] (*Geophilus*; characters; notes; USA: Virginia).
- Bollman, 1893: 125 (*Geophilus*; catalogue).
- Attems, 1903b: 237 (*Geophilus*; list).
- Attems, 1929: 169-170 (*Geophilus*; characters; USA: North Carolina, Virginia).
- Causey, 1940: 62 (*Geophilus*; synonym of *mordax*).
- Crabill, 1952b, 152-153, pl. 5; Fig. 59 (*Geophilus*; synonym of *mordax*).
- Hoffman, 1995a: 24 (*Geophilus*; list; notes; USA: Virginia). [**Note:** Although Causey (1940) and Crabill (1952b) considered this to be a synonym of *Geophilus mordax* Hoffman believes this to be a valid species].

Distribution: USA: ?North Carolina, Virginia. [**Note:** To my knowledge, the only reference to this species being found in North Carolina is in Attems, 1929].

HOLOTYPE: ?NMNH; Virginia: Natural Bridge, by L. M. Underwood. [**Note:** Type specimen not in NMNH database].

♂♀***G. vittatus*** (Rafinesque, 1820)
- Rafinesque, 1820: 8 (*Mycotheres*; characters; USA: New York).
- Hoffman & Crabill, 1953: 79 (*Geophilus*; subgenus *Nemopleura*; list).
- Crabill, 1954b: 177; Fig. 4 (*Geophilus*; characters; distribution; habitat; notes; USA: Connecticut, Illinois, Indiana, Kentucky, Maryland, Massachusetts, New York, Ohio, Pennsylvania, Virginia).
- Crabill, 1955b: 37 (*Geophilus*; range; USA: Missouri).
- Crabill, 1955f: 258 (*Geophilus*; checklist; USA: Kentucky).
- Crabill, 1955h: 158 (*Geophilus*; notes; USA: Missouri).
- Crabill, 1958b: 95 (*Geophilus*; habitat; notes; USA: Wisconsin).
- Chamberlin, 1961: 98 (*Geophilus*; note; USA: Utah).
- Branson & Batch, 1967: 81 (*Geophilus*; characters; habitat; note; USA: Kentucky).
- Summers, 1979: 697, 699 (*Geophilus*; checklist; key).
- Summers & Uetz, 1979; Tab. 1, 2: 348, 349, 350 (*Geophilus*; microhabitat; USA: Illinois).
- Summers et al., 1980: 247; Fig. 14 (*Geophilus*; checklist; habitat (Table 1, p. 242-243); USA: Illinois).

— Summers et al., 1981: 60 (*Geophilus*; checklist; USA: Illinois).

— Kevan, 1983: 2946 (*Geophilus*; list; CANADA: Quebec).

— Snider, 1991: 187–188 (*Geophilus*; list; habitat; USA: Michigan).

— Behan-Pelletier, 1993: 25 (*Geophilus*; list).

— Hoffman, 1995a: 24 (*Geophilus*; list; habitat; notes; USA: Virginia).

— Watermolen, 1997: 3 (*Geophilus*; checklist, notes; USA: Wisconsin).

— Jeekel, 2005: 92 (*Mycotheres*; list).

— Regier et al., 2005: 149; Tab. 1 (*Geophilus*; table of taxon classification with GenBank accession numbers).

cephalicus Wood, 1862: 44 (*Geophilus*; synonym of *rubens*; characters; USA: Maryland).

— Wood, 1865: 178-179 (*Geophilus*; synonym of *rubens*).

— Bollman, 1893: 125, 148 (*Geophilus*; synonym of *rubens*).

— Kenyon, 1893a: 16 (*Geophilus*; synonym of *rubens*; USA: Nebraska).

— Kenyon, 1893b: 161 (*Geophilus*; synonym of *rubens*; USA: Nebraska).

— Cook, 1896: 239 (*Geophilus*; synonym of *rubens*).

— Chamberlin, 1912b: 66 (*Geophilus*; cited as synonym of *rubens*).

— Gunthorp, 1920: 113 (*Geophilus*; synonym of *rubens*; list).

— Attems, 1929: 181 (*Geophilus*; synonym of *rubens*).

— Crabill, 1952b: 143 (*Geophilus*; synonym of *rubens*).

— Kevan, 1983: 2946 (synonym of *vittatus*).

deducens Chamberlin, 1909b: 180, pl. 25; Figs. 1-3 (*Geophilus*; characters; notes; Type locality: USA: New York, Long Island, Sea Cliff).

— Bailey, 1928a: 1083 (*Geophilus*; list).

— Bailey, 1928b: 19, 39 (*Geophilus*; list; notes; USA: New York).

— Attems, 1929: 322 (*Geophilus*; characters; distribution).

— Crabill, 1952b: 143 (*Geophilus*; cited as a new synonym of *rubens*).

— Kevan, 1983: 2946 (synonym of *vittatus*; CANADA: Quebec).

laevis Wood, 1862: 44 (*Geophilus*; synonym of *rubens*; characters; USA: Georgia). [**Note:** Wood notes there are 2 specimens at the ANSP collected by Dr. J. Le Conte].

— Wood, 1865: 180 (*Geophilus*; synonym of *rubens*; characters; notes; USA: Georgia).

— Wood, 1867b: 128(*Geophilus*; synonym of *rubens*; notes; USA: California).

— Bollman, 1893: 125 (*Geophilus*; synonym of *rubens*).

— Chamberlin, 1912b: 66 (*Geophilus*; synonym of *rubens*).

— Gunthorp, 1920: 113 (*Geophilus*; list).
— Crabill, 1952b: 143 (*Geophilus*; synonym of *rubens*).
— Kevan, 1983: 2946 (synonym of *vittatus*).

okolonae Bollman, 1888 (*Geophilus*; USA: Arkansas).
— Bollman, 1887c: 624 [=1893: 41] (*Geophilus*; key). [**Note:** Bollman included this yet undescribed species in a key of *Geophilus*, but the description appeared in a later publication noted in the footnote].
— Bollman, 1888b: 5 [= 1893: 77] (*Geophilus*; characters; note; USA: Arkansas).
— Bollman, 1893: 125 (*Geophilus*; catalogue).
— Attems, 1903b: 262 (*Geophilus*; list).
— Chamberlin, 1912b: 66 (*Geophilus*; synonym of *rubens*).
— Attems, 1929: 324 (*Geophilus*; characters; distribution; *species inquirenda*).
— Kevan, 1983: 2946 (synonym of *vittatus*).

rubens Say, 1821: 113 (*Geophilus*; characters; habitat; USA). [**Note:** A type specimen of this species is at the NMNH].
— Gervais, 1837: 52 (*Geophilus*;). [**Note:** reference not obtained].
— Newport, 1856: 87 (*Geophilus*; characters; "Americâ Boreali").
— Wood, 1865: 182 (*Strigamia*; synonym of *vittatus*).
— Bollman, 1888e: 408 [= 1893: 109] (*Geophilus*; note; USA: Indiana).
— Bollman, 1893: 125, 148 (*Geophilus*; catalogue; taxonomic notes).
— Attems, 1903b: 236 (*Geophilus*; list).
— Chamberlin, 1909b: 182 (*Geophilus*; synonym of *vittatus*; USA: California, Michigan).
— Chamberlin, 1912a: 656-657 (*Geophilus*; synonym of *vittatus*; characters; habitat; notes; USA: California).
— Chamberlin, 1912b: 66 (*Geophilus*; synonym of *vittatus*; notes; USA: Illinois, Iowa, Michigan).
— Chamberlin, 1912c: 412 (*Geophilus*; note; USA: North Carolina, Virginia).
— Chamberlin, 1914b: 301 (*Geophilus*; habitat (p. 301); USA: Michigan).
— Chamberlin, 1920b: 166 (*Geophilus*; CANADA: Ontario).
— Anonymous, 1920: 35 (*Geophilus*; USA: California).
— Chamberlin, 1925a: 54 (*Geophilus*; habitat; notes; USA: Arizona, California). [**Note:** This is an interesting geophilid in that on the east coast it tends to have fewer segments than on the west coast (see Chamberlin, 1925a). Unless Chamberlin was not aware that the leg count varies depending on the sex].

— Chamberlin, 1925e: 59 (*Geophilus*; notes; USA: Utah).
— Holmquist, 1926: 411 (*Geophilus*; notes on hibernation such as: activity, date collected, developmental stage, ecological association/niche, quantity found).
— Williams & Hefner, 1928: 132-133 (*Geophilus*; characters; USA: Ohio).
— Bailey, 1928a: 1083 (*Geophilus*; list).
— Bailey, 1928b: 19, 39 (*Geophilus*; list; notes; USA: New York).
— Attems, 1929: 181-182 (*Geophilus*; characters; distribution).
— Matthews, 1935: 13-14; Map: 15 (*Geophilus*; characters; USA: Wisconsin).
— Causey, 1940: 63; Fig. 35 (*Geophilus*; characters; habitat; USA: North Carolina).
— Chamberlin, 1942b: 15 (*Geophilus*; USA: Iowa).
— Chamberlin, 1944a: 33 (*Geophilus*; USA: Georgia).
— Chamberlin, 1944d: 191 (*Geophilus*; USA: Arkansas, Illinois, Indiana, Texas, Wisconsin).
— Auerbach, 1946: 14, (*Geophilus*; list; USA: Illinois).
— Rapp, 1946: 666 (*Geophilus*; habitat; USA: Illinois).
— Crabill, 1950d: 253 (*Geophilus*; Crabill suggests that *Mecistocephalus melanonotus* Wood may be *Geophilus rubens* (= *G. vittatus*).
— Auerbach, 1951a: 100, 101, 105, 109, 112, 120, 121, 122; Tab. 1, 3, 8, 9, 14; Fig. 1 (*Geophilus*; desiccation; food habits; habitat; heart beat; list; notes; oviposition; population percentage; USA: Illinois).
— Crabill, 1952b: 143-145, pl. 4; Figs. 45, 51 (*Geophilus*; characters; ethology; distribution).
— Crabill, 1952b: 143 (*Strigamia*; cited as synonym of *Geophilus rubens*).
— Johnson, 1952: 79-103, tabs. 6-9, 28; Map 4; Fig. 20 (*Geophilus*; autecology; characters; distribution; natural history; USA: Michigan). [**Note:** Johnson gave extensive natural history coverage of this species].
— Chamberlin, 1958c: 133 (*Geophilus*; checklist; note).
— Chamberlin, 1961: 98 (*Geophilus*; synonym of *vittatus*).
— Wray, 1967: 156 (*Geophilus*; list).
— Kevan, 1979: 298 (*Geophilus*; CANADA: Ontario). [**Note:** Kevan considered this an introduced species to Canada, however, this may be indigenous].
— Kevan, 1983: 2946 (synonym of *vittatus*).

Distribution: CANADA: Ontario, Quebec; USA: Arkansas, Arizona, California, Connecticut, Georgia, Illinois, Indiana, Iowa, Kentucky, Maryland, Massachusetts, Michigan, Missouri, North Carolina, Nebraska, New York, Ohio, Pennsylvania, Texas, Utah, Virginia, Wisconsin.

HOLOTYPE: BMNH, Pennsylvania, probably Philadelphia.

G. whitei Newport, 1845
- — Newport, 1845b: 436 (*Geophilus*; characters).
- — Newport, 1856: 88 (*Geophilus*; characters; "Americâ Boreali").
- — Gervais, 1847: 321 (*Geophilus*; characters; note).
- — Wood, 1862: 47 (*Strigamia*; characters).
- — Wood, 1865: 184 (*Strigamia*; characters).
- — Bollman, 1893: 126 (*Linotaenia*; catalogue).
- — Attems, 1929: 349 (*Geophilus*; *incertae sedis*).

Distribution: USA: In America Boreali. (see Newport, 1856)

HOLOTYPE: ?BMNH, ("*v. in* Mus. Brit.", see Newport, 1845)

G. winnetui Attems, 1947
- — Attems, 1947: 57-58, taf. 4 & 5; Figs. 27-29 (*Geophilus*; characters; notes; USA: Iowa).

Distribution: USA: Iowa.

HOLOTYPE: NMW, Iowa, Providence.

♂*G. yavapainus* (Chamberlin, 1941)
- — Chamberlin, 1941d: 778-779 (*Brachygeophilus*; characters; USA: Arizona).
- — Foddai, et al., 1995: 9 (*Brachygeophilus* transferred to *Geophilus*).

Distribution: USA: Arizona.

HOLOTYPE: NMNH, ♂, Arizona: Grand Canyon, Yavapi Point, 7,000 ft., 27 January 1911, by W.M. Wheeler.

Genus *Gosipina* Chamberlin, 1940
[1 species; Southern Central USA]

Gosipina Chamberlin, 1940d: 56 (characters). **Type Species:** *Gosipina bexara* Chamberlin, 1940, fixed by original designation and monotypy.
— Crabill, 1968b: 331 (catalogue).
— Jeekel, 2005: 82 (list).

♂*G. bexara* Chamberlin, 1940
— Chamberlin, 1940d: 56 (*Gosipina*; characters; USA: Texas). [**Note:** This species was collected from Texas but was obtained by Chamberlin from the U.S. Bureau of Entomology and Plant Quarantine suggesting it could possibly have foreign origins].
— Jeekel, 2005: 82 (*Gosipina*; list).
— Bonato, Zapparoli & Minelli, 2008: 349; Tab. 1 (*Gosipina*; "Main diagnostic characters differentiating *Diphyonyx* gen. n. from all other genera tentatively recognized in the subgroup of Geophilidae to which *Diphyonyx* belongs.").

Distribution: USA: Texas.

HOLOTYPE: NMNH, ♂, Texas: Bexar Co., from U. S. Bureau of Entomology and Plant Quarantine, 9 November 1936. [**Note:** NMNH may also have paratypes].

Genus *Harmostela* Chamberlin, 1941
[1 species; Western USA]

Harmostela Chamberlin, 1941d: 784 (characters). **Type Species:** *Harmostela hespera* Chamberlin, 1941, fixed by original designation and monotypy.

♀*H. hespera* Chamberlin, 1941
— Chamberlin, 1941d: 784-785 (*Harmostela*; characters; USA: California).
— Bonato, Zapparoli & Minelli, 2008: 349; Tab. 1 (*Harmostela*; "Main diagnostic characters differentiating *Diphyonyx* gen. n. from all other genera tentatively recognized in the subgroup of Geophilidae to which *Diphyonyx* belongs.").

Distribution: USA: California.

HOLOTYPE: NMNH, ♀, California: Monterey Co., Hastings Reservation, 1 or 10 March 1941, by Dr. J.M. Linsdale.

Genus *Leptodampius* Chamberlin, 1938
[1 species; Northwestern USA]

Leptodampius Chamberlin, 1938a: 254-255 (characters). Type Species: *Leptodampius lamprus* Chamberlin, 1938, fixed by original designation and monotypy.

♀*L. lamprus* Chamberlin, 1938
— Chamberlin, 1938a: 255 (*Leptodampius*; characters; USA: Oregon).

Distribution: USA: Oregon.

HOLOTYPE: NMNH, ♀, Oregon: Boyer, by J.A. Macnab.

Genus *Lionyx* Chamberlin, 1960
[1 species; Western USA]

Lionyx Chamberlin, 1960a: 100 (characters). Type Species: *Linoyx hedgpethi* Chamberlin, 1960, fixed by original designation and monotypy.
— Pereira & Hoffman, 1993: 1 (transferred from the family Schendylidae). [**Note:** This genus has been transferred to the family Geophilidae by Pereira & Hoffman, 1993].
— Shelley, 2006a: 13 (list).

♂*L. hedgpethi* Chamberlin, 1960
— Chamberlin, 1960a: 100-101; Fig. 1-2 (*Lionyx*; characters; USA: California). [**Note:** This species exhibits possible littoral habits: "... under stones between the tide levels ..."].

Distribution: USA: California.

HOLOTYPE: NMNH, California: Marin Co., Tomales Bay, Nick's Cove, near Dillon Beach, 8 July 1959, by J.W. Hedgepeth and students.

Genus *Nesogeophilus* Verhoeff, 1924
[1 species; Western USA]

Nesogeophilus Verhoeff, 1924: 413 (originally cited as a subgenus;) [**Note:** Reference not obtained]. Type Species: *Geophilus laticollis* Attems, 1903, fixed by original designation.

— Attems, 1929: 184 (characters; distribution; key to species; species included: *bäckströmi*, *hartmeyeri*, *laticollis*, *ormanyensis*, *palpiger*, *xylophagus*).

(Subgenus ***Dyodesmophilus*** Verhoeff, 1938)

♂*N. (**D.**) **longissimus*** Verhoeff, 1938
— Verhoeff, 1938b: 376-377; Tab. 7; Figs. 43-45 (*Nesogeophilus*; subgenus *Dyodesmophilus*; characters; USA: California).
— Attems, 1947: 119 (*Nesogeophilus*; subgenus *Dyodesmophilus*; list).

Distribution: USA: California.

HOLOTYPE: ?Unknown. ♂, California: near Bishop, at the end of March, by Herr Michelbacher.

Genus ***Nesomerium*** Chamberlin, 1953
[1 species; ?USA: Hawaii]

Nesomerium Chamberlin, 1953b: 83-84 (characters). **TYPE SPECIES:** *Nesomerium hawaiiense* Chamberlin, 1953, fixed by original designation and monotypy.
— Jeekel, 2005: 92 (list).

♂*N. **hawaiiense*** Chamberlin, 1953
— Chamberlin, 1953b: 84 (*Nesomerium*; characters). [**Note:** No locality is given for this new species but it has been assumed by the author that this species has been found somewhere in the Hawaiian Islands and is therefore tentatively included as a Hawaiian species].
— Bonato, Foddai, Minelli & Shelley, 2004: 24-25 (*Nesomerium*; characters; notes).
— Nishida, 1994, 26 (*Nesomerium*; checklist).
— Nishida, 1997, 195 (*Nesomerium*; checklist).
— Jeekel, 2005: 92 (*Nesomerium*; list).

Distribution: Unknown, but assumed to be Hawaii based on the specific name.

HOLOTYPE: ?Unknown.

Genus *Poaphilus* Chamberlin, 1912
[1 species; Northern Central USA]

Poaphilus Chamberlin, 1912b: 70 (characters). TYPE SPECIES: *Poaphilus*
kewinus Chamberlin, 1912, fixed by original designation.
— Attems, 1929: 332 (characters; distribution).
— Jeekel, 2005: 99 (list).

♀*P. kewinus* Chamberlin, 1912
— Chamberlin, 1912b: 70-71 (*Poaphilus*; characters; notes; USA: Iowa
[**Note:** Chamberlin mentions that in addition to this species there
is another known from New Mexico which I have been unable to
locate in the literature]).
— Attems, 1929: 333 (*Poaphilus*; characters; distribution).
— Auerbach, 1946: 14, 15 (*Poaphilus*; list; note; USA: Illinois).
— Auerbach, 1949: 225 (*Poaphilus*; note; USA: Illinois).
— Jeekel, 2005: 99 (*Poaphilus*; list).

Distribution: USA: Illinois, Iowa.

HOLOTYPE: NMNH, ♀, Iowa, Marshalltown.

Genus *Sepedonophilus* Attems, 1909
[1 species; USA: Hawaii]

Sepedonophilus Attems, 1909: 34 (characters). TYPE SPECIES: *Geophilus concolor*
Gervais, 1847, var. *perforatus* Haase, 1887, fixed by original designation.
— Jeekel, 2005: 104 (list).

♂*S. hodites* Chamberlin, 1940
— Chamberlin, 1940d: 55 (*Sepedonophilus*; characters; USA: Hawaii).
[**Note:** A species intercepted by a quarantine inspector and is not
likely to be established].
— Bonato, Foddai, Minelli & Shelley, 2004: 27 (*Mecistocephalus*; notes).

Distribution: USA: ?Hawaii.

HOLOTYPE: NMNH, Hawaii: Honolulu, 25 June 1938, in soil about
Cymbidium lowianum from Australia.

Genus *Serrona* Chamberlin, 1941
[1 species; Western USA]

Serrona Chamberlin, 1941d: 781 **TYPE SPECIES:** *Serrona kernensis* Chamberlin, 1941, fixed by original designation and monotypy.
— Jeekel, 2005: 104 (list).

S. kernensis Chamberlin, 1941
— Chamberlin, 1941d: 781 (*Serrona*; characters; USA: California).
— Jeekel, 2005: 104 (*Serrona*; list).
— Bonato, Zapparoli & Minelli, 2008: 349; Tab. 1 (*Serrona*; "Main diagnostic characters differentiating *Diphyonyx* gen. n. from all other genera tentatively recognized in the subgroup of Geophilidae to which *Diphyonyx* belongs.").

Distribution: USA: California.

HOLOTYPE: NMNH, California: 2 mi. W of Kernville, 18 March 1941, by S. & D. Mulaik.

Genus *Sogona* Chamberlin, 1912
[3 species; Southeastern USA]

Sogona Chamberlin, 1912c: 431 (characters). **TYPE SPECIES:** *Sogona minima* Chamberlin, 1912, fixed by original designation.
— Attems, 1929: 345 (characters; distribution).
— Crabill, 1968b: 331 (catalogue).
— Jeekel, 2005: 105 (list).
?*Pycnona* Chamberlin, 1943a: 18 (characters). **TYPE SPECIES:** *Pycnona pujola* Chamberlin, 1943, fixed by original designation.
— Crabill, 1968b: 331 (synonym of *Garrina*).
— Jeekel, 2005: 101 (list).
Nannocrix Chamberlin, 1918b: 376 (characters). **TYPE SPECIES:** *Nannocrix porethus* Chamberlin, 1918, fixed by original designation.
— Attems, 1929: 216 (characters; distribution).
— Crabill, 1961b: 125, 126 (synonym of *Sogona*).
— Jeekel, 2005: 91 (list).

♂*S. kerrana* Chamberlin, 1940
— Chamberlin, 1940e: 66 (*Sogona*; characters; notes; USA: Texas).

Distribution: USA: Texas.

HOLOTYPE: NMNH, ♂, Texas: Kerr Co., Turtle Creek, December 1939, D. & S. Mulaik.

♂♀ *S. minima* Chamberlin, 1912
— Chamberlin, 1912c: 431-432, pl. 3; Figs. 1-5, 9 (*Sogona*; characters; note; USA: South Carolina, Tennessee).
— Attems, 1929: 345 (*Sogona*; characters; distribution).
— Chamberlin, 1944a: 33 (*Sogona*; USA: Georgia).
— Crabill, 1950c: 199 (*Sogona*; notes; USA: South Carolina).
— Crabill, 1961b: 127 (*Sogona*; characters).
— Hoffman, 1995a: 24 (*Sogona*; list; notes). [**Note:** Hoffman suggests *minima* lives in Virginia].
— Jeekel, 2005: 105 (list).
— Bonato, Zapparoli & Minelli, 2008: 349; Tab. 1 (*Sogona*; "Main diagnostic characters differentiating *Diphyonyx* gen. n. from all other genera tentatively recognized in the subgroup of Geophilidae to which *Diphyonyx* belongs.").

Distribution: USA: Georgia, South Carolina, Tennessee, ?Virginia.

HOLOTYPE: MCZ, South Carolina: Taylor's or Tennessee: Johnson City.

♂♀ *S. poretha* (Chamberlin, 1918)
— Chamberlin, 1918b: 376-377 (*Nannocrix*; characters; USA: Louisiana).
— Attems, 1929: 216 (*Nannocrix*; characters; distribution).
— Crabill, 1952b: 120, 121 (*Nannocrix*; characters; key; notes).
— Crabill, 1961b: 126, 127, 129-132; Figs. 1-3 (*Sogona*; characters; lectotype designation; morphological notes; transferred to *Sogona*; USA: Arkansas). [**Note:** Crabill seemed to question the two specimens (may be smaller or younger females compared to the lectotype) he claimed were from Arkansas, but they have been included as official records until further study].
— Jeekel, 2005: 91 (*Nannocrix*; list).
— Bonato, Zapparoli & Minelli, 2008: 349; Tab. 1 (*Nannocrix*; "Main diagnostic characters differentiating *Diphyonyx* gen. n. from all other genera tentatively recognized in the subgroup of Geophilidae to which *Diphyonyx* belongs.").

Distribution: USA: Arkansas, Louisiana.

LECTOTYPE: MCZ, (Unique No. 14523; TC-38); ♀; Louisiana: Creston, March 9, 1915, K.P. Schmidt. [**Note:** Crabill note in vial reads "VII.21.59 . . . found undissected. Dissected ex. for first time this date. Found it is referable to *Sogona*, probably to *S. minima* Ch."].

PARATYPE: NMNH.

Genus *Synthophilus* Chamberlin, 1946
[1 species; USA: Alaska]

Synthophilus Chamberlin, 1946c: 180 (characters). TYPE SPECIES: *Synthophilus boreus* Chamberlin, 1946, fixed by original designation and monotypy.
— Jeekel, 2005: 106 (list).

S. boreus Chamberlin, 1946
— Chamberlin, 1946c: 180, pl. 1; Fig. 5; pl. 2, figs 6-7 (*Synthophilus*; characters; USA: Alaska).
— Behan-Pelletier, 1993: 25 (*Synthophilus*; list).
— Jeekel, 2005: 106 (*Synthophilus*; list).
— Bonato, Zapparoli & Minelli, 2008: 349; Tab. 1 (*Synthophilus*; "Main diagnostic characters differentiating *Diphyonyx* gen. n. from all other genera tentatively recognized in the subgroup of Geophilidae to which *Diphyonyx* belongs.").

Distribution: USA: Alaska.

HOLOTYPE: NMNH, Alaska: Juneau, 28-29 April 1945.

PARATYPES: NMNH, 3 specimens with slide mount, Alaska: Juneau, 28-29 April 1945, J.C. Chamberlin. NMNH, 2 specimens with slide mounts, Alaska: Haines, 22 August 1945, J.C. Chamberlin. [**Note:** Specimens at the NMNH may be labeled with the genus *Chilenophilus* Attems (1909b: 27) instead of *Synthophilus*].

Genus *Tabiphilus* Chamberlin, 1912
[1 species; Western USA]

Tabiphilus Chamberlin, 1912a: 665 (characters). TYPE SPECIES: *Tabiphilus rex* Chamberlin, 1912, fixed by monotypy.
— Attems, 1929: 332 (characters; distribution).
— Jeekel, 2005: 106 (list).

T. rex Chamberlin, 1912

— Chamberlin, 1912a: 665-666 (*Tabiphilus*; characters; USA:
California). [**Note:** Only the type specimen is known and there is
no mention of its sex].
— Anonymous, 1920: 35 (*Tabiphilus*; USA: California).
— Attems, 1929: 332 (*Tabiphilus*; characters; distribution).
— Jeekel, 2005: 106 (*Tabiphilus*; list).

Distribution: USA: California.

HOLOTYPE: Unknown?, California: Claremont, by Prof. Baker.

Genus *Timpina* Chamberlin, 1912
[1 species; South Central USA]

Timpina Chamberlin, 1912c: 432-433 (characters). **TYPE SPECIES:** *Timpina
texana* Chamberlin, 1912, fixed by original designation and monotypy.
— Attems, 1929: 345-346 (characters; distribution).
— Crabill, 1968b: 331 (catalogue).
— Jeekel, 2005: 108 (list).

♂ *T. texana* Chamberlin, 1912
— Chamberlin, 1912c: 433-434, pl. 3; Figs. 6-8 (*Timpina*; characters;
USA: Texas).
— Attems, 1929: 346 (*Timpina*; characters; distribution).
— Chamberlin & Mulaik, 1940: 125 (*Timpina*; note; USA: Texas).
— Crabill, 1968b: 331 (*Timpina*; catalogue; cited as type species).
— Jeekel, 2005: 108 (*Timpina*; list).
— Bonato, Zapparoli & Minelli, 2008: 349; Tab. 1 (*Timpina*; "Main
diagnostic characters differentiating *Diphyonyx* gen. n. from all other
genera tentatively recognized in the subgroup of Geophilidae to
which *Diphyonyx* belongs.").

Distribution: USA: Texas.

HOLOTYPE: MCZ, No. 1005, ♂, Texas: Austin, by Prof. T.H. Montgomery.
[Note: Crabill note in vial says "N.B. when found July, 1956, cephalic plate
was missing. Rest of specimen intact"].

Genus *Tuoba* Chamberlin, 1920
[1 species (Introduced); Hawaii]

Tuoba Chamberlin, 1920c: 35 (characters). **TYPE SPECIES:** *Tuoba curticeps* Chamberlin, 1920 [= *Tuoba sydneyensis* (Pocock, 1891)], fixed by original designation.
— Jeekel, 2005: 109 (list).
Honuaphilus Chamberlin, 1926b: 93-94 (characters).
— Chamberlin, 1953: 85 (list).
— Bonato, Foddai, Minelli & Shelley, 2004: 24 (synonym of *Tuoba*).
— Jeekel, 2005: 84 (list).

T. sydneyensis (Pocock, 1891)
— Pocock, 1891: 219 (*Geophilus*; characters; Australia).
— Attems, 1914: 133 (*Geophilus*; list)
— Attems, 1929: 326 (*Geophilus*; characters).
— Chamberlin, 1920: 54 (*Geophilus*; list).
— Jones, 1998: 334; Figs. 1-11 (*Tuoba*;). [**Note:** Reference not obtained].
— Bonato, Foddai, Minelli & Shelley, 2004: 23-24, 27, 28; Figs. 2, 6, 10; Tab. 1 (*Tuoba*; characters; key; notes; USA: Hawaii).
alohanus Chamberlin, 1926b: 93-94 (characters; USA: Hawaii).
— Attems, 1938b: 365, 369 (list; note).
— Chamberlin, 1953: 85 (*Honuaphilus*; notes).
— Nishida, 1994: 26 (*Honuaphilus*; checklist).
— Nishida, 1997: 195 (*Honuaphilus*; checklist).
— Nishida, 2002: 235 (*Honuaphilus*; checklist).
— Bonato, Foddai, Minelli & Shelley, 2004: 24 (synonym of *sydneyensis*).
— Jeekel, 2005: 84 (*Honuaphilus*; list).

Distribution: USA: Hawaii (Origin: Australian Region).

HOLOTYPE: ? Unknown.

Genus *Tylonyx* Cook, 1899
[1 species; Southeastern USA]

Tylonyx Cook, 1899: 308 (characters, notes). **TYPE SPECIES:** *Tylonyx tampae* Cook, 1899, fixed by monotypy. [**Note: Preoccupied** by *Tylonyx* Schulze, 1897].
— Jeekel, 2005: 67, 110, 112, 121 (list, note).

♀*T. tampae* Cook, 1899
— Cook, 1899: 308 (*Tylonyx*, characters; USA: Florida).

— Jeekel, 2005: 110 (list, note).

Distribution: USA: Florida.

HOLOTYPE: NMNH: Florida, Tampa.

Genus *Zygomerium* Chamberlin, 1943
[1 species; Western USA]

Zygomerium Chamberlin, 1943: Type Species: *Zygomerium euphanum*
Chamberlin, 1943, [=*Arenophilus watsingus* Chamberlin, 1912] fixed by
original designation.
— Crabill, 1952b: 129 (nomenclatorial note). [**Note:** Crabill
synonymized the genus *Zygomerium* with *Arenophilus* because
upon examination of the paratype of *euphanum* (Type Species of
Zygomerium) he found it to be conspecific with *Arenophilus watsingus*.
The genus *Zygomerium* has been kept separate for the time being
because Crabill stated it was not known if *rotarium* is conspecific with
watsingus and the matter deserves further investigation].
— Crabill, 1969a: 7 (synonym of *Arenophilus*).
?*Arenophilus* Chamberlin, 1912 (see note above). Type Species: *Arenophilus*
attenuatus Say, 1821, fixed by original designation.
— Crabill, 1969a: 7 (senior synonym of *Zygomerium*).

♀*Z. rotarium* Chamberlin, 1943
— Chamberlin, 1943b: 101 (*Zygomerium*; characters; USA: Utah).
— Crabill, 1952b: 129 (*Zygomerium*; note).
— Chamberlin, 1961: 97 (*Zygomerium*; note).

Distribution: USA: Utah.

HOLOTYPE: NMNH, ♀, Utah: City Creek Canyon, Rotary Park, 21 May 1941.

Genus *Zygona* Chamberlin, 1960
[1 species; Southwestern USA]

Zygona Chamberlin, 1960b: 242-243 (characters). Type Species: *Zygona*
duplex Chamberlin, 1960, fixed by original designation and monotypy.
[**Note:** This genus has been transferred to the family Geophilidae by
Pereira & Hoffman, 1993].
— Shelley, 2006a: 15 (list).

371

Z. duplex Chamberlin, 1960
— Chamberlin, 1960b: 243 (*Zygona*; characters; USA: Arizona).
— Bonato, Zapparoli & Minelli, 2008: 349; Tab. 1 (*Zygona*; "Main diagnostic characters differentiating *Diphyonyx* gen. n. from all other genera tentatively recognized in the subgroup of Geophilidae to which *Diphyonyx* belongs.").

Distribution: USA: Arizona.

HOLOTYPE: NMNH, Arizona: Santa Rita Mountains, Madison Canyon, 10 September 1941.

Family **Himantariidae** Cook, 1895
Subfamily **Bothriogastrinae** Attems, 1929

Genus *Garriscaphus* Chamberlin, 1941
[3 species; Western USA]

Garriscaphus Chamberlin, 1941d: 789-790 (characters). Type Species: *Garriscaphus oreines* Chamberlin, 1941, fixed by original designation.
— Jeekel, 2005: 80 (list).

♀ *G. amplus* Chamberlin, 1941
— Chamberlin, 1941d: 790 (*Garriscaphus*; characters; USA: California).
— Chamberlin, 1944d: 186 (*Garriscaphus*; notes; USA: California).
— Crabill, 1959a: 155 (*Garriscaphus*; list; cited as possibly belonging to *Gosiphilus*).
— Crabill, 1969c: 494, footnote 4 (*Garriscaphus*; cited as probably referable to *Chomatobius* Humbert & Saussure).

Distribution: USA: California.

HOLOTYPE: NMNH, ♀, California: 2 mi. W of Kerrville, 18 March 1941, by S. & D. Mulaik.

♂♀ *G. oreines* Chamberlin, 1941
— Chamberlin, 1941d: 790 (*Garriscaphus*; characters; USA: California).
— Crabill, 1959a: 155 (*Garriscaphus*; list).

— Jeekel, 2005: 80 (*Garriscaphus*; list).

Distribution: USA: California.

HOLOTYPE: NMNH, California: 12 mi. NE of Hammond, 22 March 1941, by S. & D. Mulaik.

♀ *G. tytthus* Crabill, 1969
  — Crabill, 1969c: 494-496; Figs. 1-4 (*Garriscaphus*; characters; USA: California).

Distribution: USA: California.

HOLOTYPE: NMNH; ♀, California, Los Angeles Co., Arroyo Seco, 10 May 1958, by R.M. Bohart & A.S. Menke.

Subfamily **Himantariinae** Verhoeff, 1908

Genus *Chomatobius* Humbert & Saussure, 1870
[7 species; Western & Southern USA]

Chomatobius Humbert & Saussure, 1870b: 205 (characters). Tʏᴘᴇ Sᴘᴇᴄɪᴇꜱ: *Geophilus mexicanus* Saussure, 1860, fixed by subsequent designation of Cook, 1895.
  — Cook, 1895b: 69 (list; designation of type species).
  — Crabill, 1968a: 108, 109, 110 (distinguishing characters; taxonomic discussion). [**Note:** Crabill gives characters of the type species *Chomatobius mexicanus* (Saussure) from Mexico].
  — Jeekel, 2005: 74 (list).
  — Bonato, Pereira & Minelli, 2007: 2–3 (*Chomatobius*; taxonomic and nomenclatural notes).
?*Californiphilus* Verhoeff, 1938: 370–371 (characters). Tʏᴘᴇ Sᴘᴇᴄɪᴇꜱ: *Californiphilus michaelbacheri* Verhoeff, 1938, fixed by monotypy.
  — Crabill, 1953e: 88 (notes). [**Note:** Crabill seems relatively certain that Verhoeff's *Californiphilus* is referable to Chamberlin's *Gosiphilus* (= *Chomatobius*)].
  — Crabill, 1959a: 155 (cited as suspected synonym of *Gosiphilus*). [**Note:** Crabill lists *Californiphilus* as a suspected synonym of *Gosiphilus* (= *Chomatobius*) and is therefore listed here as a suspected synonym of *Chomatobius*].

?*Garriscaphus* Chamberlin, 1941d: 789-790 (characters). TYPE SPECIES: *Garriscaphus oreines* Chamberlin, 1941, fixed by original designation.
— Crabill, 1959a: 155 (Crabill lists *Garriscaphus amplus* as possibly belonging to *Gosiphilus* (= *Chomatobius*) and is therefore listed here as a suspected synonym of *Chomatobius*).

Gosiphilus Chamberlin, 1912a: 671 (characters; key to species of California). TYPE SPECIES: *Strigamia laticeps* Wood, 1862, fixed by subsequent designation of Crabill, 1953e: 88. [**Note:** Chamberlin thought that *Gosiphilus* was closely allied with *Haplophilus*].
— Attems, 1929: 53-54 (characters; distribution; key to species; species included: *bakeri, laticeps, minor*).
— Crabill, 1953e: 88 (taxonomic notes). [**Note:** Crabill suspects that *Gosiphilus* (= *Chomatobius*) is very closely allied to, if not identical to, Latzel's *Stigmatogaster*].
— Crabill, 1968a: 108, 109, 110 (taxonomic discussion; synonym of *Chomatobius*).
— Jeekel, 2005: 82 (list).
— Bonato, Pereira & Minelli, 2007: 2 (synonym of *Chomatobius*).

♂*C. auximus* (Chamberlin, 1938)
— Chamberlin, 1938a: 254 (*Gosiphilus*; characters; USA: Texas).
— Chamberlin & Mulaik, 1940: 109 (*Gosiphilus*; list).
— Crabill, 1959a: 156 (*Gosiphilus*; list).
— Crabill, 1968a: 109-110 (*Gosiphilus*; transferred to *Chomatobius*; senior subjective synonym of *Gosiphilus*).
— Bonato, Pereira & Minelli, 2007: 2 (*Chomatobius*; list; taxonomic and nomenclatural notes; transferred to *Chomatobius*).

Distribution: USA: Texas.

HOLOTYPE: NMNH, ♂, Texas: Edinburg, 16 October 1936, by S. Mulaik.

C. bakeri (Chamberlin, 1912)
— Chamberlin, 1912a: 672 (*Gosiphilus*; characters; USA: California). [**Note:** Only known from the type specimen, and the sex is not mentioned in the description].
— Attems, 1929: 54 (*Gosiphilus*; characters; distribution).
— Anonymous, 1920: 35 (*Gosiphilus*; USA: California).
— Crabill, 1959a: 156 (*Gosiphilus*; list).
— Crabill, 1968a: 109-110 (*Gosiphilus*; transferred to *Chomatobius*; senior subjective synonym of *Gosiphilus*).

— Bonato, Pereira & Minelli, 2007: 2 (*Chomatobius*; list; taxonomic and nomenclatural notes; transferred to *Chomatobius*).

Distribution: USA: California.

HOLOTYPE: NMNH, California: Claremont, by W.C. Spencer.

♂♀ *C. euphorion* (Crabill, 1953)
— Crabill, 1952b: 74-77, pl. 1; Fig. 6-10, 12a, 13a-i, 14a-b, 15-16 (*Gosiphilus*; characters (immature); USA: Alabama, Kentucky, Tennessee). [**Note:** This reference is not the official publication of this species because it is from Crabill's PhD thesis which is not a peer reviewed document (see Crabill, 1953e: 85-88)].
— Crabill, 1953e: 85-88, pl. 5, figs, 1-8 (*Gosiphilus*; characters (immature); taxonomic notes).
— Crabill, 1955f: 259 (*Gosiphilus*; checklist; USA: Kentucky).
— Crabill, 1968a: 109-110 (*Gosiphilus*; transferred to *Chomatobius*; senior subjective synonym of *Gosiphilus*).
— Bonato, Pereira & Minelli, 2007: 2 (*Chomatobius*; list; taxonomic and nomenclatural notes; transferred to *Chomatobius*).

Distribution: USA: Alabama, Kentucky, Tennessee.

HOLOTYPE: NMNH; Crabill's collection, ♂, C-1362; Alabama: Tuscaloosa, 15 April 1949, by B.D. Valentine.

ALLOTYPE: NMNH, Crabill's collection, ♀, C-1364; Kentucky: Rockcastle Co., Cumberland National Forest, near Livingston, 3 September 1950, by T.J. Spilman.

PARATYPE: NMNH, Crabill's collection, ♂, C-1363; Tennessee: Sevier Co., Chilhowee Mountain, 1 October 1950, by E.O. Wilson.

♂ *C. laticeps* (H. C. Wood, 1862)
— Wood, 1862: 49 (*Strigamia*; characters; USA: Texas).
— Wood, 1865: 186 (*Strigamia*; characters; USA: Texas).
— Bollman, 1893: 125 (*Himantarium*; catalogue).
— Cook, 1896: 241-242 (*Himantarium*; characters; note).
— Chamberlin, 1909b: 177 (*Haplophilus*; notes; USA: California, Texas).
— Chamberlin, 1912a: 672 (*Gosiphilus*; characters; habitat; notes; USA: California).

— Chamberlin, 1912c: 434-435 (*Gosiphilus*; notes; USA: California, Nevada, Texas).
— Anonymous, 1920: 35 (*Gosiphilus*; USA: California).
— Gunthorp, 1920: 113 (*Strigamia*; list).
— Attems, 1929: 54 (*Gosiphilus*; characters; distribution).
— Chamberlin & Mulaik, 1940: 108-109 (*Gosiphilus*; USA: Texas).
— Chamberlin, 1944d: 185 (*Gosiphilus*; USA: Texas).
— Crabill, 1959a: 156, 158 (*Strigamia*; list) & (*Gosiphilus*; characters of holotypical specimen).
— Reddell, 1965: 166 (*Gosiphilus*; checklist; note; USA: Texas).
— Crabill, 1968a: 109-110 (*Gosiphilus*; transferred to *Chomatobius*; senior subjective synonym of *Gosiphilus*).
— Jeekel, 2005: 82 (*Strigamia*; list).
— Bonato, Pereira & Minelli, 2007: 2 (*Chomatobius*; list; taxonomic and nomenclatural notes; transferred to *Chomatobius*).

Distribution: USA: California, Nevada, Texas.

HOLOTYPE: MCZ, No. 867, ♂, Texas, Chas. Stolley. [**Note:** Label by Crabill reads "Holotype ♂, *Strigamia laticeps* Wood, 1862, TC-5(867), See slides of anterior and posterior ends of body and mouthparts. Parchment label is Wood's original and is in his handwriting. Pencil label (*Himantarium*) is Meinert's, and *Gosiphilus* label is Chamberlin's"].

♂*C. mexicanus* (Saussure, 1858)
— Saussure, 1858: 545 (*Geophilus*; characters; Mexico). [**Note:** Reference not obtained].
— Saussure, 1859: 390–391 pl. 7, Figs. 49, 49c, 49d [= Saussure, 1860: 132-133], (*Geophilus*; characters; Mexico).
— Humbert & Saussure, 1870b: 205 (*Chomatobius*; list).
— Saussure & Humbert, 1872: 145, 147, 207 (*Chomatobius*;). [**Note:** Reference not obtained].
— Sseliwanoff, 1881: 24–25; Figs. 9–16 (*Chomatobius*; ?redescription). [**Note:** Reference not obtained].
— Cook, 1895b: 69 (cited as type of *Chomatobius*).
— Pocock, 1896: 40 pl. 3, Figs. 13, 13a-13d (*Chomatobius*; list; distribution).
— Cook, 1899: 303 (note).
— Attems, 1903b: 204 (*Orphnaeus*; list).
— Attems, 1928: 124 (*Orphnaeus*; key).
— Attems, 1929: 112, 114 (*Orphnaeus*; characters; key; Mexico; USA: California "San Diego", Texas).
— Attems, 1947: 92 (*Orphnaeus*; key).

— Crabill, 1960c: 87 (*Chomatobius*; notes).
— Crabill, 1968a: 108, 109, 110–112 (*Chomatobius*; taxonomic notes; characters; Mexico).
— Foddai, Pereira & Minelli, 2000: 160–161 (*Chomatobius*; catalog, distribution).
— Jeekel, 2005: 74 (*Geophilus*; list).
— Bonato, Pereira & Minelli, 2007: 2, 3 (*Chomatobius*; list; taxonomic and nomenclatural notes; transferred to *Chomatobius*).

Distribution: USA: ?California, ?Texas.

HOLOTYPE: ?Unknown. Mexico: Veracruz, Cordoba.

♂♀ *C. minor* (Chamberlin, 1912)
— Chamberlin, 1912a: 671-672; Fig. 220g (*Gosiphilus*; characters; USA: California).
— Attems, 1929: 54 (*Gosiphilus*; characters; distribution).
— Crabill, 1959a: 156 (*Gosiphilus*; list).
— Crabill, 1968a: 109-110 (*Gosiphilus*; transferred to *Chomatobius*; senior subjective synonym of *Gosiphilus*).
— Bonato, Pereira & Minelli, 2007: 2 (*Chomatobius*; list; taxonomic and nomenclatural notes; transferred to *Chomatobius*).

Distribution: USA: California.

SYNTYPE: NMNH; California: Berkeley, April 1911, by R.V. Chamberlin.

C. minor arizonicus (Chamberlin, 1925)
— Chamberlin, 1925a: 54 (*Gosiphilus*; nominal taxon *minor*; cited as subspecies of *minor*; notes; USA: Arizona).
— Attems, 1947: 83 (*Gosiphilus*; nominal taxon *minor*; list).
— Chamberlin, 1958c: 133 (*Gosiphilus*; nominal taxon *minor*; checklist).
— Crabill, 1959a: 156 (*Gosiphilus*; nominal taxon *minor*; list).
— Crabill, 1968a: 109-110 (*Gosiphilus*; transferred to *Chomatobius*; senior subjective synonym of *Gosiphilus*).
— Bonato, Pereira & Minelli, 2007: 2 (*Chomatobius*; list; taxonomic and nomenclatural notes; transferred to *Chomatobius*).

Distribution: USA: Arizona.

HOLOTYPE: NMNH; Arizona: Fish Creek, February 1924, by H.F. Loomis.

Genus *Ephemerozaster* Crabill, 1959
[1 species; Southwestern USA]

Ephemerozaster Crabill, 1959a: 118-120 **TYPE SPECIES:** *Ephemerozaster antaeus* Crabill, 1959, fixed by original designation and monotypy.
— Shelley, 2006a: 10 (list). [**Note:** Misspelled *Empherozoster*].

♀*E. antaeus* Crabill, 1959
— Crabill, 1959a: 120-126; Figs. 1-9; p. 155 (*Ephemerozaster*; characters; list; USA: New Mexico).
— Barr & Reddell, 1967: 263 (*Ephemerozaster*; habitat; notes).

Distribution: USA: New Mexico.

HOLOTYPE: NMNH; No. 2529; ♀, New Mexico: Eddy Co., Carlsbad Caverns National Park, Spider Cave, from dry silt beds about 100 feet inside entrance.

Genus *Nothobius* Cook, 1899
[1 species; Western & ?Southeastern USA]

Nothobius Cook, 1899: 303 (characters). **TYPE SPECIES:** *Nothobius californicus* Cook, fixed by monotypy. [**Note:** Cook creates the genus *Nothobius* for a single species *californicus* which was regarded as distinct from the genus *Chomatobius* Saussure; see footnote on page 303].
— Attems, 1929: 54-55 (characters; distribution; notes).
— Crabill, 1959a: 156 (list).
— Crabill, 1960c: 97 (summary of diagnostic features).
— Chamberlin, 1912a: 669 (characters). [**Note:** Misspelled *Notobius*].
— Crabill, 1960c: 88 (comment on misspelling of *Nothobius*).
— Jeekel, 2005: 93 (list).

♂♀*N. californicus* Cook, 1899
— Cook, 1899: 303, footnote (*Nothobius*; characters; notes). [**Note:** I am unclear about the locality].
— Attems, 1903b: 290 (*Nothobius*; list). [**Note:** Misspelled *Notobius*].
— Chamberlin, 1912a: 669 (*Nothobius*; synonym of *taeniopsis*). [**Note:** Misspelled *Notobius*].
— Attems, 1929: 55 (*Nothobius*; characters; distribution; notes).

— Crabill, 1959a: 156 (*Nothobius*; list).
— Crabill, 1960c: 87, 89-99; Figs. 1-13 (*Nothobius*; characters; notes on variation; redescription; USA: California).
— Jeekel, 2005: 93 (*Nothobius*; list).

mexicanus Selivanov, 1881: 24 (*Chomatobius*;). [sensu Selivanov nec Saussure] [**Note:** reference not obtained].
— Bollman, 1893: 126 (*Chomátobius*; catalogue).
— Attems, 1903b: 290 (*Chomatobius*; synonym of *californicus*).
— Chamberlin, 1912a: 669 (*Chomatobius*; synonym of *taeniopsis*).

taeniopsis Wood, 1862: 48 (*Strigamia*; characters; notes; original spelling *tæniopsis*; USA: Georgia). [**Note:** Crabill (1960c) proposed that *Strigamia taeniopsis* Wood, a name frequently associated with *N. californicus*, be set aside and regarded as a *species inquirenda*].
— Wood, 1865: 185 (*Strigamia*; characters; USA: Georgia).
— Meinert, 1886a: 229 (*Himantarium*; characters; USA: California).
— Bollman, 1890: 211 (*Himantarium*; note; Mexico). [**Note:** Bollman records the number of legs as 148 from a young female].
— Bollman, 1893: 125 (*Himantarium*; catalogue).
— Cook, 1896: 241 (*Himantarium*; characters).
— Attems, 1903b: 177 (*Himantarium*; list).
— Chamberlin, 1909b: 177 (*Haplophilus*; misspelling of *taeniopsis*; subjective synonym of *californicus*; USA: California).
— Chamberlin, 1912a: 669, 671 (*Nothobius*, characters; USA: California). [**Note:** Misspelled *Notobius teniopsis*].
— Anonymous, 1920: 35 (*Nothobius*; note; USA: California). [**Note:** Misspelled *Notobius*].
— Gunthorp, 1920: 113 (*Strigamia*; list).
— Attems, 1929: 329 (*Strigamia*; distribution; *incertae sedis*).
— Crabill, 1959a: 156. (*Strigamia*; list). [**Note:** Crabill lists the locality as Georgia, ? = California].
— Crabill, 1960c: 88, 97 (*Strigamia*; species inquirenda). [**Note:** Crabill proposes " . . . that the name be set aside and that it be considered provisionally as a *species inquirenda* of uncertain generic assignment."].

Distribution: USA: California, ?Florida, ?Georgia.

HOLOTYPE: Unknown.

PLESIOTYPE: NMNH, No. 2566, ♀, California: Placer Co., 4 mi. W of Newcastle, 15 April 1958, by L.M. Smith & R.O. Schuster.

Genus *Stenophilus* Chamberlin, 1946

[6 species; Western and Southeastern USA]

Stenophilus Chamberlin, 1946a: 35 (characters). **TYPE SPECIES:** *Stenophilus coloradanus* Chamberlin, 1946, fixed by original designation and monotypy.
— Chamberlin, 1953a: 37-38 (notes).
— Crabill, 1959a: 156 (list).
— Crabill, 1960e: 88-89 (character discussion; key to species of North America).
— Jeekel, 2005: 105 (list).

S. audacior (Chamberlin, 1909)
— Chamberlin, 1909b: 177-178, pl. 26; Figs. 1-3 (*Haplophilus*; characters; USA: Idaho).
— Attems, 1929: 41 (*Stigmatogaster*; questioned subgenus; characters; distribution; note).
— Chamberlin, 1946a: 35 (*Stenophilus*). [**Note:** Chamberlin transferred *audacior* from *Stigmatogaster* to *Stenophilus*].
— Crabill, 1959a: 157 (*Haplophilus*; list). [**Note:** Crabill considered the generic assignment of this species uncertain].

Distribution: USA: Idaho.

HOLOTYPE: NMNH; Idaho: McKendrick.

♂*S. californicus* (Chamberlin, 1930)
— Chamberlin, 1930b: 297-299; Figs. 1-6 (*Meinertophilus*; characters; USA: California).
— Chamberlin, 1946a: 35 (*Stenophilus*). [**Note:** Chamberlin transferred *californicus* from *Meinertophilus* to *Stenophilus*).
— Attems, 1947: 78 (*Meinertophilus*; list).
— Crabill, 1959a: 157 (*Meinertophilus*; list).

Distribution: USA: California.

HOLOTYPE: NMNH, ♂, California, Potter Creek Cave, collected by Dr. J.C. Merriam. [**Note:** Chamberlin states that the type is in the Department of Zoology, University of California, however, at the NMNH there is a ♂ holotype from Potter Cave with W.J. Sinclair 1903 written on the label and not as above from the literature].

♀*S. coloradanus* Chamberlin, 1946

— Chamberlin, 1946a: 35-38; Figs. 1-6 (*Stenophilus*; characters; note; USA: Colorado).
— Crabill, 1959a: 157 (*Stenophilus*; list).
— Jeekel, 2005: 105 (*Stenophilus*; list).

Distribution: USA: Colorado.

HOLOTYPE: NMNH, ♀, Colorado: Mesa Verde, 29 June 1944, by Professor V.E. Shelford.

♀*S. grenadae* (Chamberlin, 1912)
— Chamberlin, 1912c: 435-436, pl. 3; Figs. 10-13 (*Haplophilus*; characters; USA: Mississippi).
— Attems, 1929: 42 (*Stigmatogaster*; characters; distribution). [**Note:** The distribution indicates a locality of "Grenada, Misiones", this is a lapsus calami].
— Crabill, 1959a: 157. (*Haplophilus*; list). [**Note:** Crabill considered the generic assignment of this species uncertain].
— Crabill, 1960e:89-92; Figs. 1-3 (*Stenophilus*; characters; USA: Arkansas). [**Note:** This female specimen exists at the NMNH; from Arkansas: Pulaski County, Little Rock, April, 1953, collected by B. Johnson].

Distribution: USA: Arkansas, Mississippi.

HOLOTYPE: MCZ, No. 866, ♀, Mississippi, Grenada, 15 July 1912. [**Note:** Original label has 1910 as the year collected but Chamberlin's publication states 1912].

♀*S. hesperus* (Chamberlin, 1928)
— Chamberlin, 1928d: 309-310 (*Haplophilus*; characters; USA: Utah).
— Attems, 1947: 83 (*Haplophilus*; list).
— Chamberlin, 1958c: 133 (*Gosiphilus*; checklist).
— Crabill, 1959a: 157 (*Haplophilus*; list). [**Note:** Crabill considered the generic assignment of this species uncertain].
— Chamberlin, 1961: 99 (*Stenophilus*; characters of mouthparts; notes).

Distribution: USA: Utah.

HOLOTYPE: NMNH, Utah: San Juan Co., Devil's Canyon, 18 April 1928.

♂♀***S. rothi*** Chamberlin, 1953
— Chamberlin, 1953a: 38-39; figs. 1-3 (*Stenophilus*; characters; USA: Montana, Oregon).
— Crabill, 1959a: 157 (*Stenophilus*; list).

Distribution: USA: Montana, Oregon.

HOLOTYPE: NMNH or ?Oregon State College, ♂, Oregon: Marion Co., Silver Creek Falls, 12 May 1951, by V. Roth.

ALLOTYPE: NMNH or ?Oregon State College, ♀, Montana: St. Regis, 23 September 1950, by V. Roth.

PARATYPE: NMNH.

Genus ***Stigmatogaster*** Latzel, 1880
[1 species (Introduced); Eastern CANADA & USA]

Stigmatogaster Latzel, 1880: TYPE SPECIES: *Himantarium gracile* Meinert, 1870, fixed by subsequent designation of Cook, 1896.
— Crabill, 1952b: 70-72 (characters; notes).
— Jeekel, 2005: 105 (list, notes).
Haplogaster Verhoeff, 1896: 76–78 (characters). [**Note: Preoccupied** by *Haplogaster* Chaudoir, 1879]. TYPE SPECIES: *Himantarium dimidiatum* Meinert, 1870, fixed by monotypy.
— Attems, 1929: 36 (synonym of *Stigmatogaster*).
— Jeekel, 2005: 83 (list, notes).
Haplophilus Cook, 1896: 6 (reference not obtained). TYPE SPECIES: *Himantarium dimidiatum* Meinert, 1870, fixed by direct substitution.
— Chamberlin, 1912a: 669 (note).
— Attems, 1929: 36 (synonym of *Stigmatogaster*).
— Jeekel, 2005: 83 (list, notes).
Meinertophilus Silvestri, 1897: 241 (characters). TYPE SPECIES: *Himantarium superbum* Meinert, 1870, fixed by original designation.
— Foddai, et al. 1995: 8 (synonym of *Stigmatogaster*).
Nesoporogaster Verhoeff, 1924: 99 (characters). TYPE SPECIES: *Nesoporogaster excavatum* (recte-*vata*) Verhoeff, 1924, fixed by monotypy.
— Minelli & Bonato, unpublished: ? (synonym of *Stigmatogaster*).
[**Note:** Unpublished as of December 2008, pers. comm.].

— Jeekel, 2005: 93 (list).

♂♀*S. subterranea* (Shaw, 1794)
  — Shaw, 1794: 7 (Scolopendra;). [Note: Reference not obtained].
  — Leach, 1817: 44 (*Geophilus*; characters; habitat).
  — Newport, 1845: 436-437; Tab. 33, Fig. 10 (*Geophilus*; characters; notes; England).
  — Bergsöe & Meinert, 1866: 108 (*Himantarium*; characters; notes).
  — Meinert, 1870: 31-32, 107, 119; Tab. 2; Fig. 3-5 (*Himantarium*; characters; notes).
  — Chalande & Ribaut, 1909: 256-258; fig. 20 (*Haplophilus*; characters).
  — Attems, 1929: 39; fig. 56 (*Stigmatogaster*; characters; distribution).
  — Crabill, 1952b: 71, 72 (*Stigmatogaster*; characters). [**Note:** Only taken in New York at quarantine in plant from England].
  — Palmén, 1954: 131, 133; Fig. 1 (*Haplophilus*; notes; CANADA: Newfoundland).
  — Eason, 1964: 46-51, pl. 2; Figs. 13-23 (*Haplophilus*; characters (adult, adolescens); distribution; habitat).
  — Kevan, 1979: 298 (*Haplophilus*; CANADA: Newfoundland).
  — Kevan, 1983: 2945 (*Haplophilus*; checklist).
  — Behan-Pelletier, 1993: 25 (*Haplophilus*; list).

Distribution: CANADA: Newfoundland; USA: ?New York.

HOLOTYPE: Unknown?

Family **Mecistocephalidae** Verhoeff, 1908
Subfamily **Arrupinae** Chamberlin, 1920

Genus *Arrup* Chamberlin, 1912
[1 species (?Introduced); Western USA]

Arrup Chamberlin, 1912a: 654 (characters). Type Species: *Arrup pylorus* Chamberlin, fixed by original designation and monotypy.
  — Attems, 1929: 154 (characters; distribution).
  — Crabill, 1964: 163-165 (characters; notes on systematic position).
  — Foddai, Bonato, Pereira & Minelli, 2003: 1261 (biogeography; characters; distribution).
  — Jeekel, 2005: 70 (list).

Prolamnonyx Silvestri, 1919: 47, 84–85; Fig. 25, 26 (characters, key; notes).
TYPE SPECIES: *Geophilus holstii* Pocock, 1895, fixed by original designation.

— Attems, 1929: 154 (species included: *holstii* and *sauteri*; characters; distribution; key to species).
— Crabill, 1964: 163 (junior subjective synonym of *Arrup*).
— Foddai, Bonato, Pereira & Minelli, 2003: 1261 (cited as synonym of *Arrup*).
— Jeekel, 2005: 100 (list).

Nodocephalus Attems, 1928(?): 115 (?). [**Note:** Reference not obtained].
TYPE SPECIES: *Mecistocephalus edentulus* Attems, 1904, fixed by original designation.

— Crabill, 1964: 163 (junior subjective synonym of *Arrup*).
— Foddai, Bonato, Pereira & Minelli, 2003: 1261 (synonym of *Arrup*).

♂*A. pylorus* Chamberlin, 1912
— Chamberlin, 1912a: 654; Fig. 219 c-e, 220a (*Arrup*; characters; USA: California).
— Chamberlin, 1920e: 187 (*Arrup*; list).
— Attems, 1929: 154 (*Arrup*; characters; distribution).
— Crabill, 1964: 161, 162, 163, 164, 165, 166-168; Figs. 1-3 (*Arrup*; composite description; terminology; USA: California).
— Jeekel, 2005: 70 (*Arrup*; list).

?*holstii* Pocock, 1895: 352 (*Geophilus*; possible synonym of *pylorus*; Japan). [**Note:** Holotype at the British Museum of Natural History].
— Silvestri, 1919: 85-87; Fig. 25 (*Prolamnonyx*; characters; Japan).
— Attems, 1929: 155; fig. 165(*Prolamnonyx*; characters; distribution).
— Crabill, 1964: 162, 163 (*Prolamnonyx*; notes; possible synonym of *pylorus*). [**Note:** Crabill suggests a very close relationship with *A. holstii* (Pocock) and *A. pylorus* Chamberlin. Because *holstii* was described from Japan and is congeneric with *pylorus*; is this an introduced species from Japan that has established itself or as Crabill points out is it a result of " . . . migrations across the Bering Strait during interglacial times"?].
— Foddai, Bonato, Pereira & Minelli, 2003: 1261 (*Arrup*; list).
— Jeekel, 2005: 100 (*Geophilus*; list).
— Uliana, Bonato & Minelli, 2007: 227-229 (*Arrup*; designation of neotype). [**Note:** Reference not obtained].

Distribution: USA: California.

HOLOTYPE: ?Unknown; ♂, California: Sausalito and Berkeley, April 1911, by R.V. Chamberlin. [**Note:** According to Crabill (1964: 161) the types could not be located].

Genus *Nannarrup* Foddai, Bonato, Pereira & Minelli, 2003
[1 species (probably Introduced); Northeastern USA]

Nannarrup Foddai, Bonato, Pereira & Minelli, 2003: 1256 TYPE SPECIES: *Nannarrup hoffmani* Foddai, Bonato, Pereira & Minelli, 2003, fixed by original designation and monotypy.
— Shelley, 2006a: 13 (list).

♀*N. hoffmani* Foddai, Bonato, Pereira & Minelli, 2003
— Foddai, Bonato, Pereira & Minelli, 2003: 1256-1260; Figs. 2-27 (*Nannarrup*; characters; USA: New York).

Distribution: USA: New York.

HOLOTYPE: AMNH, ♀, New York: Manhattan, Central Park, North Woods, 21 April 1998, K. Catley [sic].

PARATYPES: AMNH, 7 juveniles, New York: Manhattan, Central Park, North Woods, 21 April 1998, K. Catley [sic]; UP, 1 juvenile, New York: Manhattan, Central Park, North Woods, 21 April 1998, K. Catley [sic].

Genus *Tygarrup* Chamberlin, 1914
[2 species (Introduced); Eastern USA & Hawaii]

Tygarrup Chamberlin, 1914c: 210-212 (characters; notes). TYPE SPECIES: *Tygarrup intermedius* Chamberlin, 1914, fixed by original designation and monotypy.
— Attems, 1929: 151 (species included: *intermedius* and *javanicus*; characters; distribution; key to species).
— Foddai, Pereira & Minelli, 2000: 62 (catalogue).
— Jeekel, 2005: 109 (list).

T. intermedius Chamberlin, 1914
— Chamberlin, 1914c: 212-214. (*Tygarrup*; characters; USA: District of Columbia). [**Note:** Four specimens were taken from pots of plants from British Guiana. The possibilities of this species establishing itself in the Washington D.C. area are probably slim].

— Attems, 1929: 151-152 (*Tygarrup*; characters; distribution).
— Foddai, Pereira & Minelli, 2000: 63 (*Tygarrup*; catalogue; notes).
— Jeekel, 2005: 109 (*Tygarrup*; list).

Distribution: USA: ?District of Columbia.

HOLOTYPE: MCZ, British Guiana.

♂♀*T. javanicus* Attems, 1907
— Attems, 1907: 95, Figs. 8-9, [non *Mecistocephalus spissus* Wood, 1862] (*Mecistocephalus*; ?Java). [**Note:** Reference not obtained].
— Attems, 1929: 152 (*Tygarrup*; characters; Java).
— Verhoeff, 1937: 236 (*Tygarrup*;). [**Note:** Reference not obtained].
— Attems, 1938a: 330; Figs. 294-298 (*Tygarrup*; characters).
— Wang & Mauries, 1996: 89 (*Tygarrup*; list).
— Bonato, Foddai, Minelli & Shelley, 2004: 14-15, 27, 28; Figs. 8, 16, 19; Tab. 1 (*Tygarrup*; characters; key; notes; USA: Hawaii).

Distribution: USA: Hawaii (Origin: southeastern Asia).

HOLOTYPE: ?Unknown.

Subfamily **Mecistocephalinae** Verhoeff, 1901

Genus *Dicellophilus* Cook, 1895
[2 species (1 or both Introduced); Western USA]

Dicellophilus Cook, 1895a: 61 (characters). TYPE SPECIES: *Mecistocephalus limatus* Wood, 1862, fixed by original designation.
— Cook, 1895b: 74 (distribution; list; type species cited).
— Attems, 1929: 149 (characters; distribution; key to species; species included: *carniolensis*, *limatus*).
— Jeekel, 2005: 76 (list).

♀*D. carniolensis* (C. L. Koch, 1847)
— Koch, 1847: 185 (*Clinopodes*; characters; "Herzogthum Krain").
— Meinert, 1870: 94-96 (*Mecistocephalus*; characters; notes).
— Latzel, 1880: 162; t. 6; Fig. 53-54; t. 7; Fig. 55-62 (*Mecistocephalus*;
— Daday, 1889: 90 (*Mecistocephalus*). [**Note:** reference not obtained].
— Cook, 1895a: 61 (*Lamnonyx*; transferred from *Mecistocephalus*).

— Attems, 1929: 149-150; Figs. 160-163 (*Dicellophilus*; characters; distribution).

— Kevan, 1979: 298 (*Dicellophilus*; ?CANADA). [**Note:** Kevan hypothesized that because this species was widespread in the U.S. that it probably occurred in Canada].

— Bonato & Minelli, 2002: 193-198 (*Dicellophilus*; parental care; reproductive behavior).

anomalus Chamberlin, 1904: 655-656 (*Mecistocephalus*; characters; notes; USA: California). [**Note:** Holotype: ?NMNH, California: Pacific Grove].

— Silvestri, 1919: 83; Fig. 24 (*Dicellophilus*; characters; USA: Oregon, California).

— Chamberlin, 1920e: 186 (*Dicellophilus*; list).

— Attems, 1929: 149 (*Mecistocephalus* + *Dicellophilus*; synonym of *carniolensis*).

Distribution: ?CANADA; USA: California, Oregon.

HOLOTYPE: ?Unknown "Vaterland. Herzogthum Krain; aus der Sammlung des Herrn Kaufmann Schmidt in Laibach."

D. limatus (Wood, 1862)

— Wood, 1862: 42; Fig. 9 (*Mecistocephalus*; characters; notes; USA: California).

— Cook, 1895b: 74 (*Dicellophilus*; cited as type species of *Dicellophilus*).

— Bollman, 1893: 125 (*Mecistocephalus*; catalogue).

— Attems, 1903b: 210 (*Mecistocephalus*; list).

— Chamberlin, 1912a: 661 (*Mecistocephalus*; synonym of *melanonotus*).

— Anonymous, 1920: 35 (*Clinopodes*; note; USA: California).

— Chamberlin, 1920e: 186 (*Dicellophilus*; list).

— Gunthorp, 1920: 113 (*Mecistocephalus*; list).

— Chamberlin, 1946c: 182 (*Mecistocephalus*; synonym of *melanonotus*).

— Jeekel, 2005: 76 (*Mecistocephalus*; list).

breviceps Meinert, 1886a: 214 (*Mecistocephalus*; characters; USA: Massachusetts). [**Note:** Two specimens deposited at the MCZ, (No. 905, 906) are labeled as holotypes; one of the vials (906) had two determined labels in Meinert's handwriting that read "*Mecistocephalus breviceps* n. sp. No. 151 [and the other] No. 152". Neither of these vials have a locality label].

— Cook, 1896: 241 (*Mecistocephalus*; notes).

— Bollman, 1893: 125 (*Mecistocephalus*; catalogue).
— Attems, 1903b, 210 (*Mecistocephalus*; list).
— Chamberlin, 1920e: 186 (*Dicellophilus*; synonym of *limatus*; list).

Distribution: USA: California. [**Note:** The holotype locality for breviceps is considered erroneous and has therefore been left out of the distribution].

HOLOTYPE: ?Unknown. [**Note:** NMNH has a specimen labeled as a holotype identified by Crabill as *Dicellophilus* aff. *limatus*. The MCZ has numerous specimens from California with catalog numbers as follows: 903, 1023, 1028, 1030, 1033, 1040, 31768].

Genus *Mecistocephalus* Newport, 1843
[6 species (4 or all Introduced); USA & Hawaii]

Mecistocephalus Newport, 1843: 177. TYPE SPECIES: *Mecistocephalus punctifrons* Newport, 1843. [**Note:** Bonato & Minelli (2004) give a redescription of *punctifrons* and provide great figures].
— Cook, 1895a: 60-62 (characters; distribution). [**Note:** Cook cites *Mecistocephalus attenuatus* (Say) as the type species, and he also refers *punctifrons* to *Lamnonyx*].
— Cook, 1895b: 72 (distribution; list).
— Morse, 1902: 187 [**Note:** Morse indicates a *Mecistocephalus* sp. from Vinton, Ohio and may have been *Geophilus strigosus* (McNeill) which may be a junior synonym of *Pachymerium ferrugineum*].
— Chamberlin, 1920e: 185-186 (list of species).
— Attems, 1929: 127-128 (characters; distribution; key to subgenera).
— Chamberlin, 1953a: 77-78 (key to species).
— Crabill, 1957a: 344 (notes on type species fixation).
— Crabill, 1970b: 231, 234-237 (morphological notes).
— Foddai, Pereira & Minelli, 2000: 63 (notes).
— Bonato & Minelli, 2004: 20 (characters).
— Jeekel, 2005: 89 (list; notes).
— Bonato & Minelli, 2007: 166-173 (designation of *M. punctifrons* as type species).
Fusichila Chamberlin, 1953a: 81 (characters). TYPE SPECIES: *Fusichila waipaheenas* Chamberlin, 1953, fixed by original designation and monotypy.
— Bonato, Foddai, Minelli & Shelley, 2004: 26-27 (monotypic genus suspected of being based on a juvenile *Mecistocephalus* with anomalies). [**Note:** The author has placed this one species

under *Mecistocephalus* based on the observations of the above authors. Although this may not be an endemic species to Hawaii it remains a valid species until further investigation].
— Jeekel, 2005: 80 (list).

Lamnonyx Cook, 1895a: 61 (characters; notes). TYPE SPECIES: *Mecistocephalus punctifrons* Newport, 1843, fixed by subsequent designation of Attems, 1903 (see, Jeekel, 2005).
— Attems, 1929: 127 (synonym of *Mecistocephalus*).
— Chamberlin, 1953b: 77 (synonym of *Mecistocephalus*).
— Jeekel, 2005: 86 (list; note).

Pauroptyx Chamberlin, 1920e: 188 (characters). TYPE SPECIES: *Pauroptyx himalayanus* Chamberlin, 1920, fixed by original designation.
— Bonato & Minelli, 2004: 20 (synonym of *Mecistocephalus*).

M. (M.) guildingii Newport, 1845

— Newport, 1845b: 429-430 (*Mecistocephalus*; characters; notes; West Indies).
— Meinert, 1870: 96-97 (*Mecistocephalus*; characters; notes).
— Chamberlin, 1920e: 185 (*Mecistocephalus*; list; cited as synonym of *maxillaris*).
— Crabill, 1959b: 188-192; Figs. 1-6 (*Mecistocephalus*; notes; redescription; USA: Florida).
— Foddai, Pereira & Minelli, 2000: 63-64 (*Mecistocephalus*; catalogue).

?*punctilabratus* Newport, 1845a: 302; Figs. 18, 19 (*Mecistocephalus*; head illustrations).

Distribution: USA: Florida.

HOLOTYPE: West Indies: St. Vincent?, (v. in Mus. D. Hope.) see Newport, 1845.

♂♀M. (M.) maxillaris (Gervais, 1837)

— Gervais, 1837: 52 (*Geophilus*). [**Note:** reference not obtained].
— Cook, 1895a: 61 (*Lamnonyx*; transferred from *Mecistocephalus*).
— Chamberlin, 1920e: 185 (*Mecistocephalus*; list; cited as tropicopolitan in distribution).
— Attems, 1929: 134-135 (*Mecistocephalus*; subgenus *Mecistocephalus*; characters; distribution).
— Chamberlin, 1930c: 67 (*Mecistocephalus*; USA: Hawaii). [**Note:** "Several specimens of this species, carried widely by ships in

tropical regions, were taken at Hawaii in 1928 in soil about plants from Panama"].
— Attems, 1938b: 365 (*Mecistocephalus*; list).
— Chamberlin, 1940c: 56 (*Mecistocephalus*; notes; USA: North Carolina).
— Chamberlin, 1953b: 78, 80 (*Mecistocephalus*; notes). [**Note:** Chamberlin reports that this species is found in hothouses of America].
— Chamberlin, 1958a: 14 (*Mecistocephalus*; USA: Florida).
— Crabill, 1959b: 188 (*Mecistocephalus*; notes).
— Wray, 1967: 156 (*Mecistocephalus*; list).
— Nishida, 1994: 26 (*Mecistocephalus*; checklist).
— Nishida, 1997: 195 (*Mecistocephalus*; checklist).
— Nishida, 2002: 235 (*Mecistocephalus*; checklist).
— Foddai, Pereira & Minelli, 2000: 64-65 (*Mecistocephalus*; catalogue).
— Bonato, Foddai, Minelli & Shelley, 2004: 16, 27, 28; Fig. 22; Tab. 1 (*Mecistocephalus*; characters; notes; USA: Hawaii).

Distribution: USA: Florida, ?Hawaii, North Carolina.

HOLOTYPE: Unknown?

♂♀*M. spissus* Wood, 1862
— Wood, 1862: 43 (*Mecistocephalus*; characters; USA: Hawaii).
— Haase, 1887: 101 (*Mecistocephalus*;). [**Note:** Reference not obtained].
— Cook, 1895a: 61 (*Lamnonyx*; transferred from *Mecistocephalus*).
— Silvestri, 1904: 326-327, pl. 11; Figs. 5-7 (*Lamnonyx*; characters; note; USA: Hawaii).
— Attems, 1914: 48, 62, 130 (*Mecistocephalus*; distribution; list).
— Silvestri, 1919: 75, Fig. 19 (*Lamnonyx*; characters; USA: Hawaii).
— Chamberlin, 1920c: 63, 243 (*Mecistocephalus*; note; USA: Hawaii). [**Note:** It is not clear if this is truly indigenous to Hawaii, but is has been reported by Pocock in Burma and Sumatra].
— Chamberlin, 1920e: 186 (*Mecistocephalus*; list).
— Attems, 1929: 129; Fig. 142 (*Mecistocephalus*; subgenus *Mecistocephalus*; characters; distribution).
— Attems, 1938b: 365, 369 (*Mecistocephalus*; list; note).
— Chamberlin, 1953b: 77, 80 (*Mecistocephalus*; note). [**Note:** Chamberlin claims it is an endemic to the Hawaiian Islands].
— Nishida, 1994: 26 (*Mecistocephalus*; checklist).

— Nishida, 1997: 195 (*Mecistocephalus*; checklist).
— Nishida, 2002: 235 (*Mecistocephalus*; checklist).
— Bonato, Foddai, Minelli & Shelley, 2004: 15-16, 27, 28; Fig. 17, 20; Tab. 1 (*Mecistocephalus*; characters; key; notes; USA: Hawaii).

Distribution: USA: ?Hawaii.

HOLOTYPE: NMNH, No. ?343, Hawaii: Ohau or Kaiu, by Commodore Perry.

M. tridens Chamberlin, 1922
— Chamberlin, 1922c: 178-179 (*Mecistocephalus*; characters; notes). [**Note:** Chamberlin described this species as originating from Java but Attems (1929; 1938) considered this species to occur in Hawaii].
— Attems, 1929: 136-137 (*Mecistocephalus*; subgenus *Mecistocephalus*; characters; distribution).
— Attems, 1938b: 365, 369 (*Mecistocephalus*; list; note).
— Chamberlin, 1953b: 78, 80. (*Mecistocephalus*; note). [**Note:** Chamberlin questions the occurrence of this species in Hawaii, but it certainly shows the possibility of this species to become established].
— Bonato, Foddai, Minelli & Shelley, 2004: 27 (*Mecistocephalus*; notes).

Distribution: USA: ?Hawaii.

COTYPES: MCZ, No. 2190, & FHB, numerous specimens (MCZ vial with one specimen), mostly young, Hawaii: Honolulu, from soil of a shipment of plants from the Botanical Gardens of Buitenzorg, Java, Mar. 8, 1922. [**Note:** Original label states to see collection card and mount of head. A separate label in MCZ vial labeled as holotype reads "Head, mouthparts removed by Chamberlin prior to 1956"].

♂♀*M. waikaneus* Chamberlin, 1953
— Chamberlin, 1953b: 81 (*Mecistocephalus*; characters; habitat; USA: Hawaii).
— Nishida, 1994: 26 (*Mecistocephalus*; list).
— Nishida, 1997: 195 (*Mecistocephalus*; list).
— Nishida, 2002: 235 (*Mecistocephalus*; checklist).
— Bonato, Foddai, Minelli & Shelley, 2004: 17, 27, 28; Fig. 18, 21; Tab. 1 (*Mecistocephalus*; characters; key; notes; USA: Hawaii).

Distribution: USA: Hawaii.

SYNTYPE: NMNH: Hawaii: Waikane Island. Two specimens taken in rotten log on Dec. 28, 1928.

?*M. waipaheenas* (Chamberlin, 1953)
— Chamberlin, 1953b: 81-82 (*Fusichila*; characters; habitat; USA: Hawaii).
— Nishida, 1994: 26 (*Mecistocephalus*; list).
— Nishida, 1997: 195 (*Mecistocephalus*; list).
— Bonato, Foddai, Minelli & Shelley, 2004: 26-27 (*Fusichila*; notes; suspected juvenile specimen of *Mecistocephalus* with anomalies).
— Jeekel, 2005: 80 (*Fusichila*; list).

Distribution: USA: Hawaii.

HOLOTYPE: NMNH, ♀, Hawaii: Waipahee, Kauai. One female taken Jan. 13, 1944 under bark of dead tree by N.L.H. Krauss.

Family **Oryidae** Cook, 1895
Subfamily **Oryinae** Attems, 1914

Genus ***Orphnaeus*** Meinert, 1870
[1 species (Introduced); Southeastern USA & Hawaii]

Orphnaeus Meinert, 1870: 17-19; Tab. 2, Figs. 6-12 (characters; notes). **TYPE SPECIES:** *Orphnaeus lividus* Meinert, 1870, [= *Orphnaeus brevilabiatus* (Newport, 1845)], fixed by subsequent designation.
— Attems, 1929: 112 (characters; key to species).
— Jeekel, 2005: 95 (list; notes).

♂♀*O. brevilabiatus* (Newport, 1845)
— Newport, 1845b: 436 (*Geophilus*; characters; "Orâ Tenasserim Peninsulae Indiae Ulterioris").
— Haase, 1887: 111 (*Orphnaeus;*). [**Note:** reference not obtained].
— Pocock, 1896: 40, t. 3, f. 14 (*Orphnaeus*; figures; notes).
— Attems, 1914: 48 (*Orphnaeus*; list).
— Chamberlin, 1920c: 38-39, 239 (*Orphnaeus*; USA: Hawaii).
— Attems, 1929: 112-113 (*Orphnaeus*; characters; distribution).
— Attems, 1938b: 365 (*Orphnaeus*; list; note).

— Crabill, 1957c: 265 (*Orphnaeus*; notes). [**Note:** Crabill suspects it has likely been introduced with cargo from the tropics].
— Chamberlin, 1958a: 14 (*Orphnaeus*; USA: Florida).
— Nishida, 1994: 26 (*Orphnaeus*; checklist).
— Nishida, 1997: 195 (*Orphnaeus*; checklist).
— Foddai, Pereira & Minelli, 2000: 114-117 (*Orphnaeus*; catalogue).
— Bonato, Foddai, Minelli & Shelley, 2004: 22-23, 27, 28; Figs. 1, 5; Tab. 1 (*Orphnaeus*; characters; key; notes).
lividus Meinert, 1870: 19-20 (*Orphnaeus*; characters; distribution, notes).
— Haase, 1887: ?111 (synonym of *brevilabiatus*). [**Note:** reference not obtained].
— Jeekel, 2005: 95 (*Orphnaeus*; list).

Distribution: USA: Florida, ?Hawaii. [**Note:** Author may have seen a specimen from New York City, Manhattan, on the upper west side].

HOLOTYPE: ?BMNH; "Orâ Tenasserim Peninsulae Indiae Ulterioris" (*v. in Mus. Brit.*). [**Note:** MCZ has a specimen that could be from Florida, catalog number 891].

Genus *Orya* Meinert, 1870
[1 species (Introduced); Southeastern USA]

Orya Meinert, 1870: 14-16; Tab. 1, Figs. 1-12 (characters). TYPE SPECIES: *Geophilus barbarica* Gervais, 1835, fixed by original designation.
— Attems, 1929: 110 (characters; distribution).
— Jeekel, 2005: 95 (list).
Parorya Cook, 1896: 33 (characters). TYPE SPECIES: *Parorya valida* Cook, 1896, fixed by monotypy.
— Crabill, 1957c: 266 (synonymy of *Orya*).
— Jeekel, 2005: 96 (list).

O. barbarica (Gervais, 1835)
— Gervais, 1835: 10-11; Fig. 3 (*Geophilus*; characters).
— Lucas, 1840: 551 (*Geophilus*; characters).
— Meinert, 1870: 16, 104-105, 115-117; Tab. 1; Fig. 1-12 (*Orya*; characters; notes).
— Attems, 1929: 110-111; Fig. 121-123 (*Orya*; characters; distribution).
— Crabill, 1957c: 266, 267 (*Orya*; synonymy of *Parorya valida* Cook with *barbarica*).
— Jeekel, 2005: 95 (*Orya*; list).

valida Cook, 1896: 33 (*Parorya*; synonym of *barbarica*). [**Note:** Holotype: NMNH; No. 2363; ♀; USA: Louisiana or Texas, by Schufelt].
— Crabill, 1957c: 266-267 (*Parorya*; synonymy of *valida* with *barbarica*). [**Note:** Crabill states that the holotype of *valida* must have been introduced with produce from the Mediterranean].
— Jeekel, 2005: 96 (*Parorya*; list).

Distribution: USA: ?Louisiana, ?Texas.

HOLOTYPE: ?Europe.

Family **Schendylidae** Cook, 1895

Genus *Apunguis* Chamberlin, 1947
[1 species (Introduced); South Central USA]

Apunguis Chamberlin, 1947b: 260 **Type Species:** *Apunguis prosoicus* Chamberlin, fixed by original designation and monotypy.
— Crabill, 1961a: 68 (catalogue).

♀*A. prosoicus* Chamberlin, 1947
— Chamberlin, 1947b: 260 (*Apunguis*; USA: Texas). [**Note:** This species was intercepted on fruit from Mexico and is not known to occur naturally in the United States].
— Crabill, 1961a: 68 (*Apunguis*; catalogue).

Distribution: USA: ?Texas.

HOLOTYPE: ?NMNH, ♀, Texas: Eagle Pass, June 2, 1947, intercepted at quarantine on fruit from Mexico.

Genus *Escaryus* Cook and Collins, 1891
[8 species; CANADA; USA & Alaska]

Escaryus Cook and Collins, 1891: 391-392 (characters; notes). **Type Species:** *Escaryus sibericus* Cook, 1899 fixed by subsequent designation of Cook, 1895: 71.
— Cook, 1895b: 71 (distribution; list).
— Bailey, 1928b: 43 (key to species of New York; notes; species included: *liber, urbicus*).

— Attems, 1929: 94-95 (species included: *albus, japonicus, latzeli, liber, phyllophilus, retusidens* and *sibiricus*; characters; key to species).
— Chamberlin, 1946c: 178 (key to species).
— Chamberlin, 1947a: 37 (key to species).
— Crabill, 1952b: 63-65 (characters; key to species of northeastern United States).
— Crabill, 1953c: 96 (characters; key to species of northeastern United States).
— Crabill, 1961a: 35, 68 (key, reference to keys).
— Carter & Brown, 1973: 1066, Tab. 1, Fig. 1 (data from study on soil arthropods; CANADA: New Brunswick).
— Pereira & Hoffman, 1993: 8-13 (characters; distribution; key to species; notes).
— Jeekel, 2005: 78 (list).

♂♀*E. cryptorobius* Pereira & Hoffman, 1993
— Pereira & Hoffman, 1993: 30-41; Figs. 38-83; Maps 2, 3 (Escaryus; characters; distribution; notes; USA: Virginia).
— Hoffman, 1995a: 22 (*Escaryus*; list; notes).

Distribution: USA: Virginia.

HOLOTYPE: VMNH, ♂, Virginia: Grayson Co., Whitetop Mountain, ca. 5,400 ft. ASL, 15 December 1984, R.L. Hoffman.

PARATYPES: NMNH, ♂; VMNH 2♀, 2 juv. ♂; Virginia: Grayson Co., Whitetop Mountain, ca. 5,400 ft. ASL, 15 December 1984, R.L. Hoffman.

TOPOPARATYPE: NMNH ♀; MCZ ♀; Virginia: Grayson Co., Whitetop Mountain, ca. 5,400 ft. ASL, 12 May 1988, L.A. Pereira & R.L. Hoffman.

♂♀*E. ethopus* (Chamberlin, 1920)
— Chamberlin, 1920d: 43 (*Geophilus*; characters; notes; USA: Alaska).
— Attems, 1929: 328 (*Geophilus*; distribution; *incertae sedis*).
— Chamberlin, 1946c: 179 (*Geophilus*; list).
— Crabill, 1961a: 69-70 (*Escaryus*; catalogue; notes).
— Kevan, 1983: 2945 (*Escaryus*; checklist).
— Behan-Pelletier, 1993: 25 (*Escaryus*; list).
— Pereira & Hoffman, 1993: 50-56; Figs. 102-116; Map 1 (*Escaryus*; characters; distribution; CANADA: Yukon Territory; USA: Alaska).

delus Chamberlin, 1946c: 178, pl. 1; Figs. 1-2 (*Escaryus*; characters; USA: Alaska). [Type: deposited at the NMNH].
— Crabill, 1961a: 69 (*Escaryus*; catalogue).
— Kevan, 1983: 2945 (*Escaryus*; checklist).
— Behan-Pelletier, 1993: 25 (*Escaryus*; list).
— Pereira & Hoffman, 1993: 50 (*Escaryus*; synonym of *ethopus*).
japonicus [nec Attems, 1903): Chamberlin, 1952c: 209 (*Escaryus*; USA: Alaska). {misidentification}
— Attems, 1929: 96-97; figs. 112-115 (*Escaryus*; characters; distribution).
— Pereira & Hoffman, 1993: 50 (*Escaryus*; synonym of *ethopus*).

Distribution: CANADA: Yukon Territory; USA: Alaska.

HOLOTYPE: MCZ, (Unique No. 14326; TC-39), ♂, Alaska: Iditarod Island, (?River 63N, 158W), June 1918, A.H. Twitchell.

♂♀***E. liber*** Cook & Collins, 1891
— Cook & Collins, 1891: 394-395; pl. 35; Figs. 16-17 (*Escaryus*; characters; habitat; USA: New York).
— Attems, 1903b: 196 (*Escaryus*; list).
— Cook, 1904: 77 (*Escaryus*; characters; key).
— Attems, 1927: 303 (*Escaryus*; key).
— Bailey, 1928a: 1083 (*Escaryus*; list).
— Bailey, 1928b: 19, 43 (*Escaryus*; characters; habitat; list; notes; USA: New York).
— Attems, 1929: 98 (*Escaryus*; characters; distribution).
— Crabill, 1952b: 65 (*Escaryus*; characters; USA: District of Columbia, Maryland, New York, Ohio).
— Crabill, 1953c: 97 (*Escaryus*; characters; USA: District of Columbia, Maryland, New York, Ohio).
— Crabill, 1961a: 70 (*Escaryus*; catalogue).
— Summers, 1979: 697, 700 (*Escaryus*; checklist; key).
— Pereira & Hoffman, 1993: 41-46; Figs. 84-92; Map 2 (*Escaryus*; characters; distribution; notes; USA: New York).
— Hoffman, 1995a: 22 (*Escaryus*; list; notes). [**Note:** Hoffman suggests *liber* lives in Virginia].

Distribution: USA: District of Columbia, Maryland, New York, Ohio, ?Virginia.

HOLOTYPE: Unknown, New York: Onondaga Co., Kirkville, April 1890, Cook and Collins.

NEOTYPE: NMNH, ♀ from Rochester, New York, 8 December 1956.

♂♀ *E. missouriensis* Chamberlin, 1942
— Chamberlin, 1942d: 185 (*Escaryus*; characters; USA: Missouri).
— Crabill, 1952b: 67 (*Escaryus* [sic]; characters; notes).
— Crabill, 1953c: 96-97 (*Escaryus*; characters; distribution).
— Crabill, 1955b: 38 (*Escaryus*; notes; USA: Missouri).
— Crabill, 1961a: 70 (*Escaryus*; catalogue). [**Note:** Crabill includes Illinois as a locality without specifics for this species but this should be confirmed].
— Summers, 1979: 697, 700 (*Escaryus*; checklist; key).
— Summers et al., 1980: 249 (*Escaryus*; checklist; habitat (Table 1, p. 242-243); note; USA: Illinois).
— Summers et al., 1981: 60 (*Escaryus*; checklist; note; USA: Illinois).
— Pereira & Hoffman, 1993: 56-67; Figs. 117-154; Map 2. (*Escaryus*; characters; notes; USA: Kansas, Missouri). [**Note:** According to Pereira & Hoffman the location of the holotype is unknown. Pereira & Hoffman are doubtful that this species exists in Indiana but this also needs to be confirmed].

Distribution: USA: ?Illinois, ?Indiana, Kansas, Missouri.

HOLOTYPE: NMNH; ♂, Missouri: St. Louis Co., 4.3 mi. NW Glencoe, 11 March 1936, by Leslie Hubricht.

♀ *E. monticolens* Chamberlin, 1947
— Chamberlin, 1947a: 37-38 (*Escaryus*; habitat; USA: Utah).
— Chamberlin, 1961: 100 (*Escaryus*; note).
— Crabill, 1961a: 70 (*Escaryus*; catalogue).
— Pereira & Hoffman, 1993: 26-29, figs. 27-37; Map 1 (*Escaryus*; characters; distribution; notes).

Distribution: USA: Utah.

HOLOTYPE: NMNH, ♀, Utah: Davis Co., Mill Creek Canyon, ~7,500 ft, dug up in soil, by A.M. Woodbury.

♂♀ *E. orestes* Pereira & Hoffman, 1993
— Pereira & Hoffman, 1993: 20-25; Figs. 11-26; Maps 2, 3 (*Escaryus*; characters; distribution, habitat, notes; USA: Virginia).
— Hoffman, 1995a: 22 (*Escaryus*; list; notes).

Distribution: USA: Virginia.

HOLOTYPE: VMNH; ♂, Virginia: Washington Co., Whitetop Mountain, 5400 ft ASL, 12 May 1988, L.A. Pereira & R.L. Hoffman.

PARATYPES: VMNH; 2♀; NMNH; ♂; MCZ; ♂; all from Virginia: Washington Co., Whitetop Mountain, 5400 ft ASL, 12 May 1988, L.A. Pereira & R.L. Hoffman.

♂*E. paucipes* Chamberlin, 1946
   — Chamberlin, 1946c: 179, pl. 1; Figs. 3-4 (*Escaryus*; characters; USA: Alaska).
   — Chamberlin, 1949a: 14 (*Escaryus*; notes; USA: Alaska).
   — Crabill, 1961a: 70 (*Escaryus*; catalogue).
   — Kevan, 1983: 2945 (*Escaryus*; checklist).
   — Behan-Pelletier, 1993: 25 (*Escaryus*; list).
   — Pereira & Hoffman, 1993: 47-50; Figs. 93-101; Map 1 (*Escaryus*; characters; distribution; notes).

Distribution: USA: Alaska.

HOLOTYPE: NMNH; ♂, Alaska: Haines, (see Pereira & Hoffman, 1993).

♂♀*E. urbicus* (Meinert, 1886)
   — Meinert, 1886a: 218-219 (*Geophilus*; characters; USA: Massachusetts).
   — Cook & Collins, 1891: 393 (*Geophilus*; note).
   — Cook, 1896: 239-240 (*Geophilus*; notes).
   — Cook, 1896: 240 (*Escaryus*; note).
   — Bollman, 1893: 125, 184 (*Geophilus*; catalogue; notes; USA: Minnesota).
   — Attems, 1903b: 197 (*Geophilus*; list).
   — Cook, 1904: 76 (*Escaryus*; key).
   — Chamberlin, 1912b: 66 (*Escaryus*; note).
   — Bailey, 1928a: 1083 (*Escaryus*; list).
   — Bailey, 1928b: 19, 43, 44 (*Escaryus*; characters; habitat; list; notes; USA: New York).
   — Williams & Hefner, 1928: 135 (*Escaryus*; characters; USA: Ohio).
   — Attems, 1929: 327 (*Geophilus*; distribution; *species inquirenda*).
   — Crabill, 1952b: 66 (*Escaryus*; characters; note; USA: Massachusetts, New York, Ohio, Virginia).

— Crabill, 1953c: 97-98 (*Escaryus*; characters; distribution note; USA:
Massachusetts, New York, Ohio, Virginia).

— Crabill, 1958b: 93, 94 (*Escaryus*; notes; USA: Wisconsin).

— Crabill, 1961a: 71 (*Escaryus*; catalogue).

— Branson & Batch, 1967: 82-83 (*Escaryus*; characters; habitat; USA:
Kentucky).

— Summers, 1979: 697, 700 (*Escaryus*; checklist; key).

— Kevan, 1983: 2945 (*Escaryus*; checklist).

— Behan-Pelletier, 1993: 25 (*Escaryus*; list).

— Pereira & Hoffman, 1993: 14-20, figs 1-10; Maps 2, 3 (*Escaryus*;
characters; distribution; notes; USA: Virginia).

— Hoffman, 1995a: 22 (*Escaryus*; list; notes).

— Watermolen, 1997: 2 (*Escaryus*; checklist, notes; USA: Wisconsin).

phyllophilus Cook and Collins, 1891: 392-394; pl. 34; Figs. 9-11 (*Escaryus*;
characters; habitat; notes; USA: New York).

— Cook, 1895: 71 (*Escaryus*; cited as type species of *Escaryus*).

— Cook, 1896: 240 (*Escaryus*; note).

— Cook, 1899: 304 (*Escaryus*; cited as type species of *Escaryus*; note).

— Attems, 1903b: 196 (*Escaryus*; list).

— Attems, 1904: 124 (*Escaryus*; note).

— Chamberlin, 1909b: 177 (*Escaryus*; habitat; USA: New York).

— Attems, 1927, 303 (*Escaryus*; key).

— Attems, 1929: 95, 97 (*Escaryus*; characters; distribution).

— Pereira & Hoffman, 1993: 14 (synonym of *urbicus*).

— Jeekel, 2005: 78 (*Escaryus*; list).

Distribution: USA: Kentucky, Massachusetts, Minnesota, New York, Ohio,
Virginia, Wisconsin.

HOLOTYPE: MCZ, No. 966, ♀, Massachusetts, Cambridge, Winter
1872-1873, Schwartz. [Note: Original determined label reads "*Geophilus
urbicus* n. sp. No. 160"

Genus *Gosendyla* Chamberlin, 1960
[1 species; Western USA]

Gosendyla Chamberlin, 1960b: 241-242 (characters; note). TYPE SPECIES:
Gosendyla socarina Chamberlin, 1960, fixed by original designation and
monotypy.

♂*G. socarina* Chamberlin, 1960

— Chamberlin, 1960b: 242 (*Gosendyla*; characters; USA: Utah).
— Chamberlin, 1961: 100 (*Gosendyla*; note).

Distribution: USA: Utah.

HOLOTYPE: ?NMNH, ♂, Utah: Dry Canyon, near University of Utah campus.

Genus *Holitys* Cook, 1899
[1 species; Southwestern USA]

Holitys Cook, 1899: 304 (characters; notes). Type Species: *Holitys neomexicanus* Cook, fixed by monotypy.
— Cook, 1904: 76 (key to genus within key to Schendylidae).
— Attems, 1929: 99 [**Note:** Attems comments that there has been an inadequate diagnosis of this genus].
— Crabill, 1961a: 71 (notes).
— Hoffman & Pereira, 1991: 45 (questioned synonym of *Marsikomerus*).
— Jeekel, 2005: 84 (list).
?*Mexiconyx* Chamberlin, 1922d: 9 (characters; Mexico). Type Species: *Mexiconyx hidalgoensis* Chamberlin, 1922, fixed by original designation.
— Crabill, 1961a: 71 (cited as possible synonym of *Holitys*).
— Hoffman & Pereira, 1991: 45 (questioned synonym of *Marsikomerus*).
— Jeekel, 2005: 90 (list).
Simoporus (?in part) Chamberlin & Mulaik, 1940: 109 (characters; note). Type Species: *Simoporus texanus* Chamberlin & Mulaik, 1940, fixed by original designation.
— Crabill, 1961a: 71 (cited as possible synonym of *Holitys*).
— Jeekel, 2005: 104 (list).

♀*H. neomexicanus* Cook, 1899
— Cook, 1899: 304; pl. 4; Figs. 4a-b (*Holitys*; characters; notes; USA: New Mexico).
— Attems, 1929: 99 (*Holitys*; distribution; notes). [**Note:** Attems comments that there has been no further description].
— Chamberlin, 1943b: 99 (*Holitys*; USA: New Mexico).
— Crabill, 1961a: 72 (*Holitys*; catalogue).
— Jeekel, 2005: 84 (*Holitys*; list).

Distribution: USA: New Mexico.

HOLOTYPE: ?Unknown, New Mexico: Organ Mountains, Dripping Spring.

Genus *Marsikomerus* Attems, 1938
[3 species; south Central USA & Hawaii]

Marsikomerus Attems, 1938b: 371 (characters). **TYPE SPECIES:** *Marsikomerus pacificus* Attems, 1938, fixed by original designation and monotypy.
— Chamberlin, 1953b: 85 (notes).
— Hoffman & Pereira, 1991: 45-48 (characters; distribution; key to species; taxonomy).
— Jeekel, 2005: 88 (list).
Lanonyx Chamberlin, 1953b: 75 (characters). **TYPE SPECIES:** *Lanonyx lanaius* Chamberlin, 1953, fixed by original designation and monotypy.
— Hoffman & Pereira, 1991: 47 (synonym of *Marsikomerus*).
— Jeekel, 2005: 86 (list).
Simoporus Chamberlin & Mulaik, 1940: 109 (characters). **TYPE SPECIES:** *Simoporus texanus* Chamberlin & Mulaik, 1940, fixed by original designation.
— Crabill, 1961a: 71 (cited as possible synonym of *Holitys*).
— Hoffman & Pereira, 1991: 45, 47 (synonym of *Marsikomerus*).
?*Mexiconyx* Chamberlin, 1922d: 9 (characters; Mexico). **TYPE SPECIES:** *Mexiconyx hidalgoensis* Chamberlin, 1922, fixed by original designation.
— Crabill, 1961a: 71 (cited as possible synonym of *Holitys*).
— Hoffman & Pereira, 1991: 45 (cited as questioned synonym of *Marsikomerus*).
?*Holitys* Cook, 1899: 304 (characters; notes). **TYPE SPECIES:** *Holitys neomexicana* Cook, 1899, fixed by monotypy.
— Hoffman & Pereira, 1991: 45 (cited as questioned synonym of *Marsikomerus*).
?*Morunguis* Chamberlin, 1922:. **TYPE SPECIES:** *Morunguis morelus* Chamberlin, 1943, fixed by original designation.
— Hoffman & Pereira, 1991: 45 (cited as questioned synonym of *Marsikomerus*).

♂*M. arcanus* Crabill, 1961
— Crabill, 1961a: 32-35; figs. 1-4 (*Simoporus*; characters; USA: Arkansas).
— Hoffman & Pereira, 1991: 48, 54 (*Marsikomerus*; diagnosis; key, notes).

Distribution: USA: Arkansas.

HOLOTYPE: NMNH, No. 2598, ♂, Arkansas: Washington Co., 4 mi. W of Farmington, 16 June 1950, by Nell B. Causey.

PARATYPE: NMNH, ♂, Arkansas: Washington Co., 4 mi. W of Farmington, 16 June 1950, by Nell B. Causey.

♂♀*M. bryanus* (Chamberlin, 1926)
— Chamberlin, 1926b: 92-93 (*Nyctunguis*; characters; habitat; note; USA: Hawaii).
— Attems, 1938b: 365, 369 (*Nyctunguis*; list; note).
— Chamberlin, 1953b: 75 (*Nyctunguis*; USA: Hawaii).
— Nishida, 1994: 27 (*Nyctunguis*; checklist).
— Nishida, 1997: 195 (*Nyctunguis*; checklist).
— Nishida, 2002: 236 (*Marsikomerus*; checklist).
— Bonato, Foddai, Minelli & Shelley, 2004: 18-22, 27, 28; Figs. 3, 7, 9, 11; Tab. 1 (*Marsikomerus*; characters; key; notes; transferred from *Nyctunguis*; USA: Hawaii).
lanaius—Chamberlin, 1953b, 76 (*Lanonyx*; characters; USA: Hawaii).
  HOLOTYPE: NMNH; ♂, Hawaii: Lanai Island, Lanai Mtns., Nov. 1, 1947, N.L.H. Krauss.
— Hoffman & Pereira, 1991: 48, 52-54, 59; Figs. 38-42 (*Marsikomerus*; diagnosis; description of holotype, notes).
— Nishida, 1994: 27 (*Marsikomerus*; checklist).
— Nishida, 1997: 195 (*Marsikomerus*; checklist).
— Shelley, 2000b: 39 (*Marsikomerus*; note).
— Nishida, 2002: 236 (*Marsikomerus*; checklist).
— Bonato, Foddai, Minelli & Shelley, 2004: 18 (*Lanonyx*; synonym of *bryanus*).
— Jeekel, 2005: 86 (*Lanonyx*; list).
pacificus—Attems, 1938b: 365, 369, 371, 372-374; Figs. 1-6 (*Marsikomerus*; characters; list; note; USA: Hawaii). **HOLOTYPE:** NMW, Hawaii: Island of Hawaii, Nauhi Gulch. (see Hoffman & Pereira, 1991).
— Attems, 1947: 128 (*Marsikomerus*; list). [**Note:** Misspelled *Marsicomerus*].
— Chamberlin, 1953b: 85 (*Marsikomerus*; notes). [**Note:** Misspelled *Marsukomerus*].
— Hoffman & Pereira, 1991: 45, 48-52, 59; Figs. 1-37 (*Marsikomerus*; diagnosis; description of holotype; key; notes).
— Nishida, 1994: 27 (*Marsikomerus*; checklist).

— Nishida, 1997: 195 (*Marsikomerus*; checklist).
— Nishida, 2002: 236 (*Marsikomerus*; checklist).
— Shelley, 2000b: 39 (*Marsikomerus*; note).
— Bonato, Foddai, Minelli & Shelley, 2004: 18 (*Marsikomerus*; synonym of *bryanus*).
— Jeekel, 2005: 88 (*Marsikomerus*; list).

Distribution: USA: Hawaii.

HOLOTYPE: NMNH, Hawaii: Necker Island.

♂♀*M. texanus* Chamberlin & Mulaik, 1940
— Chamberlin & Mulaik, 1940: 109 (*Simoporus*; characters; USA: Texas).
— Crabill, 1961a: 79 (*Simoporus*; catalogue).
— Hoffman & Pereira, 1991: 45, 48, 54-56; Figs. 43-59 (*Marsikomerus*; diagnosis; description of male allotype; key; notes).

Distribution: USA: Texas.

HOLOTYPE: NMNH, ♀, Texas: Bandera Co., 2 mi. N of Medina, 16 December 1939, Stanely and Dorothea Mulaik.

ALLOTYPE: NMNH, ♂, Texas: Bandera Co., 2 mi. N of Medina, 16 December 1939, Stanely and Dorothea Mulaik.

PARATYPES: NMNH, 4♂, 3♀, Texas: Bandera Co., 2 mi. N of Medina, 16 December 1939, Stanely and Dorothea Mulaik.

Genus *Nyctunguis* Chamberlin, 1914
[13 species (1 Introduced); Eastern and Western USA]

Nyctunguis Chamberlin, 1914c: 201 (transfer of species [*montereus* and *heathii* (sic)] from *Pectiniunguis*). TYPE SPECIES: *Pectiniunguis montereus* Chamberlin, 1904, fixed by original designation.
— Attems, 1929: 87 (characters; distribution; notes).
— Crabill, 1961a: 75 (catalogue; notes).
— Chamberlin, 1962: 135-136 (notes; "tentative" key to species; species included: *apachus, archochilus, auxus, bryanus, catalinae, dampfi, danzantinus, glendorus, heathi, libercolens, mirus, molinor, montereus, pholeter, stenus*).
— Jeekel, 2005: 94 (list).

403

♀*N. apachus* Chamberlin, 1941
— Chamberlin, 1941d: 786-787 (*Nyctunguis*; characters; notes; USA: Arizona).

Distribution: USA: Arizona.

HOLOTYPE: NMNH, ♀, Arizona: 38 mi. S of Ajo, 4 January 1941. [**Note:** The author could not find a vial labeled as the holotype].

PARATYPE: NMNH, ♀, Arizona: 20 mi. S of Ajo, 4 January 1941. [**Note:** The specimen at the NMNH has a locality of Arizona: N. Sasaba, 4-41, S.-D. Mulaik].

♀*N. archochilus* Chamberlin, 1941
— Chamberlin, 1941d: 785 (*Nyctunguis*; characters; USA: Texas). [**Note:** This specimen was imported on a bromeliad from Mexico and is not indigenous to the United States].
— Crabill, 1961a: 76 (*Nyctunguis*; catalogue).

Distribution: USA: ?Texas.

HOLOTYPE: ?NMNH, ♀, Texas: Laredo, 13 July 1938.

N. auxus Chamberlin, 1941
— Chamberlin, 1941d: 787 (*Nyctunguis*; characters; notes; USA: California).
— Crabill, 1961a: 76 (*Nyctunguis*; catalogue).

Distribution: USA: California.

HOLOTYPE: ?NMNH, California: Coyote Wells, 8 January 1941, by D. & S. Mulaik.

N. catalinae (Chamberlin, 1912)
— Chamberlin, 1912a: 669 (*Pectiniunguis*; nominal taxon *heathii*; characters; USA: California).
— Anonymous, 1920: 35 (*Nyctunguis*; nominal taxon *heathii*; USA: California).
— Attems, 1929: 89 (*Nyctunguis*; characters; distribution).
— Crabill, 1961a: 76 (*Nyctunguis*; catalogue).

Distribution: USA: California.

HOLOTYPE: ?Unknown; California: Catalina Island, Claremont.

♀*N. glendorus* Chamberlin, 1946
— Chamberlin, 1946d: 69-70; Figs. 1-4 (*Nyctunguis*; characters; habitat; notes; USA: California).
— Crabill, 1961a: 77 (*Nyctunguis*; catalogue).

Distribution: USA: California.

HOLOTYPE: NMNH, ♀, California: Los Angeles National Forest, Glendora, July 15, 1944, by Dr. V.E. Shelford in growth of golden oak.

♀*N. heathii* (Chamberlin, 1909)
— Chamberlin, 1909b: 176, pl. 26; Figs. 4-6 (*Pectiniunguis*; habitat; notes; [**Note:** Chamberlin failed to indicate what species his figures 4-6 on plate 26 refer to and it is likely that these are diagrams for *heathii*]. USA: California).
— Brölemann & Ribaut, 1912: 104 (*Pectiniunguis*; characters; list).
— Chamberlin, 1914c: 201 (*Nyctunguis*). [**Note:** Chamberlin mentions the two species from the southwestern United States comparing the genus *Pectiniunguis* to *Adenoschendyla*. Chamberlin then transfers the two mentioned species {*montereus* and *heathii* [sic]} to a new genus *Nyctunguis*].
— Attems, 1929: 88 (*Nyctunguis*; characters; distribution).
— Chamberlin, 1944d: 186-187 (*Nyctunguis*; USA: California).
— Crabill, 1961a: 77 (*Nyctunguis*; catalogue).

Distribution: USA: California.

HOLOTYPE: ?NMNH; ♀, California: Monterey Co., near Cypress Point, found in an Indian shell mound some distance below the surface.

N. libercolens Chamberlin, 1923
— Chamberlin, 1923b: 395, 397; Fig. 12 (*Nyctunguis*; habitat; USA: California).
— Attems, 1929: 89 (*Nyctunguis*; characters; distribution).
— Crabill, 1961a: 77 (*Nyctunguis*; catalogue).

Distribution: USA: California.

HOLOTYPE: NMNH, California: Stanford and environs, numerous specimens taken under bark of Eucalyptus trees in December, 1920. [**Note:**

Other specimens are from Los Angeles, 1909, probably also deposited in the NMNH; MCZ (Unique No. 14403) has a single specimen labeled as the holotype with a TC-52 identification number.].

♂♀*N. molinor* Chamberlin, 1925
— Chamberlin, 1925e: 58-59 (*Nyctunguis*; characters; USA: Utah).
— Attems, 1929: 89 (*Nyctunguis*; distribution).
— Crabill, 1961a: 78 (*Nyctunguis*; catalogue).
— Chamberlin, 1961: 100 (*Nyctunguis*; note).

Distribution: USA: Utah.

HOLOTYPE: NMNH, Utah: Salt Lake Co., Mill Creek Canyon.

N. montereus (Chamberlin, 1904)
— Chamberlin, 1904: 653-654 (*Pectiniunguis*; characters; key to species; notes; USA: California).
— Chamberlin, 1909b: 176 (*Pectiniunguis*; USA: California).
— Brölemann & Ribaut, 1912: 104 (*Pectiniunguis*; list).
— Chamberlin, 1914c: 201 (*Nyctunguis*). [**Note:** Chamberlin mentions the two species from the southwestern United States comparing the genus *Pectiniunguis* to *Adenoschendyla*. Chamberlin then transfers the two mentioned species {*montereus* and *heathii* (sic)} to a new genus *Nyctunguis*].
— Attems, 1929: 88 (*Nyctunguis*; characters; distribution).
— Jeekel, 2005: 94 (*Pectiniunguis*; list).

Distribution: USA: California.

HOLOTYPE: ?NMNH; California: Monterey Bay, Pacific Grove.

♀*N. pholeter* Crabill, 1958
— Crabill, 1958c: 154-159; Figs. 1-14 (*Nyctunguis*; characters; notes; USA: Tennessee). [**Note:** Crabill provides information on using Hoyer's Mountant for slide preparation].
— Crabill, 1961a: 78 (*Nyctunguis*; catalogue).

Distribution: USA: Tennessee.

HOLOTYPE: NMNH; No. 2453; ♀, Tennessee: DeKalb Co., Cripps' Mill. Cripps' Mill Cave, 27 December 1956, by Thomas C. Barr.

N. stenus Chamberlin, 1962

— Chamberlin, 1962: 134-135 (*Nyctunguis*; characters; USA: Nevada).

Distribution: USA: Nevada.

HOLOTYPE: ?University of Utah, Nevada: Clark Co., Mercury, Nevada Test Area. [**Note:** Sex of holotype is not mentioned in the description].

♂*N. vallis* Chamberlin, 1941
— Chamberlin, 1941d: 786 (*Nyctunguis*; characters; note; USA: California).
— Crabill, 1961a: 78 (*Nyctunguis*; catalogue).

Distribution: USA: California.

HOLOTYPE: NMNH, ♂, California: Monterey Co., Carmel Valley, Hastings Reservation, 9 March 1940, by Dr. Linsdale.

Genus ***Parunguis*** Chamberlin, 1941
[1 species; Western USA]

Parunguis Chamberlin, 1941d: 788 (characters). **Type Species:** *Parunguis kernensis* Chamberlin, 1941, fixed by original designation and monotypy.
— Crabill, 1961a: 73 (catalogue).
— Jeekel, 2005: 97 (list).

♂♀*P. kernensis* Chamberlin, 1941
— Chamberlin, 1941d: 788-789 (*Parunguis*; USA: California).
— Crabill, 1961a: 73 (*Parunguis*; catalogue).
— Jeekel, 2005: 97 (*Parunguis*; list).

Distribution: USA: California.

HOLOTYPE: NMNH, California: Kern Co., 4 mi. E of Glenville, 19 March 1941, by S. & D. Mulaik.

PARATYPE: NMNH.

Genus ***Pectiniunguis*** Bollman, 1890
[3 species; Southeastern and Western USA]

Pectiniunguis Bollman, 1890: 212-213 [= 1893: 113] (characters; notes).
TYPE SPECIES: *Pectiniunguis americanus* Bollman, 1890, fixed by original designation.
— Cook & Collins, 1891: 388-389 (characters; notes).
— Bollman, 1893: 167, see footnote (key to subgenera; subgenera included: *Nannopus, Pectiniunguis*).
— Cook, 1895b: 71 (distribution, list).
— Chamberlin, 1904: 653-654 (key to known species).
— Brölemann, 1909: 334-335 (characters).
— Attems, 1929: 80-81 (characters; distribution; key to species).
— Jeekel, 2005: 97 (list).
Adenoschendyla Brölemann & Ribaut, 1911: 192 (characters; key). **TYPE SPECIES:** *Adenoschendyla geayi* Brölemann & Ribaut, 1911, fixed by subsequent monotypy.
— Chamberlin, 1923b: 391 (cited as synonym of *Pectiniunguis*).
— Crabill, 1959c: 324-326 (notes).
— Crabill, 1961a: 73 (catalogue; cited as synonym of *Pectiniunguis*).
Litoschendyla Chamberlin, 1923b: 391 (name change). **TYPE SPECIES:** *Litoschendyla insulanus* Brölemann & Ribaut, 1911, fixed by original designation.
— Crabill, 1959c: 324-326 (junior subjective synonym of *Pectiniunguis*).
— Crabill, 1961a: 73 (synonym of *Pectiniunguis*).

♂♀*P. americanus* Bollman, 1890
— Bollman, 1890: 212-213 (*Pectiniunguis*; characters; Mexico). [**Note:** Bollman originally described this species from the southern portion of Baja California, Mexico. Cook (1899) mentions this species from Florida but according to Crabill (1959c) it is a misidentification and in the same paper Crabill describes it as the new species *P. halirrhytus*].
— Cook & Collins, 1891: 389-391, plate 33 figs. 1-5, plate 34; Figs. 6-8 (*Pectiniunguis*; characters; notes).
— Bollman, 1893: 113 (*Pectiniunguis*; characters; notes; Mexico).
— Cook, 1895b: 71 (*Pectiniunguis*; cited as type species of *Pectiniunguis*).
— Attems, 1903b: 194 (*Pectiniunguis*; list).
— Chamberlin, 1904: 653 (*Pectiniunguis*; key to species; note)
— Brölemann & Ribaut, 1912: 101 (*Pectiniunguis*; characters; United States).
— Chamberlin, 1912a: 668 (*Pectiniunguis*; characters; range).
— Attems, 1929: 81 (*Pectiniunguis*; characters; distribution).

— Crabill, 1959c: 324, 325, 326. (*Pectiniunguis*; discussion of misidentification).
— Crabill, 1961a: 74 (*Pectiniunguis*; catalogue).
— Foddai, Pereira & Minelli, 2000: 126-127 (*Pectiniunguis*; catalogue).
— Jeekel, 2005: 97 (*Pectiniunguis*; list).

americanus Cook (not Bollman), 1899: 305; pl. 4; Fig. 3a (*Pectiniunguis*; habitat; notes; USA: Florida). [**Note:** misidentification, see note under Bollman, 1890]

Distribution: USA: Not found in the USA (see note).

HOLOTYPE: NMNH, No. 598, ♀, Mexico: Baja Sur: Gulf of California, Pichilinque Bay, near La Paz.

♂*P. catalinensis* Chamberlin, 1941
— Chamberlin, 1941d: 787-788 (*Pectiniunguis*; characters; habitat; USA: California).
— Crabill, 1961a: 74 (*Pectiniunguis*; catalogue).

Distribution: USA: California.

HOLOTYPE: ?NMNH, ♂, California: Catalina Island, 11 November 1927, under stone near Black Jack.

♂♀*P. halirrhytus* Crabill, 1959
— Crabill, 1959c: 326-330; Figs. 1-14 (*Pectiniunguis*; characters; USA: Florida).
— Crabill, 1961a: 74 (*Pectiniunguis*; catalogue).
— Foddai, Pereira & Minelli, 2000: 129 (*Pectiniunguis*; catalogue).

americanus Cook (not Bollman), 1899: 305, pl. 4; Fig. 3a (*Pectiniunguis*; habitat; notes; USA: Florida). [**Note:** Misidentification: Cook, 1899: 305 referred these specimens from Florida to *americanus*, however, Crabill, 1959 discovered that they were a new species]

Distribution: USA: Florida.

HOLOTYPE: NMNH No. 2548, ♀, Florida: Monroe Co., Big Pine Key, 30 December 1957, in beach seaweed, by H.V. Weems, Jr.

ALLOTYPE: NMNH, ♂, Florida: Monroe Co., Big Pine Key, 30 December 1957, in beach seaweed, by H.V. Weems, Jr.

PARATYPES: NMNH, 1♂, 1♀, Florida: Monroe Co., Big Pine Key, 30 December 1957, in beach seaweed, by H. V. Weems, Jr.; 1♀, Florida: Monroe Co., Flamingo, 3 May 1958, R.S. Swanson & C.F. Dowling.

Genus *Schendyla* Bergsöe and Meinert, 1866
[1 species (Introduced); Eastern Canada, Northwestern and Eastern USA]

Schendyla Bergsöe and Meinert, 1866 TYPE SPECIES: *Geophilus nemorensis* C. L. Koch, 1837, fixed by monotypy.
— Cook & Collins, 1891: 386 (characters).
— Cook, 1895b: 70 (distribution; list).
— Brölemann & Ribaut, 1912: 145 (key to subgenera; subgenera included: *Echinoschendyla, Schendyla*).
— Bailey, 1928b: 44 (characters; note).
— Crabill, 1952b: 61-62 (characters, notes).
— Crabill, 1953c: 94 (characters; notes).
— Crabill, 1961a: 75 (catalogue).
— Jeekel, 2005: 101 (list).

♂♀*S. nemorensis* (C. L. Koch, 1837)
— C. L. Koch, 1837: unnumbered page (142. 4.) (*Geophilus*; characters).
— C. L. Koch, 1863: 26-27; Tab. 72, Fig. 148 (*Linotaenia*; characters; "Baiern").
— Cook & Collins, 1891: 386-388 (*Schendyla*; characters; habitat; notes; USA: New York).
— Cook, 1895b: 70 (list; distribution).
— Attems, 1903b: 188 (*Schendyla*; distribution; USA: New York).
— Chamberlin, 1909b: 175-176 (*Schendyla*; habitat; notes; USA: Utah).
— Brölemann & Ribaut, 1912: 154-155, pl. 9; Figs. 159-166 (*Schendyla*, subgenus *Schendyla*; characters; notes; United States).
— Bailey, 1928a: 1083 (*Schendyla*; list).
— Bailey, 1928b: 19, 44 (*Schendyla*; characters; list, USA: New York).
— Williams & Hefner, 1928: 135 (*Schendyla*; characters; USA: Ohio).
— Attems, 1929: 60 (*Schendyla*; subgenus *Schendyla*; characters; distribution).
— Chamberlin, 1943b: 98 (*Schendyla*; note; USA: Utah).
— Chamberlin, 1944d: 186 (*Schendyla*; USA: Illinois).
— Chamberlin, 1951b: 35 (*Schendyla*; note; USA: Utah).

— Crabill, 1952b: 62-63 (*Schendyla*; characters; habitat; notes; USA: Connecticut, Illinois, Massachusetts, Michigan, New Hampshire, New York, Ohio).

— Johnson, 1952: 73-74; Tab. 4 (*Schendyla*; autecology; characters; distribution; USA: Michigan).

— Crabill, 1953c: 94-95 (*Schendyla*; characters; habitat; notes; USA: Connecticut, Illinois, Massachusetts, Michigan, New Hampshire, New York, Ohio).

— Palmén, 1954: 133; Fig. 1(2,3,6) (*Schendyla*; notes; CANADA: Newfoundland).

— Crabill, 1958b: 94 (*Schendyla*; habitat; notes; USA: Wisconsin).

— Crabill, 1961a: 75 (*Schendyla*; catalogue; range).

— Chamberlin, 1961: 100 (*Schendyla*; USA: Utah).

— Chamberlin, 1962: 134 (*Schendyla*; USA: Oregon).

— Eason, 1964: 68, 70-73; Figs. 56-69 (*Schendyla*; characters; distribution; habitat).

— Kevan, 1979: 298 (*Schendyla*; CANADA: "widely distributed").

— Kevan, 1983: 2945 (*Schendyla*; list; notes; CANADA: Newfoundland).

— Summers, 1979: 697, 699 (*Schendyla*; checklist; key).

— Summers & Uetz, 1979: 347; Tab. 1 (*Schendyla*; USA: Illinois).

— Summers et al., 1980: 249; Fig. 15 (*Schendyla*; checklist; habitat (Table 1, p. 242-243); USA: Illinois).

— Summers et al., 1981: 60 (*Schendyla*; checklist; USA: Illinois).

— Snider, 1991: 186 (*Schendyla*; list).

— Behan-Pelletier, 1993: 25 (*Schendyla*; list).

— Watermolen, 1997: 2 (*Schendyla*; checklist, notes; USA: Wisconsin).

— Jeekel, 2005: 101 (*Geophilus*; list).

bistriatus C. L. Koch, 1847: 183-184 (*Poabius*; characters; Type Locality: "Pola").

— Latzel, 1880: 198 (*Poabius*; synonym of *nemorensis*).

Distribution: CANADA: Newfoundland; USA: Connecticut, Illinois, Massachusetts, Michigan, New Hampshire, New York, Ohio, Oregon, Utah, Wisconsin.

HOLOTYPE: ?Unknown.

Genus *Serrunguis* Chamberlin, 1941
[1 species; Western USA]

Serrunguis Chamberlin, 1941d: 789 (characters). **Type Species:** *Serrunguis paroicus* Chamberlin, 1941, fixed by original designation and monotypy.
— Crabill, 1961a: 75 (catalogue).
— Jeekel, 2005: 104 (list).

♂*S. paroicus* Chamberlin, 1941
— Chamberlin, 1941d: 789 (*Serrunguis*; characters; USA: California).
— Crabill, 1961a: 75 (*Serrunguis*; catalogue).
— Jeekel, 2005: 104 (*Serrunguis*; list).

Distribution: USA: California.

HOLOTYPE: ?NMNH, ♂, California: Mountain Springs, 8 January 1941, by S. & D. Mulaik.

PARATYPE: NMNH.

Family **Tampiyidae** Chamberlin, 1912

Genus *Eremorus* Chamberlin, 1963
[1 species; Western USA]

Eremorus Chamberlin, 1963b: 33-34 (characters). **Type Species:** *Eremorus becki* Chamberlin, 1963, fixed by original designation and monotypy.
— Shelley, 2006a: 10 (list).

E. becki Chamberlin, 1963
— Chamberlin, 1963b: 34-35; Fig. 1-4 (*Eremorus*; characters; USA: Nevada). [**Note:** Two specimens known and sex of these were not mentioned in the description].
— Bonato, Zapparoli & Minelli, 2008: 348; Tab. 1 (*Eremorus*; "Main diagnostic characters differentiating *Diphyonyx* gen. n. from all other genera tentatively recognized in the subgroup of Geophilidae to which *Diphyonyx* belongs.").

Distribution: USA: Nevada.

HOLOTYPE: ?University of Utah, Nevada: Nevada Test Area, 6 March 1961.

Genus *Tampiya* Chamberlin, 1912
[1 species; Western USA]

Tampiya Chamberlin, 1912a: 655 (characters). TYPE SPECIES: *Tampiya pylorus* Chamberlin, fixed by original designation and monotypy.
— Attems, 1929: 365 (characters; distribution; notes).

T. pylorus Chamberlin, 1912
— Chamberlin, 1912a: 655; Fig. 220 c-e (*Tampiya*; characters; USA: California).
— Attems, 1929: 365 (*Tampiya*; characters; distribution).
— Bonato, Zapparoli & Minelli, 2008: 349; Tab. 1 (*Tampiya*; "Main diagnostic characters differentiating *Diphyonyx* gen. n. from all other genera tentatively recognized in the subgroup of Geophilidae to which *Diphyonyx* belongs.").

Distribution: USA: California.

HOLOTYPE: NMNH, California: Sausalito.

Species Inquirendae

[**Note:** The following names are either of uncertain status; are suspected synonyms, have unknown or vague localities, or may be from other countries].

LITHOBIOMORPHA

Gosibius saccharogeus Chamberlin, 1941 [**Note:** Weaver (1982) suspected that *G. louisianus* Chamberlin, 1942 is a junior synonym to *saccharogeus*].

Gosibius arizonensis Chamberlin, 1917 [**Note:** Weaver (1982) suspected that *G. mulaiki* Chamberlin, 1938 is a junior synonym to *arizonensis*].

Lithobius cantabrigensis Meinert, 1885 (?= *Paitobius zinus* Chamberlin, 1911)

Lithobius planus Newport, 1845 [**Note:** Vague Locality].

Lithobius punctulatus C.L. Koch, 1847 [**Note:** According to Eason (1972b: 111) this species " . . . should be rejected as a *nomen dubium* . . ." and the type locality is Triest, Italy. I have included It here because I may have come across this name in the literature and it may have been reported as being found in the USA as *Eulithobius punctulatus*. See Eason (1972b) for further information about this species].

Lithobius spinipes Say, 1821 [**Note:** ?=*Neolithobius mordax* Koch, 1862; or ?= *Neolithobius transmarnius* Koch, 1862].

Neolithobius mordax (L. Koch, 1862) (?= *Neolithobius transmarinus* Koch, 1862)

Planobius aletes Chamberlin & Wang, 1952 [**Note:** Vague Locality].

SCOLOPENDROMORPHA

Cryptomera lunularis Rafinesque, 1820
— Rafinesque, 1820: 7 (*Cryptomera*; characters; habitat; USA: "near Baltimore and Philadelphia"].
— Hoffman & Crabill, 1953: 75-76 (*Cryptomera*; *species inquirenda*; taxonomic discussion).

Mycotheres (*Exocera*) *oligopoda* Rafinesque, 1820
— Rafinesque, 1820: 8 (*Mycotheres*; characters; USA: New York).
— Hoffman & Crabill, 1953: 77 (*Mycotheres*; *species inquirenda*; taxonomic discussion).
— Jeekel, 2005: 16 (list).

GEOPHILOMORPHA

Arctogeophilus fulvus (Wood, 1862) [**Note:** Attems (1929) listed this species as *incertae sedis*].

Arctogeophilus glaber (Bollman, 1887) [**Note:** Attems (1929) listed this species as *incertae sedis*].

Escaryus albus Cook, 1904 [**Note:** Pereira & Hoffman (1993) considered this species as a species inquirenda].

Geophilus attenuatus Say, 1821 [**Note:** Attems (1929) listed this species as *incertae sedis*].

G. brunneus McNeill, 1887 [**Note:** Attems (1929) considered this species incertae sedis & Crabill (1954b) cited it as a species inquirenda].

Geophilus californiensis Bollman, 1887 [**Note:** Attems (1929) listed this species as *incertae sedis*].

Geophilus setiger Bollman, 1887 [?=*Geophilus rupestris* (Crabill, 1949)]

Geophilus whitei Newport, 1845 [**Note:** Attems (1929) listed this species as *incertae sedis*].

Strigamia chionophila Wood, 1862 [?= *Strigamia acuminata* (Leach, 1815)]

♂*Strigamia exul* (Meinert, 1886)

— Meinert, 1886a: 224-225 (*Scolioplanes*; characters; no locality).
— Cook, 1896: 241 (*Scolioplanes*; characters; notes).
— Attems, 1929: 227 (*Scolioplanes*; characters; ?North America).
— Crabill, 1960b: 190 (*Scolioplanes*; synonym of *Strigamia*).

Distribution: Unknown. [**Note:** Attems (1929) suspected it is North America].

HOLOTYPE: MCZ, No. 988, no locality. [**Note:** Meinert's original determined label reads "*Scolioplanes exul* n. sp. No. 170". Vial of holotype has a label from Crabill that reads "Has the 3-partite ult. pretergite of some - but not all - of the large *Tomotaenia* exx. from Pacific Coast. If this is interspecif. significant, the species may take the name *exul*. Currently, this species would be called *T. parviceps* (Wood) by R. V. Chamberlin. - R. Crabill, I.13.60"].

Strigamia gracilis H. C. Wood, 1867 [**Note:** Attems (1929:329) listed this species as *incertae sedis*].

Strigamia inermis Wood, 1867 [**Note:** Attems (1929) listed this species as *incertae sedis*].

Strigamia longicornis (Meinert, 1886)
— Meinert, 1886a: 226-227 (*Scolioplanes*; characters; notes; no locality).
— Cook, 1896: 241 (*Scolioplanes*; characters; notes).
— Attems, 1929: 229 (*Scolioplanes*; listed as a species of inadequate description).
— Crabill, 1960b: 190 (*Scolioplanes*; synonym of *Strigamia*).

Distribution: Unknown. [**Note:** The author has included this species as there is a possibility that it originated from North America].

HOLOTYPE: ?MCZ.

Strigamia taeniopsis Wood, 1862
— Wood, 1862: 48 (*Strigamia*; characters; notes; USA: Georgia).
— Wood, 1865: 185 (*Strigamia*; characters; notes; USA: Georgia).
— Attems, 1929: 55 (synonym of *Nothobius californicus*).

Distribution: USA: Georgia.

HOLOTYPE: ?ANSP; USA: Georgia: collected in the mountains by Dr. J. Le Conte.

Nomina Nuda

[**Note:** The following names are not available according to the ICZN].

SCUTIGEROMORPHA

Cermatia coleoptrata var. *floridensis* Newport, 1844
— Newport, 1844b: 95 (*Cermatia*; nominal taxon *coleoptrata*; cited as variety *floridensis*; appears in list of centipedes as one of forty-seven new species without a description; USA: Florida).
— Minelli,—Chilobase (*Cermatia*; in list of unavailable names as a *nomen nudum* according to ICZN art. 12). [**Note:** Author accessed website 4/2/2009].

LITHOBIOMORPHA

Agothothus carolinus Chamberlin, ?
— Chamberlin, ? unpublished.
— Brimley, 1938: 501 (*Agathothus*; USA: North Carolina).
— Causey, 1940: 61 (*Agathothus*; USA: North Carolina).
— Wray, 1967: 156 (*Agathothus*; list).
— Crabill, 1952b: 85 (note; *nomen nudum*).
— Hoffman, 1995a: 22 (*Agathothus*; list; notes).
— Minelli,—Chilobase (*Agathotus* [misspelled]; in list of unavailable names as a *nomen nudum* according to ICZN art. 13.1; cited as authored by Wray (1967)). [**Note:** Author accessed website 4/2/2009].

Neolithobius audacior Chamberlin, 1942
— Chamberlin, 1942d: 188 (*Neolithobius*; USA: Arkansas). [**Note:** No known description for this species. Locality data as follows: Arkansas: Jackson County, 1.5 miles southwest of Oliphant, April 10, 1936, four specimens].

Paitobius diversus Chamberlin, ?

— Chamberlin, ? unpublished.
— Causey, 1940: 75 (*Paitobius*; habitat; USA: North Carolina). [**Note:** Causey (1940) lists this species as being described by Chamberlin but doesn't give a year of publication and I have yet to find it published anywhere. Causey notes that it was "Collected from a pine log in October in the Duke Forest (Causey)."
— Wray, 1967: 155 (*Paitobius*; list).
— Minelli,—Chilobase (*Paitobius*; in list of unavailable names as a *nomen nudum* according to ICZN art. 13.1; cited as authored by Wray (1967)). [**Note:** Author accessed website 4/2/2009].

Pokabius yukus (Chamberlin, 1912)
— Chamberlin, 1912d: 153. (*Poabius*; cited as new species). [**Note:** No description, but mentioned as a new species in note under the genus name *Poabius* (= *Pokabius*)].

Sigibius reductus Chamberlin, ? or Starling, 1944
— Chamberlin, ? unpublished. [**Note:** According to Chilobase this was published by Starling, 1944, however, Wray (1967) states it was described by Chamberlin].
— Wray, 1967: 155 (*Sigibius*; list; USA: North Carolina). [**Note:** Wray (1967) lists this species as being described by Chamberlin but does not give a publication date. Locality as "Duke Forest, Starling."].
— Minelli,—Chilobase (*Sigibius*; in list of unavailable names as a *nomen nudum* according to ICZN art. 13.1; cited as authored by Starling, 1944). [**Note:** Author accessed website 4/2/2009].

Sonibius yanikans Chamberlin, 1912
— Chamberlin, 1912e: 177 (no description). [**Note:** This species is mentioned by Chamberlin as a new species while listing other species that belong to the new genus *Sonibius* but a description has not been found].

Genus *Startobius* Starling, 1944

Startobius Starling, 1944: 300 (no description). [**Note:** Reference not obtained].
— Minelli,—Chilobase (in list of unavailable names as a *nomen nudum* according to ICZN art. 13.1). [**Note:** Author accessed website 4/2/2009].

Startobius gracilis Starling, 1944

— Starling, 1944: 300 (*Startobius*; no description or locality). [**Note:** Reference not obtained].

— Minelli,—Chilobase (*Startobius*; in list of unavailable names as a *nomen nudum* according to ICZN art. 13.1). [**Note:** Author accessed website 4/2/2009].

Tidabius lilacius Starling, 1944

— Starling, 1944: 301 (*Tidabius*; no description or locality). [**Note:** Reference not obtained].

— Minelli,—Chilobase (*Tidabius*; in list of unavailable names as a *nomen nudum* according to ICZN art. 13.1). [**Note:** Author accessed website 4/2/2009].

GEOPHILOMORPHA

?*G.* (*G.*) *claremontus* Chamberlin, 1917

— Chamberlin, 1917: 208. (*Gosibius*; subgenus *Gosibius*; listed in key to species). [**Note:** Chamberlin included this new species in his key but never actually described the species, therefore it should be considered a *nomen nudum*. It is assumed this species was collected in Claremont, California. In any case, it may be conspecific with *G. monicus*].

Distribution: USA: ?California.

SYNTYPES: MCZ, No. 583, 4 specimens, "Cal. Mts., 1891". [**Note:** Original locality label reads "*Gosibius*, Cal. Mts., 1891" and original determined label reads "*Gosibius claremontus* sp. n., (near *monicus*)"].

List of Reports to Generic Level

LITHOBIOMORPHA

Helembius sp.—Crabill, 1950c: 202 (*Helembius*; possible report from USA: South Carolina). [**Note:** Crabill hesitatingly reports of a *Helembius* sp. from South Carolina and does not refer the specimens to *nannus* due to a U-shaped diastema instead of a V-shaped].

Lamyctes sp.—Chamberlin, 1930c: 68 (*Lamyctes*; USA: Hawaii). [**Note:** Chamberlin reports "A specimen of uncertain species was taken in 1928 in soil about plants in baggage from Japan".

Lamyctes sp.—Chamberlin & Wang, 1952b: 181 (*Lamyctinus*; USA: Hawaii). [**Note:** Taken in a cargo from Japan at quarantine at Honolulu, 18 March 1938. Originally described as a *Lamyctinus* sp. but transferred to *Lamyctes* based on Edgecombe & Giribet, 2003].

Lithobius sp.—Chamberlin, 1930c: 68 (*Lithobius*; USA: Hawaii). [**Note:** "Specimens too young for identification were taken in 1927 and 1928 in soil about plants from China"].

Lithobius sp.—Palmén 1954: 143 (Palmén reports a *Lithobius* sp. from Newfoundland, Canada and suggests that it may belong to the genus *Garibius*. The only species of *Garibius* that is known to exists so far north is *Garibius opicolens* (Chamberlin) and this may be worth investigating).

♀*Lithobius* (*Lithobius*) sp.—Zapparoli & Shelley, 2000: 40–42; Figs. 1–4; Tab. 1 (*Lithobius*; subgenus *Lithobius*; characters; plectrotaxy; notes; USA: Hawaii).

Nadabius sp.—Carter & Brown, 1973: 1066, Tab. 1 (data from study on soil arthropods; CANADA: New Brunswick).

Nampabius sp.—Holsinger & Culver, 1988: 57 (*Nampabius*; cave species; USA: Virginia). [**Note:** Holsinger & Culver report a *Nampabius* sp. from Virginia: Montgomery Co.: Erharts Cave but do not indicate why they could not identify it to species; maybe due to immaturity, missing legs or possibly a new species].

Neolithobius sp.—Chamberlin, 1920d: 44 (notes; possible occurrence of this genus in Montana). [**Note:** Chamberlin reports two specimens of a *Neolithobius* sp. from St. Xavier, Montana, on the 31 of May, 1917, that was collected from the crop of an eared grebe, *Colymbus nigricollis californicus* (Brehm). This is the only report of this genus occurring in Montana so it is worth mentioning. Further collecting efforts should be made in this state to establish the actual species occurring here because it is a considerable distance from the known biogeographic distribution of the genus].

Neolithobius sp.—Reddell, 1965: 166 (USA: Texas).
— Reddell, 1994: ? (*Neolithobius* sp.; notes; USA: Texas).

♀*Oabius* sp.—Chamberlin, 1949a: 14 (*Oabius*; habitat, notes; CANADA: Yukon Territory).

Sozibius sp.—Crabill, 1958b: 98 [**Note:** Crabill reports of finding a *Sozibius* sp. in Wisconsin but was unable to identify species due to the male specimen being in poor condition].

Sozibius sp.—Branson & Batch, 1967: 88 [**Note:** Branson & Batch report of a *Sozibius* sp. from Kentucky but were not sure of it's status].

Tidabius sp.—Lee, 1980: 5; Tab. 1 (*Tidabius*; habitat; USA: Ohio). [**Note:** Lee reports of a *Tidabius* sp. from Ohio but apparently was not able to identify the 3 specimens to species. It is probable that these specimens were referable to *tivius* as it is the only species known from Ohio at this time].

Zygethobius sp.—Bailey, 1928b: 19, 32-33 (list; notes). [**Note:** Bailey reports of a *Zygethobius* sp. from New York at the McLean Bogs Reservation, 24 April 1924, by C. R. Crosby.]

SCOLOPENDROMORPHA

Newportia sp.—Chamberlin, 1930c: 66 (*Newportia*; USA: Hawaii). [**Note:** The genus *Newportia* is not indigenous to Hawaii. "A specimen of a species of this genus was taken at Hawaii in 1928 in soil about plants in the cargo of a vessel from Panama. The anal legs are missing, making specific determination impossible." Recently, a contribution to the taxonomic revision of the genus *Newportia* has been completed by Schileyko & Minelli, 1998].

GEOPHILOMORPHA

Escaryus sp.—Carter & Brown, 1973: 1066, Tab. 1, Fig. 1 (data from study on soil arthropods; CANADA: New Brunswick).

Himantarium sp.—Chamberlin & Wang, 1952b: 177 (*Himantarium*; USA: Pennsylvania). [**Note:** This specimen was intercepted in a cargo shipment from Japan in Philadelphia, 30 April 1926. This genus is not normally found in North America and has probably not established itself although it should be kept in mind that this genus can find its way here].

Mecistocephalus sp.—Morse, 1902: 187 [**Note:** Morse indicates a *Mecistocephalus* sp. from Vinton, Ohio and may have been *Geophilus strigosus* (McNeill) which may be a junior synonym of *Pachymerium ferrugineum*].

Catalog of Extinct Centipedes

Note: Scudder (1890a) described seven species of what he thought were centipedes but Mundel, (1979) suggests that only one of these is a true centipede. Some of Scudder's drawings apparently represent geophilomorph centipedes but these have not been included until further study verifies they are true centipedes.

Order **Scutigeromorpha** Leach, 1815
Family **Crussolidae** Shear, Jeram & Selden, 1998

Genus *Crussolum* Shear, Jeram & Selden, 1998
[1 species (extinct); Northeastern USA]

Crussolum Shear, Jeram & Selden, 1998: 10 (characters). Type Species: Crussolum crusseratum Shear, Jeram & Selden, 1998, fixed by original designation.
— Giribet & Edgecombe, 2006: 531 (notes).

C. crusseratum Shear, Jeram & Selden, 1998
— Shear, Jeram & Selden, 1998: 10-13, tab, 1; Figs. 8-19, 26, 28 (*Crussolum*; characters; notes; USA: New York).

Distribution: USA: New York.

HOLOTYPE: AMNH, 334/1b/AR78, New York, Schoharie County, near Gilboa, in the upper part of the Panther Mountain Formation on the west flank of Brown Mountain.

Family **Latzeliidae** (Scudder, 1890)

Genus *Latzelia* Scudder, 1890
[1 species (extinct); Central Northern USA]

Latzelia Scudder, 1890: 418-419 (characters; note). **TYPE SPECIES:** *Latzelia primordialis* Scudder, 1890, fixed by monotypy.
— Hoffman, 1969: R605 (cited as *incertae sedis*).

L. primordialis Scudder, 1890
— Scudder, 1890a: 419, pl. 38; fig. 3 (*Latzelia*; characters; USA: Illinois).
— Scudder, 1890b: (*Latzelia*;). [**Note:** reference not obtained].
— Hoffman, 1969: R605 (*Latzelia*; cited as *incertae sedis*).
— Mundel, 1979: 361, 366-374; figs. 3-11 (*Latzelia*; characters; notes; USA: Illinois).
— Ross & Briggs, 1993: 360 (*Latzelia*; list).
— Shear et al., 1998: 4, 5, 6; Figs. 4-7 (*Latzelia*; leg illustration; photograph of CMNH 008672; notes). [**Note:** The leg illustrations are of uncertain identity but are thought to belong to *primordialis*).

Distribution: USA: Illinois.

HOLOTYPE: NMNH, No. 38003, Illinois, Mazon Creek.

OTHER MATERIAL: Eleven FMNH specimens (PE 29003, PE 25597, PE 29004, PE 29005, PE 29006, PE 25596, PE 32253, PE 22925, PE 28604, PE 24537, PE 29006) from Illinois: Will and Kankakee Counties, Peabody Coal Company, Pit 11. Francis Creek Shale, Westphilian D., Desmoinesian, Middle Pennsylvanian.

Order **Devonobiomorpha** Shear & Bonamo, 1988
[Extinct Order]

Family **Devonobiidae** Shear & Bonamo, 1988

Genus *Devonobius* Shear & Bonamo, 1988
[1 species (extinct); Northeastern USA]

Devonobius Shear & Bonamo, 1988: 20 (characters are of the order). **TYPE SPECIES:** *Devonobius delta* Shear & Bonamo, 1988, fixed by original designation.
— Edgecombe & Giribet, 2004: 108, 119, 124-125; Fig. 14 ; Tab. 1 (cladogram; systematic position discussion).
— Giribet & Edgecombe, 2006: 531 (notes).

D. delta Shear & Bonamo, 1988
- — Shear & Bonamo, 1988: 20-28; Figs. 1-75; Tab. 1-4 (*Devonobius*; characters; notes; USA: New York).
- — Ross & Briggs, 1993: 360 (*Devonobius*; list).

Distribution: USA: New York.

HOLOTYPE: AMNH, slide 411-15-AR18, complete head with 15 or 16 trunk segments.

PARATYPES: AMNH, slides: 41-7-AR97, head with 11 antennal segments, maxillipeds; slide 329-AR4, maxillipedes; slide 411-2-AR1, telescoped, folded, and compressed exuvium including posterior legs.

Order **Scolopendromorpha** Newport, 1844

Genus ***Mazoscolopendra*** Mundel, 1979
[1 species (extinct); Northern Central USA]

Mazoscolopendra Mundel, 1979: 363 (characters). **TYPE SPECIES:** *Mazoscolopendra richardsoni* Mundel, 1979, fixed by original designation. [**Note:** According to Mundel the family placement is uncertain].

M. richardsoni Mundel, 1979
- — Mundel, 1979: 363-366; Figs. 1-2 (*Mazoscolopendra*; characters; notes; USA: Illinois).
- — Ross & Briggs, 1993: 361 (*Mazoscolopendra*; list).

Distribution: USA: Illinois.

HOLOTYPE: FMNH, PE 22936, Illinois: Will and Kankakee Counties, Peabody Coal Company, Pit 11. Francis Creek Shale, Westphilian D., Desmoinesian, Middle Pennsylvanian.

PARATYPES: FMNH, PE 29002, PE 28606, PE 32244.

Order **Geophilomorpha** Pocock, 1895
Family **Geophilidae** Leach, 1815

Genus *Calciphilus* Chamberlin, 1949
[1 species (extinct); Southwestern USA]

Calciphilus Chamberlin, 1949b: 118 (characters). **Type Species:** *Calciphilus abboti* Chamberlin, 1949, fixed by original designation and monotypy.
— Jeekel, 2005: 72 (list).

C. abboti Chamberlin, 1949
— Chamberlin, 1949b: 118 (*Calciphilus*; characters; USA: Arizona).
— Ross & Briggs, 1993: 360 (*Calciphilus*; list; note).
— Jeekel, 2005: 72 (*Calciphilus*; list).

Distribution: USA: Arizona.

HOLOTYPE: SDSNH, No. 109737, Arizona: Bonner Quarry, near Ashfork.
[**Note:** The holotype is preserved in a thin sectioned piece of travertine of which there are two individual pieces mounted on slides because according to Chamberlin (1949b), "The pen base as received was broken in such a way as to divide the elongate specimen into two parts." The Bonner Quarry locality has been assigned to SDSNH locality 2628].

Bibliography

Albert, A.M. 1979. Chilopoda as part of the predatory macroarthropod fauna in forests: abundance, life-cycle, biomass, and metabolism. *In*: Myriapod Biology. Ch. 22, (Camatini Marina ed.) p. 215-231.

Almond, J.E. 1985. The Silurian-Devonian Fossil Record of the Myriapoda. Philosophical Transactions of the Royal Society of London (B), 309: 227-237.

Andersson, G. 1978. An investigation of the post-embryonic development of the Lithobiidae—some introductory aspects. Abhandlungen und Verhandlungen des Naturwissenschaftlichen Vereins in Hamburg (NF) 21/22: 63-71.

Andersson, G. 1979. On the use of larval characters in the classification of lithobiomorph centipedes (Chilopoda: Lithobiomorpha). *In*: Myriapod Biology. Ch. 8, (Camatini Marina ed.) p. 73-81.

Anonymous. 1920. Centipedes and Millipedes from near Claremont. Pomona College Journal of Entomology and Zoology 12: 35.

Attems, C. 1901. Neue, durch den Schiffsverkehr in Hamburg eingeschleppte. Mitteilungen aus dem Naturhistorischen Museum in Hamburg. 18: 111-116.

Attems, C. 1903a. Beiträge zur Myriopodenkunde. Zoologische Jahrbucher (Syst.) 18: 63-154.

Attems, C. 1903b. Synopsis der Geophiliden., Zoologische Jahrbucher (Syst.) 18: 155-302.

Attems, C. 1909a. Die Myriapoden der Vega-Expedition. Arkiv För Zoologi. Stockholm, Band 5 No. 3: 1-84.

Attems, C. 1909b. Myriapoda. Forschungsreise im Westlichen und Zentralen Südafrica (ed. by L. Schultze). Denkschriften der Medizinisch-naturwissenschaftlichen Gesellschaft zu Jena. 14: 1-52.

Attems, C. 1914. Die indo-australischen Myriapoden. Archiv für Naturgeschichte, Abteilung A. 80, 4: 1-398.

Attems, C. 1926. Chilopoda. *In*: Handbuch der Zoologie (Eds. W. Kukenthal & T. Krumbach), 4: 239-402, Berlin: Walter de Gruyter.

Attems, C. 1927. Zoologischer Anzeiger. 72: 303.

Attems, C. 1928. The myriapoda of South Africa. Annals of the South African Museum 26: 1-431.

Attems, C. 1929. Myriapoda 1. Geophilomorpha. Das Tierreich 52: 1-388.

Attems, C. 1930. Myriapoda 2. Scolopendromorpha. Das Tierreich 54: 1-308.

Attems, C. 1934. Einige neue Geophiliden und Lithobiiden des Hamburger Museums. Zoolgischer Anzeiger. 107: p. 313.

Attems, C. 1938a. Die von Dr. C. Dawydoff in franzosisch Indochina gesammelten Myriapoden. Mémoires du Muséum National d'Histoire Naturelle., Paris. N.S. 6:187-353.

Attems, C. 1938b. Myriapoden von Hawai. Proceedings of the Zoological Society, London, (B) 108: 365-387.

Attems, C. 1947. Neue Geophilomorpha des Wiener Museums. Annalen des Naturhistorischen Museums in Wien. 55: 50-149.

Auerbach, S.I. 1946. Ecological relationships of deciduous forest chilopods. University of Illinois, Ph.D. thesis. 110 pp. Info on Centipedes from Illinois in a park study.

Auerbach, S.I. 1949. A preliminary ecological study on certain deciduous forest centipedes. American Midland Naturalist 42(1): 220-227.

Auerbach, S.I. 1950. A new species of centipede from a tree hole in California. Natural History Miscellanea 76: 1-5.

Auerbach, S.I. 1951a. The centipedes of the Chicago Area with special reference to their ecology. Ecological Monographs, 21(1): 97-124.

Auerbach, S.I. 1951b. A Key to the Centipedes of the Chicago Area. Bulletin of the Chicago Academy of Sciences, 9(6): 109-114.

Auerbach, S.I. 1952a. Centipedes in the diet of salamanders. Natural History Miscellanea 103: 1-2.

Auerbach, S.I. 1952b. Taxonomic notes on an exotic centipede taken from Illinois. Annals of the Entomological Society of America. 45: 413-414.

Auerbach, S.I. 1954. A new centipede from Illinois (Chilopoda; Geophilomorpha: Geophilidae). Natural History Miscellanea 136: 1-5.

Back, E.A. 1939. Centipedes and Millipedes in the House. United States Department of Agriculture, Leaflet No. 192: 1-6.

Bailey, J.W. 1928a. Class Chilopoda. In: A list of the insects of New York with a list of the spiders and certain other allied groups. Mem. Cornell Univ. Agric. Exp. Stn. (Ithaca), 101: 1081-1083.

Bailey, J.W. 1928b. The Chilopoda of New York State with Notes on the Diplopoda. New York State Museum Bulletin 276: 3-50.

Baker, A.C. 1920. The house centipede in Canada. Canadian Entomologist 52: 93.

Barber, A.D. and E.H. Eason, 1986. A redescription of *Lithobius peregrinus* Latzel, a centipede new to Britain (Chilopoda: Lithobiomorpha). Journal of Natural History 20: 431-437.

Barker, G.M. 2004. Millipedes (Diplopoda) and Centipedes (Chilopoda) (Myriapoda) as Predators of Terrestrial Gastropods. Ch. 7 *In*: Natural Enemies of Terrestrial Mollusks. CABI Publishing, Biddles Ltd., England.

Barr, T. C., and J.R. Reddell 1967. The arthropod cave fauna of the Carlsbad Caverns Region, New Mexico. Southwestern Naturalist 12: 253-274.

Behan-Pelletier, V. M. 1993. Diversity of soil arthropods in Canada: systematic and ecological problems. Memoirs of the Entomological Society of Canada 165: 11-50.

Bergsöe, J. and F. Meinert 1866. Danmarks Geophiler. Naturhistorisk Tidsskrift 4(3): 81-108.

Böcher, J. and H. Enghoff 1984. A Centipede in Greenland: *Lamyctes fulvicornis* Meinert, 1868 (Chilopoda: Lithobiomorpha, Henicopidae). Entomologiske Meddeleser 52: 49-50.

Bollman, C.H. 1887a. Preliminary Descriptions of Ten New North American Myriapods. American Naturalist 21: 81-82.

Bollman, C.H. 1887b. Notes on the North American Lithobiidae and Scutigeridae. Proceedings of the United States National Museum 10: 254-266.

Bollman, C.H. 1887c. Descriptions of Fourteen New Species of North American Myriapods. Proceedings of the United States National Museum : 617-627.

Bollman, C.H. 1887d. Description of new genera and species of North American Myriapoda (Julidae). Entomologica Americana 1: 225-229.

Bollman, C.H. 1887e. New North American myriapods. Entomologica Americana 3: 81-83.

Bollman, C.H. 1888a. Notes Upon a Collection of Myriapoda from East Tennessee, with Description of a New Genus and Six New Species. Annals of the New York Academy of Sciences 4(3-4): 106-112.

Bollman, C.H. 1888b. A Preliminary List of the Myriapoda of Arkansas with Descriptions of New Species. Entomologica Americana 4 (1): 1-8.

Bollman, C.H. 1888c. Notes on a Collection of Myriapoda from Mossy Creek, Tenn., with a Description of a New Species. Proceedings of the United States National Museum p. 339-342.

Bollman, C.H. 1888d. Notes on Some Myriapods Belonging to the U. S. National Museum. Proceedings of the United States National Museum: 343-350.

Bollman, C.H. 1888e. Catalogue of the Myriapods of Indiana. Proceedings of the United States National Museum : 403-410.

Bollman, C.H. 1890. Myriapoda. Proceedings of the United States National Museum 12: 211-216.

Bollman, C.H. 1893. The Myriapoda of North America, Bulletin of the United States National Museum, No. 46: 1-210. (Published posthumously, this work contains all of the published and unpublished works of Bollman).

Bonato, L., D. Foddai, A. Minelli and R. Shelley, 2004. The Centipede Order Geophilomorpha in the Hawaiian Islands (Chilopoda). Bishop Museum Occasional Papers 78: 13-32.

Bonato, L. and A. Minelli, 2002. Parental care in *Dicellophilus carniolensis* (C. L. Koch, 1847): New behavioral evidence with implications for the higher phylogeny of centipedes (Chilopoda). Zoologische Anzeiger 241: 193-198.

Bonato, L. and A. Minelli, 2004. The centipede genus *Mecistocephalus* Newport, 1843 in the Indian Peninsula (Chilopoda: Geophilomorpha: Mecistocephalidae). Tropical Zoology 17: 15-63.

Bonato, L. and A. Minelli, 2007. *Mecistocephalus* Newport, 1843 and *Pachymerium* Koch, 1847 (Chilopoda): proposed conservation of current usage by designation of *Mecistocephalus punctifrons* Newport, 1843 as the type species of *Mecistocephalus* Newport, 1843. Bulletin of Zoological Nomenclature, 64(3): 166-173.

Bonato, L. and A. Minelli, 2008. *Stenotaenia* Koch, 1847: a hitherto unrecognized lineage of western Palaearctic centipedes with unusual diversity in body size and segment number (Chilopoda: Geophilidae). Zoological Journal of the Linnean Society 153: 253-286.

Bonato, L., L.A. Pereira and A. Minelli, 2007. Taxonomic and Nomenclatural notes on the centipede genera *Chomatobius*, *Ityphilus*, *Hapleurytion*, *Plateurytion*, and *Steneurytion* (Chilopoda: Geophilomorpha). Zootaxa 1485: 1-12.

Bonato, L., M. Zapparoli and A. Minelli, 2008. Morphology, taxonomy and distribution of *Diphyonyx* gen. n., a lineage of geophilid centipedes with unusually shaped claws (Chilopoda: Geophilidae). European Journal of Entomology 105: 343-354.

Borucki, H. 1996. Evolution und phylogenetisches System der Chilopoda (Mandibulata, Tracheata), Verhandlungen des Naturwissenschaftlichen Vereins in Hamburg 35: 95-226.

Bouchard, N.C., G.M. Chan and R.S. Hoffman, 2004. Vietnamese centipede envenomation. Veterinary and Human Toxicology 46(6): 312-313.

Branson, B.A., and D.L. Batch, 1967. Valley centipedes (Chilopoda; Symphyla) from northern Kentucky. Transactions of the Kentucky Academy of Science 28: 77-89.

Brimley, C.S. 1938. Insects of North Carolina. North Carolina Department of Agriculture, Division of Entomology. Raleigh, 560 pp.

Brodie, W. and J.E. White 1883. Myriapoda. p. 67. *In*: Check List of Insects of the Dominion of Canada. Natural History Society of Toronto, C. Blackett Robinson Co., Toronto.

Brölemann, H.W. 1889. Contributions á la faune myriopodologique méditerranéenne. Trois espéces nouvelles. Annales de la Societe Linnéenne de Lyon 35: 271-282.

Brölemann, H.W. 1896. Liste de Myriapodes des États-Unis, et principalemente de la Carolina du Nord, faisant partie des collections de M. Eugène Simon. Annales de la Société Entomologique de France 65: 43-70.

Brölemann, H.W. 1900. Myriapodes d'Amerique. Mémoires de la Société Zoologique de France 8: 89-131.

Brölemann, H.W. 1904. Catalogue des Scolopendrides des collections du Muséum d'Histoire Naturelle de Paris. Bulletin du Muséum National d'Histoire Naturelle 5-6: 243-250, 316-324.

Brölemann, H.W. 1909. A Propos d'un Systéme des Géophilomophes. *Arch. Zool. Exper.* Gen., Ser. 5, Tome 3: 303-340.

Brölemann, H.W. 1924. Trois géophiliens (myriapodes) nouveaux ou peu connus. Bulletin de la Société d'Histoire Naturelle de Toulouse 52: 14-20.

Brölemann, H.W. 1930. Elements d'une faune des Myriapodes de France, Chilopodes. Faune de France, 25: 1-405.

Brölemann, H.W. 1932. Chilopodes. Faune de France 25. 405 pp. [Information on introduced species]

Brölemann, H.W. and Ribaut, H. 1911. Note préliminare sur les genres de *Schendylina* (Myriapoda: Geophilomorpha). Bulletin of the Society of Entomologists France 8: 191-193.

Brölemann, H.W. and Ribaut, H. 1912. Nouvelles Archives du Muséum d'Histoire Naturelle Paris (5) 4: 53-183.

Bücherl, W. 1939. Os Quilopodos do Brasil. Memórias do Instituto de Butantan. 13: 43-362.

Bücherl, W. 1941. Catálogo dos quilopopdos da zona Neotropica. Memórias do Instituto de Butanan 15: 251-372.

Bücherl, W. 1971. Venomous chilopods or centipedes. (p. 169-196) *In*: Bücherl, W. & Buckley, E. Venomous animals and their venoms, Academic Press, 537 pp.

Bücherl, W. 1974. Die Scolopendramorpha der Neotropischen Region. Symposia of the Zoological Society of London. 32: 99-133.

Buchsbaum, R. 1948. Animals without backbones, an introduction to the invertebrates. University of Chicago Press. 405 pp.

Burke, R. and R. Mercurio 2002. Food Habits of a New York Population of Italian Wall Lizards, *Podarcis sicula* (Reptilia, Lacertidae). American Midland Naturalist 147: 368-375.

Butler, G.D. and Usinger R.L. 1963. Insects and other arthropods from Kure Island. Proceedings of the Hawaiian Entomological Society 18: 237-244.

Cameron, J.A. 1926. Regeneration in *Scutigera forceps*. The Journal of Experimental Zoology 46: 169-179.

Carlson, A.J. 1904. Contributions to the Physiology of the Ventral Nerve-Cord of Myriapoda (Centipedes and Millipedes). Journal of Experimental Zoology. 1: 269-287.

Carpenter, C.C. and J.C. Gillingham, 1984. Giant Centipede (*Scolopendra alternans*) attacks marine toad (*Bufo marinus*). Carribean Journal of Science. 20: 71-72.

Carter, N.E. and N.R. Brown 1973. Seasonal abundance of certain soil arthropods in a fenitrothion-treated red spruce stand. Canadian Entomologist 105: 1065-1073.

Causey, N.B. 1940. Ecological and systematic studies on North Carolina myriapods. Unpublished Ph. D. thesis, Zoology Dept., Duke University, Durham, NC, 181 pp.

Causey, N.B. 1942. New Lithobiid Centipedes from North Carolina. Journal of the Mitchell Society: 79-83.

Chalande, J. and H. Ribaut 1909. Étude sur la systématique de la famille des Himantariidae (Myriapodes). Archives de Zoologie Expérimentale et Generale. ser. 5, v. I: 197-275.

Chamberlin, R.V. 1901. (in 1902 vol.) List of the Myriapod family Lithobiidae of Salt Lake County, Utah, with descriptions of five new species. Proceedings of the United States National Museum, 24 (1242): 21-25.

Chamberlin, R.V. 1902a. A new genus and three new species of chilopods. Proceedings of the Academy of Natural Sciences of Philadelphia: 39-43.

Chamberlin, R.V. 1902b. *Henicops Dolichopus*, a new chilopod from Utah. Proceedings of the United States National Museum, 24 (1270): 797-800.

Chamberlin, R.V. 1902c. Utah chilopods of the Geophilidae. American Naturalist 36 (426): 473-480.

Chamberlin, R.V. 1903a. Myriopods from Beulah, New Mexico. Proceedings of the Academy of Natural Sciences of Philadelphia : 35-40.

Chamberlin, R.V. 1903b. New Lithobii from California and Oregon. Proceedings of the Academy of Natural Sciences of Philadelphia: 152-160.

Chamberlin, R.V. 1903c. *Henicops*. Entomological News 14(10): 335.

Chamberlin, R.V. 1904. New chilopods, Proceedings of the Academy of Natural Sciences of Philadelphia 56: 651-657.

Chamberlin, R.V. 1906. A new *Lithobius* from Colorado. Proceedings of the Academy of Natural Sciences of Philadelphia 58: 3-4.

Chamberlin, R.V. 1909a. A *Newportia* from Utah., The Canadian Entomologist 41: 27-30.

Chamberlin, R.V. 1909b. Some records of North American Geophilidae and Lithobiidae with descriptions of new species., Annals of the Entomological Society of America 2: 175-195.

Chamberlin, R.V. 1910. The chilopoda of California I. Pomona College Journal of Entomology : 363-374.

Chamberlin, R.V. 1911a. The Lithobiomorpha of Colorado. The Canadian Entomologist 43(2): 67-70.

Chamberlin, R.V. 1911b. The Lithobiomorpha of the southeastern States. Annals of the Entomological Society of America 4(1): 32-48.

Chamberlin, R.V. 1911c. The Lithobiomorpha of Wisconsin and neighboring States, The Canadian Entomologist 43: 98-104.

Chamberlin, R.V. 1911d. The Chilopoda of California II. Pomona College Journal of Entomology 3(2): 470-479.

Chamberlin, R.V. 1911e. Notes on Myriapods from Alaska and Washington. The Canadian Entomologist : 260-264.

Chamberlin, R.V. 1911f. Some Lithobiomorpha from the region of San Francisco Bay. The Canadian Entomologist : 377-384.

Chamberlin, R.V. 1912a. The Chilopoda of California III. Pomona College Journal of Entomology 4(1): 651-672.

Chamberlin, R.V. 1912b. Notes on Geophiloidea from Iowa and some neighbouring States. The Canadian Entomologist 44(3): 65-71.

Chamberlin, R.V. 1912c. The Geophiloidea of the southeastern States. Bulletin of the Museum of Comparative Zoology Harvard Cambridge, Massachusetts 54(13): 407-436.

Chamberlin, R.V. 1912d. New North American chilopods and diplopods. Annals of the Entomological Society of America 5: 141-172.

Chamberlin, R.V. 1912e. New genera of North American Lithobiidae. The Canadian Entomologist 44: 173-178, 204.

Chamberlin, R.V. 1912f. Change of name. (*Poabius* to *Pokabius*) The Canadian Entomologist 44: 316.

Chamberlin, R.V. 1912g. The Henicopidae of America North of Mexico. Bulletin of the Museum of Comparative Zoology Harvard 57(1): 1-36.

Chamberlin, R.V. 1913. The Lithobiid genera *Nampabius*, *Garibius*, *Tidabius*, and *Sigibius*. Bulletin of the Museum of Comparative Zoology Harvard 57(2): 39-104.

Chamberlin, R.V. 1914a. The Genus *Watobius*. Bulletin of the Museum of Comparative Zoology Harvard 57(3): 107-112.

Chamberlin, R.V. 1914b. Notes on myriapods from Douglas Lake, Michigan. The Canadian Entomologist 46: 301-306.

Chamberlin, R.V. 1914c. The Stanford Expedition to Brazil, 1911, John C. Branner, Director. The Chilopoda of Brazil. Bulletin of the Museum of Comparative Zoology, Harvard 58(3): 151-221.

Chamberlin, R.V. 1915. New chilopods from Mexico and the West Indies. Bulletin of the Museum of Comparative Zoology Harvard 54: 495-541.

Chamberlin, R.V. 1916. The Lithobiid genera *Oabius*, *Kiberbius*, *Paobius*, *Arebius*, *Nothembius*, and *Tigobius*. Bulletin of the Museum of Comparative Zoology Harvard 57(4): 115-201.

Chamberlin, R.V. 1917. The Gosibiidae of America North of Mexico., Bulletin of the Museum of Comparative Zoology Harvard 57: 209-255.

Chamberlin, R.V. 1918a. Myriapods from Nashville, Tennessee. Psyche 25 (2): 23-30.

Chamberlin, R.V. 1918b. Myriopods from Okefenokee Swamp, GA., and from Natchitoches Parish, Louisiana. Annals of the Entomological Society of America 11(4): 369-380.

Chamberlin, R.V. 1919. The Chilopoda collected by the Canadian Arctic Expedition, 1913-1918. Report of the Canadian Arctic Expedition 1913-1918 Volume III: Insects. Part H: Spiders, Mites, and Myriapods: 15H-22H.

Chamberlin, R.V. 1920a. Some Records of Canadian myriopods., The Canadian Entomologist 52: 94-95.

Chamberlin, R.V. 1920b. Canadian myriapods collected in 1882-1883 by J. B. Tyrrell, with additional records. The Canadian Entomologist 52: 166-168.

Chamberlin, R.V. 1920c. The Myriapoda of the Australian region. Bulletin of the Museum of Comparative Zoology Harvard 64: 1-269.

Chamberlin, R.V. 1920d. A new diplopod from Texas and a new chilopod from Alaska. Proceedings of the Biological Society of Washington 33: 41-44.

Chamberlin, R.V. 1920e. On chilopods of the family Mecistocephalidae., The Canadian Entomologist 52: 184-189.

Chamberlin, R.V. 1920f. Corrections to Mr. Gunthorp's summary of Wood's myriapod papers. The Canadian Entomologist 52: 202-203.

Chamberlin, R.V. 1920g. The Myriapod Fauna of the Bermuda Islands, with notes on variation in *Scutigera*. Annals Entomological Society of America. 13(3): 271-302.

Chamberlin, R.V. 1921a. On some chilopods and diplopods from Knox Co., Tennessee. The Canadian Entomologist 53: 230-233.

Chamberlin, R.V. 1921b. The Centipedes of Central America. Proceedings of the United States National Museum. No. 2402, 60(7): 1-17.

Chamberlin, R.V. 1922a. A new Lithobiid of the genus *Paobius*. The Canadian Entomologist 54: 47-48.

Chamberlin, R.V. 1922b. Further studies on North American Lithobiidae. Bulletin of the Museum of Comparative Zoology Harvard 57(6): 259-382.

Chamberlin, R.V. 1922c. A new Javan chilopod of the genus Mecistocephalus. Psyche 29(4): 178-179.

Chamberlin, R.V. 1922d. A new Schendyloid chilopod from Mexico. Psyche 29: 9-10.

Chamberlin, R.V. 1923a. Chilopoda" In a biological Survey of the Priblof Islands, Alaska. Part II, Insects, Arachnids, and Chilopods of the Priblof Islands, Alaska North American Fauna 46: 240-244.

Chamberlin, R.V. 1923b. On chilopods and diplopods from islands in the Gulf of California. Proceedings of the. California Academy of Science 12(18): 389-407.

Chamberlin, R.V. 1925a. Notes on a small collection of chilopods from Arizona and adjacent parts. Journal Entomology and Zoology : 53-54.

Chamberlin, R.V. 1925b. Notes on chilopods and diplopods from the Barro Colorado Id., and other parts of the canal zone, with diagnoses of new species. Proceedings of the Biological Society of Washington 38: 35-44.

Chamberlin, R.V. 1925c. The Ethopolidae of America North of Mexico. Bulletin of the Museum of Comparative Zoology Harvard 57(7): 385-437.

Chamberlin, R.V. 1925d. The genera *Lithobius*, *Neolithobius*, *Gonibius*, and *Zinapolys* in America North of Mexico. Bulletin of the Museum of Comparative Zoology Harvard 57(8): 441-504.

Chamberlin, R.V. 1925e. Notes on some centipeds and millipeds from Utah. Pan-Pacific Entomologist 2(2): 55-63.

Chamberlin, R.V. 1925f. A new species of Lithobiid, genus *Nampabius*, from Tennessee, The Canadian Entomologist 57: 291.

Chamberlin, R.V. 1926a. Two new American chilopods. Proceedings of the Biological Society of Washington 39: 9-10.

Chamberlin, R.V. 1926b. Chilopoda. Bernice P. Bishop Museum Bulletin 31: 92-94.

Chamberlin, R.V. 1928a. On three chilopods from the La Sal Mountains of Utah. Entomological News 39(3): 93-96.

Chamberlin, R.V. 1928b. Some chilopods and diplopods from Missouri. Entomological News 39: 153-155.

Chamberlin, R.V. 1928c. Three new lithobiomorphous chilopods from Washington and Oregon. Pan-Pacific Entomologist 5(2): 85-86.

Chamberlin, R.V. 1928d. Notes on chilopods and diplopods from southeastern Utah. Entomological News 39: 307-311.

Chamberlin, R.V. 1929. Two new lithobioid chilopods. The Canadian Entomologist 61: 37-38.

Chamberlin, R.V. 1930a. On some centipeds and millipeds from Utah and Arizona. Pan-Pacific Entomologist 6(3): 111-122.

Chamberlin, R.V. 1930b. A new Geophiloid chilopod from Potter Creek Cave, California. University of California Publications in Entomology, 33 (14): 297-300.

Chamberlin, R.V. 1930c. On some chilopods immigrant at Hawaii. Pan-Pacific Entomologist 7(2) 65-69.

Chamberlin, R.V. 1931a. On three new chilopods. Pan-Pacific Entomologist 7 (4): 189-191.

Chamberlin, R.V. 1931b. On a collection of chilopods and diplopods from Oklahoma. Entomological News 42(4): 97-104.

Chamberlin, R.V. 1938a. Three new geophiloid chilopods. Entomological News 49: 254-255.

Chamberlin, R.V. 1938b. On eighteen new lithobiomorphous chilopods, Annals and Magazine of Natural History Ser. 11, 2: 625-635.

Chamberlin, R.V. 1939. Four new centipeds of the genus *Cryptops*. Pan-Pacific Entomologist 15 (2): 63-65.

Chamberlin, R.V. 1940a. A new *Geophilus* from San Nicholas Island, California. Pan-Pacific Entomologist 16(1): 4.

Chamberlin, R.V. 1940b. Two new lithobiid chilopoda from burrows of the Florida Pocket Gopher. Entomological News 51: 48-50.

Chamberlin, R.V. 1940c. On some chilopods and diplopods from North Carolina. The Canadian Entomologist 72: 56-59.

Chamberlin, R.V. 1940d. Diagnoses of ten new chilopods with a new genus of Sogonidae and a key to the species of *Lophobius*. Pan-Pacific Entomologist 16(2): 49-56.

Chamberlin, R.V. 1940e. Two new Geophiloid chilopods from Mexico and Texas. Proceedings of the Biological Society of Washington 53: 65-66.

Chamberlin, R.V. 1940f. On six new lithobiid centipeds from North Carolina. Proceedings of the Biological Society of Washington 53: 75-78.

Chamberlin, R.V. 1941a. New genera and species of American lithobiid centipeds. Bulletin of the University of Utah 31(13): 1-23.

Chamberlin, R.V. 1941b. Three new centipeds of the genus *Cryptops*. Pomona College Journal of Entomology and Zoology : 41-42.

Chamberlin, R.V. 1941c. New chilopods from Mexico. Pan-Pacific Entomologist 17(4): 184-188.

Chamberlin, R.V. 1941d. New genera and species of North American geophiloid centipeds. Annals of the Entomological Society of America 34(4): 773-790.

Chamberlin, R.V. 1942a. A new American centiped of the genus *Scutigera* (Chilopoda: Scutigeridae). Entomological News 53: 10-11.

Chamberlin, R.V. 1942b. On a collection of myriopods from Iowa. The Canadian Entomologist 74: 15-17.

Chamberlin, R.V. 1942c. On ten new centipeds from Mexico and Venezuela. Proceedings of the Biological Society of Washington 55: 17-24.

Chamberlin, R.V. 1942d. Notes on a collection of centipeds chiefly from Louisiana, Arkansas and Missouri (Chilopoda). Entomological News 53: 184-188.

Chamberlin, R.V. 1943a. On Mexican Centipedes. Bulletin of the University of Utah. 33(6): 1-55.

Chamberlin, R.V. 1943b. Some records and descriptions of American chilopods. Proceedings of the Biological Society of Washington 56: 97-108.

Chamberlin, R.V. 1944a. Some centipedes from Georgia. Entomological News 55: 32-35.

Chamberlin, R.V. 1944b. Two new centipeds. Entomological News 55(3): 64-66.

Chamberlin, R.V. 1944c. Some records of myriopods collected by W. M. Pearce in California. Pan-Pacific Entomologist 20: 79-80.

Chamberlin, R.V. 1944d. Chilopods in the collections of Field Museum of Natural History. Field Museum of Natural History, Zoological Series 28(4): 175-216.

Chamberlin, R.V. 1945a. A new henicopid centiped from Utah. Entomological News 56 (6) 153-154.

Chamberlin, R.V. 1945b. On three lithobioid chilopods. Entomological News 56 (8): 197 199.

Chamberlin, R.V. 1945c. Occurrence of a European centiped in Utah. Entomological News 56(8): 199.

Chamberlin, R.V. 1945d. On some centipedes from Georgia. The Canadian Entomologist 77: 215-216.

Chamberlin, R.V. 1946a. A new American genus in the chilopod family Himantariidae. Proceedings of the Biological Society of Washington 59: 35-38.

Chamberlin, R.V. 1946b. A new centiped of the genus *Guambius* from Mississippi. Entomological News 57(8): 194-195.

Chamberlin, R.V. 1946c. On the chilopods of Alaska. Annals of the Entomological Society of America 39(2): 177-189.

Chamberlin, R.V. 1946d. A new schendyloid chilopod from California. Pan-Pacific Entomologist 22(2): 64-70.

Chamberlin, R.V. 1946e. A new chilopod genus of the family Sogonidae. The Canadian Entomologist 78(4): 69-71.

Chamberlin, R.V. 1946f. A new Texan *Lithobius* (Chilopoda). Pan-Pacific Entomologist 22: 20-21.

Chamberlin, R.V. 1947a. On four new American chilopods. Pan-Pacific Entomologist 23(1): 37-39.

Chamberlin, R.V. 1947b. A new geophiloid centiped taken at the Mexican Border. Entomological News 58(10): 260.

Chamberlin, R.V. 1949a. On some centipeds from northern Alaska. Entomological News 55: 12-15.

Chamberlin, R.V. 1949b. A new fossil chilopod from the late cenozoic. Transactions of the San Diego Society of Natural History 11(7) 117-120.

Chamberlin, R.V. 1950. Some Chilopods from Puerto Rico. Proceedings of the Entomological Society of Washington 63: 155-162.

Chamberlin, R.V. 1951a. A new species in the chilopod genus *Theatops*. Psyche 58(3): 100-101.

Chamberlin, R.V. 1951b. Records of American Millipeds and Centipeds Collected by Dr. D. Elden Beck in 1950. Great Basin Naturalist 11(1-2): 27-35.

Chamberlin, R.V. 1951c. On five new American lithobiid centipedes. Great Basin Naturalist 11(3-4) 115-118.

Chamberlin, R.V. 1952a. A new geophiloid centiped from the littoral of southeast Alaska. Proceedings of the Biological Society of Washington 65: 83-84.

Chamberlin, R.V. 1952b. The centipedes (Chilopoda) of South Bimini, Bahama Islands, British West Indies. American Museum Novitates 1576: 1-8.

Chamberlin, R.V. 1952c. Occurrence of a Japanese centipede in Alaska. Entomological News 63: 209.

Chamberlin, R.V. 1953a. Two new Oregon chilopods of the order Geophilida. Psyche 60(1): 37-39.

Chamberlin, R.V. 1953b. Geophiloid chilopods of the Hawaiian and other oceanic islands of the Pacific. The Great Basin Naturalist 13(3-4): 75-85.

Chamberlin, R.V. 1954. Notes on the chilopod genera *Linotaenia* and *Tomotaenia* with description of a new *Korynia*. Entomological News : 117-122.

Chamberlin, R.V. 1955a. Four new American chilopods. Proceedings of the Biological Society of Washington 68: 179-182.

Chamberlin, R.V. 1955b. The Chilopoda of the Lunds University and California Academy of Science Expeditions. Lunds Universitets Årsskrift N. F. Avd. 2. 51(5): 1-61.

Chamberlin, R.V. 1958a. Some records of chilopods from Florida. Entomological News 69(1): 13-14.

Chamberlin, R.V. 1958b. Occurrence of the chilopod genus *Ethmostigmus* in America. Proceedings of the Biological Society of Washington 71: 185-186.

Chamberlin, R.V. 1958c. Checklist of arthropods (Diplopoda and Chilopoda). p. 131-133, *In*: Anthropol. Pap. No. 31 (Glen Canyon Reservoir).

Chamberlin, R.V. 1960a. A new marine centiped from the California littoral. Proceedings of the Biological Society of Washington 73: 99-102.

Chamberlin, R.V. 1960b. Five new western geophilid chilopods. Proceedings of the Biological Society of Washington 73: 239-244.

Chamberlin, R.V. 1961. Notes on the geophilid chilopods of Utah. Entomological News 72(5): 96-100.

Chamberlin, R.V. 1962. New records and species of chilopods from Nevada and Oregon. Entomological News 73(5): 134-138.

Chamberlin, R.V. 1963a. Further observations on the chilopod genus *Tomotaenia*. Entomological News 74(4): 88-90.

Chamberlin, R.V. 1963b. A New Genus in the Chilopod Family Tampiyidae. Proceedings of the Biological Society of Washington 76: 33-36.

Chamberlin, R.V. 1966. A new genus in the chilopod family Dignathodontidae with proposal of two subfamilies (Chilopoda: Geophilomorpha). Proceedings of the Entomological Society of Washington 79: 215-220.

Chamberlin, J.C. and R.V. Chamberlin. 1945-1946. Biological Series. Contents of Volume IX. Systematic Index to Volume IX. Bulletin of the University of Utah IX: i-vii.

Chamberlin, R.V. and S. Mulaik, 1940. On a collection of centipeds from Texas, New Mexico and Arizona (Chilopoda). Entomological News 51: 107-110, 125-128, 156-158.

Chamberlin, R.V. and Y.-h.M. Wang. 1952a. Miscellaneous New North American Centipeds of The Order Lithobiida. Proceedings of the Biological Society of Washington 65: 55-62.

Chamberlin, R.V. and Y.-h.M. Wang. 1952b. Some records and descriptions of chilopods from Japan and other oriental areas. Proceedings of the Biological Society of Washington 65: 177-188.

Chipman, A.D., W. Arthur, and M. Akam 2004. A double segment periodicity underlies segment generation in centipede development. Current Biology (14): 1250-1255.

Clark, D. B. 1979. A centipede preying on a nestling rice rat (*Oryzomys bauri*). Journal of Mammology 60(3): 654.

Cloudsley-Thompson, J. L. 1945. Behavior of the common Centipede *Lithobius forficatus*. Nature 156(3966): 537-538.

Cloudsley-Thompson, J.L. and C.S. Crawford, 1970a. Lethal temperatures of some arthropods of the southwestern United States. Entomologist's Monthly Magazine 106: 26-29.

Cloudsley-Thompson, J.L. and C.S. Crawford, 1970b. Water and temperature relations, and diurnal rhythms of scolopendromorph centipedes. Entomologia Experimentalis et Applicata 13: 187-193.

Cockerell, T.D.A. 1893. The entomology of the mid-alpine zone of Custer county, Colorado. Transactions of the American Entomological Society 20: 305-370.

Cole, L.C. 1946. A study of the cryptozoa of an Illinois woodland. Ecological Monographs 16(1): 49-86.

Cook, O.F. 1895a. On *Geophilus attenuatus*, Say, of the class Chilopoda Proceedings of the United States National Museum 18(1038): 59-62.

Cook, O.F. 1895b. An Arrangement of the Geophilidae, a Family of Chilopoda, Proceedings of the United States National Museum 18(1039); 63-75.

Cook, O.F. 1895c. On the Generic Names *Strigamia*, *Linotaenia* and *Scolioplanes*., American Naturalist 29: 864-866.

Cook, O.F. 1896. On Certain Geophilidae Described by Meinert, American Naturalist : 239-242.

Cook, O.F. 1899. The Geophiloidea of the Florida Keys, Proceedings of the Entomological Society of Washington 4(3): 303-312.

Cook, O.F. 1904. Myriapoda of Northwestern North America In: Harriman Alaska Expedition, 8(Insects pt. 1). Doubleday, Page and Co., New York: 47-77.

Cook, O. F. and G. N. Collins 1891. Notes on North American Myriapoda of the family Geophilidae, with descriptions of three genera. Proceedings of the United States National Museum 13: 383-396.

Cope, E.D. 1869. Synopsis of the extinct Mammalia of the cave formations of the United States with observations on some Myriapoda found in and near the same etc. Proceedings of the American Philosophical Society 11: 171-192.

Cornwell, W.S. 1934. Notes on the egg-laying and nesting habits of certain species of North Carolina myriapods, and various phases of their life histories. Journal of the Elisha Mitchell Scientific Society 48: 289-291.

Costa, O. 1834. see Newport 1845. for *Rhysida = Branchistoma*.

Crabill, R.E. Jr. 1949a. Presence of a European centipede in New York State. Entomological News 55(4): 101.

Crabill, R.E. Jr. 1949b. A new centipede from the eastern United States (Chilopoda: Geophilidae). Entomological News 55(8): 210-213.

Crabill, R.E. Jr. 1950a. A review of the genus *Pseudolithobius* (Chilopoda: Gosibiidae). Entomological News 61(1): 8-12.

Crabill, R.E. Jr. 1950b. A new centipede record for Canada. The Canadian Entomologist 82(5): 101.

Crabill, R.E. Jr. 1950c. On a collection of centipedes from western South Carolina. Entomological News 61(7): 199-202.

Crabill, R.E. Jr. 1950d. On the true identity of *Arctogeophilus fulvus* (Wood), with some remarks concerning the status of *Mecistocephalus melanonotus* Wood (Chilopoda: Geophilomorpha: Geophilidae). The Canadian Entomologist 82(12): 253-256.

Crabill, R.E. Jr. 1951. On the true identity of *Geophilus huronicus* Meinert and the presence of *Geophilus longicornis* Leach in North America (Chilopoda: Geophilomorpha: Geophilidae).The Canadian Entomologist 83(11):314-315.

Crabill, R.E. Jr. 1952a. A New Subspecies of *Otocryptops gracilis* (Wood) from the eastern United States, together with remarks on the status of *Otocryoptops nigridius* (McNeill) and a key to the species of the genus known to occur east of the Rocky Mountains (Chilopoda: Scolopendromorpha: Cryptopidae). Entomological News 63(5): 123-129.

Crabill, R.E. Jr. 1952b. The centipedes of northeastern North America. Ithaca, Cornell University: 450 + 9 plates.

Crabill, R.E. Jr. 1952c. A new cavernicolous *Nampabius* with a key to its northeastern American congeners (Chilopoda: Lithobiidae). Entomological News 63(8): 203-206.

Crabill, R.E. Jr. 1953a. A New *Zygethopolys* from Kentucky and a key to the members of the genus. (Chilopoda: Lithobiomorpha: Lithobiidae: Ethopolyinae), The Canadian Entomologist 85(3): 119-120.

Crabill, R.E. Jr. 1953b. Concerning a new genus, *Dinocryptops*, and the nomenclatorial status of *Otocryptops* and *Scolopocryptops* (Chilopoda: Scolopendromorpha: Cryptopidae). Entomological News 64(4): 96.

Crabill, R.E. Jr. 1953c. The Schendylidae of northeastern North America (Chilopoda: Geophilomorpha: Schendylidae). Journal of the New York Entomological Society 61(2): 93-98.

Crabill, R.E. Jr. 1953d. The genotypes of *Strigamia*, *Linotaenia*, and *Scolioplanes* (Chilopoda: Geophilomorpha: Dignathodontidae). Entomological News 64(7): 169-172.

Crabill, R.E. Jr. 1953e. A new Himantariid from the eastern United States (Chilopoda: Geophilomorpha: Himantariidae). Bulletin of the Brooklyn Entomological Society 48(4): 85-88.

Crabill, R.E. Jr. 1954a. Concerning the true identity of *Strigamia fulva* Sager and *Strigamia bothriopa* Wood (Chilopoda: Geophilomorpha: Dignathodontidae), Entomological News 65(2): 40-46.

Crabill, R.E. Jr. 1954b. A conspectus of the northeastern North American species of *Geophilus* (Chilopoda: Geophilomorpha: Geophilidae), Entomological Society of Washington 56(4): 172-188.

Crabill, R.E. Jr. 1954c. Concerning *Tomotaenia* and *Paraplanes*, with the description of a new Dignathodontid centipede from Missouri (Chilopoda: Geophilomorpha: Dignathodontidae). The Canadian Entomologist 86(9): 416-419.

Crabill, R.E. Jr. 1955a. Proposed use of the plenary powers to designate for the genus "Scolopendra" Linnaeus, 1758 (Class Myriapoda) a type species in harmony with accustomed usage. Bulletin of Zoological Nomenclature 11(4): 134-136.

Crabill, R.E. Jr. 1955b. A preliminary Report on the Chilopoda of Missouri. Entomological News 66(2): 36-41.

Crabill, R.E. Jr. 1955c. Concerning the genotypes of *Bothropolys*, *Polybothrus* and *Eupolybothrus* (Chilopoda: Lithobiomorpha: Lithobiidae). Entomological News 66(4): 107-110.

Crabill, R.E. Jr. 1955d. On the identity of the Gunthorp Types (Chilopoda: Geophilomorpha: Geophilidae). The Canadian Entomologist 87(5): 221-223.

Crabill, R.E. Jr. 1955e. Report of another European chilopod in eastern North America (Chilopoda: Geophilomorpha). Entomological News 66(9): 248-249.

Crabill, R.E. Jr. 1955f. A checklist of the Chilopoda known to occur in Kentucky. Entomological News 66(10): 257-261.

Crabill, R.E. Jr. 1955g. On the reapperance of a possible ancestral characteristic in a modern chilopod (Chilopoda: Scolopendromopha: Cryptopidae). Bulletin of the Brooklyn Entomological Society 50(5): 133-136.

Crabill, R.E. Jr. 1955h. New Missouri chilopod records with remarks concerning geographical affinities. Journal of the New York Entomological Society 63: 153-159.

Crabill, R.E. Jr. 1957a. On the Newport chilopod genera. Journal of the Washington Academy of Sciences 47(10): 343-345.

Crabill, R.E. Jr. 1957b. A new *Garibius* from Virginia, with a key to the North American congeners (Chilopoda: Lithobiomorpha: Lithobiidae). Journal of the Washington Academy of Sciences 47(11): 375-377.

Crabill, R.E. Jr. 1957c. A note on the identity of *Parorya valida* Cook (Chilopoda: Geophilomorpha: Oryidae). Entomological News 68(10) 265-267.

Crabill, R.E. Jr. 1958a. Report of *Geophilus proximus* in North America, with a key to its eastern North American congeners (Chilopoda: Geophilomorpha: Geophilidae). Entomological News 69(1): 15-17.

Crabill, R.E. Jr. 1958b. On a collection of centipedes from Wisconsin (Chilopoda). Entomological News 69(4): 93-99.

Crabill, R.E. Jr. 1958c. A new Schendylid from the eastern United States, with notes on distribution and morphology (Chilopoda: Geophilomorpha: Schendylidae). Entomological News 69(6): 153-160.

Crabill, R.E. Jr. 1958d. A new *Eulithobius*, with a key to the known American species (Chilopoda: Lithobiidae). Journal of the Washington Academy of Sciences 48(8): 260-262.

Crabill, R.E. Jr. 1958e. A new *Kethops* from New Mexico, with a key to its congeners (Chilopoda: Scolopendromorpha: Cryptopidae). Proceedings of the Entomological Society of Washington 60(5): 235-238.

Crabill, R.E. Jr. 1959a. A synonymical list of American Himantariidae, with a generic key and description of a new genus (Chilopoda: Geophilomorpha: Himantariidae). Entomological News 70(6): 117-126; 153-159.

Crabill, R.E. Jr. 1959b. Notes on *Mecistocephalus* in the Americas, with a redescription of *Mecistocephalus guildingi* Newport (Chilopoda: Geophilomorpha: Mecistocephalidae). Journal of the Washington Academy of Sciences 49(6): 188-192.

Crabill, R.E. Jr. 1959c. A new Floridan *Pectiniunguis*, with re-appraisal of its type species and comments on the status of *Adenoschendyla* and *Litoschendyla* (Chilopoda: Geophilomorpha: Schendylidae). Journal of the Washington Academy of Sciences 49(9): 324-330.

Crabill, R.E. Jr. 1960a. A new American genus of Cryptopid centipede, with an annotated key to the Scolopendromorph genera from America North of Mexico. Proceedings of the United States National Museum 111(3422): 1-15.

Crabill, R.E. Jr. 1960b. Centipedes of the Smithsonian Bredin Expeditions to the West Indies. Proceedings of the United States National Museum 111(3427): 167-195.

Crabill, R.E. Jr. 1960c. Concerning the aberrant genus *Nothobius*, with a redescription of its type species. (Chilopoda: Geophilomorpha: Himantariidae). Entomological News 71(4): 87-99.

Crabill, R.E. Jr. 1960d. A remarkable form of sexual dimorphism in a centipede (Chilopoda: Lithobiomorpha: Lithobiidae). Entomological News 71(6): 156-161.

Crabill, R.E. Jr. 1960e. On the identity of *Stenophilus grenadae* (Chamberlin) with a key to the known North American congeners (Chilopoda: Geophilomorpha: Himantariidae). Proceedings of the Biological Society of Washington 73: 87-94.

Crabill, R.E. Jr. 1960f. A new *Nuevobius* with review of the genus (Chilpoda: Lithobiomorpha: Lithobiidae). Bulletin of the Brooklyn Entomological Society 55(5): 121-133.

Crabill, R.E. Jr. 1961a. A catalogue of the Schendylinae of North America including Mexico, with a generic key and proposal of a new *Simoporus* (Chilopoda: Geophilomorpha: Schendylinae). Entomological News 72(2): 29-36, 72(3): 67-80.

Crabill, R.E. Jr. 1961b. Concerning the identities of *Nannocrix* and *Sogona*, with pertinent morphological notes (Chilopoda: Geophilomorpha: Sogonidae). Proceedings of the Entomological Society of Washington 63(2): 125-135.

Crabill, R.E. Jr. 1962a. A new *Damothus* and a key to the North American Dignathodontid genera (Chilopoda: Geophilomorpha: Dignathodontidae). Psyche 69(2): 81-86.

Crabill, R.E. Jr. 1962b. Plectrotaxy as a systematic criterion in lithobiomorphic centipedes (Chilopoda: Lithobiomorpha), Proceedings of the United States National Museum 113(3459): 399-412.

Crabill, R.E. Jr. 1962c. A new interpretation of some troublesome Dignathodontid species and genera (Chilopoda: Geophilomorpha). Entomological News 73(7): 179-186.

Crabill, R.E. Jr. 1963. Concerning chilopod types in the British Museum, Part I Annals and Magazine of Natural History (13) 5(56): 505-510.

Crabill, R.E. Jr. 1964. A revised interpretation of the primitive centipede genus *Arrup*, with redescription of its type-species and list of known species. Proceedings of the Biological Society of Washington 77: 161-169.

Crabill, R.E. Jr. 1968a. Concerning the true identities of *Gosiphilus* and *Chomatobius*, with redescription of the latter's type-species (Chilopoda: Geophilomorpha: Himantariidae). Entomological News 79(4): 108-112.

Crabill, R.E. Jr. 1968b. On the true identity of *Chomatophilus* with description of a new species, and with key and catalogue of all sogonid genera. Proceedings of the Entomological Society of Washington 70(4): 323-331.

Crabill, R.E. Jr. 1969a. Review of *Arenophilus* and key to all species. Entomological News 80(1): 7-11.

Crabill, R.E. Jr. 1969b. A new Floridian *Cryptops* with key to the State's species (Chilopoda: Scolopendromorpha: Cryptopidae). Proceedings of the Biological Society of Washington 82: 201-204.

Crabill, R.E. Jr. 1969c. A new Californian *Garriscaphus*: report of the Bothriogastrinae in the New World. Proceedings of the Entomological Society of Washington 71(4): 494-496.

Crabill, R.E. Jr. 1970a. A new family of centipedes from Baja California with introductory thoughts on ordinal revision. Proceedings of the Entomological Society of Washington 72(1): 112-118.

Crabill, R.E. Jr. 1970b. Concerning Mecistocephalid morphology and the true identity of the type-species of *Mecistocephalus*. Journal of Natural History 4: 231-237.

Crabill, R.E. Jr. 1976. A new *Watophilus* from Utah, including a list of all known species (Chilopoda: Geophilomorpha: Chilenophilidae). Proceedings of the Biological Society of Washington 88(37): 395-398.

Crabill, R.E. Jr. 1981a. On the identity of *Dysmesus* Chambelrin: a new generic synonomy (Chilopoda: Geophilomorpha: Geophilidae) Proceedings of the Entomological Society of Washington 83(1): 174.

Crabill, R.E. Jr. 1981b. Synonomy by way of teratology (Chilopoda: Lithobiomorpha: Lithobiidae) Proceedings of the Entomological Society of Washington 83(2): 359.

Crabill, R.E. Jr. 1981c. On the true identity of *Zygethobius pontis* Chambelrin (Chilopoda: Lithobiomorpha: Henicopidae) Proceedings of the Entomological Society of Washington 83(2): 360.

Crabill, R.E. Jr. and M.A. Lorenzo 1957. On the identity of the Gunthorp Types, Part II, and some notes on plectrotaxic criteria (Chilopoda: Lithobiomorpha: Lithobiidae). The Canadian Entomologist 89(9): 428-432.

Cragin, F.W. 1885. First contribution to a knowledge of the myriopoda of Kansas. Bulletin of the Washburn College Laboratoryt of Natural History 4:143-145.

Crawford, C.S. and W.A. Riddle 1974. Cold hardiness in centipedes and scorpions in New Mexico. Oikos 25: 86-92.

Daday, J. 1889. A Magyarországi Myriopodák Magánrajza.-Myriopoda Regni Hungariae. Budapest, 126 pp, 3 plates.

Daday, J. 1889. Myriopoda extranea musei nationalis Hungarici. Termésezetrajzi Füzetek, 12: 115-156.

Daday, J. 1891a. A Heidelbergi egyetum zoologiai gyujtcmcnycnck idcgenfoldi myriopodai. Termésezetrajzi Füzetek, 14(1-2): 135-154?.

Daday, J. 1891b. Ausländische myriapoden der zoologischen collection der universität zu Heidelberg, Termésezetrajzi Füzetek, 14(3-4): 135-?.

Daday, J. 1893. Myriopoda extranea nova vel minus cognita in collectione Musaei nationalis hungarici. Termésezetrajzi Füzetek, 16: 98-113.

Dearolf, K. 1953. The invertebrates of 75 Caves in the United States. Proceedings of the Pennsylvania Acedemy of Science 27: 225-241.

Dohle, W. 1985. Phylogenetic pathways in the Chilopoda. Bijdragen tot de Dierkunde 55(1): 55-66.

Donovan, E. 1810. The natural history of British insects 14: 23-24.

Dowdy, W.W. 1944. A community study of a disturbed deciduous forest area near Cleveland, Ohio with special reference to invertebrates. Ecological Monographs, 14: 193-222.

Duboscq, O. 1898. Recherches sur les Chilopodes. Archives de Zoologie Experimentale et generale 3(6): 481-650.

Dubuisson, M. 1928. Recherches sur la ventilation trachéenne chez lez chilopodes et sur la circulation sanguine chez les Scutigères. Archives de Zoologie Expérimentale et Générale 67: 49-63.

Eason, E. H. 1961. On the synonymy of some British Centipedes. Annals and Magazine of Natural History 13(4): 385-391.

Eason, E.H. 1964. Centipedes of the British Isles. London: Warne. p. 1-294.

Eason, E.H. 1972a. The type specimens and identity of the species described in the genus *Lithobius* by George Newport in 1844, 1845 and 1849 (Chilopoda: Lithobiomorpha). Bulletin of the British Museum of Natural History (Zool.) 21: 297-311.

Eason, E.H. 1972b. The type specimens and identity of the species described in the genus *Lithobius* by C.L. Koch and L. Koch from 1841 to 1878 (Chilopoda: Lithobiomorpha) Bulletin of the British Museum of Natural History (Zool.) 22(4): 105-150.

Eason, E.H. 1973. The type specimens and identity of the species described in the genus Lithobius by R. I. Pocock from 1890 to 1901 (Chilopoda: Lithobiomorpha). Bulletin of the British Museum of Natural History (Zool.) 25: 41-83.

Eason, E.H. 1974a. On certain aspects of the generic classification of the lithobiidae, with special reference to geographical distribution. Symposia of the Zoological Society of London 32: 65-73.

Eason, E.H. 1974b. The type specimens and identity of the species described in the genus Lithobius by F. Meinert, and now preserved in the Zoological Museum, Copenhagen University (Chilopoda: Lithobiomorpha). Zoological Journal of the Linnean Society 55: 1-52.

Eason, E.H. 1976a. The type specimens and identity of the Siberian species described in the genus *Lithobius* by Anton Stuxberg in 1876 (Chilopoda: Lithobiomorpha). Zoological Journal of the Linnean Society 58: 91-127.

Eason, E.H. 1976b. On *Lithobius melanops* Newport (Chilopoda: Lithobiomorpha) in North America. Entomologist's Monthly Magazine 112: 65-66.

Eason, E.H. 1977. A redescription of the type specimens of *Lithobius hawaiiensis* Silvestri, 1904 with a note on the Hawaiian Lithobiidae. Journal of Natural History 11: 485-492.

Eason, E.H. 1983. The indentity of the European and Mediteranean species of Lithobiidae (Chilopoda) described by K. W. Verhoeff and now represented by material preserved in the British Museum (Natural History). Zoological Journal of the Linnean Society. 77: 111-144.

Eason, E.H. 1991. Distribution of the centipedes *Lithobius obscurus* Meinert, and *L. peregrinus* Latzel (Chilopoda, Lithobiomorpha). Entomologist's Monthly Magazine 127: 23.

Eason, E.H. 1992. On the taxonomy and geographical distribution of the Lithobiomorpha. *Ber. Nat.-Med.* Verein Innsbruck, Supplement 10: 1-9.

Eason, E.H. 1993. Descriptions of four new species of Lithobius from the Oriental Region and a redescription of *Australobius palnis* (Eason, 1973) from Sri Lanka. Bolletino del Museo Civico di Storia Naturale di Verona. 17[1990]: 181-200.

Eason, E.H. 1996. Lithobiomorpha from Sakhalin Island, the Kamchatka Peninsula, and the Kurile Islands (Chilopoda). Arthropoda Selecta 6(1/2): 117-123.

Easterla, D.A. 1975. Giant desert centipede preys upon snake. The Southwestern Naturalist 20: 411.

Edgecombe, G.D. 2004a. The henicopid centipede Haasiella (Chilopoda: Lithobiomorpha): new species from Australia, with a morphology-based phylogeny of Henicopidae.

Edgecombe, G.D. 2004b. Monophyly of the Lithobiomorpha (Chilopoda): New Characters from the Pretarsal Claws. Insect Systematics and Evolution 35: 1: 29-41.

Edgecombe, G. and G. Giribet 2003a. A new blind *Lamyctes* (Chilopoda: Lithobiomorpha) from Tasmania with an analysis of molecular sequence data for the *Lamyctes-Henicops* Group. Zootaxa 152: 1-23.

Edgecombe, G. and G. Giribet 2003b. Relationships of Henicopidae (Chilopoda: Lithobiomorpha): New molecular data, classification and biogeography. African Invertebrates 44(1): 13-38.

Edgecombe G. and G. Giribet 2004. Adding mitochondrial sequence data (16S rRNA and cytochrome *c* oxidase subunit I) to the phylogeny of centipedes (Myriapoda: Chilopoda): an analysis of morphology and four molecular loci Journal of Zoology and Systematic Evolutionary Rsearch 42: 89-134.

Edgecombe, G.D., G. Giribet and W.C. Wheeler 1999. Phylogeny of Chilopoda: Analysis of 18S and 28S rDNA Sequences and Morphology. *In*: A. Melic, (Eds.), Evolución y Filogenia de Arthropoda. Boletin S. E. A. 26: 293-331.

Edgecombe, G.D, G. Giribet and W.C. Wheeler 2002. Phylogeny of Henicopidae (Chilopoda: Lithobiomorpha): a combined analysis af morphology and five molecular loci. Systematic Entomology 27: 31-64.

Enghoff, H. 1975. Notes on *Lamyctes coeculus* (Brölemann), a cosmopolitic, parthenogenetic centipede (Chilopoda: Henicopidae). Entomologica Scandinavica 6: 45-46.

Filinger, G.A. 1928. Observations on the habits and control of the garden centipede Scutigerella immaculata, a pest in greenhouses. Journal of Economic Entomology 21: 351-360.

Foddai, D., L.A. Pereira, and A. Minelli 2000. A catalogue of the geophilomorph centipedes (Chilopoda) from Central and South America including Mexico. Amazoniana 16(1/2): 59-185.

Foddai, D., L. Bonato, L.A. Pereira, and A. Minelli 2003. Phylogeny and systematics of the Arrupinae (Chilopoda: Geophilomorpha: Mecistocephalidae) with the description of a new dwarfed species. Journal of Natural History 37(10): 1247-1267.

Frank, J.H., S. Sreenivasan, P.J. Benshoff, M.A. Deyrup, G.B. Edwards, S.E. Halbert, A.B. Hamon, M.D. Lowman, E.L. Mockford, R.H. Scheffrahn, G.J. Steck, M.C. Thomas, T.J. Walker and W.C. Welbourn 2004. Invertebrate animals extracted from native Tillandsia (Bromeliales: Bromeliaceae) in Sarasota County, Florida. Florida Entomologist 87(2): 176-185.

Gardner, J.E. 1986. Invertebrate fauna from Missouri caves and springs. Conservation Commision of the State of Missouri, Natural History Series no. 3: 1-72.

Geoffroy, M. 1762. Histoire abregée des insectes: qui se trouvent aux environs de Paris : dans laquelle ces animaux sont rangés suivant un ordre méthodique.

Gervais: 1835. Sur les Myriapodes du genere Géophile, *Geophilus*, Leach, et Description de trios espéces nouvelles. Magasin de Zoologie. vol. 5, cl. 9, t. 133, f. 3: 1-12.

Gervais: 1847. Myriapodes. *In*: Histoire naturelle des Insectes Aptères. (Suites a Buffon). *Edited by*: C.A. de Walckenaer and P. Gervais, Roret, Paris. 4: 1-623, Atlas Plates 37-45.

Girard, C. 1853. Myriapods. *In*: Marcy, Report on exploration of the Red River of Louisiana expedition in 1852, Appendix F: 243-246? 272-274.

Giribet, G. and G. Edgecombe, 2006. Conflict between datasets and phylogeny of centipedes: an analysis based on seven genes and morphology. Proceedings of the Royal Society B 273: 531-538.

Gray, 1832. Griffith's Animal Kingdom, (Insects), pl. 1; Fig. 2.

Gray, J. 1842. Myriapoda In: W. Todd (ed.), The Cyclopaedia of Anatomy and Physiology. 3: 547.

Grimaldi, D. 1996. Amber window to the past. Harry N. Abrams, Inc., Publishers, in association with the American Museum of Natural History. 216 pp.

Grimaldi, D. and M.S. Engle 2005. Evolution of the insects. Cambridge 755 pp.

Guarisco, H., Cameron Liggett and Rowland Shelley 2007. Rediscovery of the centipede *Scolopendra heros* (Chilopoda: Scolopendromorpha; Scolopendridae) in southeastern Colorado. Transactions of the Kansas Academy of Science 110(3/4) 274-275.

Gunn, A. and J.M. Cherrett 1993. The exploitation of food resources by meso and macro invertebrates. Pedobiologia 37: 303–320.

Gunthorp, H. 1913. Annotated list of the Diplopoda and Chilopoda, with a key to the Myriapoda of Kansas. Kansas University Science Bulletin 7: 161-182.

Gunthorp, H. 1920. Summary of Wood's Myriapoda Papers. The Canadian Entomologist 52: 112-114.

Gunthorp, H. 1921. Cragin's collection of Kansas Myriapoda. The Canadian Entomologist 53: 87-91.

Haase, E. 1880. Schlesiens Chilopoda I. Chilopoda-Anamorpha (Inaugural Dissertation) Breslau, 4: 1-45.

Haase, E. 1881. Beitrag zur phylogenie und ontogenie der chilopoden. Zeitschrift für Entomologie 8(2): 93-115.

Haase, E. 1881. Schlesiens Chilopoda II., Chilopoda epimorpha, Zeitschrift für Entomologie 8: 66-92.

Haase, E. 1887. Die Indisch-Australischen Myriopoden. Pt. I Chilopoden. Abhand. Ber. Zool. Anthropol. Mus. Dresden No. 5: 1–118. 6 plates.

Harger, O. 1872. Description of new North American Myriapods. The American Journal of Science and Arts (3rd series) 4: 116-121.

Hatch, M. 1939. The house centipede (*Scutigera forceps* Raf.) in Washington. Pan-Pacific Entomologist 15(4): 189.

Hewitt, C.G. 1914. The occurrence of the House centipede, *Scutigera forceps* Raf., in Canada. The Canadian Entomologist 46: 219.

Hickerson, C.M., C.D. Anthony and J.A. Wicknick 2004. Behavioral interactions between salamanders and centipedes: competition in divergent taxa. Behavioral Ecology 15(4) 679-686.

Hilken, G., C. Brockmann, and L. Nevermann 2003a. Hemocytes of the centipede *Scutigera coleoptrata* (Chilopoda: Notostigmophora) with notes on their interactions with the tracheae. Journal of Morphology 257: 181-189.

Hilken, G., C. Brockmann and L. Nevermann 2003b. Exocytosis of fibrous material from plasmocytes in *Scutigera coleoptrata* (Chilopoda, Notostigmophohora) in relation to wound healing. African Invertebrates 44(1): 169-173.

Hilken, G., C. Brockmann, and J. Rosenberg 2003c. The maxillary organ gland: Description of a new head gland in *Scutigera coleoptrata* (Chilopoda: Notostigmophora). African Invertebrates 44(1) 175-184.

Hindle, E. 1948. Reginald Innes Pocock. 1863-1947. Obituary Notices of Fellows of the Royal Society 6(17): 189-211.

Hoffman, R.L. 1993. *Serrobius pulchellus* Causey, a Poorly Known Centiped, Rediscovered in Virginia (Lithobiomorpha: Lithobiidae). Banisteria 2: 18-20.

Hoffman, R.L. 1994. The Occurrence of *Hemiscolopendra punctiventris* (Newport), an Ecophilus Centiped, in Virginia (Chilopoda: Scolopendromorpha). Banisteria 3: 33-34.

Hoffman, R.L. 1995a. The Centipeds (Chilopoda) of Virginia: A First List. Banisteria 5: 20-32.

Hoffman, R.L. 1995b. The Type Locality of *Lithobius latzelii* Meinert (Chilopoda: Lithobiomorpha; Lithobiidae). Banisteria 6: 30-32.

Hoffman, R.L. and R.E. Jr. Crabill 1953. C.S. Rafinesque as the real father of American myriapodology: an analysis of his hitherto unrecognized species. Florida Entomologist 36(2) 73-82.

Hoffman, R.L. and L. Hubricht 1954. Distributional records for two species of Plethodon in the Southern Appalachians. Herpetologica 10: 191-193.

Hoffman, R.L. and L.A. Pereira 1991. Systematics and biogeography of Marsikomerus Attems, 1938, a misunderstood genus of centipedes (Geophilomorpha: Schendylidae) Insecta Mundi 5(1): 45-60.

Hoffman, R.L. and Shelley, R.M. 1996. The Identity of *Scolopendra marginata* Say (Chilopoda: Scolopendromorpha: Scolopendridae). Myriapodologica 4(5) 35-42.

Hölldobler, B. and E.O. Wilson 1990. The Ants. Harvard University Press, Cambridge, Massachusetts, 732 pp.

Hollington, L.M. & G.D. Edgecombe 2004. Two new species of the Henicopid centipede *Henicops* (Chilopoda: Lithobiomorpha) from Queensland and Victoria, with revision species from Western Australia and synoptic classification of Henicopidae. Records of the Australian Museum 56: 1-28.

Holmquist, A.M. 1926. Studies in arthropod hibernation. Annals Entomological Society of America 19: 395-426.

Holmquist, A.M. 1928. Notes on the life history and habits of the mound-building ant, *Formica ulkei* Emery. Ecology 9(1): 70-87.

Holsinger, J.R. and D.C. Culver 1988. The invertebrate cave fauna of Virginia and part of eastern Tennessee: Zoogeography and Ecology. Brimleyana 14: 1-162.

Holsinger, J.R. and S.B. Peck 1971. The invertebrate cave fauna of Georgia. Bulletin of the National Speleological Society 33: 23-44.

Humbert, A. and H. de Saussure, 1869. Myriapoda nova Americana. Revue et magasin de zoologie pure et appliquée. (2)22: 149-159.

Humbert, A. and H. de Saussure, 1870a. Expedition Scient. du Mexique. Zoologie 6(2): 1-211.

Humbert, A. and H. de Saussure, 1870b. Myriapoda nova Americana [description de divers Myriapodes nouveaux de musée de Vienne]. Revue et magasin de zoologie pure et appliquée. (2)22: 202-205.

ICZN - International Commision on Zoological Nomenclature, 1999. International Code of Zoological Nomenclature. Fourth Edition. International Trust for Zoological Nomenclature, London, 29: 1-306.

ICZN Opinion 454. 1957. Designation under the plenary powers of a type species in harmony with accustomed usage for "Scolopendra" Linnaeus, 1758 (Class Chilopoda). Opinions and declarations rendered by the International Commission on Zoological Nomenclature. 15: 357-378.

ICZN Opinion 1228. 1982. Henicopidae Pocock, 1901 given nomenclatural precedence over Cermatobiidae Haase, 1885 (Myriapoda, Chilopoda). Bulletin of Zoological Nomenclature 39(4): 235-237.

ICZN Opinion 1880. 1997. Plutoniinae Bollman, 1893 (Arthropoda, Chilopoda): spelling emended to Plutoniuminae, so removing the homonymy with Plutoniinae Cockerell, 1893 (Mollusca, Gastropoda). The Bulletin of Zoological Nomenclature. 54: 197-199.

Illiger, J.K.W. 1807. Caractères des espèces de papillonides et d'hespérides : suivi de traductions de trois articles par Karl Illiger. (*In*: P. Rossi, Fauna Etrusca sistens Insecta quae in Provinciis Florentina et Pisana praesertim collegit Petrus Rossius (ed. 2), 2:198).

Jameson, E.W., Jr. 1944. Food of the red-backed salamander. Copeia 3: 145-147.

Jeekel C.A.W. 1963. The generic and subgeneric names of the European Lithobiidae generally referred to *Polybothrus* Latzel 1880 (Chilopoda Lithobiida). Entomologische Berichten 23: 193-195.

Jeekel, C.A.W. 2005. Nomenclator generum et familiarum Chilopodorum: A list of the genus and family-group names in the class CHILOPODA from the 10th edition of Linnaeus, 1758, to the end of 1957. Myriapoda Memoranda, Supplement 1: 1-130.

Jensen, D.B. 2003. The biodiversity of Greenland - a country study. 165 pp. http://www.natur.gl/filer/Biodiversity_of_Greenland.pdf

Johnson, B.M. 1952. The centipedes and millipedes of Michigan. Ph. D. Dissertation, Univiversity of Michigan.

Jones, D.T. 1944. List of publications of Dr. Ralph V. Chamberlin to June 1st, 1944. Vintonia 4: 1-22.

Jones, R.E. 1998. On the species of *Tuoba* (Chilopoda: Geophilomorpha) in Australia, New Zealand, New Caledonia, Solomon Islands, and New Britain. Records of the Western Australian Museum 18: 333-346.

Judd, W.W. 1957. The food of Jefferson's salamander *Ambystoma jeffersonianum*, in Rondeau Park, Ontario. Ecology 38: 77-81.

Judd, W.W. 1964. Studies on the Byron Bog in southwestern Ontario XXI. Distribution of centipedes (Chilopoda) and millipedes (Diplopoda). Natural History Papers of the National Museum of Canada 26: 1-4.

Karsch, F. 1884. Ueber einige neue und minder bekannte Arthropoden des Bremer Museums. Abhandlugen herausgegeben vom naturwissenschaftlichen Vereine zu Bremen. Band 9, Mit 8 Tafeln und 6 Holzschnitten, 65-71.

Kenyon, F.C. 1893a. A preliminary list of the Myriapoda of Nebraska, with descriptions of new species. Publications of the Nebraska Academy of Sciences 3: 14-18.

Kenyon, F.C. 1893b. Nebraska Myriapoda. The Canadian Entomologist 25: 161-162. [misspelled Kenyan]

Kevan, D.K. McE. 1983. Chilopoda, Memoirs of the Entomological Society of Canada Ch. 17: 296-298.

Kevan, D.K. McE. 1983. A Preliminary Survey of Known and Potentially Canadian and Alaskan Centipedes (Chilopoda). Candian Journal of Zoology 61: 2938-2955.

Kevan, D.K. McE., and V.R. Vickery 1978. The orthopteroid insects of the Magdalen Islands with notes from adjacent regions. Annales de la Société Entomologique du Québec 22: 193-204.

Koch, C.L. 1835. Deutsch Crust. Myr. Arach., vol. 3.

Koch, C.L. 1837, Crust., Myr., u. Arachnida vol. 4, fasc. 9: ?

Koch, C.L. 1847. System der Myriapoden., Krit. Revis. Insektenfaun. Deutschl., 3.

Koch, C.L. 1862. Die Myriapodengattung *Lithobius*. pp. 94. two plates, Nurnberg.

Koch, C.L. 1863. Die Myriapoden getreu nach der Natur abgebildet und beschrieben., Vol. 2. Halle, Regensburg: 1-112.

Koch, L. 1878. Japanesische Arachniden und Myriapoden. Verh. Z. b. Wien, 27: 787-797.

Kohlrausch, 1879. Beitrage zur Kenntnis der Scolopendriden., Journal des Museum Godeffroy 14: 51-74, pl. vi. [Inaug. Diss.: 1-23, Marburg]

Kohlrausch, 1881. Gattungen und Arten der Scolopendriden. Archiv des Vereins der Freunde der Naturgeschichte in Mecklenburg. 47: 50-132.

Kraepelin, 1903. Revision der Scolopendren. Mitteilungen aus dem Naturhistorischen Museum in Hamburg 20: 1-276.

Kraus, O. 1998. Phylogenetic Relationships between Higher Taxa of Tracheate Arthropods. p. 295-303. *In*: Forety, R. A., & R. H. Thomas (eds.) *Arthropod Relationships*, Chapman and Hall; London, UK; xii+383pp.

Kraus, O. 2001. "Myriapoda" and the ancestry of the hexapoda. Annals de la Société Entomolgique de France 37: 105-127.

Kraus, O. and M. Kraus. 1994. Phylogenetic system of the Tracheata (Mandibulata): On "Myriapoda"-Insect interrelationships, phyogentic age and primary ecological niches. Verhandlungen des Naturwissenschaftlichen Vereins in Hamburg 35: 95-226.

Lamarck, J.B.P. 1801. Système des animaux sans vertèbres; ou, tableau général des classes, des ordres, et des genres de ces animaux . . . précdeé du discours d'ouverture du cours de zoologie, donné dans le Muséum national d'histoire naturelle l'an 8 de la République. 8: 432pp.

Latzel, R. 1880. Mit Bestimmungs-Tabellen aller bisher aufgestellten Myriapoden-Gattungen, und zahlreichen, die morphologischen Verhältnisse dieser Thiere illustrirenden, Abbildungen. Erste Hälfte. Die Chilopoden. Die Myriapoden der Österreichisch-Ungarischen Monarchie. Wein, 1: 1-228.

Leach W.E. 1814. Crustaceology, in D. Brewster (ed.) The Edinburgh Encyclopaedia Blackwood, Edinburgh. 7(2): 383–437.

Leach, W.E. 1815. A tabular view of the external characters of four classes of animals which Linné arranged under Insecta; with the distribution of the genera composing three of these classes into orders, etc. and descriptions of several new genera and species. Transactions of the Linnean Society of London Ser. 1, 11: 306-400.

Leach, W.E. 1817. Characters of the genera of the class Myriapoda, with descriptions of some species. Zoological Miscellany; being descriptions of new or interesting animals 3: 38-45.

Lee, R.E. 1980. Summer microhabitat distribution of some centipedes in a deciduous and coniferous community of central Ohio (Chilopoda). Entomological News 91(1): 1-6.

Lewis, J.G.E. 1965. The food and reproductive cycles of the centipedes *Lithobius variegates* and *Lithobius forficatus* in a Yorkshire Woodland. Proceedings of the Zoological Society of London. 144: 269–283.

Lewis, J.G.E. 1980. Swimming in the centipede *Scolopendra subspinipes* Leach (Chilopoda, Scolopendromorpha). Entomologist's Monthly Magazine 116: 121-122.

Lewis, J.G.E. 1986. The genus *Trachycormocephalus* a junior synonym of *Scolopendra*, with remarks on the validity of other genera of the tribe Scolopendrini (Chilopoda: Scolopendromorpha). Journal of Natural History 20: 1083-1088.

Lewis, J.G. E. 2003. The problems involved in the characterisation of scolopendromorph species (Chilopoda: Scolopendromorpha). African Invertebrates 44(1): 61-69.

Linnaeus, C. 1758. Systema Naturae Vol. 1, ed. 10, Stockholm, 824 pp.

Lucas, H. 1840. Histoire Naturelle des Crustaces des arachnides et des myriapodes. p. 534-551, + 3 plates.

Marlatt, C.L. 1914. The House Centipede. United States Department of Agriculture, Farmer's Bulletin [Reprint of Bureau of Entomology Circular 48] No. 627: 1-4.

Martin, A.C., H.S. Zim, A.L. Nelson 1961. American Wildlife and Plants: A guide to wildlife food habits. McGraw-Hill Book Company, Inc. New York.

Matic, Z. 1960. Die Cryptopiden (Myriapoda, Chilopoda) der Sammlung des speologischen Institutes "E. Gh. Racovitza" aus Cluj. Zoologischer Anzeiger 165: 442-447.

Matic, Z. 1966. Clasa Chilopoda, Subclasa Anamorpha, Fauna Republicii Socialiste România. Bucuresti. 6(1): 1-272.

Matic Z. 1970 Contributo alla conoscenza dei chilopodi di Turchia. Fragmenta Entomologica 7: 5-13.

Matic, Z. 1972. Clasa Chilopoda, Subclasa Epimorpha. Fauna Republicii Socialiste România. Buchresti. 6(2): 1-204.?

Matic, Z. 1973. Pseudolithobiidae fam. nov.: Una nuova famiglia dell'ordine dei Lithobiomorpha (Chilopoda, Anamorpha). Fragmenta Entomologica 9(3): 135-142.

Matic Z. 1974. Contribution à la connaissance du genre Bothropolys Wood 1863 (Lithobiomorpha Ethopolidae). Annales Zoologici, Instytut Zoologiczny, Polska Akademia Nauk 31: 329-341.

Matic, Z., S. Negrea and F. Martinez 1977. Reserches sur les Chilopodes hypogés de Cuba. *In*: Resultas de Expeditions Biospeleogiques Cubano-Roumaines à Cuba. II Editura Academici Republicii Socialiste Romania, Bucurest 2: 277-301.

Matthewman, W.G. and D.P. Pielou 1971. Arthropods inhabiting the sporophores of *Fomes fomentarius* (Polyporaceae) in Gatineau Park, Quebec. Canadian Entomologist 103: 775-847.

Matthews, D.C. 1935. The Chilopoda of Wisconsin. University of Wisconsin - Madison, Ph.D. Thesis, 119 pp.

McAllister, C.T., R.M. Shelley, and J.T. McAllister 2003. Geographic distribution records for scolopendromorph centipedes (Arthropoda: Chilopoda) from Arkansas, Oklahoma, and Texas. Journal of the Arkansas Academy of Science 57:111-114.

McAllister, C.T., R.M. Shelley, and M.L. Cameron 2004. Significant new distribution records for the centipede, *Theatops posticus* (Say) (Chilopoda: Scolopendromorpha: Cryptopidae), from Oklahoma, with four new records from the Ark-La-Tex. *Proc. Okla. Acad. Sci.* 84:73-74.

McFee, R.B., T.R. Caraccio, H.C. Mofenson and M.A. McGuigan, 2002. Envenomation by the Vietnamese centipede in a Long Island pet store. Journal of Toxicology—Clinical Toxicology. 40(5): 573-574.

McNeill, J. 1887a. List of the Myriapods found in Escambia County, Florida, with descriptions of six new species. Proceedings of the United States National Museum 10: 323-327.

McNeill, J. 1887b. Descriptions of twelve new species of myriapoda, cheifly from Indiana. Proceedings of the United States National Museum 10: 328-334.

McNeill, J. 1888. A list, with breif descriptions of all the species, including one new to science, of myriapoda of Franklin Co., Ind. Bulletin of the Brookville Society of Natural History 3: 1-20.

Meinert, F. 1868. Danmarks Scolopender og Lithobier., Naturhistorisk Tidsskrift (3)5: 241-268.

Meinert, F. 1870. Myriapoda Musaei Hauniensis I. Geophili. Naturhistorisk Tidsskrift 3: 1-128.

Meinert, F. 1872. Myriapoda Musaei Hauniensis II. Lithobini. Naturhistorisk Tidsskrift (3)8: 281-344.

Meinert, F. 1886a. Myriapoda musei cantabrigensis, Part I. Chilopoda. Proceedings of the American Philosophical Society 23: 161-233.

Meinert, F. 1886b. Myriapoda Musaei Hauniensis III. Chilopoda. Videnskabelige meddelelser fra Dansk naturhistorisk forening 38: 100-150.

Mercurio, R.J. 2005. New centipede records from Long Island, New York. Journal of the New York Entomological Society 113(3-4): 232-233.

Mercurio, R.J. (Unpublished). First report of a centipede from North Dakota.

Metz, R. 1987. Recent traces by invertebrates in aquatic nonmarine environments. The Bulletin of the New Jersey Academy of Science, 32(1): 19-24.

Meyer-Rochow, V.B., C.H.G. Müller and M. Lindström, 2006. Spectral sensitivity of the eye of *Scutigera coleoptrata* (Linnaeus, 1758) (Chilopoda: Scutigeromorpha: Scutigeridae). Applied Entomology and Zoology 41(1): 117-122.

Minelli, A. (ed.)—Chilobase: A World Catalogue of Centipedes (Chilopoda) for the Web—*http://chilobase.bio.unipd.it*, used as a reference intermitently since inception.

Molinari, J., Gutiérrez, E., De Ascencão, A., Nasar, J., Arends, A. & Márquez, R. 2005. Predation by giant centipedes, *Scolopendra gigantia*, on three species of bats in a Venezuelan cave. Carribean Journal of Science 41(2): 340-346.

Morse, M. 1902. Myriapods from Vinton, Ohio. Ohio Naturalist 2(3): 187.

Müller, C.H.G., J. Rosenberg, S. Richter & V.B. Meyer-Rochow 2003. The compund eye of *Scutigera coleoptrata* (Linnaeus, 1758) (Chilopoda: Notostigmophora): an ultrastructural reinvestigation that adds support to the Mandibulata concept. Zoomorphology 122: 191-209.

Müller, Carsten H. G., J. Rosenberg, & V. B. Meyer-Rochow. 2003. Hitherto undescribed interommatidial exocrine glands in Chilopoda. African Invertebrates, 44 (1): 185-197.

Mundel: 1979. The centipedes (Chilopoda) of the Mazon Creek. In: Mazon Creek Fossils: 361-378.

Mundel: 1981. A review of the Lithobiomorph centipedes of Mexico. Ph. D. thesis, University of Wisconsin-Madison.

Neck, R.W. 1985. Comparative behavior and external color patters of two sympatric centipedes (Chilopoda: Scolopendra) from central Texas. The Texas Journal of Science 37: 253-255.

Newport, G. 1844a. Untitled. Proceedings of the Linnean Society of London 1: 191-196.

Newport, G. 1844b. A list of the species of Myriapoda, Order Chilopoda, contained in the cabinets of the British Museum, with synoptic descriptions of forty-seven new species. The Annals and Magazine of Natural History ser. 1, 8: 94-101.

Newport, G. 1844-45. Monograph of the Class Myriapoda, Order Chilopoda; with observations on the general arrangement of the Articulata. Transactions of the Linnaen Society of London 19: 265-302 (1844), 349-340 (1845).

Newport, G. 1856. Catalogue of the myriapoda in the collection of the British Museum, Part I. Chilopoda. British Museum (Natural History), London, 96pp.

Nishida, G.M. 1994. Hawaiian terrestrial arthropod checklist. Hawaii Biological Survey. Second Edition. Bishop Museum Technical Report 4: 26-27.

Nishida, G.M. 1997. Hawaiian terrestrial arthropod checklist. Hawaii Biological Survey. Third Edition. Bishop Museum Technical Report 12: 195.

Nishida, G.M. 2002. Hawaiian terrestrial arthropod checklist. Fourth Edition. Bishop Museum Technical Report 22: 313 p.

Okeden, W.P. 1903. A centipede eating a snake. The Journal of the Bombay Natural History Society 15: 135.

Packard, A.S. Jr. 1870. New or rare Neuroptera, Thysanunra and Myriapoda Proceedings of the Boston Society of Natural History 13: 405-411.

Palmén, E. 1949. The chilopoda of Eastern Fennoscandia. Annales Zoologicae Societatis Zoologicae Botanicae Fennicae. 'Vanamo' Tom. 13, no. 4: 1-46.

Palmén, E. 1954. Survey of the Chilopoda of Newfoundland. Archives Societatis Zoologicae Botanicae Fennicae 'Vanamo' 8: 2: 131-149.

Pereira, L. A. and R.L. Hoffman, 1993. The American species of *Escaryus*, a genus of Holarctic centipedes (Geophilomorpha: Schendylidae), Jeffersoniana 3: 1-72.

Peters, E.H. 1954. A centipede (*Arenophilus bipuncticeps* Wood): Ontario. Canadian Insect Pest Review. 32: 135.

Plateau, F. 1872. Matériaux pour la Faune Belge. 2me note. Myriapodes. Bulletin de l'Académie Royale des Sciences, des Lettres et des Beaux-Arts de Belgique (2)33: 409-418.

Pocock, R.I. 1888a. On the genus *Theatops*. The Annals and Magazine of Natural History. ser. 6, 1: 283-290.

Pocock, R.I. 1888b. Contributions to our knowledge of the myriapoda of Dominica. The Annals and Magazine of Natural History. ser. 6, 2: 472-483.

Pocock, R.I. 1890. Contributions to our knowledge of the Chilopoda of Liguria. Annali Mus. Civ. Stor. Nat. Giacomo Doria, 29: 59-68.

Pocock, R.I. 1891a. Notes on the synonomy of some species of Scolopendridae, with descriptions of new genera and species of the group. The Annals and Magazine of Natural History ser. 6, 7: 51-68, 221-231.

Pocock, R.I. 1891b. Descriptions of some new species of Chilopoda. The Annals and Magazine of Natural History ser. 6, 8: 152-164.

Pocock, R.I. 1893. Report upon the myriapoda of the 'Challenger' Expedition, with remarks upon the fauna of Bermuda. The Annals and Magazine of Natural History. ser. 6, 11: 121-142.

Pocock, R.I. 1894a. Contributions to our knowledge of the arthropod fauna of the West Indies. Part II. Chilopoda. The Journal of the Linnean Society of London. Zoology. 24: 454-473.

Pocock, R.I. 1894b. In: Weber, Ergeb. Reise Niederl. O.-Ind., 3: p. 316.

Pocock, R.I. 1895-1910. Chilopoda and diplopoda. Biologia Centrali-Americana, 217 pp. [Fascicles treating North American Scolopenromorpha distributed in December 1895 and January 1896].

Pocock, R.I. 1898. List of the Arachnida and "Myriopoda" obtained in Funafuti by Prof. W.J. Sollas & Mr. Stanley Gardiner, and in Rotuma by Mr. Stanley Gardiner. Ann. Nat. Hist. 7(1): 321-329.

Poinar, G. Jr. and Poinar, R. 1999. The amber forest. A reconstruction of a vanished world. Princeton University Press, Princeton, New Jersey, 239 pp.

Porath. C.O. von. 1876. Om några exotiska myriapoder. Bihang Till Kongl. Svenska Vetenskaps-Akademiens Handlingar. 4: 1-48.

Provancher, L. 1873. Myriapodes. Le Naturliste Canadien (Qué.) 5: 410-419.

Rafinesque, C. S. 1820. Annals of Nature, or Annual synopsis of new genera and species of animals and plants discovered in North America. First Annual Number for 1820: 1-20.

Rapp, J.L.C. 1946. List of Myriapoda taken in Champaign County, Illinois, during the fall and winter of 1944-45. 36: 666-667.

Reddell, J.R. 1965. A checklist of the cave fauna of Texas. I. The Invertebrata (exclusive of Insecta). The Texas Journal of Science 17: 143-187.

Reddell, J.R. 1970. A checklist of the cave fauna of Texas. IV. Additional records of Invertebrata (exclusive of Insecta). The Texas Journal of Science 21: 389-415.

Reddell, J.R. 1994. The Cave Fauna of Texas with Special Reference to the Western Edwards Plateau. Pp. 31-50, in: The Caves and Karst of Texas (W.R. Elliot, and G. Venti, eds.). National Speleological Society, Huntsville, Alabama. 252 pp.

Regier, J.C., H.M. Wilson and J.W. Shultz 2005. Phylogenetic Analysis of Myriapoda Using Three Nuclear Protein-Coding Genes. Molecular Phylogeny and Evolution 34: 147-158.

Remington, C.L. 1950. The bite and habits of a giant centipede (*Scolopendra subspinipes*) in the Philippine Islands. The American Journal of Tropical Medicine 30: 453-455.

Ribaut, H. 1911. Sur un genere nouveau de la sous-tribu des Ribautiina Brol. Toulouse Bul. Soc. Hist. Nat. 43: 105–126.

Ribaut, H. 1921. L'armement des pattes chez les Lithobies. Bulletin de la Société d'histoire naturelle de Toulouse. 49: 312-319.

Rilling, G. 1968. *Lithobius forficatus*. Grosses Zoologisches Praktikum, Heft 13b, Mit 52 Abbildungen, Gustav Fischer Verlag • Stuttgart, 136pp. [In German but has excellent figures of musculature.]

Risso, 1826. Histoire naturelle des principales productions de l'Europe méridionale et particulièrement de celles des environs de Nice et des Alpes Maritimes: 153.

Roberts, H. 1956. An ecological study of the arthropods of a mixed beech-oak woodland, with particular reference to Lithobiidae. Ph. D. thesis, University of Southhampton.

Ross, A.J. and D.E.G. Briggs 1993. Arthropoda (Euthycarcinoidea and Myriapoda). *In*: M.J. Benton (Editor). The Fossil Record 2, Chapman & Hall, London.

Sager, Ab. 1856. Descriptions of three Myriapoda. Proceedings of the Academy of Natural Sciences of Philadelphia. 8: 109.

Salt, G., F.S. Hollick, F. Raw and M.V. Brian 1948. The arthropod population of pasture soil. The Journal of Animal Ecology 17(2): 139-150.

Sandefer, C. 1998. The Giant Centipedes of the genus *Scolopendra*: Their Captive Care and Husbandry. Privately printed. Pages unnumbered.

Saussure, H. 1859. Essai d'une faune des Myriopodes du Mexique, avec la description de quelques espèces des autres parties de l'Amérique. Mém. Soc. Phys. Hist. Nat. Genève, 15: 259-393.

Saussure, H. 1860. Essai d'une faune des Myriopodes du Mexique, avec la description de quelques espèces des autres parties de l'Amérique. Mém. Soc. Phys. Hist. Nat. Genève, 121-135.

Saussure H. and A. Humbert, 1872. Études sur les myriapodes. 212 p.

Say, T. 1821. Descriptions of the Myriapodae of the United States. Journal of the Academy of Natural Sciences of Philadelphia 2(1): 102-114.

Schileyko, A. 1992. Scolopenders of Viet-Nam and some aspects of the system of Scolopendromorpha (Chilopoda Epimorpha). Part 1. Arthropoda Selecta 1(1): 5-19.

Schileyko, A. 1995. The scolopendromorph centipedes of Vietnam (Chilopoda Scolopendromorpha). Part 2. Arthropoda Selecta 4(2): 73-87.

Schileyko, A. & Minelli, A. 1998. On the genus *Newportia* Gervais, 1847. Arthropoda Selecta 7(4): 265-299.

Scudder, S.H. 1890a. New Carboniferous Myriapoda from Illinois. *In:* Fossil Insects. Memoirs of the Boston Society of Natural History 4: 417-442.

Scudder, S.H. 1890b. The fossil insects of North America, with notes on some European species, vol. 1, the Pre-Tertiary insects. New York: MacMillan, 455 pp.

Selander, R.B. and P. Vaurie 1962. A gazetteer to accompany the "Insecta" volumes of the "Biologia Centrali-Americana" American Museum Novitates 2099: 1-70.

Selivanoff, 1881. Geophilidae Muzeia Imperatorskoi Akademii Nauk. Imperatorskoi Akademii Nauk St. Petersburg 40(7): 1-27.

Shaw, 1794. Linn. Trans, ii. p. 7Info on Stigmatogaster subterraneus.

Shear, W.A. and P.M. Bonamo 1988. Devonobiomorpha, A New Order of Centipeds (Chilopoda) from the Middle Devonian of Gilboa, New York State, USA, and the Phylogeny of Centiped Orders. American Museum Novitates 2927: 1-30; Figs. 1-75, tables 1-4.

Shear, W.A., J.T. Hannibal, and J. Kukalová-Peck 1992. Terrestrial arthropods from the Upper Pennsylvanian rocks at the Kinney Brick Quarry, New Mexico. New Mexico Bureau of Mines and Mineral Resources Bulletin 138: 135-141.

Shear, W.A., A.J. Jeram and P.A. Selden 1998. Centipede legs (Arthropoda, Chilopoda, Scutigeromorpha) from the Silurian and Devonian of Britain and the Devonian of North America. American Museum Novitates 3231: 1-16.

Shelley, R.M. 1978. Class Chilopoda p. 221, *In*: Zingmark, Richard G., ed., An Annotated Checklist of the Biota of the Coastal Zone of South Carolina. University of South Carolina Press, Columbia, 364 pp.

Shelley, R.M. 1987. The Scolopendromorph centipedes of North Carolina, with a taxonomic assesment of *Scolopocryptops gracilis peregrinator* (Crabill) (Chilopoda: Scolopendromorpha). Florida Entomologist 70(4): 498-512.

Shelley, R.M. 1990. The Centipede *Theatops posticus* (Say) (Scolopendromorpha: Cryptopidae) in the southwestern United States and Mexico. Canadian Journal of Zoology 68: 2637-2644.

Shelley, R.M. 1991. Deletion of the centipede *Theatops spinicaudus* (Wood) from the Hawaiian fauna (Scolopendromorpha: Cryptopidae). Bishop Museum Occasional Papers 31: 182-184.

Shelley, R.M. 1992. Distribution of the centipede *Scolopocryptops sexspinosus* (Say) in Alaska and Canada (Scolopendromorpha: Cryptopidae). Insecta Mundi 6(1): 23-27.

Shelley, R.M. 1997. The Holarctic Centipede Subfamily Plutoniuminae (Chilopoda: Scolopendromorpha: Cryptopidae) (Nomen Correctum Ex Subfamily Plutoniinae Bollman, 1893). Brimleyana 24: 51-113.

Shelley, R.M. 2000a. Occurrence of the centipede, *Dinocryptops miersii* (Newport) (Scolopendromorpha: Scolopocryptopidae), in Tobago, Trinidad and Tobago. Caribbean Journal of Science 36(1-2): 155-156.

Shelley, R.M. 2000b. The centipede order Scolopendromorpha in the Hawaiian Islands (Chilopoda). Bishop Museum Occasional Papers 64: 39-48.

Shelley, R.M. 2000c. Neotype designation for the centipede *Mycotheres leucopoda* Rafinesque (Scolopendromorpha: Cryptopidae). Myriapodologica 7(2): 15-17.

Shelley, R.M. 2002. A synopsis of the North American centipedes of the order Scolopendromorpha (Chilopoda), Virginia Museum Natural History, Memoir 5, 108 pgs.

Shelley, R.M. 2003. *Otocryptops gracilis berkeleyensis* Verhoeff, 1938, a synonym of *Scolopocryptops gracilis* Wood, 1862 (Chilopoda: Scolopendromorpha: Scolopocryptopidae). Entomological News 114(1): 57.

Shelley, R.M. 2006a. Nomenclator generum et familiarum Chilopodorum II: A list of the genus—and family-group names in the Class Chilopoda from 1958 through 2005. Zootaxa 1198: 1-20.

Shelley, R.M. 2006b. A chronological catalog of the New World species of *Scolopendra* L., 1758 (Chilopoda: Scolopendromorpha; Scolopendridae). Zootaxa 1253: 1-50.

Shelley, R.M. 2008. Revision of the Centipede Genus *Hemiscolopendra* Kraepelin, 1903: Description of *H. marginata* (Say, 1821) and possible misidentifications as *Scolopendra* spp.; proposal of *Akymnopellis*, n. gen., and redescriptions of its South American components (Scolopendromorpha: Scolopendridae: Scolopendrinae). International Journal of Myriapodology. 2: 171-204.

Shelley, R.M. and T. Backeljau 1995. Plutoniinae Bollman, 1893 (Arthropoda, Chilopoda) and Plutoniinae Cockerell, 1893 (Mollusca, Gastropoda): Proposed Removal of Homonymy. Bulletin of Zoological Nomenclature 52(2) 150-152.

Shelley, R.M. and G.B. Edwards 1987. The Scolopendromorph centipedes of Florida, with an introduction to the common myriapodous arthropods. Florida Dept. Agric. Cons. Servs., Div. Plant Ind., Ent. Circ. No. 300: 1-4.

Shelley, R.M. and G.B. Edwards 2004. A fourth Floridian record of the centipede genus *Rhysida* Wood, 1862; potential establishment of *R. L. Longipes* (Newport, 1845) in Miami-Dade County (Scolopendromorpha: Scolopendridae: Otostigminae). Entomological News 115(2) 116-119.

Shelley, R.M. and A. Chagas Jr. 2004. The centipede genus *Arthrorhabdus* Pocock, 1891, in the western hemisphere: potential occurrence of *A. pygmaeus* (Pocock, 1895) in Belize (Scolopendromorpha: Scolopendridae: Scolopendrinae). Western North American Naturalist 64(4): 532-537.

Shelley, R.M. and D.L. Six 2004. Discovery of the centipede *Scolopocryptops gracilis* Wood in Montana (Scolopendromorpha: Scolopocryptopidae) Western North American Naturalist 64(2): 257-258.

Shelley, R.M., G.B. Edwards, and A. Chagas Jr. 2005. Introduction of the centipede *Scolopendra morsitans* L., 1758, into northeastern Florida, the first authentic North American record, and a review of its global occurences (Scolopendromorpha: Scolopendridae: Scolopendrinae). Entomological News 116(1): 39-58.

Shinohara, K. 1990. A new species of the genus *Scolopocryptops* (Chilopoda: Cryptopidae) from Japan. Proceedings of the Japanese Society of Systematic Zoology. 41: 62-65.

Shugg, H.B. 1961. Predation on mouse by centipede. The Western Australian Naturalist 8(2): 52.

Silvestri F. 1897. Contributo alla conoscenza dei chilopodi e diplopodi della Sicilia Bullettino della Società Entomologica Italiana 29(4): 233-261.

Silvestri, F. 1904. (1913). Fauna Hawaiiensis, Myriapoda, 3, Part 4: 323-338, + 2 Plates.

Silvestri, F. 1909. Contribuzioni alla conoscenza dei Chilopodi III-IV. III Descrizioni di alcuni generi e specie die Henicopidae. Bolletino del Laboratorio di Zoologia Portici. 4: 38-50.

Silvestri, F. 1919. Contributions to a knowledge of the Chilopoda Geophilomorpha of India. Rec. Ind. Mus. Calcutta 16: 45-107.

Snider, R.M. 1991: Updated species lists and distribution records for the Diplopoda and Chilopoda of Michigan. Michigan Acad. 24(1) 177-194.

Spencer, G.J. 1942. Insects and other arthropods in buildings in British Columbia. Proceedings of the Entomological Society of British Columbia. 39: 23-29.

Stoev and Jean-Jacques Geoffroy, 2004. An annotated catalogue of the scutigeromorph centipedes in the collection of the Muséum National d'Histoire Naturelle, Paris (France) (Chilopoda: Scutigeromorpha). Zootaxa 635: 1-12.

Stuxberg, A. 1871. Öfversight af Kongl. Vetenskaps-Akademiens Förhandlingar 28: 504.

Stuxberg, A. 1875a. Nya Nordamerikanska Lithobier. Öfversight af Kongl. Vetenskaps-Akademiens Förhandlingar 32(2): 65-72.

Stuxberg, A. 1875b. Genera et species Lithobiodarum disposuit. Öfversight af Kongl. Vetenskaps-Akademiens Förhandlingar 3: 5-22.

Stuxberg, A. 1875c. Lithobioidae Americae Borealis: Öfversight af Nord-Amerikas hittills kända Lithobiider. Öfversight af Kongl. Vetenskaps-Akademiens Förhandlingar 3: 23-32.

Stuxberg, A. 1875d. Descriptions of some new North-American Lithobioidae. Annals and Magazine of Natural History 4(15): 188-192.

Stuxberg, A. 1876. Myriopoder fran Sibirien och Waigatsch on samlade under Nordenskioldska expeditionen 1875. Öfversight af Kongl. Vetenskaps-Akademiens Förhandlingar No. 2, Tafl. 2. p. 11-38.

Stuxberg, A. 1877. Lithobioidae Americae Borealis: Preliminary Report on the Lithobii of North America. Proceedings of the California Academy of Sciences 7: 132-139. (This is the English translation of Stuxberg, 1875c)

Summers, G. 1979. An Illustrated Key to the Chilopods of the North-Central Region of the United States. Journal of the Kansas Entomological Society 52: 690-700.

Summers, G., J.A. Beatty, and N. Magnuson 1980. A checklist of Illinois Centipedes (Chilopoda). Great Lakes Entomologist 13: 241-257.

Summers, G., J.A. Beatty, and N. Magnuson 1981. A Checklist of Illinois Centipedes (Chilopoda): Supplement. Great Lakes Entomologist 14: 59-62.

Summers, G. and G.W. Uetz 1979. Microhabitats of woodland centipedes in a streamside forest. The American Midland Naturalist 102: 346-352.

Uliana, M., L. Boanato and A. Minelli, 2007. *Geophilus holstii* Pocock, 1895 (currently *Arrup holstii*; Chilopoda; Mecistocephalidae): replacement of holotype by designation of neotype. Bulletin of the International Commission on Zoological Nomenclature, 64(4): 227-229.

Underwood, L.M. 1885. The North American Myriapods., Entomologica Americana 1: 141-151.

Underwood, L.M. 1887. The Scolopendridae of the United States., Entomologica Americana 3: 61-65.

Verhoeff, K.W. 1900. Üeber Schendyla und Petiniunguis. Zoologischer Anzeiger 23(624): 483-486.

Verhoeff K.W. 1905a. Über Scutigeriden. 5 Aufsatz. Zoologischer Anzeiger Band 29(2/3): 73-119.

Verhoeff K.W. 1905b. Über die Entwickelungsstufen der Steinläufer Lithobiiden und Beiträge zur Kenntnis der Chilopoden. Zoologische Jahrbücher, Abteilung für Systematik 8: 195-298.

Verhoeff, K.W. 1907. (1902-1925). Chilopoda: 1-725, In: Bronn, H. G., Klassen und Ordnungen des Tier-Reichs, wissenschaftlich dargestellt in Wort und Bild, Band 5, Abt. 2, Lief. 63-101.

Verhoeff, K.W. 1924. Über Myriapoden von Juan Fernandez und der Osterinsel, p. 403-418. *In*: Skottsberg K., ed., The Natural History of Juan Fernandez and Easter Island. Vol. 3, Zoology. Almqvist & Wiksells, Uppsala. 688 p.

Verhoeff, K.W. 1925. Beitrage zur Kenntis der Steinlaufer, Lithobiiden., Arkiv. f. Naturg. A9: 124-158.

Verhoeff, K.W. 1925. Klasse Chilopoda. *Bronn's Klass. U. Ordnung.* Tier-reich, 5(2) 1: 1-725.

Verhoeff, K.W. 1928. Geophilomorpha Beitrage und ein *Lithobius* form., Mitt. Zool. Mus. Berlin, 14: 229-286.

Verhoeff, K.W. 1928. Über Chilopoden aus Bulgarien gesammelt von Herrn Dr. Iw. Buresch 1. Aufsatz-Mitteilungen der Bulgarischen Entomologischen Gesellschaft 4: 115-124.

Verhoeff, K.W. 1934. Beitrage zur Systematik und Geographie der Chilopoden. Zoologische Jahrbucher (Syst.) 66: 1-112.

Verhoeff, K.W. 1935. Über *Scolioplanes* (Chilopoda). Zoologischer Anzeiger. 111: 10-23.

Verhoeff, K.W. 1937. Zur Kenntnis der Lithobiiden., Arkiv. f. Naturg., (n. F.) 6: 169-257.

Verhoeff, K.W. 1938a. Über einige Amerikanische Myriapoden. Zoologischer Anzeiger. 122: 273-284.

Verhoeff, K.W. 1938b. Chilopoden-Studien, zur Kenntnis der Epimorphen., Zoologische Jahrbucher (Syst.) 71: 339-388.

Viosca: 1918. Myriapoda of Louisiana. Annual Report of the Louisiana State Museum : 42-47.

Wang D. and Mauriès J.-P. 1996. Review and Perspective Study on Myriapodology in China. *In*: Geoffroy j.-J., Mauriès J.-P. & Nguyen Duy-Jacquemin M. (eds.), Acta Myriapodologica. Mémoires du Muséum National d'Histoire Naturelle, Paris 169: 81-99.

Waterhouse, J.S. 1968. An evaluation of a new predaceous centipede *Lamyctes* sp., on the Garden symphylan *Scutigerella immaculata*. Canadian Entomologist 101: 1081-1083.

Watermolen, D.J. 1996. The centipede Lithobius celer (Chilopoda: Lithobiidae) in Wisconsin. University of Wisconsin-Milwaukee Field Station Bulletin 29(2): 11-13.

Watermolen, D.J. 1997. Updated Checklist of Wisconsin Centipedes (Chilopoda). Wisconsin Entomological Society Special Publication No. 4, March: 8 pp.

Wheeler, W.C.: Cartwright, and C.Y. Hayashi. 1993. Arthropod Phylogeny: A combined approach. Caldistics 9: 1-39.

Wild, A. 2005. Observations on larval cannibalism and other behaviors in a captive colony of *Amblyopone oregonensis*. *http://www.notesfromunderground. org/issue11-1/features/amblyopone.html*.

Willey, A. and E.M. Walker 1920. The house centipede, *Cermatia forceps* Raf. In Montreal [and Toronto]. Canadain Entomologist 52: 8.

Williams, F.X. (Com.) 1931. The Myriapoda: Classes Diplopoda (Millipedes) and Chilopoda (Centipedes). pgs. 313-352. *In*: Handbook of The Insects and Other Invertebrates of Hawaiian Sugar Cane fields. Experiment Station of the Hawaiian Sugar Planters' Association. 400pp.

Williams, S.R. and R.A. Hefner 1928. The Millipedes and Centipedes of Ohio. Ohio Biological Survey, The Ohio State University Bulletin 4: 91-146.

Wood, D.M. and A.G. Wheeler 1972. First record in North America of the centipede parasite *Loewia foeda* (Diptera: Tachinidae). Canadian Entomologist 104: 1363-1369.

Wood, H.C. 1861. Descriptions of new species of *Scolopendra* in the collection of the Academy. Proceedings of the Academy of Natural Sciences of Philadelphia 13: 10-15.

Wood, H.C. 1862. On the Chilopoda of North America, with a Catalogue of all the Specimens in the Collection of the Smithsonian Institution. Journal of the Academy of Natural Sciences of Philadlphia 6: 5-52.

Wood, H.C. 1865. The Myriapoda of North America., Transactions of the American Philosophical Society 13: 137-248.

Wood, H.C. 1867a. Descriptions of new species of Texan Myriapoda. Proceedings of the Academy of Natural Sciences of Philadelphia 19: 42-44.

Wood, H.C. 1867b. Notes on a collection of California Myriapoda, with the descriptions of new eastern species. Proceedings of the Academy of Natural Sciences of Philadelphia 19: 127-130.

Wray, D.L. 1950. Insects of North Carolina, Second Supplement. North Carolina Department of Agriculture, Division of Entomology, Raleigh, 181 pp.

Wray, D.L. 1967. Insects of North Carolina, Third Supplement. North Carolina Department of Agriculture, Division of Entomology, Raleigh, 181 pp. [**Note:** The Third Supplement is essentially the same as the Second therefore in most cases only the Third is referenced].

Würmli, M. 1973. Die Scutigeromorpha (Chilopoda) von Costa Rica. Ueber Dendrothereua arborum Verhoeff, 1944. Studies on the Neotropical Fauna 8: 75-80."

Zapparoli, M. and R.M. Shelley 2000. The centipede order lithobiomorpha in the Hawaiian Islands (Chilopoda). I. The Epigean Fauna. Bishop Museum Occasional Papers 63: 35-49.

Appendix I

List of Indigenous Extant Centipedes

1. *Agathothus gracilis* (Bollman, 1888)
2. *Alaskobius adlatus* Chamberlin, 1946
3. *Alaskobius josephus* Chamberlin, 1946
4. *Alaskobius parvior* Chamberlin, 1946
5. *Archethopolys gosobius* Chamberlin, 1928
6. *Archethopolys kaibabus* Chamberlin, 1930
7. *Archethopolys parowanus* (Chamberlin, 1925)
8. *Arctogeophilus atopus* (Chamberlin, 1902)
9. *Arctogeophilus corvallis* Chamberlin, 1941
10. *Arctogeophilus fulvus* (Wood, 1862)
11. *Arctogeophilus glacialis* (Attems, 1909)
12. *Arctogeophilus insularis* Attems, 1947
13. *Arctogeophilus melanonotus* (H. C. Wood, 1862)
14. *Arctogeophilus shelfordi* Chamberlin, 1946
15. *Arctogeophilus umbraticus* (McNeill, 1887)
16. *Arebius agamus* Chamberlin, 1941
17. *Arebius cherosus* Chamberlin, 1941
18. *Arebius convergens* Chamberlin, 1943
19. *Arebius crenius* Chamberlin, 1941
20. *Arebius diplonyx* Chamberlin, 1916
21. *Arebius dolius* Chamberlin, 1916
22. *Arebius elysianus* Chamberlin, 1916
23. *Arebius epelus* Chamberlin, 1941
24. *Arebius fremontus* Chamberlin, 1943
25. *Arebius integrior* Chamberlin, 1949
26. *Arebius kochii* (Stuxberg, 1875)
27. *Arebius medius* Chamberlin, 1916
28. *Arebius montivagus* Chamberlin, 1943

29. *Arebius navajo* Chamberlin, 1943
30. *Arebius obesus* (Stuxberg, 1875)
31. *Arebius oregonensis* Chamberlin, 1916
32. *Arebius petrovius* Chamberlin, 1947
33. *Arebius platypus* Chamberlin, 1941
34. *Arebius sequens* Chamberlin, 1947
35. *Arebius sequoius* Chamberlin, 1941
36. *Arebius tetonus* Chamberlin, 1943
37. *Arebius tridens* Chamberlin, 1941
38. *Arenobius manegitus* (Chamberlin, 1911)
39. *Arenophilus bipuncticeps* (H. C. Wood, 1862)
40. *Arenophilus iugans* Chamberlin, 1944
41. *Arenophilus osborni* Gunthorp, 1913
42. *Arenophilus psednus* Crabill, 1969
43. *Arenophilus unaster* (Chamberlin, 1909)
44. *Arenophilus watsingus* Chamberlin, 1912
45. *Arkansobius lamprus* Chamberlin, 1938
46. *Arrup pylorus* Chamberlin, 1912
47. *Arthrorhabdus pygmaeus* Pocock, 1895
48. *Banobius tener* Chamberlin, 1938
49. *Bothropolys columbiensis* Chamberlin, 1925
50. *Bothropolys dasys* Chamberlin, 1941
51. *Bothropolys ethus* Chamberlin, 1946
52. *Bothropolys hoples* (Brölemann, 1896)
53. *Bothropolys multidentatus* (Newport, 1845)
54. *Bothropolys permundus* (Chamberlin, 1902)
55. *Bothropolys victorianus* Chamberlin, 1925
56. *Brachygonarea borealis* Attems, 1934
57. *Buethobius arizonicus* Chamberlin, 1925
58. *Buethobius coniugans* Chamberlin, 1911
59. *Buethobius huestoni* Williams & Hefner, 1928
60. *Buethobius oabitus* Chamberlin, 1911
61. *Buethobius translucens* Williams & Hefner, 1928
62. *Calcibius calcarifer* Chamberlin & Wang, 1952
63. *Californiphilus michaelbacheri* Verhoeff, 1938
64. *Caliphilus alamedanus* Chamberlin, 1941
65. *Cheiletha alaska* Chamberlin, 1946
66. *Cheiletha kincaidi* Chamberlin, 1955
67. *Cheiletha phoenix* Chamberlin, 1955
68. *Chomatobius auximus* (Chamberlin, 1938)
69. *Chomatobius bakeri* (Chamberlin, 1912)

70. *Chomatobius euphorion* (Crabill, 1953)
71. *Chomatobius laticeps* (H. C. Wood, 1862)
72. *Chomatobius mexicanus* (Saussure, 1858)
73. *Chomatobius minor* (Chamberlin, 1912)
74. *Chomatobius minor arizonicus* (Chamberlin, 1925)
75. *Condylona isabella* Chamberlin, 1941
76. *Condylona sontipes* Chamberlin, 1941
77. *Cryptops floridanus* Chamberlin, 1925
78. *Cryptops leucopodus* (Rafinesque, 1820)
79. *Cryptops parisi* Brölemann, 1920
80. *Damothus alastus* Crabill, 1962
81. *Damothus montis* Crabill, 1960
82. *Dicellophilus limatus* (Wood, 1862)
83. *Enarthrobius bullifer* Chamberlin, 1926
84. *Enarthrobius covenus* Chamberlin, 1944
85. *Enarthrobius dybasi* Chamberlin, 1944
86. *Enarthrobius fumans* Chamberlin, 1944
87. *Enarthrobius litus* Chamberlin, 1944
88. *Enarthrobius oblitus* Chamberlin & Wang, 1952
89. *Ephemerozaster antaeus* Crabill, 1959
90. *Eremerium apachum* Chamberlin, 1941
91. *Eremorus becki* Chamberlin, 1963
92. *Erithophilus neopus* Cook, 1899
93. *Escaryus cryptorobius* Pereira & Hoffman, 1993
94. *Escaryus ethopus* (Chamberlin, 1920)
95. *Escaryus liber* Cook & Collins, 1891
96. *Escaryus missouriensis* Chamberlin, 1942
97. *Escaryus monticolens* Chamberlin, 1947
98. *Escaryus orestes* Pereira & Hoffman, 1993
99. *Escaryus paucipes* Chamberlin, 1946
100. *Escaryus urbicus* (Meinert, 1886)
101. *Escimobius cryophilus* Chamberlin, 1949
102. *Ethopolys bipunctatus* (Wood, 1862)
103. *Ethopolys bipunctatus insulatus* Chamberlin, 1955
104. *Ethopolys calibius* Chamberlin, 1951
105. *Ethopolys californicus* (Daday, 1889)
106. *Ethopolys integer alaskanus* Chamberlin, 1919
107. *Ethopolys integer* Chamberlin, 1919
108. *Ethopolys monticola* (Stuxberg, 1875)
109. *Ethopolys positivus* Chamberlin, 1941
110. *Ethopolys pusio* (Stuxberg, 1875)

111. *Ethopolys spectans* Chamberlin, 1951
112. *Ethopolys timpius* Chamberlin, 1951
113. *Ethopolys xanti* (Wood, 1862)
114. *Eulithobius fattigi* Chamberlin, 1945
115. *Eulithobius hypogeus* Chamberlin, 1940
116. *Eulithobius sphactes* Crabill, 1958
117. *Garibius alabamae* Chamberlin, 1913
118. *Garibius branneri* (Bollman, 1888)
119. *Garibius catawbae* Chamberlin, 1913
120. *Garibius georgiae* Chamberlin, 1913
121. *Garibius mississippiensis* Chamberlin, 1913
122. *Garibius monticolens* Chamberlin, 1913
123. *Garibius opicolens* Chamberlin, 1913
124. *Garibius pagoketes* Chamberlin, 1913
125. *Garibius psychrophilus* Crabill, 1957
126. *Garrina alicea* Chamberlin, 1943
127. *Garrina ochra* Chamberlin, 1915
128. *Garrina parapodus* (Chamberlin, 1941)
129. *Garriscaphus amplus* Chamberlin, 1941
130. *Garriscaphus oreines* Chamberlin, 1941
131. *Garriscaphus tytthus* Crabill, 1969
132. *Geophilus admarinus* (Chamberlin, 1952)
133. *Geophilus ampyx* Crabill, 1954
134. *Geophilus anonyx* (Chamberlin, 1941)
135. *Geophilus atopodon* Chamberlin, 1903
136. *Geophilus becki* Chamberlin, 1951
137. *Geophilus brevicornis* Wood, 1862
138. *Geophilus brunneus* McNeill, 1887
139. *Geophilus cayugae* Chamberlin, 1904
140. *Geophilus claremontus* Chamberlin, 1909
141. *Geophilus compactus* (Attems, 1934)
142. *Geophilus delotus* (Chamberlin, 1941)
143. *Geophilus embius* (Chamberlin, 1912)
144. *Geophilus fruitanus* Chamberlin, 1928
145. *Geophilus geronimo* (Chamberlin, 1912)
146. *Geophilus glyptus* (Chamberlin, 1902)
147. *Geophilus indianae* McNeill, 1887
148. *Geophilus leionyx* (Chamberlin, 1938)
149. *Geophilus mordax* Meinert, 1886
150. *Geophilus nasintus* Chamberlin, 1909
151. *Geophilus nealotus* Chamberlin, 1902

152. *Geophilus nicolanus* Chamberlin, 1940
153. *Geophilus oregonus* (Chamberlin, 1941)
154. *Geophilus orites* (Chamberlin, 1944)
155. *Geophilus oweni* Bollman, 1887
156. *Geophilus parki* (Auerbach, 1954)
157. *Geophilus phanus* Chamberlin, 1943
158. *Geophilus piedus* Chamberlin, 1930
159. *Geophilus regnans* Chamberlin, 1904
160. *Geophilus rupestris* (Crabill, 1949)
161. *Geophilus secundus* (Chamberlin, 1912)
162. *Geophilus sequoia* Chamberlin, 1941
163. *Geophilus setiger* Bollman, 1887
164. *Geophilus shoshoneus* Chamberlin, 1925
165. *Geophilus smithi* Bollman, 1888
166. *Geophilus strigosus* (McNeill, 1887)
167. *Geophilus tampophor* (Chamberlin, 1953)
168. *Geophilus terrae-novae* Palmén, 1954
169. *Geophilus transitus* (Chamberlin, 1941)
170. *Geophilus varians* McNeill, 1887
171. *Geophilus virginiensis* Bollman, 1888
172. *Geophilus vittatus* (Rafinesque, 1820)
173. *Geophilus whitei* Newport, 1845
174. *Geophilus winnetui* Attems, 1947
175. *Geophilus yavapainus* (Chamberlin, 1941)
176. *Georgibius georgiae* Chamberlin, 1944
177. *Gonibius glyptocephalus* (Chamberlin, 1903)
178. *Gonibius rex* (Bollman, 1888)
179. *Gosendyla socarina* Chamberlin, 1960
180. *Gosibius aberrantus* Chamberlin, 1943
181. *Gosibius ameles* Chamberlin, 1940
182. *Gosibius angelicus* Chamberlin, 1944
183. *Gosibius arizonensis* Chamberlin, 1917
184. *Gosibius atopops* Chamberlin, 1941
185. *Gosibius auxodontus* Chamberlin, & Wang, 1952
186. *Gosibius benespinosus* Chamberlin, 1941
187. *Gosibius brevicornis* Chamberlin, 1917
188. *Gosibius claremontus* Chamberlin, 1917
189. *Gosibius escabosanus* Chamberlin, 1943
190. *Gosibius fusatus* Chamberlin, 1941
191. *Gosibius intermedius* Chamberlin, 1917
192. *Gosibius louisianus* Chamberlin, 1942

193. *Gosibius monicus* Chamberlin, 1912
194. *Gosibius montereus* Chamberlin, 1912
195. *Gosibius mulaiki* Chamberlin, 1938
196. *Gosibius paucidens* (Wood, 1862)
197. *Gosibius saccharogeus* Chamberlin, 1941
198. *Gosibius sequens* Chamberlin, 1941
199. *Gosibius submarginis* Chamberlin & Wang, 1952
200. *Gosibius texicolens* Chamberlin, 1938
201. *Gosipina bexara* Chamberlin, 1940
202. *Guambius christianus* Chamberlin, 1946
203. *Guambius coloradanus* (Chamberlin, 1912)
204. *Guambius euthus* (Chamberlin, 1904)
205. *Guambius hesperus* Chamberlin, 1941
206. *Guambius hubrichti* Chamberlin, 1942
207. *Guambius mississippiensis* (Chamberlin, 1912)
208. *Guambius oedipes* (Bollman, 1888)
209. *Guambius pinguis* (Bollman, 1888)
210. *Harmostela hespera* Chamberlin, 1941
211. *Helembius nannus* Chamberlin, 1918
212. *Hemiscolopendra marginata* (Say, 1821)
213. *Holitys neomexicanus* Cook, 1899
214. *Horonia bella* Chamberlin, 1966
215. *Ityphilus lilacinus* Cook, 1899
216. *Ityphilus nemoides* Chamberlin, 1943
217. *Juanobius eremus* Chamberlin, 1928
218. *Kethops atypus* Chamberlin, 1943
219. *Kethops utahensis* (Chamberlin, 1909)
220. *Kiberbius cayoteus* Chamberlin, 1941
221. *Kiberbius dyscritus* Chamberlin, 1941
222. *Kiberbius gosobius* Chamberlin, 1941
223. *Kiberbius nannus* Chamberlin, 1916
224. *Kiberbius ogmopus* Chamberlin, 1916
225. *Kiberbius remex* (Chamberlin, 1903)
226. *Kiberbius robles* Chamberlin, 1941
227. *Korynia auxa* Chamberlin, 1954
228. *Korynia carmela* Chamberlin, 1941
229. *Korynia texensis* Chamberlin, 1941
230. *Korynia tripora* Chamberlin, 1941
231. *Korynia urania* (Crabill, 1954)
232. *Lamyctes caducens* Chamberlin, 1938
233. *Lamyctes diffusus* Chamberlin & Mulaik, 1940

234. *Lamyctes pinampus* Chamberlin, 1910
235. *Lamyctes pius* Chamberlin, 1911
236. *Lamyctes tivius* Chamberlin, 1911
237. *Leptodampius lamprus* Chamberlin, 1938
238. *Liobius mimus* Chamberlin & Mulaik, 1940
239. *Lionyx hedgpethi* Chamberlin, 1960
240. *Lithobius annectus* Chamberlin, 1941
241. *Lithobius apheles* Chamberlin, 1940
242. *Lithobius atkinsoni* Bollman, 1887
243. *Lithobius aureus* McNeill, 1887
244. *Lithobius bellulus* Chamberlin, 1903
245. *Lithobius beulae* Chamberlin, 1903
246. *Lithobius cantabrigensis* Meinert, 1886
247. *Lithobius cardinalis* (Bollman, 1887)
248. *Lithobius celer* Bollman, 1888
249. *Lithobius chumasanus* Chamberlin, 1903
250. *Lithobius cockerelli* Chamberlin, 1904
251. *Lithobius eucnemis* Stuxberg, 1875
252. *Lithobius hardyi* Chamberlin, 1946
253. *Lithobius intermontanus* Chamberlin, 1901
254. *Lithobius lindrothi* Palmén, 1954
255. *Lithobius minnesotae* Bollman, 1887
256. *Lithobius paradoxus* Stuxberg, 1875
257. *Lithobius pinetorum* Harger, 1872
258. *Lithobius rapax* Meinert, 1872
259. *Lithobius spinipes* Say, 1821
260. *Lithobius stejnegeri* Bollman, 1893
261. *Lithobius tricalcaratus* (Attems, 1909)
262. *Lithobius watovius* Chamberlin, 1911
263. *Llanobius chamberlini* Causcy, 1942
264. *Llanobius paucispinus* Chamberlin & Mulaik, 1940
265. *Llanobius santus* Chamberlin & Mulaik, 1940
266. *Lophobius apachus* Chamberlin, 1940
267. *Lophobius carinipes* (Daday, 1889)
268. *Lophobius collium* (Chamberlin, 1901)
269. *Lophobius francisae* Chamberlin, 1925
270. *Lophobius lasalanus* Chamberlin, 1928
271. *Lophobius sororis* Chamberlin, 1940
272. *Malochora linsdalei* Chamberlin, 1941
273. *Marsikomerus arcanus* (Crabill, 1961)
274. *Marsikomerus bryanus* (Chamberlin, 1926)

275. *Marsikomerus texanus* (Chamberlin & Mulaik, 1940)
276. *Mecistocephalus waikaneus* Chamberlin, 1953
277. *Mecistocephalus waipaheenas* (Chamberlin, 1953)
278. *Monotarsobius tricalcaratus* Attems, 1909
279. *Nadabius ameles* Chamberlin, 1944
280. *Nadabius aristeus* Chamberlin, 1922
281. *Nadabius caducipes* Chamberlin, 1946
282. *Nadabius cherokeenus* Chamberlin, 1947
283. *Nadabius coloradensis* (Cockerell, 1893)
284. *Nadabius eigenmanni* (Bollman, 1888)
285. *Nadabius eremites* Chamberlin, 1944
286. *Nadabius holzingeri* (Bollman, 1887)
287. *Nadabius jowensis* (Meinert, 1886)
288. *Nadabius mesechinus* (Chamberlin, 1903)
289. *Nadabius oreinus* Chamberlin, 1922
290. *Nadabius phanus* Chamberlin, 1941
291. *Nadabius pluto* Chamberlin & Wang, 1952
292. *Nadabius pullus* (Bollman, 1887)
293. *Nadabius saphes* Chamberlin, 1940
294. *Nadabius vaquens* Chamberlin & Wang, 1952
295. *Nadabius waccamanus* Chamberlin, 1940
296. *Nampabius carolinensis* Chamberlin, 1913
297. *Nampabius embius* Chamberlin, 1913
298. *Nampabius fungiferopes* (Chamberlin, 1904)
299. *Nampabius georgianus* Chamberlin, 1913
300. *Nampabius inimicus* Chamberlin, 1913
301. *Nampabius longiceps* Chamberlin, 1913
302. *Nampabius lulae* Chamberlin, 1913
303. *Nampabius lundii* (Meinert, 1886)
304. *Nampabius major* Chamberlin, 1925
305. *Nampabius mycophor* Chamberlin, 1940
306. *Nampabius parienus* Chamberlin, 1913
307. *Nampabius perspinosus* Chamberlin, 1928
308. *Nampabius pinus* Causey, 1942
309. *Nampabius tennesseensis* Chamberlin, 1913
310. *Nampabius turbator* Crabill, 1952
311. *Nampabius virginiensis* Chamberlin, 1913
312. *Nannarrup hoffmani* Foddai et al., 2003
313. *Navajona miuropus* Chamberlin, 1930
314. *Neolithobius arkansensis* Chamberlin, 1944
315. *Neolithobius devorans* (Chamberlin, 1912)

316. *Neolithobius entonus* Chamberlin, 1942
317. *Neolithobius ethopus* Chamberlin, 1945
318. *Neolithobius helius* Chamberlin, 1918
319. *Neolithobius latzeli* (Meinert, 1886)
320. *Neolithobius mordax* (L. Koch, 1862)
321. *Neolithobius suprenans* Chamberlin, 1925
322. *Neolithobius transmarinus* (L. Koch, 1862)
323. *Neolithobius tyrannus* (Bollman, 1887)
324. *Neolithobius underwoodi* (Bollman, 1888)
325. *Neolithobius voracior* (Chamberlin, 1912)
326. *Neolithobius vorax* (Meinert, 1872)
327. *Neolithobius xenopus* (Bollman, 1888)
328. *Nesidiphilus marginalis* (Meinert, 1886)
329. *Nesogeophilus longissimus* Verhoeff, 1938
330. *Nesomerium hawaiiense* Chamberlin, 1953
331. *Nothembius aberrans* Chamberlin, 1916
332. *Nothembius amplus* Chamberlin, 1941
333. *Nothembius insulae* Chamberlin, 1916
334. *Nothembius nampus* Chamberlin, 1916
335. *Nothobius californicus* Cook, 1899
336. *Nuevobius cottus* Crabill, 1960
337. *Nyctunguis apachus* Chamberlin, 1941
338. *Nyctunguis archochilus* Chamberlin, 1941
339. *Nyctunguis auxus* Chamberlin, 1941
340. *Nyctunguis catalinae* (Chamberlin, 1912)
341. *Nyctunguis glendorus* Chamberlin, 1946
342. *Nyctunguis heathii* (Chamberlin, 1909)
343. *Nyctunguis libercolens* Chamberlin, 1923
344. *Nyctunguis molinor* Chamberlin, 1925
345. *Nyctunguis montereus* (Chamberlin, 1904)
346. *Nyctunguis pholeter* Crabill, 1958
347. *Nyctunguis stenus* Chamberlin, 1962
348. *Nyctunguis vallis* Chamberlin, 1941
349. *Oabius adjacens* Chamberlin, 1946
350. *Oabius aiolus* Chamberlin, 1938
351. *Oabius ajonus* Chamberlin, 1941
352. *Oabius alaskanus* Chamberlin, 1946
353. *Oabius arktaus* Chamberlin, 1946
354. *Oabius boyeranus* Chamberlin, 1940
355. *Oabius decipiens* Chamberlin, 1916
356. *Oabius dissimulans* Chamberlin, 1916

357. *Oabius eugenus* Chamberlin, 1928
358. *Oabius fratris* Chamberlin, 1916
359. *Oabius ineptus* Chamberlin, 1916
360. *Oabius kernensis* Chamberlin, 1941
361. *Oabius mercurialis* Chamberlin, 1962
362. *Oabius mimosus* Chamberlin, 1938
363. *Oabius oreinus* Chamberlin, 1925
364. *Oabius paiutus* Chamberlin, 1925
365. *Oabius parvior* Chamberlin, 1938
366. *Oabius patonius* (Chamberlin, 1911)
367. *Oabius patonius flavus* Chamberlin, 1916
368. *Oabius patonius micrus* Chamberlin, 1916
369. *Oabius pelotes* Chamberlin, 1941
370. *Oabius pylorus* Chamberlin, 1916
371. *Oabius rodocki* Chamberlin, 1940
372. *Oabius sanjuanus* Chamberlin, 1928
373. *Oabius sastianus* (Chamberlin, 1903)
374. *Oabius tabiphilus* Chamberlin, 1916
375. *Oabius tiganus* (Chamberlin, 1910)
376. *Oabius uleorus* Chamberlin, 1916
377. *Oabius wamus* Chamberlin, 1962
378. *Pachymerium idium* Chamberlin, 1960
379. *Paitobius adelus* Chamberlin, 1922
380. *Paitobius arienus* (Chamberlin, 1911)
381. *Paitobius atlantae* Chamberlin, 1922
382. *Paitobius carolinae* (Chamberlin, 1911)
383. *Paitobius eutypus* Chamberlin, 1922
384. *Paitobius exceptus* Chamberlin, 1922
385. *Paitobius exiguus* (Meinert, 1886)
386. *Paitobius juventus* (Bollman, 1887)
387. *Paitobius naiwatus* (Chamberlin, 1911)
388. *Paitobius simitus* (Chamberlin, 1911)
389. *Paitobius tabius* (Chamberlin, 1911)
390. *Paitobius watsuitus* (Chamberlin, 1911)
391. *Paitobius zinus* (Chamberlin, 1911)
392. *Paitobius zygethus* Chamberlin & Wang, 1952
393. *Pampibius paitus* (Chamberlin, 1911)
394. *Paobius albertanus* Chamberlin, 1922
395. *Paobius berkeleyensis* (Verhoeff, 1937)
396. *Paobius boreus* Chamberlin, 1916
397. *Paobius columbiensis* Chamberlin, 1916

398. *Paobius orophilus* Chamberlin, 1916
399. *Paobius vagrans* Chamberlin, 1916
400. *Parunguis kernensis* Chamberlin, 1941
401. *Pectiniunguis catalinensis* Chamberlin, 1941
402. *Pectiniunguis halirrhytus* Crabill, 1959
403. *Pholobius goffi* Chamberlin, 1940
404. *Pholobius mundior* Chamberlin & Mulaik, 1940
405. *Planobius aletes* Chamberlin & Wang, 1952
406. *Pleotarsobius heterotarsus* (Silvestri, 1904)
407. *Poaphilus kewinus* Chamberlin, 1912
408. *Pokabius aethes* Chamberlin, 1951
409. *Pokabius arizonae* (Chamberlin, 1922)
410. *Pokabius bilabiatus* (Wood, 1867)
411. *Pokabius castellopes* (Chamberlin, 1903)
412. *Pokabius centurio* (Chamberlin, 1905)
413. *Pokabius clavigerens* (Chamberlin, 1903)
414. *Pokabius disantus* Chamberlin, 1922
415. *Pokabius eremus* Chamberlin, 1922
416. *Pokabius gilae* Chamberlin, 1922
417. *Pokabius helenae* Chamberlin, 1922
418. *Pokabius hopianus* Chamberlin, 1938
419. *Pokabius iginus* (Chamberlin, 1912)
420. *Pokabius iosemiteus* Chamberlin & Wang, 1952
421. *Pokabius liber* Chamberlin, 1941
422. *Pokabius linsdalei* Chamberlin, 1941
423. *Pokabius loganus* (Chamberlin, 1925)
424. *Pokabius oreines* Chamberlin, 1941
425. *Pokabius piedus* Chamberlin, 1930
426. *Pokabius pitophilus* (Chamberlin, 1903)
427. *Pokabius praefectus* Chamberlin, 1938
428. *Pokabius pungonius* (Chamberlin, 1912)
429. *Pokabius simplex* Chamberlin, 1941
430. *Pokabius socius* (Chamberlin, 1901)
431. *Pokabius sokovus* (Chamberlin, 1909)
432. *Pokabius stenenus* Chamberlin, 1938
433. *Pokabius utahensis* (Chamberlin, 1901)
434. *Pokabius utahensis tidus* Chamberlin, 1962
435. *Pokabius vaquero* Chamberlin, 1941
436. *Pokabius verdescens* (Chamberlin, 1912)
437. *Pseudolithobius festinatus* Crabill, 1950
438. *Pseudolithobius megaloporus* (Stuxberg, 1875)

439. *Queenslandophilus elongatus* Verhoeff, 1938
440. *Scolopendra alternans* Leach, 1813
441. *Scolopendra heros* Girard, 1853
442. *Scolopendra polymorpha* H. C. Wood, 1861
443. *Scolopendra viridis* Say 1821
444. *Scolopocryptops gracilis* Wood, 1862
445. *Scolopocryptops nigridius* McNeill, 1887
446. *Scolopocryptops peregrinator* (Crabill, 1952)
447. *Scolopocryptops rubiginosus* Koch, 1878
448. *Scolopocryptops sexspinosus* (Say, 1821)
449. *Scolopocryptops spinicaudus* Wood, 1862
450. *Scutigera buda* Chamberlin, 1944
451. *Scutigera dorothea* Chamberlin, 1943
452. *Scutigera homa* Chamberlin, 1942
453. *Scutigera linceci* (Wood, 1867)
454. *Scutigera phana* Chamberlin, 1943
455. *Serrobius pulchellus* Causey, 1942
456. *Serrona kernensis* Chamberlin, 1941
457. *Serrunguis paroicus* Chamberlin, 1941
458. *Shosobius cordialis* Chamberlin & Wang, 1952
459. *Sigibius nidicolens* Chamberlin, 1938
460. *Sigibius siopius* Chamberlin & Wang, 1952
461. *Sigibius starlingi* Causey, 1942
462. *Sigibius urbanus* Chamberlin, 1944
463. *Simobius gardneri* Auerbach, 1950
464. *Simobius ginampus* (Chamberlin, 1909)
465. *Simobius lobophor* Chamberlin, 1941
466. *Simobius opibius* Chamberlin & Wang, 1952
467. *Sogona kerrana* Chamberlin, 1940
468. *Sogona minima* Chamberlin, 1912
469. *Sogona poretha* (Chamberlin, 1918)
470. *Sonibius bius* (Chamberlin, 1911)
471. *Sonibius politus* (McNeill in Bollman, 1887)
472. *Sonibius scepticus* Chamberlin & Wang, 1952
473. *Sozibius carolinus* (Causey, 1942)
474. *Sozibius mullanua* Chamberlin & Wang, 1952
475. *Sozibius paurops* Chamberlin, 1944
476. *Sozibius pennsylvanicus* Chamberlin, 1922
477. *Sozibius proridens* (Bollman, 1887)
478. *Sozibius texanus* Chamberlin, 1938
479. *Sozibius tuobukus* (Chamberlin, 1911)

480. *Steneurytion hawaiiensis* (Chamberlin, 1953)
481. *Stenophilus audacior* (Chamberlin, 1909)
482. *Stenophilus californicus* (Chamberlin, 1930)
483. *Stenophilus coloradanus* Chamberlin, 1946
484. *Stenophilus grenadae* (Chamberlin, 1912)
485. *Stenophilus hesperus* (Chamberlin, 1928)
486. *Stenophilus rothi* Chamberlin, 1953
487. *Strigamia bidens* Wood, 1862
488. *Strigamia bothriopa* Wood, 1862
489. *Strigamia branneri* (Bollman, 1888)
490. *Strigamia chionophila* Wood, 1862
491. *Strigamia fulva* Sager, 1856
492. *Strigamia gracilis* Wood, 1867
493. *Strigamia inermis* Wood, 1867
494. *Strigamia kerrana* Chamberlin & Mulaik, 1940
495. *Strigamia maculaticeps* Wood, 1862
496. *Strigamia parviceps* Wood, 1862
497. *Synthophilus boreus* Chamberlin, 1946
498. *Tabiphilus rex* Chamberlin, 1912
499. *Taiyubius angelus* (Chamberlin, 1903)
500. *Taiyubius dux* Chamberlin, 1940
501. *Taiyubius purpureus* (Chamberlin, 1901)
502. *Taiyubius satanus* (Chamberlin, 1911)
503. *Taiyuna claremontus* Chamberlin, 1912
504. *Taiyuna idahoana* Chamberlin, 1941
505. *Taiyuna isantus* (Chamberlin, 1909)
506. *Taiyuna moderata* Chamberlin, 1941
507. *Taiyuna occidentalis* (Meinert, 1885)
508. *Taiyuna opita* Chamberlin, 1912
509. *Tampiya pylorus* Chamberlin, 1912
510. *Texobius unicus* Chamberlin & Mulaik, 1940
511. *Thalkethops grallatrix* Crabill, 1960
512. *Theatops californiensis* Chamberlin, 1902
513. *Theatops phanus* Chamberlin, 1951
514. *Theatops posticus* (Say, 1821)
515. *Theatops spinicaudus* (H. W. Wood, 1862)
516. *Tidabius aberrans* Chamberlin, 1929
517. *Tidabius anderis* Chamberlin, 1913
518. *Tidabius bonvillensis* (Chamberlin, 1909)
519. *Tidabius kansensis* (Gunthorp, 1913)
520. *Tidabius nasintus* Chamberlin, 1913

521. *Tidabius opiphilus* Chamberlin, 1913
522. *Tidabius pallidus alabamensis* Chamberlin, 1913
523. *Tidabius pallidus* Chamberlin, 1913
524. *Tidabius plesius* Chamberlin, 1945
525. *Tidabius poaphilus* Chamberlin, 1913
526. *Tidabius suitus* (Chamberlin, 1911)
527. *Tidabius tivius* (Chamberlin, 1909)
528. *Tidabius zionicus* Chamberlin, 1925
529. *Tigobius paralus* Chamberlin, 1916
530. *Timpina texana* Chamberlin, 1912
531. *Tylonyx tampae* Cook, 1899
532. *Typhlobius coecus* (Bollman, 1888)
533. *Typhlobius kebus* Chamberlin, 1922
534. *Watobius anderisus* Chamberlin, 1911
535. *Watophilus alabamae* Chamberlin, 1912
536. *Watophilus dolicocephalus* (Gunthorp, 1913)
537. *Watophilus errans* Chamberlin, 1912
538. *Watophilus knowltoni* Crabill, 1976
539. *Watophilus laetus* Chamberlin, 1912
540. *Watophilus utus* Chamberlin, 1928
541. *Yobius haywardi* Chamberlin, 1945
542. *Zantaenia idahona* Chamberlin, 1960
543. *Zinapolys uticola* Chamberlin & Wang, 1952
544. *Zinapolys zipius* Chamberlin, 1912
545. *Zygethobius columbiensis* Chamberlin, 1912
546. *Zygethobius dolichopus* (Chamberlin, 1901)
547. *Zygethobius ecologus* Chamberlin, 1938
548. *Zygethobius pontis* Chamberlin, 1911
549. *Zygethobius sokarienus* Chamberlin, 1911
550. *Zygethopolys atrox* Crabill, 1953
551. *Zygethopolys nothus* Chamberlin, 1925
552. *Zygethopolys pugetensis* Chamberlin, 1928
553. *Zygethopolys pugetensis tiganus* Chamberlin & Wang, 1952
554. *Zygomerium rotarium* Chamberlin, 1943
555. *Zygona duplex* Chamberlin, 1960

Appendix II

List of Non-indigenous Centipedes

1. *Apunguis prosoicus* Chamberlin, 1947
2. *Australobius vians* Chamberlin, 1938
3. *Bothropolys epelus* Chamberlin, 1931
4. *Bothropolys imaharensis* Verhoeff, 1937
5. *Bothropolys maluhianus* Attems, 1914
6. *Bothropolys rugosus* (Meinert, 1872)
7. *Chaetechelyne vesuviana* (Newport, 1845)
8. *Cruzobius verus* Chamberlin, 1942
9. *Cryptops anomalans* Newport, 1844
10. *Cryptops hortensis* (Donovan, 1810)
11. *Cryptops melanotypus* Chamberlin, 1941
12. *Cryptops nautiphilus* Chamberlin, 1939
13. *Cryptops navis* Chamberlin, 1930
14. *Cryptops positus* Chamberlin, 1939
15. *Cryptops vector* Chamberlin, 1931.
16. *Cryptops venezuelae* Chamberlin, 1939
17. *Cryptops watsingus* Chamberlin, 1939
18. *Dicellophilus carniolensis* (C. L. Koch, 1847)
19. *Dinocryptops miersii* (Newport, 1845)
20. *Elattobius simplex* Chamberlin, 1941
21. *Garrina cruzanus* (Chamberlin, 1942)
22. *Geophilus carpophagus* Leach, 1815
23. *Geophilus electricus* (Linnaeus, 1758)
24. *Geophilus flavus* (De Geer, 1778)
25. *Geophilus proximus* C. L. Koch, 1847
26. *Geophilus seurati* Brölemann, 1924
27. *Labrobius investigans* Chamberlin, 1938
28. *Lamyctes africanus* (Porat, 1871)

29. *Lamyctes coeculus* (Brölemann, 1889)
30. *Lamyctes emarginatus* (Newport, 1844)
31. *Lamyctes* sp. Chamberlin, 1930
32. *Lamyctes* sp. Chamberlin & Wang, 1952
33. *Lithobius australis* (Chamberlin, 1944)
34. *Lithobius borealis* Meinert, 1868
35. *Lithobius bullatus* Eason, 1993
36. *Lithobius calcaratus* Koch, 1844
37. *Lithobius cepeus* (Chamberlin, 1940)
38. *Lithobius crassipes* (L. Koch, 1862)
39. *Lithobius forficatus* (Linnaeus, 1758)
40. *Lithobius hawaiiensis* Silvestri, 1913
41. *Lithobius melanops* Newport, 1845
42. *Lithobius microps* Meinert, 1868
43. *Lithobius migrans* (Chamberlin, 1925)
44. *Lithobius moananus* (Chamberlin, 1926)
45. *Lithobius obscurus* Meinert, 1872
46. *Lithobius peregrinus* Latzel, 1880
47. *Lithobius sinensis* (Chamberlin, 1930)
48. *Mecistocephalus guildingii* Newport, 1845
49. *Mecistocephalus maxillaris* (Gervais, 1837)
50. *Mecistocephalus spissus* Wood, 1862
51. *Mecistocephalus tridens* Chamberlin, 1922
52. *Neolithobius aztecus* (Humbert & Saussure, 1869)
53. *Newportia* sp. Chamberlin, 1930
54. *Orphnaeus brevilabiatus* (Newport, 1845)
55. *Orya barbarica* (Gervais, 1835)
56. *Otostigmus mians* Chamberlin, 1930
57. *Otostigmus scaber* Porat, 1876
58. *Otostigmus sinicolens* Chamberlin, 1930
59. *Pachymerium ferrugineum* (C. L. Koch, 1835)
60. *Paracryptops inexpectus* Chamberlin, 1914
61. *Paracryptops weberi* Pocock, 1891
62. *Rhysida longipes* (Newport, 1845)
63. *Schendyla nemorensis* (C. L. Koch, 1837)
64. *Scolopendra mima* Chamberlin, 1942
65. *Scolopendra morsitans* Linnaeus, 1758
66. *Scolopendra pachygnatha* Pocock, 1895
67. *Scolopendra subspinipes* Leach, 1815
68. *Scutigera coleoptrata* (Linnaeus, 1758)
69. *Sepedonophilus hodites* Chamberlin, 1940

70. *Sigibius enans* Chamberlin, 1938
71. *Strigamia acuminata* (Leach, 1815)
72. *Stigmatogaster subterranea* (Shaw, 1794)
73. *Taiyuna australis* Chamberlin, 1914
74. *Tidabius emporus* Chamberlin, 1941
75. *Tidabius vector* Chamberlin, 1931
76. *Tuoba sydneyensis* (Pocock, 1891)
77. *Tygarrup intermedius* Chamberlin, 1914
78. *Tygarrup javanicus* Attems, 1907

Appendix III

List of Centipedes by Country and State or Province

Alphabetical list of centipedes by country, state, province, district and by genus and species. Foreign species that have only been taken at quarantine have also been included for future possibilities that it may be encountered again. If there was reasonable doubt as to the presence of a species (indigenous or non-indigenous) in a certain area because it is not known or of ambiguity in the literature it has been included, however, with a question mark in front of the entry until further investigation clarifies the distribution. If there is no questionmark the species is known to be established. Asterisk (*) indicates an introduced species that is known to be established, a question mark (?) indicates there is uncertainty to the existence of this species in this state, a question mark followed by an asterisk (?*) indicates this foreign species is probably not established, and † indicates an extinct species.

CANADA

Geophilus carpophagus Leach, 1815 [**Note:** In general, no specific locality (see Brodie & White, 1883)].

?*Lithobius spinipes* Say, 1821 [**Note:** In general, no specific locality (see Brodie & White, 1883)].

Pachymerium ferrugineum (C. L. Koch, 1835) [**Note:** Kevan (1979) reports it as "Subarctic" with no specific province].

ALBERTA:
Paobius albertanus Chamberlin, 1922
Strigamia chionophila Wood, 1862

BRITISH COLUMBIA:
Arctogeophilus insularis Attems, 1947
?*Arctogeophilus melanonotus* (H. C. Wood, 1862)

Bothropolys columbiensis Chamberlin, 1925
Bothropolys hoples (Brölemann, 1896)
Bothropolys victorianus Chamberlin, 1925
*Cryptops hortensis (Donovan, 1810)
Ethopolys californicus (Daday, 1889)
Ethopolys spectans Chamberlin, 1951
Nadabius eigenmanni (Bollman, 1888)
Paobius columbiensis Chamberlin, 1916
Paobius orophilus Chamberlin, 1916
Pokabius eremus Chamberlin, 1922
?**Scolopocryptops sexspinosus* (Say, 1821)
Scolopocryptops spinicaudus Wood, 1862
*Scutigera coleoptrata (Linnaeus, 1758)
Simobius ginampus (Chamberlin, 1909)
Strigamia parviceps Wood, 1862
Taiyuna opita Chamberlin, 1912
Zygethobius columbiensis Chamberlin, 1912
Zygethopolys pugetensis tiganus Chamberlin & Wang, 1952

MANITOBA:
?*Scolopocryptops rubiginosus* Koch, 1878

NEW BRUNSKWICK:
Bothropolys multidentatus (Newport, 1845)
Escaryus sp. (see Carter & Brown, 1973)
Nadabius sp. (see Carter & Brown, 1973)

NEWFOUNDLAND:
Cryptops parisi Brölemann, 1920
Garibius sp. (see Kevan, 1983: 2948)
*Geophilus electricus (Linnaeus, 1758)
*Geophilus flavus (De Geer, 1778)
Geophilus terrae-novae Palmén, 1954
*Lamyctes emarginatus (Newport, 1844)
*Lithobius forficatus (Linnaeus, 1758)
Lithobius lindrothi Palmén, 1954
*Lithobius melanops Newport, 1845
*Lithobius microps Meinert, 1868
*Schendyla nemorensis (C. L. Koch, 1837)
?**Stigmatogaster subterranea* (Shaw, 1794)

NORTHWEST TERRITORIES:
Strigamia chionophila Wood, 1862

NOVA SCOTIA:
Arenophilus bipuncticeps (H. C. Wood, 1862)
*Lithobius forficatus (Linnaeus, 1758)
?**Rhysida longipes* (Newport, 1845)
*Scolopendra viridis Say, 1821

ONTARIO:
Arenophilus bipuncticeps (H. C. Wood, 1862)
Bothropolys multidentatus (Newport, 1845)
?**Chaetechelyne vesuviana* (Newport, 1845)
*Cryptops anomalans Newport, 1844
*Geophilus flavus (De Geer, 1778)
*Geophilus proximus C. L. Koch, 1847
Geophilus vittatus (Rafinesque, 1820)
*Lamyctes emarginatus (Newport, 1844)
*Lithobius forficatus (Linnaeus, 1758)
*Lithobius microps Meinert, 1868
Nadabius aristeus Chamberlin, 1922
Nadabius jowensis (Meinert, 1886)
Nampabius lundii (Meinert, 1886)
?**Scolopocryptops rubiginosus* Koch, 1878
Scolopocryptops sexspinosus (Say, 1821)
*Scutigera coleoptrata (Linnaeus, 1758)
Sonibius politus (McNeill in Bollman, 1887)
Strigamia chionophila Wood, 1862

QUEBEC:
*Cryptops anomalans Newport, 1844
Geophilus vittatus (Rafinesque, 1820)
*Lithobius forficatus (Linnaeus, 1758)
*Lithobius microps Meinert, 1868
?**Scolopendra alternans* Leach, 1813
*Scutigera coleoptrata (Linnaeus, 1758)
Sonibius politus (McNeill in Bollman, 1887)

YUKON TERRITORY:
Escaryus ethopus (Chamberlin, 1920)
Oabius sp. (see Chamberlin, 1949a)

GREENLAND

GREENLAND:
Lamyctes emarginatus (Newport, 1844)
Lithobius forficatus (Linnaeus, 1758)

UNITED STATES OF AMERICA

?*Cryptops anomalans* Newport, 1844 [**Note:** Northeastern USA in general
   (see Shelley, 2002)].
?*Lithobius spinipes* Say, 1821 [**Note:** In general, no specific locality].
Planobius aletes Chamberlin & Wang, 1952 [**Note:** North America with no
   specific locality].

ALABAMA:
Arctogeophilus umbraticus (McNeill, 1887)
Arenophilus bipuncticeps (H. C. Wood, 1862)
Arenophilus watsingus Chamberlin, 1912
Bothropolys multidentatus (Newport, 1845)
Chomatobius euphorion (Crabill, 1953)
Cryptops leucopodus (Rafinesque, 1820)
Garibius alabamae Chamberlin, 1913
Garibius branneri (Bollman, 1888)
Geophilus ampyx Crabill, 1954
Geophilus mordax Meinert, 1886
Hemiscolopendra marginata (Say, 1821)
Lamyctes tivius Chamberlin, 1911
Neolithobius devorans (Chamberlin, 1912)
Nadabius pullus (Bollman, 1887)
Neolithobius mordax (L. Koch, 1862)
Neolithobius transmarinus (L. Koch, 1862)
Neolithobius underwoodi (Bollman, 1888)
Neolithobius vorax (Meinert, 1872)
Paitobius exceptus Chamberlin, 1922
Paitobius exiguus (Meinert, 1886)
Paitobius zinus (Chamberlin, 1911)
?*Scolopendra heros* Girard, 1853
Scolopendra viridis Say 1821
Scolopocryptops nigridius McNeill, 1887
Scolopocryptops sexspinosus (Say, 1821)
Tidabius pallidus alabamensis Chamberlin, 1913

Tidabius suitus (Chamberlin, 1911)
Tidabius tivius (Chamberlin, 1909)
Theatops posticus (Say, 1821)
Theatops spinicaudus (H. W. Wood, 1862)
Watobius anderisus Chamberlin, 1911
Watophilus alabamae Chamberlin, 1912

ALASKA:
Alaskobius adlatus Chamberlin, 1946
Alaskobius josephus Chamberlin, 1946
Alaskobius parvior Chamberlin, 1946
Arctogeophilus glacialis (Attems, 1909)
?*Arctogeophilus melanonotus* (H. C. Wood, 1862)
Arebius integrior Chamberlin, 1949
Arenophilus bipuncticeps (Wood, 1862)
Bothropolys ethus Chamberlin, 1946
Bothropolys victorianus Chamberlin, 1925
Cheiletha alaska Chamberlin, 1946
Escaryus ethopus (Chamberlin, 1920)
Escaryus paucipes Chamberlin, 1946
Escimobius cryophilus Chamberlin, 1949
Ethopolys integer alaskanus Chamberlin, 1919
Geophilus admarinus (Chamberlin, 1952)
Geophilus alaskanus Cook, 1904
Lamyctes emarginatus (Newport, 1844)
Lithobius rapax Meinert, 1872
Lithobius stejnegeri Bollman, 1893
Lithobius tricalcaratus Attems, 1909
Nadabius caducipes Chamberlin, 1946
Oabius adjacens Chamberlin, 1946
Oabius alaskanus Chamberlin, 1946
Oabius arktaus Chamberlin, 1946
Oabius uleorus Chamberlin, 1916
Pachymerium ferrugineum (C. L. Koch, 1835)
Paobius boreus Chamberlin, 1916
?*Scolopocryptops rubiginosus* Koch, 1878
?*Scolopocryptops sexspinosus* (Say, 1821)
Scolopocryptops spinicaudus Wood, 1862
Simobius ginampus (Chamberlin, 1909)
?**Strigamia acuminata* (Leach, 1815)
Strigamia chionophila Wood, 1862
Synthophilus boreus Chamberlin, 1946
Zygethopolys nothus Chamberlin, 1925

ARIZONA:

Archethopolys kaibabus Chamberlin, 1930
Arctogeophilus atopus (Chamberlin, 1902)
Arthrorhabdus pygmaeus Pocock, 1895
Buethobius arizonicus Chamberlin, 1925
†*Calciphilus abboti* Chamberlin, 1949
Cheiletha phoenix Chamberlin, 1955
Chomatobius minor arizonicus (Chamberlin, 1925)
Eremerium apachum Chamberlin, 1941
Geophilus vittatus (Rafinesque, 1820)
Geophilus yavapainus (Chamberlin, 1941)
Gosibius arizonensis Chamberlin, 1917
Hemiscolopendra marginata (Say, 1821)
Kiberbius dyscritus Chamberlin, 1941
Kiberbius gosobius Chamberlin, 1941
Kiberbius robles Chamberlin, 1941
Lophobius apachus Chamberlin, 1940
Navajona miuropus Chamberlin, 1930
Nothembius nampus Chamberlin, 1916
Nyctunguis apachus Chamberlin, 1941
Oabius ajonus Chamberlin, 1941
Oabius oreinus Chamberlin, 1925
Pokabius apachus Chamberlin, 1940
Pokabius arizonae (Chamberlin, 1922)
Pokabius gilae Chamberlin, 1922
Pokabius hopianus Chamberlin, 1938
Pseudolithobius festinatus Crabill, 1950
Scolopendra heros Girard, 1853
Scolopendra polymorpha H. C. Wood, 1861
Scolopendra viridis Say 1821
Scolopocryptops gracilis Wood, 1862
Scutigera homa Chamberlin, 1942
Scutigera linceci (Wood, 1867)
Theatops posticus (Say, 1821)
Zygona duplex Chamberlin, 1960

ARKANSAS:

Arenophilus bipuncticeps (H. C. Wood, 1862)
Arenophilus watsingus Chamberlin, 1912
Arkansobius lamprus Chamberlin, 1938
Cryptops leucopodus (Rafinesque, 1820)
Bothropolys multidentatus (Newport, 1845)

Garibius branneri (Bollman, 1888)
Geophilus mordax Meinert, 1886
Geophilus vittatus (Rafinesque, 1820)
Guambius oedipes (Bollman, 1888)
Guambius pinguis (Bollman, 1888)
Hemiscolopendra marginata (Say, 1821)
*Lamyctes emarginatus (Newport, 1844)
Lithobius celer Bollman, 1888
*Lithobius forficatus (Linnaeus, 1758)
Marsikomerus arcanus Crabill, 1961
Neolithobius arkansensis Chamberlin, 1944
Neolithobius entonus Chamberlin, 1942
Neolithobius mordax (L. Koch, 1862)
Neolithobius suprenans Chamberlin, 1925
Neolithobius transmarinus (L. Koch, 1862)
Neolithobius vorax (Meinert, 1872)
Scolopendra heros Girard, 1853
Scolopendra viridis Say 1821
Scolopocryptops rubiginosus Koch, 1878
Scolopocryptops sexspinosus (Say, 1821)
*Scutigera coleoptrata (Linnaeus, 1758)
Sogona poretha (Chamberlin, 1918)
Sozibius proridens (Bollman, 1887)
Stenophilus grenadae (Chamberlin, 1912)
Strigamia bothriopa Wood, 1862
Strigamia branneri (Bollman, 1888)
Strigamia fulva Sager, 1856
Theatops posticus (Say, 1821)
Theatops spinicaudus (H. W. Wood, 1862)

CALIFORNIA:
?*Arctogeophilus melanonotus* (H. C. Wood, 1862)
Arctogeophilus shelfordi Chamberlin, 1946
Arctogeophilus umbraticus (McNeill, 1887)
Arebius agamus Chamberlin, 1941
Arebius cherosus Chamberlin, 1941
Arebius crenius Chamberlin, 1941
Arebius diplonyx Chamberlin, 1916
Arebius dolius Chamberlin, 1916
Arebius elysianus Chamberlin, 1916
Arebius epelus Chamberlin, 1941
Arebius kochii (Stuxberg, 1875)
Arebius medius Chamberlin, 1916

Arebius obesus (Stuxberg, 1875)
Arebius petrovius Chamberlin, 1947
Arebius sequens Chamberlin, 1947
Arebius sequoius Chamberlin, 1941
Arebius tridens Chamberlin, 1941
Arenophilus bipuncticeps (H. C. Wood, 1862)
Arenophilus iugans Chamberlin, 1944
Arrup pylorus Chamberlin, 1912
Bothropolys dasys Chamberlin, 1941
Buethobius coniugans Chamberlin, 1911
Californiphilus michaelbacheri Verhoeff, 1938
Caliphilus alamedanus Chamberlin, 1941
Chomatobius bakeri (Chamberlin, 1912)
Chomatobius laticeps (H. C. Wood, 1862)
?*Chomatobius mexicanus* (Saussure, 1858)
Chomatobius minor (Chamberlin, 1912)
Condylona isabella Chamberlin, 1941
Condylona sontipes Chamberlin, 1941
*Cryptops hortensis (Donovan, 1810)
Dicellophilus carniolensis (C. L. Koch, 1847)
Dicellophilus limatus (Wood, 1862)
?*Dinocryptops miersii* (Newport, 1845)
Enarthrobius oblitus Chamberlin & Wang, 1952
?*Ethmostigmus californicus* Chamberlin, 1958
?*Ethopolys bipunctatus* (Wood, 1862)
Ethopolys calibius Chamberlin, 1951
Ethopolys californicus (Daday, 1889)
Ethopolys monticola (Stuxberg, 1875)
Ethopolys positivus Chamberlin, 1941
Ethopolys pusio (Stuxberg, 1875)
Ethopolys xanti (Wood, 1862)
Garriscaphus amplus Chamberlin, 1941
Garriscaphus oreines Chamberlin, 1941
Garriscaphus tytthus Crabill, 1969
Geophilus becki Chamberlin, 1951
Geophilus claremontus Chamberlin, 1909
Geophilus delotus (Chamberlin, 1941)
Geophilus nasintus Chamberlin, 1909
Geophilus nicolanus Chamberlin, 1940
Geophilus regnans Chamberlin, 1904
Geophilus secundus (Chamberlin, 1912)
Geophilus sequoia Chamberlin, 1941
Geophilus vittatus (Rafinesque, 1820)

Gosibius angelicus Chamberlin, 1944
Gosibius atopops Chamberlin, 1941
Gosibius benespinosus Chamberlin, 1941
Gosibius brevicornis Chamberlin, 1917
?*Gosibius claremontus* Chamberlin, 1917
Gosibius fusatus Chamberlin, 1941
Gosibius intermedius Chamberlin, 1917
Gosibius monicus Chamberlin, 1912
Gosibius montereus Chamberlin, 1917
Gosibius paucidens (Wood, 1862)
Gosibius sequens Chamberlin, 1941
Guambius hesperus Chamberlin, 1941
Harmostela hespera Chamberlin, 1941
Kethops utahensis (Chamberlin, 1909)
Kiberbius cayoteus Chamberlin, 1941
Kiberbius nannus Chamberlin, 1916
Kiberbius ogmopus Chamberlin, 1916
Kiberbius remex (Chamberlin, 1903)
Korynia auxa Chamberlin, 1954
Korynia carmela Chamberlin, 1941
Korynia tripora Chamberlin, 1941
?**Labrobius investigans* Chamberlin, 1938
?*Lamyctes emarginatus* (Newport, 1844)
Lamyctes pinampus Chamberlin, 1910
Lionyx hedgpethi Chamberlin, 1960
Lithobius annectus Chamberlin, 1941
Lithobius chumasanus Chamberlin, 1903
*Lithobius melanops Newport, 1845
Lithobius paradoxus Stuxberg, 1875
Lophobius carinipes (Daday, 1889)
Malochora linsdalei Chamberlin, 1941
Nadabius oreinus Chamberlin, 1922
Nadabius phanus Chamberlin, 1941
?*Neolithobius aztecus* (Humbert & Saussure, 1869)
Nesogeophilus longissimus Verhoeff, 1938
Nothembius aberrans Chamberlin, 1916
Nothembius amplus Chamberlin, 1941
Nothembius insulae Chamberlin, 1916
Nothembius nampus Chamberlin, 1916
?*Nothobius californicus* Cook, 1899 [*species inquirenda*]
Nyctunguis auxus Chamberlin, 1941
Nyctunguis catalinae (Chamberlin, 1912)
Nyctunguis glendorus Chamberlin, 1946

Nyctunguis heathii (Chamberlin, 1909)
Nyctunguis libercolens Chamberlin, 1923
Nyctunguis montereus (Chamberlin, 1904)
Nyctunguis vallis Chamberlin, 1941
Oabius decipiens Chamberlin, 1916
Oabius dissimulans Chamberlin, 1916
Oabius fratris Chamberlin, 1916
Oabius kernensis Chamberlin, 1941
Oabius patonius (Chamberlin, 1911)
Oabius patonius flavus Chamberlin, 1916
Oabius patonius micrus Chamberlin, 1916
Oabius pelotes Chamberlin, 1941
Oabius pylorus Chamberlin, 1916
Oabius sastianus (Chamberlin, 1903)
Oabius tabiphilus Chamberlin, 1916
Oabius tiganus (Chamberlin, 1910)
Pachymerium idium Chamberlin, 1960
Paitobius zygethus Chamberlin & Wang, 1952
Paobius berkeleyensis (Verhoeff, 1937)
Parunguis kernensis Chamberlin, 1941
Pectiniunguis catalinensis Chamberlin, 1941
Pokabius aethes Chamberlin, 1951
Pokabius castellopes (Chamberlin, 1903)
Pokabius clavigerens (Chamberlin, 1903)
Pokabius disantus Chamberlin, 1922
Pokabius iosemiteus Chamberlin & Wang, 1952
Pokabius liber Chamberlin, 1941
Pokabius linsdalei Chamberlin, 1941
Pokabius oreines Chamberlin, 1941
Pokabius pitophilus (Chamberlin, 1903)
Pokabius simplex Chamberlin, 1941
Pokabius utahensis (Chamberlin, 1901)
Pseudolithobius megaloporus (Stuxberg, 1875)
Queenslandophilus elongatus Verhoeff, 1938
?**Rhysida longipes* (Newport, 1845)
?*Scolopendra heros* Girard, 1853
?**Scolopendra morsitans* Linnaeus, 1758
Scolopendra polymorpha H. C. Wood, 1861
?**Scolopendra subspinipes* Leach, 1815
?*Scolopendra viridis* Say, 1821
Scolopocryptops gracilis Wood, 1862
?*Scolopocryptops sexspinosus* (Say, 1821)
Scolopocryptops spinicaudus Wood, 1862

Scutigera coleoptrata (Linnaeus, 1758)
Serrona kernensis Chamberlin, 1941
Serrunguis paroicus Chamberlin, 1941
Simobius gardneri Auerbach, 1950
Simobius opibius Chamberlin & Wang, 1952
Stenophilus californicus (Chamberlin, 1930)
Strigamia cephalica Wood, 1862
Strigamia chionophila Wood, 1862
Strigamia fulva Sager, 1856
Strigamia gracilis Wood, 1867
Strigamia inermis Wood, 1867
Strigamia parviceps Wood, 1862
Tabiphilus rex Chamberlin, 1912
Taiyubius angelus (Chamberlin, 1903)
Taiyubius satanus (Chamberlin, 1911)
Taiyuna claremontus Chamberlin, 1912
Taiyuna isantus (Chamberlin, 1909)
Taiyuna moderata Chamberlin, 1941
Taiyuna occidentalis (Meinert, 1885)
Tampiya pylorus Chamberlin, 1912
Theatops californiensis Chamberlin, 1902
Theatops posticus (Say, 1821)
Tigobius paralus Chamberlin, 1916
Typhlobius kebus Chamberlin, 1922
Zygethobius dolichopus (Chamberlin, 1901)
Zygethobius sokarienus Chamberlin, 1911
Watophilus errans Chamberlin, 1912
Watophilus laetus Chamberlin, 1912

COLORADO:
Arctogeophilus umbraticus (McNeill, 1887)
Arebius kochii (Stuxberg, 1875)
Cryptops hortensis (Donovan, 1810)
Guambius coloradanus (Chamberlin, 1912)
Guambius oedipes (Bollman, 1888)
Lamyctes emarginatus (Newport, 1844)
Lithobius melanops Newport, 1845
Lithobius forficatus (Linnaeus, 1758)
Nadabius coloradensis (Cockerell, 1893)
Nadabius jowensis (Meinert, 1886)
Neolithobius mordax (L. Koch, 1862)
Neolithobius suprenans Chamberlin, 1925
Pokabius pungonius (Chamberlin, 1912)

?*Scolopendra heros* Girard, 1853
Scolopendra polymorpha H. C. Wood, 1861
Scolopendra viridis Say 1821
Scolopocryptops spinicaudus Wood, 1862
Stenophilus coloradanus Chamberlin, 1946
Strigamia chionophila Wood, 1862
Strigamia fulva Sager, 1856
?*Strigamia maculaticeps* Wood, 1862
Tidabius tivius (Chamberlin, 1909)

CONNECTICUT:

Bothropolys multidentatus (Newport, 1845)
*Cryptops hortensis (Donovan, 1810)
?**Geophilus seurati* Brölemann, 1924
Geophilus vittatus (Rafinesque, 1820)
?*Hemiscolopendra marginata* (Say, 1821)
*Lithobius forficatus (Linnaeus, 1758)
*Lithobius microps Meinert, 1868
Nadabius ameles Chamberlin, 1944
*Schendyla nemorensis (C. L. Koch, 1837)
Scolopocryptops sexspinosus (Say, 1821)
Sozibius pennsylvanicus Chamberlin, 1922
Theatops posticus (Say, 1821)

DELAWARE:

Bothropolys multidentatus (Newport, 1845)
?*Theatops posticus* (Say, 1821)

DISTRICT OF COLOMBIA:

Arenophilus bipuncticeps (H. C. Wood, 1862)
Bothropolys multidentatus (Newport, 1845)
*Cryptops hortensis (Donovan, 1810)
?**Cryptops venezuelae* Chamberlin, 1939
Escaryus liber Cook & Collins, 1891
Geophilus mordax Meinert, 1885
Geophilus smithi Bollman, 1888
Geophilus varians McNeill, 1887
*Lithobius forficatus (Linnaeus, 1758)
*Nadabius pullus (Bollman, 1887)
?**Paracryptops inexpectus* Chamberlin, 1940
?**Rhysida longipes* (Newport, 1845)
?**Scolopendra subspinipes* Leach, 1815
Scolopocryptops nigridius McNeill, 1887

Scolopocryptops peregrinator (Crabill, 1952)
Scolopocryptops sexspinosus (Say, 1821)
*Scutigera coleoptrata (Linnaeus, 1758)
Sozibius proridens (Bollman, 1887)
?*Strigamia acuminata* (Leach, 1815)
Strigamia chionophila Wood, 1862
?*Taiyuna australis* Chamberlin, 1914
?*Theatops posticus* (Say, 1821)
?*Tygarrup intermedius* Chamberlin, 1914

FLORIDA:

Arenophilus bipuncticeps (H. C. Wood, 1862)
Cryptops floridanus Chamberlin, 1925
Cryptops leucopodus (Rafinesque, 1820)
Erithophilus neopus Cook, 1899
Eulithobius hypogeus Chamberlin, 1940
Geophilus mordax Meinert, 1885
Hemiscolopendra marginata (Say, 1821)
Ityphilus lilacinus Cook, 1899
Lithobius aureus McNeill, 1887
?*Lithobius calcaratus* C. L. Koch, 1844
?*Mecistocephalus guildingii* Newport, 1845
*Mecistocephalus maxillaris (Gervais, 1837)
Neolithobius ethopus Chamberlin, 1945
Neolithobius mordax (L. Koch, 1862)
Neolithobius vorax (Meinert, 1872)
Neolithobius xenopus (Bollman, 1888)
Nesidiphilus marginalis (Meinert, 1886)
?*Nothobius californicus* Cook, 1899
*Orphnaeus brevilabiatus (Newport, 1845)
*Pachymerium ferrugineum (C. L. Koch, 1835)
Pectiniunguis halirrhytus Crabill, 1959
Pholobius goffi Chamberlin, 1940
*Rhysida longipes (Newport, 1845)
Scolopendra alternans Leach, 1813
?*Scolopendra heros* Girard, 1853
*Scolopendra morsitans Linnaeus, 1758
?*Scolopendra pachygnatha* Pocock, 1895
?*Scolopendra subspinipes* Leach, 1815
Scolopendra viridis Say 1821
Scolopocryptops nigridius McNeill, 1887
Scolopocryptops sexspinosus (Say, 1821)
?*Scutigera coleoptrata* (Linnaeus, 1758)

Theatops posticus (Say, 1821)
Tylonyx tampae Cook, 1899

GEORGIA:

Arctogeophilus melanonotus (H. C. Wood, 1862)
Arctogeophilus umbraticus (McNeill, 1887)
Arenophilus bipuncticeps (H. C. Wood, 1862)
Arenophilus watsingus Chamberlin, 1912
Bothropolys multidentatus (Newport, 1845)
Cryptops leucopodus (Rafinesque, 1820)
?*Dinocryptops miersii* (Newport, 1845)
Eulithobius fattigi Chamberlin, 1945
Garibius branneri (Bollman, 1888)
Garibius georgiae Chamberlin, 1913
Geophilus mordax Meinert, 1885
Geophilus vittatus (Rafinesque, 1820)
Georgibius georgiae Chamberlin, 1944
Gonibius rex (Bollman, 1888)
Helembius nannus Chamberlin, 1918
Hemiscolopendra marginata (Say, 1821)
Lamyctes pius Chamberlin, 1911
Lamyctes tivius Chamberlin, 1911
Lithobius atkinsoni Bollman, 1887
Nadabius cherokeenus Chamberlin, 1947
Nampabius georgianus Chamberlin, 1913
Nampabius lulae Chamberlin, 1913
?*Nampabius lundii* (Meinert, 1886)
Neolithobius ethopus Chamberlin, 1945
Neolithobius helius Chamberlin, 1918
Neolithobius underwoodi (Bollman, 1888)
Neolithobius xenopus (Bollman, 1888)
?*Nothobius californicus* Cook, 1899
Pachymerium ferrugineum (C. L. Koch, 1835)
Paitobius atlantae Chamberlin, 1922
Paitobius naiwatus (Chamberlin, 1911)
Paitobius watsuitus (Chamberlin, 1911)
Paitobius zinus (Chamberlin, 1911)
?*Rhysida longipes* (Newport, 1845)
?*Scolopendra heros* Girard, 1853
?*Scolopendra morsitans* Linnaeus, 1758
Scolopendra viridis Say 1821
Scolopocryptops nigridius McNeill, 1887
Scolopocryptops sexspinosus (Say, 1821)

Scutigera coleoptrata (Linnaeus, 1758)
Sogona minima Chamberlin, 1912
Sozibius paurops Chamberlin, 1944
Strigamia bidens Wood, 1862
Strigamia branneri (Bollman, 1888)
Strigamia branneri var. *miura* Chamberlin, 1912
Strigamia fulva Sager, 1856
Theatops posticus (Say, 1821)
Theatops spinicaudus (H. W. Wood, 1862)
Watobius anderisus Chamberlin, 1911
Watophilus alabamae Chamberlin, 1912

HAWAII:
[**Note:** Zapparoli & Shelley (2000) deleted 10 species of centipedes from the Hawaiian fauna that are considered invasive, however, these are still considered possible invasive species and have not been omitted in this catalog, hence they are included as reference for future ecological studies].

*Allothereua lesueurii Lucas, 1840
?**Australobius vians* Chamberlin, 1938
*Bothropolys maluhianus Attems, 1914
*Bothropolys rugosus (Meinert, 1872)
*Cryptops hortensis (Donovan, 1810)
?**Cryptops melanotypus* Chamberlin, 1941
?**Cryptops navis* Chamberlin, 1930
?**Cryptops sinesicus* Chamberlin, 1940
?**Cryptops weberi* Pocock, 1891
*Lamyctes africanus (Porat, 1871)
*Lamyctes coeculus (Brölemann, 1889)
*Lamyctes emarginatus (Newport, 1844)
?**Lamyctes* sp. Chamberlin, 1930
?**Lamyctes* sp. Chamberlin &Wang, 1952
?**Lithobius australis* Chamberlin, 1944
?**Lithobius borealis* Meinert, 1868
Lithobius bullatus Eason, 1993
?*Lithobius cepeus* Chamberlin, 1940
*Lithobius forficatus Linnaeus, 1758
?*Lithobius hawaiiensis* Silvestri, 1913
*Lithobius moananus (Chamberlin, 1926)
*Lithobius obscurus Meinert, 1872
?**Lithobius sinensis* Chamberlin, 1930
*Lithobius sp. Chamberlin & Wang, 1952
Marsikomerus bryanus (Chamberlin, 1926)

?*Mecistocephalus maxillaris (Gervais, 1837)
?*Mecistocephalus spissus Wood, 1862
?*Mecistocephalus tridens Chamberlin, 1922
Mecistocephalus waikaneus Chamberlin, 1953
Mecistocephalus waipaheenas (Chamberlin, 1953)
?Nesomerium hawaiiense Chamberlin, 1953
?*Newportia sp. Chamberlin, 1930
?*Oabius pylorus Chamberlin, 1916
?*Orphnaeus brevilabiatus (Newport, 1845)
?*Otostigmus mians Chamberlin, 1930
?*Otostigmus scaber Porat, 1876
?*Otostigmus sinicolens Chamberlin, 1930
?*Pachymerium ferrugineum (Koch, 1835)
?*Paracryptops weberi Pocock, 1891
Pleotarsobius heterotarsus (Silvestri, 1904)
*Scolopendra polymorpha Wood, 1861
*Scolopendra subspinipes Leach, 1815
?*Sepedonophilus hodites Chamberlin, 1940
Steneurytion hawaiiensis (Chamberlin, 1953)
?*Tidabius emporus Chamberlin, 1941
?*Tidabius vector Chamberlin, 1931
?*Theatops spinicaudus (H.W. Wood, 1862)
*Tuoba sydneyensis (Pocock, 1891)
*Tygarrup javanicus Attems, 1907

IDAHO:
Arctogeophilus umbraticus (McNeill, 1887)
Bothropolys permundus (Chamberlin, 1902)
Bothropolys hoples (Brölemann, 1896)
?Geophilus glyptus (Chamberlin, 1902)
*Lamyctes emarginatus (Newport, 1844)
?Lamyctes pinampus Chamberlin, 1910
*Lithobius forficatus (Linnaeus, 1758)
Nadabius jowensis (Meinert, 1886)
Nadabius mesechinus (Chamberlin, 1903)
Oabius rodocki Chamberlin, 1940
Pokabius loganus (Chamberlin, 1925)
Pokabius utahensis (Chamberlin, 1901)
Scolopendra polymorpha H. C. Wood, 1861
Scolopocryptops gracilis Wood, 1862
Shosobius cordialis Chamberlin & Wang, 1952
Sozibius mullanua Chamberlin & Wang, 1952
Stenophilus audacior (Chamberlin, 1909)

Strigamia chionophila Wood, 1862
Taiyuna idahoana Chamberlin, 1941
Zantaenia idahona Chamberlin, 1960
Zinapolys zipius Chamberlin, 1912

ILLINOIS:

Arctogeophilus umbraticus (McNeill, 1887)
Arenophilus bipuncticeps (H. C. Wood, 1862)
Bothropolys multidentatus (Newport, 1845)
Buethobius huestoni Williams & Hefner, 1928
*Cryptops hortensis (Donovan, 1810)
Cryptops leucopodus (Rafinesque, 1820)
?Escaryus missouriensis Chamberlin, 1942
Garibius opicolens Chamberlin, 1913
Geophilus ampyx Crabill, 1954
Geophilus brevicornis Wood, 1862
Geophilus embius (Chamberlin, 1912)
Geophilus mordax Meinert, 1886
Geophilus parki (Auerbach, 1954)
Geophilus rupestris (Crabill, 1949)
?Geophilus setiger Bollman, 1887
Geophilus vittatus (Rafinesque, 1820)
Hemiscolopendra marginata (Say, 1821)
*Lamyctes emarginatus (Newport, 1844)
†Latzelia primordialis Scudder, 1890
?Lithobius celer Bollman, 1888
*Lithobius crassipes (L. Koch, 1862)
*Lithobius forficatus (Linnaeus, 1758)
†Mazoscolopendra richardsoni Mundel, 1979
*Lithobius crassipes (L. Koch, 1862)
Nadabius ameles Chamberlin, 1944
Nadabius holzingeri (Bollman, 1887)
Nadabius jowensis (Meinert, 1886)
Nadabius pullus (Bollman, 1887)
Neolithobius mordax (L. Koch, 1862)
Neolithobius tyrannus (Bollman, 1887)
Neolithobius voracior (Chamberlin, 1912)
*Pachymerium ferrugineum (C. L. Koch, 1835)
Paitobius exiguus (Meinert, 1866)
Paitobius juventus (Bollman, 1887)
Poaphilus kewinus Chamberlin, 1912
Pokabius bilabiatus (Wood, 1867)
Pokabius verdescens (Chamberlin, 1912)

Schendyla nemorensis (C. L. Koch, 1837)
Scolopendra viridis Say 1821
Scolopocryptops nigridius McNeill, 1887
Scolopocryptops rubiginosus Koch, 1878
Scolopocryptops sexspinosus (Say, 1821)
Scutigera coleoptrata (Linnaeus, 1758)
Sigibius urbanus Chamberlin, 1944
Sonibius bius (Chamberlin, 1911)
Sonibius politus (McNeill in Bollman, 1887)
Sozibius proridens (Bollman, 1887)
Strigamia bidens Wood, 1862
Strigamia bothriopa Wood, 1862
Strigamia branneri (Bollman, 1887)
Strigamia chionophila Wood, 1862
Strigamia fulva Sager, 1856
Taiyuna opita Chamberlin, 1912
?*Tidabius anderis* Chamberlin, 1913
Tidabius plesius Chamberlin, 1945
Tidabius suitus (Chamberlin, 1911)
Tidabius tivius (Chamberlin, 1909)
Theatops posticus (Say, 1821)
Theatops spinicaudus (H. W. Wood, 1862)

INDIANA:
Arctogeophilus umbraticus (McNeill, 1887)
Arenophilus bipuncticeps (H. C. Wood, 1862)
Bothropolys multidentatus (Newport, 1845)
Cryptops leucopodus (Rafinesque, 1820)
?*Dinocryptops miersii* (Newport, 1845)
?*Escaryus missouriensis* Chamberlin, 1942
Geophilus brunneus McNeill, 1887
Geophilus indianae McNeill, 1887
Geophilus mordax Meinert, 1885
Geophilus oweni Bollman, 1887
?*Geophilus rupestris* (Crabill, 1949)
Geophilus setiger Bollman, 1887
Geophilus smithi Bollman, 1888
Geophilus strigosus (McNeill, 1887)
Geophilus varians McNeill, 1887
Geophilus vittatus (Rafinesque, 1820)
Hemiscolopendra marginata (Say, 1821)
Lithobius cardinalis (Bollman, 1887)
?*Lithobius celer* Bollman, 1888

Lithobius forficatus (Linnaeus, 1758)
Nadabius ameles Chamberlin, 1944
Nadabius jowensis (Meinert, 1886)
Nadabius pullus (Bollman, 1887)
Neolithobius mordax (L. Koch, 1862)
Neolithobius tyrannus (Bollman, 1887)
Pachymerium ferrugineum (C. L. Koch, 1835)
Paitobius juventus (Bollman, 1887)
Pokabius bilabiatus (Wood, 1867)
Scolopocryptops nigridius McNeill, 1887
Scolopocryptops rubiginosus Koch, 1878
Scolopocryptops sexspinosus (Say, 1821)
Scutigera coleoptrata (Linnaeus, 1758)
Sonibius politus (McNeill in Bollman, 1887)
Sozibius proridens (Bollman, 1887)
Strigamia bidens Wood, 1862
Strigamia bothriopa Wood, 1862
Strigamia chionophila Wood, 1862
Strigamia fulva Sager, 1856
Theatops posticus (Say, 1821)

IOWA:

Arenophilus bipuncticeps (H. C. Wood, 1862)
Geophilus embius (Chamberlin, 1912)
Geophilus vittatus (Rafinesque, 1820)
Geophilus winnetui Attems, 1947
?*Lithobius celer* Bollman, 1888
Lithobius forficatus (Linnaeus, 1758)
Nadabius jowensis (Meinert, 1886)
Neolithobius mordax (L. Koch, 1862)
Neolithobius suprenans Chamberlin, 1925
Pachymerium ferrugineum (C. L. Koch, 1835)
Paitobius exiguus (Meinert, 1866)
Poaphilus kewinus Chamberlin, 1912
Pokabius bilabiatus (Wood, 1867)
Pokabius verdescens (Chamberlin, 1912)
Scolopocryptops rubiginosus Koch, 1878
Scolopocryptops sexspinosus (Say, 1821)
Strigamia fulva Sager, 1856
Theatops spinicaudus (Wood, 1862)

KANSAS:

Arenophilus bipuncticeps (H. C. Wood, 1862)
Arenophilus osborni Gunthorp, 1913
*Cryptops hortensis (Donovan, 1810)
Escaryus missouriensis Chamberlin, 1942
Geophilus mordax Meinert, 1885
?*Hemiscolopendra marginata* (Say, 1821)
*Lithobius forficatus (Linnaeus, 1758)
Nadabius jowensis (Meinert, 1886)
Neolithobius mordax (L. Koch, 1862)
Neolithobius suprenans Chamberlin, 1925
Neolithobius transmarinus (L. Koch, 1862)
Pokabius bilabiatus (Wood, 1867)
Scolopendra heros Girard, 1853
?**Scolopendra morsitans* Linnaeus, 1758
Scolopendra polymorpha H. C. Wood, 1861
Scolopendra viridis Say 1821
Scolopocryptops rubiginosus Koch, 1878
Scolopocryptops sexspinosus (Say, 1821)
*Scutigera coleoptrata (Linnaeus, 1758)
Strigamia fulva Sager, 1856
Tidabius kansensis (Gunthorp, 1913)
Theatops spinicaudus (Wood, 1862)
Watophilus dolicocephalus (Gunthorp, 1913)

KENTUCKY:
Arctogeophilus umbraticus (McNeill, 1887)
Arenobius manegitus (Chamberlin, 1911)
Arenophilus bipuncticeps (H. C. Wood, 1862)
Arenophilus psednus Crabill, 1969
Arenophilus watsingus Chamberlin, 1912
Bothropolys multidentatus (Newport, 1845)
Chomatobius euphorion (Crabill, 1953)
Cryptops leucopodus (Rafinesque, 1820)
Escaryus urbicus (Meinert, 1886)
Garibius opicolens Chamberlin, 1913
Garibius pagoketes Chamberlin, 1913
Geophilus ampyx Crabill, 1954
Geophilus mordax Meinert, 1886
Geophilus parki (Auerbach, 1954)
Geophilus rupestris (Crabill, 1949)
?*Geophilus setiger* Bollman, 1887

Geophilus varians McNeill, 1887
Geophilus vittatus (Rafinesque, 1820)
Hemiscolopendra marginata (Say, 1821)
Lithobius celer Bollman, 1888
*Lithobius forficatus (Linnaeus, 1758)
Nadabius jowensis (Meinert, 1886)
Nadabius pullus (Bollman, 1887)
Nampabius lundii (Meinert, 1886)
Paitobius exiguus (Meinert, 1886)
Paitobius naiwatus (Chamberlin, 1911)
Paitobius zinus (Chamberlin, 1911)
Pokabius bilabiatus (Wood, 1867)
Pokabius verdescens (Chamberlin, 1912)
?*Scolopendra heros* Girard, 1853
?*Scolopendra viridis* Say, 1821
Scolopocryptops nigridius McNeill, 1887
Scolopocryptops peregrinator (Crabill, 1952)
Scolopocryptops sexspinosus (Say, 1821)
*Scutigera coleoptrata (Linnaeus, 1758)
Sonibius politus (McNeill in Bollman, 1887)
Sozibius proridens (Bollman, 1887)
Sozibius tuobukus (Chamberlin, 1911)
Strigamia bidens Wood, 1862
Strigamia bothriopa Wood, 1862
Strigamia branneri (Bollman, 1888)
Strigamia branneri var. *miura* Chamberlin, 1912
Strigamia chionophila Wood, 1862
Strigamia fulva Sager, 1856
Tidabius tivius (Chamberlin, 1909)
Theatops posticus (Say, 1821)
Zygethobius pontis Chamberlin, 1911
Zygethopolys atrox Crabill, 1953

LOUISIANA:
Arenophilus bipuncticeps (H. C. Wood, 1862)
Arenobius unaster (Chamberlin, 1909)
Arenophilus watsingus Chamberlin, 1912
Bothropolys multidentatus (Newport, 1845)
?**Bothropolys epelus* Chamberlin, 1931
*Cryptops hortensis (Donovan, 1810)
Cryptops leucopodus (Rafinesque, 1820)
?**Cryptops nautiphilus* Chamberlin, 1939
?**Cryptops positus* Chamberlin, 1939

?*_Cryptops vector_ Chamberlin, 1931.
?*_Cryptops watsingus_ Chamberlin, 1939
?*_Elattobius simplex_ Chamberlin, 1941
Geophilus mordax Meinert, 1885
Gosibius louisianus Chamberlin, 1942
Gosibius saccharogeus Chamberlin, 1941
Guambius hubrichti Chamberlin, 1942
Guambius oedipes (Bollman, 1888)
Hemiscolopendra marginata (Say, 1821)
Lamyctes tivius Chamberlin, 1911
?*_Lithobius peregrinus_ Latzel, 1880
Neolithobius mordax (L. Koch, 1862)
Neolithobius transmarinus (L. Koch, 1862)
Neolithobius vorax (Meinert, 1872)
?*_Orya barbarica_ (Gervais, 1835)
Scolopendra heros Girard, 1853
Scolopendra polymorpha H. C. Wood, 1861
Scolopendra viridis Say 1821
Scolopocryptops sexspinosus (Say, 1821)
Sogona poretha (Chamberlin, 1918)
Strigamia bidens Wood, 1862
Strigamia fulva Sager, 1856
Theatops posticus (Say, 1821)

MAINE:
Bothropolys multidentatus (Newport, 1845)
*_Lamyctes emarginatus_ (Newport, 1844)
*_Lithobius forficatus_ (Linnaeus, 1758)
*_Lithobius microps_ Meinert, 1868

MARYLAND:
*_Cryptops hortensis_ (Donovan, 1810)
Cryptops leucopodus (Rafinesque, 1820)
Escaryus liber Cook & Collins, 1891
Geophilus vittatus (Rafinesque, 1820)
*_Lithobius forficatus_ (Linnaeus, 1758)
*_Pachymerium ferrugineum_ (C. L. Koch, 1835)
?*_Scolopendra subspinipes_ Leach, 1815
Scolopocryptops nigridius McNeill, 1887
Scolopocryptops peregrinator (Crabill, 1952)
Scolopocryptops sexspinosus (Say, 1821)
*_Scutigera coleoptrata_ (Linnaeus, 1758)
?*_Sigibius enans_ Chamberlin, 1938

Theatops posticus (Say, 1821)

MASSACHUSETTS:
Arenophilus bipuncticeps (H. C. Wood, 1862)
Bothropolys multidentatus (Newport, 1845)
*Cryptops hortensis (Donovan, 1810)
Escaryus urbicus (Meinert, 1886)
Garibius opicolens Chamberlin, 1913
Garibius pagoketes Chamberlin, 1913
*Geophilus flavus (De Geer, 1778)
Geophilus vittatus (Rafinesque, 1820)
?*Hemiscolopendra marginata* (Say, 1821)
*Lamyctes emarginatus (Newport, 1844)
Lithobius cantabrigensis Meinert, 1886
*Lithobius crassipes (L. Koch, 1862)
*Lithobius forficatus (Linnaeus, 1758)
*Lithobius melanops Newport, 1845
*Lithobius microps Meinert, 1868
Nadabius aristeus Chamberlin, 1922
*Schendyla nemorensis (C. L. Koch, 1837)
Scolopocryptops sexspinosus (Say, 1821)
*Scutigera coleoptrata (Linnaeus, 1758)
Strigamia bothriopa Wood, 1862
Strigamia chionophila Wood, 1862
Tidabius tivius (Chamberlin, 1909)

MICHIGAN:
Arenophilus bipuncticeps (H. C. Wood, 1862)
Bothropolys multidentatus (Newport, 1845)
*Cryptops hortensis (Donovan, 1810)
Cryptops leucopodus (Rafinesque, 1820)
?*Geophilus rupestris* (Crabill, 1949)
Geophilus setiger Bollman, 1887
Geophilus varians McNeill, 1887
Geophilus vittatus (Rafinesque, 1820)
*Lamyctes emarginatus (Newport, 1844)
Lithobius celer Bollman, 1888
*Lithobius forficatus (Linnaeus, 1758)
*Lithobius microps Meinert, 1868
Nadabius ameles Chamberlin, 1944
Nadabius jowensis (Meinert, 1886)
Nadabius pullus (Bollman, 1887)
Nampabius fungiferopes (Chamberlin, 1904)

Nampabius lundii (Meinert, 1886)
*Pachymerium ferrugineum (C. L. Koch, 1835)
Pokabius bilabiatus (Wood, 1867)
*Schendyla nemorensis (C. L. Koch, 1837)
Scolopocryptops sexspinosus (Say, 1821)
*Scutigera coleoptrata (Linnaeus, 1758)
Sonibius bius (Chamberlin, 1911)
Sonibius politus (McNeill in Bollman, 1887)
Sozibius proridens (Bollman, 1887)
Strigamia branneri (Bollman, 1888)
Strigamia chionophila Wood, 1862
Strigamia fulva Sager, 1856
Taiyuna opita Chamberlin, 1912
Theatops posticus (Say, 1821)
Tidabius opiphilus Chamberlin, 1913
Tidabius tivius (Chamberlin, 1909)

MINNESOTA:
Arenophilus bipuncticeps (H. C. Wood, 1862)
*Cryptops hortensis (Donovan, 1810)
Escaryus urbicus (Meinert, 1886)
?Geophilus rupestris (Crabill, 1949)
Geophilus setiger Bollman, 1887
*Lamyctes emarginatus (Newport, 1844)
*Lithobius forficatus (Linnaeus, 1758)
Lithobius minnesotae Bollman, 1887
Nadabius holzingeri (Bollman, 1887)
Neolithobius mordax (L. Koch, 1862)
?*Pachymerium ferrugineum (C. L. Koch, 1835)
Pokabius bilabiatus (Wood, 1867)
Scolopocryptops rubiginosus Koch, 1878
?Scolopocryptops sexspinosus (Say, 1821)
?Sonibius politus (McNeill in Bollman, 1887)
Strigamia fulva Sager, 1856

MISSISSIPPI:
Arctogeophilus umbraticus (McNeill, 1887)
Arenophilus bipuncticeps (H. C. Wood, 1862)
Arenophilus watsingus Chamberlin, 1912
Bothropolys multidentatus (Newport, 1845)
Buethobius oabitus Chamberlin, 1911
*Cryptops hortensis (Donovan, 1810)
Cryptops leucopodus (Rafinesque, 1820)

Garibius branneri (Bollman, 1888)
Garibius mississippiensis Chamberlin, 1913
Geophilus mordax Meinert, 1885
Guambius christianus Chamberlin, 1946
Guambius curtior Chamberlin, 1917
Guambius euthus (Chamberlin, 1904)
Guambius mississippiensis (Chamberlin, 1912)
Guambius oedipes (Bollman, 1888)
Guambius pinguis (Bollman, 1888)
Hemiscolopendra marginata (Say, 1821)
Lamyctes tivius Chamberlin, 1911
Lithobius watovius Chamberlin, 1911
Neolithobius mordax (L. Koch, 1862)
Neolithobius transmarinus (L. Koch, 1862)
Neolithobius voracior (Chamberlin, 1912)
Neolithobius vorax (Meinert, 1872)
*Pachymerium ferrugineum (C. L. Koch, 1835)
Paitobius exiguus (Meinert, 1886)
Paitobius simitus (Chamberlin, 1911)
Pokabius bilabiatus (Wood, 1867)
Pokabius verdescens (Chamberlin, 1912)
Scolopendra viridis Say 1821
Scolopocryptops nigridius McNeill, 1887
Scolopocryptops sexspinosus (Say, 1821)
*Scutigera coleoptrata (Linnaeus, 1758)
Sozibius proridens (Bollman, 1887)
Stenophilus grenadae (Chamberlin, 1912)
Strigamia bidens Wood, 1862
Tidabius nasintus Chamberlin, 1913
Tidabius pallidus Chamberlin, 1913
Tidabius tivius (Chamberlin, 1909)
Theatops posticus (Say, 1821)

MISSOURI:

Arctogeophilus fulvus (Wood, 1862)
Arctogeophilus umbraticus (McNeill, 1887)
Arenophilus bipuncticeps (Wood, 1862)
Arenophilus watsingus Chamberlin, 1912
Bothropolys multidentatus (Newport, 1845)
*Cryptops hortensis (Donovan, 1810)
Cryptops leucopodus (Rafinesque, 1820)
Escaryus missouriensis Chamberlin, 1942
Garrina parapodus (Chamberlin, 1941)

Geophilus mordax Meinert, 1885
Geophilus oweni Bollman, 1887
Geophilus vittatus (Rafinesque, 1820)
?*Gosibius paucidens* (Wood, 1862)
Hemiscolopendra marginata (Say, 1821)
Korynia urania (Crabill, 1954)
?*Lithobius celer* Bollman, 1888
*Lithobius forficatus (Linnaeus, 1758)
Nadabius ameles Chamberlin, 1944
Nadabius jowensis (Meinert, 1886)
Neolithobius suprenans Chamberlin, 1925
Neolithobius voracior (Chamberlin, 1912)
*Pachymerium ferrugineum (C. L. Koch, 1835)
Pokabius bilabiatus (Wood, 1867)
Pokabius verdescens (Chamberlin, 1912)
Scolopocryptops rubiginosus Koch, 1878
Scolopocryptops sexspinosus (Say, 1821)
*Scutigera coleoptrata (Linnaeus, 1758)
Sigibius urbanus Chamberlin, 1944
Sonibius politus (McNeill in Bollman, 1887)
Sozibius proridens (Bollman, 1887)
Strigamia bidens Wood, 1862
Strigamia bothriopa Wood, 1862
Strigamia branneri (Bollman, 1888)
Strigamia chionophila Wood, 1862
Strigamia fulva Sager, 1856
?*Theatops posticus* (Say, 1821)
Theatops spinicaudus (H. W. Wood, 1862)

MONTANA:

Arebius fremontus Chamberlin, 1943
Arebius montivagus Chamberlin, 1943
Bothropolys hoples (Brölemann, 1896)
Geophilus phanus Chamberlin, 1943
Nadabius coloradensis (Cockerell, 1893)
Nadabius mesechinus (Chamberlin, 1903)
Nadabius pluto Chamberlin & Wang, 1952
Neolithobius sp. Chamberlin, 1920
Pokabius helenae Chamberlin, 1922
Scolopendra polymorpha Wood, 1861
Scolopocryptops gracilis Wood, 1962
Stenophilus rothi Chamberlin, 1953
Strigamia chionophila Wood, 1862

?*Zygethobius montana* Chamberlin, unpublished

NEBRASKA:
Arenophilus bipuncticeps (Wood, 1862)
Geophilus mordax Meinert, 1885
Geophilus vittatus (Rafinesque, 1820)
*Lamyctes emarginatus (Newport, 1844)
Lithobius celer Bollman, 1888
*Lithobius forficatus (Linnaeus, 1758)
Nadabius jowensis (Meinert, 1886)
Nadabius pullus (Bollman, 1887)
Neolithobius mordax (L. Koch, 1862)
Neolithobius suprenans Chamberlin, 1925
*Pachymerium ferrugineum (C. L. Koch, 1835)
Paitobius exiguus (Meinert, 1886)
Pokabius bilabiatus (Wood, 1867)
?*Scolopendra heros* Girard, 1853
Scolopendra polymorpha Wood, 1861
Scolopendra viridis Say, 1821
Scolopocryptops sexspinosus (Say, 1821)
Scolopocryptops rubiginosus Koch, 1878
*Scutigera coleoptrata (Linnaeus, 1758)
?*Strigamia bothriopa* Wood, 1862
?*Strigamia fulva* Sager, 1856
Tidabius poaphilus Chamberlin, 1913

NEVADA:
Chomatobius laticeps (Wood, 1862)
*Cryptops hortensis (Donovan, 1810)
Eremorus becki Chamberlin, 1963
Ethopolys bipunctatus (Wood, 1862)
?*Ethopolys monticola* (Stuxberg, 1875)
*Lamyctes emarginatus (Newport, 1844)
Lamyctes pinampus Chamberlin, 1910
*Lithobius forficatus (Linnaeus, 1758)
*Lithobius melanops Newport, 1845
Lophobius sororis Chamberlin, 1940
Nyctunguis stenus Chamberlin, 1962
Oabius mercurialis Chamberlin, 1962
Pokabius sokovus (Chamberlin, 1909)
Pokabius vaquero Chamberlin, 1941
Scolopendra polymorpha Wood, 1861
Scolopendra viridis Say 1821

Scolopocryptops gracilis Wood, 1862
?*Tidabius tivius* (Chamberlin, 1909)
Theatops posticus (Say, 1821)

NEW HAMPSHIRE:
Bothropolys multidentatus (Newport, 1845)
*Lithobius forficatus (Linnaeus, 1758)
*Schendyla nemorensis (C. L. Koch, 1837)

NEW JERSEY:
Bothropolys multidentatus (Newport, 1845)
*Cryptops hortensis (Donovan, 1810)
Cryptops leucopodus (Rafinesque, 1820)
Garibius opicolens Chamberlin, 1913
Lamyctes pius Chamberlin, 1911
*Lithobius forficatus (Linnaeus, 1758)
Nadabius aristeus Chamberlin, 1922
?**Scolopendra mima* Chamberlin, 1942
Scolopocryptops sexspinosus (Say, 1821)
Tidabius suitus (Chamberlin, 1911)
Tidabius tivius (Chamberlin, 1909)
Theatops posticus (Say, 1821)

NEW MEXICO:
Arctogeophilus umbraticus (McNeill, 1887)
Arebius convergens Chamberlin, 1943
Arebius diplonyx Chamberlin, 1916
Arebius navajo Chamberlin, 1943
Arenophilus watsingus Chamberlin, 1912
Arthrorhabdus pygmaeus (Pocock, 1895)
Ephemerozaster antaeus Crabill, 1959
Geophilus atopodon Chamberlin, 1903
Geophilus geronimo (Chamberlin, 1912)
Gonibius glyptocephalus (Chamberlin, 1903)
Gosibius aberrantus Chamberlin, 1943
Gosibius arizonensis Chamberlin, 1917
Gosibius escabosanus Chamberlin, 1943
Gosibius mulaiki Chamberlin, 1938
Holitys neomexicanus Cook, 1899
Horonia bella Chamberlin, 1966
Kethops utahensis (Chamberlin, 1909)
Lamyctes caducens Chamberlin, 1938
Lithobius beulae Chamberlin, 1903

Lithobius cockerelli Chamberlin, 1904
Neolithobius suprenans Chamberlin, 1925
Oabius parvior Chamberlin, 1938
Pokabius centurio (Chamberlin, 1905)
Pokabius praefectus Chamberlin, 1938
Scolopendra heros Girard, 1853
Scolopendra polymorpha Wood, 1861
Scolopendra viridis Say 1821
Taiyuna moderata Chamberlin, 1941
Thalkethops grallatrix Crabill, 1960
Theatops posticus (Say, 1821)

NEW YORK:
Arenophilus bipuncticeps (Wood, 1862)
?*Bothropolys imaharensis* (Verhoeff, 1937)
Bothropolys multidentatus (Newport, 1845)
Chaetechelyne vesuviana (Newport, 1845)
†*Crussolum crusseratum* Shear et al., 1998
Cryptops hortensis (Donovan, 1810)
Cryptops leucopodus (Rafinesque, 1820)
†*Devonobius delta* Shear & Bonamo, 1988
Escaryus liber Cook & Collins, 1891
Escaryus urbicus (Meinert, 1886)
Garibius opicolens Chamberlin, 1913
Geophilus cayugae Chamberlin, 1904
Geophilus mordax Meinert, 1885
Geophilus rupestris (Crabill, 1949)
?*Geophilus setiger* Bollman, 1887
Geophilus vittatus (Rafinesque, 1820)
?*Hemiscolopendra marginata* (Say, 1821)
Lamyctes emarginatus (Newport, 1844)
Lamyctes pius Chamberlin, 1911
Lithobius crassipes (L. Koch, 1862)
Lithobius eucnemis Stuxberg, 1875
Lithobius forficatus (Linnaeus, 1758)
Lithobius melanops Newport, 1845
Lithobius microps Meinert, 1868
Nadabius aristeus Chamberlin, 1922
Nadabius pullus (Bollman, 1887)
Nampabius fungiferopes (Chamberlin, 1904)
Nampabius lundii (Meinert, 1886)
Nannarrup hoffmani Foddai et al., 2003
?*Orphnaeus brevilabiatus* (Newport, 1845)

Pachymerium ferrugineum (C. L. Koch, 1835)
Paitobius adelus Chamberlin, 1922
Paitobius exiguus (Meinert, 1886)
?**Rhysida longipes* (Newport, 1845)
**Schendyla nemorensis* (C. L. Koch, 1837)
?*Scolopendra heros* Girard, 1853
?**Scolopendra morsitans* Linnaeus, 1758
?**Scolopendra subspinipes* Leach, 1815
Scolopocryptops peregrinator (Crabill, 1952)
Scolopocryptops sexspinosus (Say, 1821)
**Scutigera coleoptrata* (Linnaeus, 1758)
Sonibius politus (McNeill in Bollman, 1887)
Sonibius scepticus Chamberlin & Wang, 1952
Sozibius pennsylvanicus Chamberlin, 1922
Sozibius proridens (Bollman, 1887)
?**Stigmatogaster subterranea* (Shaw, 1794)
?**Strigamia acuminata* (Leach, 1815)
Strigamia bothriopa Wood, 1862
Strigamia branneri (Bollman, 1888)
Strigamia chionophila Wood, 1862
Strigamia fulva Sager, 1856
Tidabius aberrans Chamberlin, 1929
Tidabius suitus (Chamberlin, 1911)
Tidabius tivius (Chamberlin, 1909)
Theatops posticus (Say, 1821)

NORTH CAROLINA:

Arctogeophilus umbraticus (McNeill, 1887)
Arenobius manegitus (Chamberlin, 1911)
Arenophilus bipuncticeps (Wood, 1862)
Arenophilus watsingus Chamberlin, 1912
Bothropolys multidentatus (Newport, 1845)
Cryptops leucopodus (Rafinesque, 1820)
Garibius branneri (Bollman, 1888)
Garibius catawbae Chamberlin, 1913
Garibius georgiae Chamberlin, 1913
Garibius monticolens Chamberlin, 1913
Geophilus cayugae Chamberlin, 1904
Geophilus mordax Meinert, 1886
Geophilus varians McNeill, 1887
Geophilus vittatus (Rafinesque, 1820)
Hemiscolopendra marginata (Say, 1821)
Lamyctes pius Chamberlin, 1911

Lithobius apheles Chamberlin, 1940
Lithobius atkinsoni Bollman, 1887
Lithobius cantabrigensis Meinert, 1886
*_Lithobius forficatus_ (Linnaeus, 1758)
Llanobius chamberlini Causey, 1942
*_Mecistocephalus maxillaris_ (Gervais, 1837)
Nadabius pullus (Bollman, 1887)
Nadabius saphes Chamberlin, 1940
Nadabius waccamanus Chamberlin, 1940
Nampabius carolinensis Chamberlin, 1913
Nampabius longiceps Chamberlin, 1913
?*Nampabius lundii* (Meinert, 1886)
Nampabius mycophor Chamberlin, 1940
Nampabius parienus Chamberlin, 1913
Nampabius pinus Causey, 1942
Nampabius virginiensis Chamberlin, 1913
Neolithobius latzelii (Meinert, 1886)
Neolithobius mordax (L. Koch, 1862)
Neolithobius vorax (Meinert, 1872)
*_Pachymerium ferrugineum_ (C. L. Koch, 1835)
Paitobius arienus (Chamberlin, 1911)
Paitobius carolinae (Chamberlin, 1911)
Paitobius eutypus Chamberlin, 1922
?*Paitobius exiguus* (Meinert, 1886)
Paitobius naiwatus (Chamberlin, 1911)
Paitobius zinus (Chamberlin, 1911)
Pampibius paitus (Chamberlin, 1911)
?*_Rhysida longipes_ (Newport, 1845)
Sozibius carolinus (Causey, 1942)
Scolopendra viridis Say 1821
Scolopocryptops nigridius McNeill, 1887
Scolopocryptops peregrinator (Crabill, 1952)
Scolopocryptops sexspinosus (Say, 1821)
*_Scutigera coleoptrata_ (Linnaeus, 1758)
Serrobius pulchellus Causey, 1942
Sigibius starlingi Causey, 1942
Sozibius pennsylvanicus Chamberlin, 1922
Sozibius proridens (Bollman, 1887)
Sozibius tuobukus (Chamberlin, 1911)
Strigamia bidens Wood, 1862
Strigamia branneri (Bollman, 1888)
Strigamia branneri var. *miura* Chamberlin, 1912
Strigamia fulva Sager, 1856

Taiyubius dux Chamberlin, 1940
Tidabius suitus (Chamberlin, 1911)
Theatops posticus (Say, 1821)
Theatops spinicaudus (Wood, 1862)
Typhlobius coecus (Bollman, 1888)

NORTH DAKOTA:

Pokabius bilabiatus (Wood, 1867) [**Note:** Mercurio, unpublished; to my knowledge this is the first report of a centipede from North Dakota].

OHIO:

Arctogeophilus umbraticus (McNeill, 1887)
Arenophilus bipuncticeps (H. C. Wood, 1862)
Bothropolys multidentatus (Newport, 1845)
Buethobius huestoni Williams & Hefner, 1928
Buethobius translucens Williams & Hefner, 1928
*Cryptops hortensis (Donovan, 1810)
Cryptops leucopodus (Rafinesque, 1820)
Escaryus liber Cook & Collins, 1891
Escaryus urbicus (Meinert, 1886)
Garibius monticolens Chamberlin, 1913
Garibius opicolens Chamberlin, 1913
Garibius pagoketes Chamberlin, 1913
Geophilus brevicornis Wood, 1862
Geophilus brunneus McNeill, 1887
Geophilus mordax Meinert, 1885
Geophilus oweni Bollman, 1887
?Geophilus rupestris (Crabill, 1949)
Geophilus setiger Bollman, 1887
Geophilus strigosus (McNeill, 1887)
Geophilus varians McNeill, 1887
Geophilus vittatus (Rafinesque, 1820)
Hemiscolopendra marginata (Say, 1821)
*Lamyctes emarginatus (Newport, 1844)
Lithobius cardinalis (Bollman, 1887)
Lithobius celer Bollman, 1888
*Lithobius forficatus (Linnaeus, 1758)
Nadabius ameles Chamberlin, 1944
Nadabius aristeus Chamberlin, 1922
Nadabius jowensis (Meinert, 1886)
Nadabius pullus (Bollman, 1887)
Nampabius lundii (Meinert, 1886)
Neolithobius mordax (L. Koch, 1862)

Neolithobius tyrannus (Bollman, 1887)
**Pachymerium ferrugineum* (C. L. Koch, 1835)
Paitobius juventus (Bollman, 1887)
Pokabius bilabiatus (Wood, 1867)
**Schendyla nemorensis* (C. L. Koch, 1837)
Scolopocryptops nigridius McNeill, 1887
?*Scolopocryptops rubiginosus* Koch, 1878
Scolopocryptops sexspinosus (Say, 1821)
**Scutigera coleoptrata* (Linnaeus, 1758)
Sonibius politus (McNeill in Bollman, 1887)
Sozibius pennsylvanicus Chamberlin, 1922
Sozibius proridens (Bollman, 1887)
Strigamia bidens Wood, 1862
Strigamia chionophila Wood, 1862
Strigamia fulva Sager, 1856
Tidabius tivius (Chamberlin, 1909)
Theatops posticus (Say, 1821)
Zygethobius pontis Chamberlin, 1911

OKLAHOMA:
Arenophilus bipuncticeps (H. C. Wood, 1862)
Cryptops leucopodus (Rafinesque, 1820)
Eulithobius sphactes Crabill, 1958
Hemiscolopendra marginata (Say, 1821)
Neolithobius entonus Chamberlin, 1942
Neolithobius suprenans Chamberlin, 1925
Scolopendra heros Girard, 1853
Scolopendra polymorpha H. C. Wood, 1861
Scolopendra viridis Say 1821
Scolopocryptops rubiginosus Koch, 1878
Scolopocryptops sexspinosus (Say, 1821)
**Scutigera coleoptrata* (Linnaeus, 1758)
Theatops posticus (Say, 1821)
Theatops spinicaudus (H. W. Wood, 1862)

OREGON:
Arctogeophilus corvallis Chamberlin, 1941
?*Arctogeophilus melanonotus* (H. C. Wood, 1862)
Arebius kochii (Stuxberg, 1875)
Arebius oregonensis Chamberlin, 1916
Banobius tener Chamberlin, 1938
Bothropolys hoples (Brölemann, 1896)
Bothropolys victorianus Chamberlin, 1925

Cryptops hortensis (Donovan, 1810)
Dicellophilus carniolensis (C. L. Koch, 1847)
Ethopolys integer Chamberlin, 1919
Ethopolys monticola (Stuxberg, 1875)
Ethopolys californicus (Daday, 1889)
Ethopolys xanti (Wood, 1862)
Geophilus anonyx (Chamberlin, 1941)
Geophilus leionyx (Chamberlin, 1938)
Geophilus oregonus (Chamberlin, 1941)
Geophilus tampophor (Chamberlin, 1953)
Geophilus transitus (Chamberlin, 1941)
Lamyctes emarginatus (Newport, 1844)
?*Lamyctes pinampus* Chamberlin, 1910
Leptodampius lamprus Chamberlin, 1938
Lithobius bellulus Chamberlin, 1903
Lithobius pinetorum Harger, 1872
Nadabius eigenmanni (Bollman, 1888)
Nadabius mesechinus (Chamberlin, 1903)
Oabius aiolus Chamberlin, 1938
Oabius boyeranus Chamberlin, 1940
Oabius eugenus Chamberlin, 1928
Oabius ineptus Chamberlin, 1916
Oabius mimosus Chamberlin, 1938
Oabius wamus Chamberlin, 1962
Pokabius stenenus Chamberlin, 1938
Pokabius utahensis (Chamberlin, 1901)
Pokabius utahensis tidus Chamberlin, 1962
Schendyla nemorensis (C. L. Koch, 1837)
Scolopendra polymorpha H. C. Wood, 1861
?*Scolopendra viridis* Say, 1821
Scolopocryptops gracilis Wood, 1862
Scolopocryptops spinicaudus Wood, 1862
Stenophilus rothi Chamberlin, 1953
Strigamia parviceps Wood, 1862
Theatops californiensis Chamberlin, 1902
Zygethobius ecologus Chamberlin, 1938

PENNSYLVANIA:
Arctogeophilus fulvus (Wood, 1862)
?*Bothropolys rugosus* (Meinert, 1872)
Bothropolys multidentatus (Newport, 1845)
Cryptops hortensis (Donovan, 1810)
Cryptops leucopodus (Rafinesque, 1820)

Garibius opicolens Chamberlin, 1913
**Geophilus proximus* C. L. Koch, 1847
Geophilus varians McNeill, 1887
Geophilus vittatus (Rafinesque, 1820)
?*Himantarium* sp. Chamberlin & Wang, 1952
**Lamyctes emarginatus* (Newport, 1844)
Lamyctes pius Chamberlin, 1911
**Lithobius crassipes* (L. Koch, 1862)
**Lithobius forficatus* (Linnaeus, 1758)
?**Lithobius migrans* Chamberlin, 1925
Nadabius aristeus Chamberlin, 1922
Paitobius adelus Chamberlin, 1922
?**Scolopendra morsitans* Linnaeus, 1758
?**Scolopendra subspinipes* Leach, 1815
Scolopocryptops nigridius McNeill, 1887
Scolopocryptops peregrinator (Crabill, 1952)
Scolopocryptops sexspinosus (Say, 1821)
**Scutigera coleoptrata* (Linnaeus, 1758)
Sozibius pennsylvanicus Chamberlin, 1922
Strigamia bidens Wood, 1862
Strigamia bothriopa Wood, 1862
Strigamia fulva Sager, 1856
Tidabius tivius (Chamberlin, 1909)
Theatops posticus (Say, 1821)
Theatops spinicaudus (H. W. Wood, 1862)

RHODE ISLAND:
Bothropolys multidentatus (Newport, 1845)
**Cryptops hortensis* (Donovan, 1810)
**Lithobius forficatus* (Linnaeus, 1758)

SOUTH CAROLINA:
?*Agathothus gracilis* (Bollman, 1888)
Arctogeophilus umbraticus (McNeill, 1887)
Arenobius manegitus (Chamberlin, 1911)
Arenophilus bipuncticeps (H. C. Wood, 1862)
Arenophilus watsingus Chamberlin, 1912
Bothropolys multidentatus (Newport, 1845)
**Cryptops hortensis* (Donovan, 1810)
Cryptops leucopodus (Rafinesque, 1820)
Enarthrobius bullifer Chamberlin, 1926
Garibius monticolens Chamberlin, 1913
Geophilus ampyx Crabill, 1954

Geophilus mordax Meinert, 1885
?*Geophilus varians* McNeill, 1887
?*Helembius nannus* Chamberlin, 1918
Hemiscolopendra marginata (Say, 1821)
Lithobius atkinsoni Bollman, 1887
Lithobius cantabrigensis Meinert, 1886
*Lithobius forficatus (Linnaeus, 1758)
?*Nadabius pullus* (Bollman, 1887)
Nampabius carolinensis Chamberlin, 1913
Nampabius embius Chamberlin, 1913
?*Nampabius lundii* (Meinert, 1886)
Nampabius parienus Chamberlin, 1913
Neolithobius underwoodi (Bollman, 1888)
Paitobius arienus (Chamberlin, 1911)
Paitobius atlantae Chamberlin, 1922
Paitobius carolinae (Chamberlin, 1911)
Paitobius naiwatus (Chamberlin, 1911)
Paitobius zinus (Chamberlin, 1911)
?**Rhysida longipes* (Newport, 1845)
Scolopendra viridis Say 1821
Scolopocryptops nigridius McNeill, 1887
Scolopocryptops sexspinosus (Say, 1821)
Sogona minima Chamberlin, 1912
Sozibius proridens (Bollman, 1887)
Sozibius tuobukus (Chamberlin, 1911)
Strigamia branneri (Bollman, 1888)
Strigamia fulva Sager, 1856
Theatops posticus (Say, 1821)
Theatops spinicaudus (H. W. Wood, 1862)
Watophilus alabamae Chamberlin, 1912

SOUTH DAKOTA:
Scolopendra polymorpha H. C. Wood, 1861

TENNESSEE:
Agathothus gracilis (Bollman, 1888)
Arctogeophilus umbraticus (McNeill, 1887)
Arenobius manegitus (Chamberlin, 1911)
Arenophilus bipuncticeps (H. C. Wood, 1862)
Bothropolys multidentatus (Newport, 1845)
Chomatobius euphorion (Crabill, 1953)
Cryptops leucopodus (Rafinesque, 1820)
Enarthrobius covenus Chamberlin, 1944

Enarthrobius dybasi Chamberlin, 1944
Enarthrobius fumans Chamberlin, 1944
Enarthrobius litus Chamberlin, 1944
Garibius branneri (Bollman, 1888)
Garibius monticolens Chamberlin, 1913
Geophilus cayugae Chamberlin, 1904
Geophilus flavus (De Geer, 1778)
Geophilus mordax Meinert, 1885
Geophilus orites (Chamberlin, 1944)
Geophilus varians McNeill, 1887
Gonibius rex (Bollman, 1888)
Hemiscolopendra marginata (Say, 1821)
Lithobius cantabrigensis Meinert, 1886
Nadabius eremites Chamberlin, 1944
Nadabius jowensis (Meinert, 1886)
Nadabius pullus (Bollman, 1887)
Nampabius inimicus Chamberlin, 1913
Nampabius lulae Chamberlin, 1913
?*Nampabius lundii* (Meinert, 1886)
Nampabius major Chamberlin, 1925
Nampabius tennesseensis Chamberlin, 1913
Nampabius virginiensis Chamberlin, 1913
Neolithobius mordax (L. Koch, 1862)
Neolithobius transmarinus (L. Koch, 1862)
Neolithobius vorax (Meinert, 1872)
Nuevobius cottus Crabill, 1960
Nyctunguis pholeter Crabill, 1958
Pachymerium ferrugineum (Koch, 1835)
Paitobius carolinae (Chamberlin, 1911)
Paitobius exiguus (Meinert, 1886)
Paitobius juventus (Bollman, 1887)
Paitobius naiwatus (Chamberlin, 1911)
Paitobius tabius (Chamberlin, 1911)
Paitobius zinus (Chamberlin, 1911)
Pampibius paitus (Chamberlin, 1911)
?*Scolopendra viridis* Say, 1821
Scolopocryptops nigridius McNeill, 1887
Scolopocryptops sexspinosus (Say, 1821)
Scutigera coleoptrata (Linnaeus, 1758)
Sogona minima Chamberlin, 1912
Sozibius proridens (Bollman, 1887)
Sozibius tuobukus (Chamberlin, 1911)
Strigamia bidens Wood, 1862

Strigamia bothriopa Wood, 1862
Strigamia branneri (Bollman, 1888)
Strigamia branneri var. *miura* Chamberlin, 1912
Strigamia fulva Sager, 1856
Tidabius tivius (Chamberlin, 1909)
Theatops posticus (Say, 1821)
Theatops spinicaudus (H. W. Wood, 1862)
Typhlobius coecus (Bollman, 1888)
Zygethobius pontis Chamberlin, 1911

TEXAS:
?**Apunguis prosoicus* Chamberlin, 1947
Arctogeophilus umbraticus (McNeill, 1887)
Arenophilus unaster (Chamberlin, 1909)
Arthrorhabdus pygmaeus (Pocock, 1895)
Bothropolys multidentatus (Newport, 1845)
Chomatobius auximus (Chamberlin, 1938)
Chomatobius laticeps (Wood, 1862)
?*Chomatobius mexicanus* (Saussure, 1858)
?*Cruzobius verus Chamberlin, 1942
**Cryptops hortensis* (Donovan, 1810)
Cryptops leucopodus (Rafinesque, 1820)
Garrina alicea Chamberlin, 1943
?*Garrina cruzanus Chamberlin, 1942
Garrina ochra Chamberlin, 1915
Garrina parapodus (Chamberlin, 1941)
Geophilus brevicornis Wood, 1862
Geophilus oweni Bollman, 1887
Geophilus secundus (Chamberlin, 1912)
Geophilus vittatus (Rafinesque, 1820)
Gosibius succharogeus Chamberlin, 1941
Gosibius texicolens Chamberlin, 1938
Gosipina bexara Chamberlin, 1940
Guambius euthus (Chamberlin, 1904)
Hemiscolopendra marginata (Say, 1821)
Ityphilus nemoides Chamberlin, 1943
Korynia texensis Chamberlin, 1941
Lamyctes diffusus Chamberlin & Mulaik, 1940
Liobius mimus Chamberlin & Mulaik, 1940
Lithobius hardyi Chamberlin, 1946
Llanobius paucispinus Chamberlin & Mulaik, 1940
Llanobius santus Chamberlin& Mulaik, 1940
Marsikomerus texanus (Chamberlin & Mulaik, 1940)

Neolithobius mordax (L. Koch, 1862)
Neolithobius suprenans Chamberlin, 1925
Neolithobius transmarinus (L. Koch, 1862)
?**Nyctunguis archochilus* Chamberlin, 1941
?*Orya barbarica (Gervais, 1835)
**Pachymerium ferrugineum* (C. L. Koch, 1835)
Pholobius mundior Chamberlin & Mulaik, 1940
Pokabius praefectus Chamberlin, 1938
Scolopendra heros Girard, 1853
?**Scolopendra morsitans* Linnaeus, 1758
Scolopendra polymorpha H. C. Wood, 1861
?**Scolopendra subspinipes* Leach, 1815
Scolopendra viridis Say 1821
Scolopocryptops rubiginosus Koch, 1878
Scolopocryptops sexspinosus (Say, 1821)
Scutigera buda Chamberlin, 1944
**Scutigera coleoptrata* (Linnaeus, 1758)
Scutigera dorothea Chamberlin, 1943
Scutigera linceci (Wood, 1867)
Scutigera phana Chamberlin, 1943
Sogona kerrana Chamberlin, 1940
Sozibius texanus Chamberlin, 1938
Strigamia chionophila Wood, 1862
Strigamia kerrana Chamberlin & Mulaik, 1940
Strigamia maculaticeps Wood, 1862
Texobius unicus Chamberlin & Mulaik, 1940
Theatops phanus Chamberlin, 1951
Theatops posticus (Say, 1821)
Timpina texana Chamberlin, 1912

UTAH:
Archethopolys gosobius Chamberlin, 1928
Archethopolys parowanus (Chamberlin, 1925)
Arctogeophilus atopus (Chamberlin, 1902)
Arctogeophilus umbraticus (McNeill, 1887)
Arebius obesus (Stuxberg, 1875)
Bothropolys permundus (Chamberlin, 1902)
**Cryptops hortensis* (Donovan, 1810)
Damothus alastus Crabill, 1962
Damothus montis Crabill, 1960
Escaryus monticolens Chamberlin, 1947
Ethopolys bipunctatus (Wood, 1862)
Ethopolys bipunctatus insulatus Chamberlin, 1955

Ethopolys timpius Chamberlin, 1951
Ethopolys xanti (Wood, 1862)
Geophilus fruitanus Chamberlin, 1928
Geophilus glyptus (Chamberlin, 1902)
Geophilus nealotus Chamberlin, 1902
Geophilus piedus Chamberlin, 1930
Geophilus shoshoneus Chamberlin, 1925
Geophilus vittatus (Rafinesque, 1820)
Gosendyla socarina Chamberlin, 1960
Gosibius ameles Chamberlin, 1940
Gosibius arizonensis Chamberlin, 1917
?*Gosibius auxodontus* Chamberlin, & Wang, 1952
Juanobius eremus Chamberlin, 1928
Kethops atypus Chamberlin, 1943
Kethops utahensis (Chamberlin, 1909)
*_Lamyctes emarginatus_ (Newport, 1844)
?*Lamyctes pinampus* Chamberlin, 1910
*_Lithobius forficatus_ (Linnaeus, 1758)
Lithobius intermontanus Chamberlin, 1901
Lophobius collium (Chamberlin, 1901)
Lophobius francisae Chamberlin, 1925
Lophobius lasalanus Chamberlin, 1928
Nyctunguis molinor Chamberlin, 1925
Oabius paiutus Chamberlin, 1925
Oabius sanjuanus Chamberlin, 1928
Pokabius arizonae (Chamberlin, 1922)
Pokabius centurio (Chamberlin, 1905)
Pokabius loganus (Chamberlin, 1925)
Pokabius piedus Chamberlin, 1930
Pokabius socius (Chamberlin, 1901)
Pokabius utahensis (Chamberlin, 1901)
*_Schendyla nemorensis_ (C. L. Koch, 1837)
?*Scolopendra heros* Girard, 1853
?*_Scolopendra morsitans_ Linnaeus, 1758
Scolopendra polymorpha H. C. Wood, 1861
Scolopendra viridis Say 1821
Scolopocryptops gracilis Wood, 1862
Sigibius siopius Chamberlin & Wang, 1952
Stenophilus hesperus (Chamberlin, 1928)
Strigamia chionophila Wood, 1862
Strigamia fulva Sager, 1856
Taiyubius purpureus (Chamberlin, 1901)
Tidabius bonvillensis (Chamberlin, 1909)

Tidabius tivius (Chamberlin, 1909)
Tidabius zionicus Chamberlin, 1925
Theatops posticus (Say, 1821)
Watophilus knowltoni Crabill, 1976
Watophilus utus Chamberlin, 1928
Yobius haywardi Chamberlin, 1945
Zinapolys uticola Chamberlin & Wang, 1952
Zygethobius dolichopus (Chamberlin, 1901)
Zygomerium rotarium Chamberlin, 1943

VERMONT:

Bothropolys multidentatus (Newport, 1845)
*Cryptops hortensis (Donovan, 1810)
*Lithobius forficatus (Linnaeus, 1758)
*Lithobius crassipes (L. Koch, 1862)
Nampabius fungiferopes (Chamberlin, 1904)
Paobius vagrans Chamberlin, 1916
Sonibius politus (McNeill in Bollman, 1887)

VIRGINIA:

Arctogeophilus fulvus (Wood, 1862)
Arctogeophilus umbraticus (McNeill, 1887)
Arenobius manegitus (Chamberlin, 1911)
Arenophilus bipuncticeps (H. C. Wood, 1862)
Arenophilus watsingus Chamberlin, 1912
Bothropolys multidentatus (Newport, 1845)
*Cryptops hortensis (Donovan, 1810)
Cryptops leucopodus (Rafinesque, 1820)
?**Dinocryptops miersii* (Newport, 1845)
Escaryus cryptorobius Pereira & Hoffman, 1993
?*Escaryus liber* Cook & Collins, 1891
Escaryus orestes Pereira & Hoffman, 1993
Escaryus urbicus (Meinert, 1886)
Garibius opicolens Chamberlin, 1913
Garibius pagoketes Chamberlin, 1913
Garibius psychrophilus Crabill, 1957
Geophilus ampyx Crabill, 1954
Geophilus cayugae Chamberlin, 1904
Geophilus mordax Meinert, 1885
Geophilus rupestris (Crabill, 1949)
Geophilus setiger Bollman, 1887

Geophilus varians McNeill, 1887
Geophilus virginiensis Bollman, 1888
Geophilus vittatus (Rafinesque, 1820)
Hemiscolopendra marginata (Say, 1821)
Lamyctes pius Chamberlin, 1911
Lithobius cantabrigensis Meinert, 1886
**Lithobius forficatus* (Linnaeus, 1758)
Nadabius aristeus Chamberlin, 1922
Nadabius pullus (Bollman, 1887)
?*Nampabius lundii* (Meinert, 1886)
Nampabius mycophor Chamberlin, 1940
Nampabius parienus Chamberlin, 1913
Nampabius turbator Crabill, 1952
Nampabius virginiensis Chamberlin, 1913
Neolithobius latzelii (Meinert, 1886)
Neolithobius vorax (Meinert, 1872)
**Pachymerium ferrugineum* (C. L. Koch, 1835)
Paitobius adelus Chamberlin, 1922
?*Paitobius exceptus* Chamberlin, 1922
Paitobius exiguus (Meinert, 1886)
Paitobius naiwatus (Chamberlin, 1911)
Paitobius tabius (Chamberlin, 1911)
Paitobius watsuitus (Chamberlin, 1911)
Paitobius zinus (Chamberlin, 1911)
Pampibius paitus (Chamberlin, 1911)
?*Scolopendra viridis* Say, 1821
Scolopocryptops nigridius McNeill, 1887
Scolopocryptops peregrinator (Crabill, 1952)
Scolopocryptops sexspinosus (Say, 1821)
**Scutigera coleoptrata* (Linnaeus, 1758)
Serrobius pulchellus Causey, 1942
Sigibius nidicolens Chamberlin, 1938
?*Sogona minima* Chamberlin, 1912
Sozibius carolinus (Causey, 1942)
Sozibius pennsylvanicus Chamberlin, 1922
Sozibius proridens (Bollman, 1887)
Sozibius tuobukus (Chamberlin, 1911)
Strigamia bidens Wood, 1862
Strigamia bothriopa Wood, 1862
Strigamia branneri (Bollman, 1888)
Strigamia chionophila Wood, 1862

Strigamia fulva Sager, 1856
Tidabius suitus (Chamberlin, 1911)
Tidabius tivius (Chamberlin, 1909)
Theatops posticus (Say, 1821)
Theatops spinicaudus (H. W. Wood, 1862)
Typhlobius coecus (Bollman, 1888)
Zygethobius pontis Chamberlin, 1911

WASHINGTON:

?*Arctogeophilus melanonotus* (H. C. Wood, 1862)
Arebius platypus Chamberlin, 1941
Bothropolys hoples (Brölemann, 1896)
Brachygonarea borealis Attems, 1934
Calcibius calcarifer Chamberlin & Wang, 1952
Cheiletha kincaidi Chamberlin, 1955
Cryptops hortensis (Donovan, 1810)
Ethopolys integer Chamberlin, 1919
Ethopolys monticola (Stuxberg, 1875)
Ethopolys californicus (Daday, 1889)
Geophilus compactus (Attems, 1934)
Geophilus glyptus (Chamberlin, 1902)
Gosibius submarginis Chamberlin & Wang, 1952
Nampabius perspinosus Chamberlin, 1928
Pokabius iginus (Chamberlin, 1912)
Scolopendra polymorpha H. C. Wood, 1861
Scolopendra viridis Say, 1821
Scolopocryptops gracilis Wood, 1862
?*Scolopocryptops sexspinosus* (Say, 1821)
Scolopocryptops spinicaudus Wood, 1862
Scutigera coleoptrata (Linnaeus, 1758)
Simobius ginampus (Chamberlin, 1909)
Simobius lobophor Chamberlin, 1941
?*Strigamia parviceps* Wood, 1862
Tidabius anderis Chamberlin, 1913
Tidabius tivius (Chamberlin, 1909)
Zygethopolys pugetensis Chamberlin, 1928

WEST VIRGINIA:
Arctogeophilus umbraticus (McNeill, 1887)
Bothropolys multidentatus (Newport, 1845)
Cryptops hortensis (Donovan, 1810)

Cryptops leucopodus (Rafinesque, 1820)
Garibius opicolens Chamberlin, 1913
Geophilus mordax Meinert, 1885
Hemiscolopendra marginata (Say, 1821)
*Lithobius forficatus (Linnaeus, 1758)
Nadabius pullus (Bollman, 1887)
Scolopocryptops nigridius McNeill, 1887
Scolopocryptops peregrinator (Crabill, 1952)
Scolopocryptops sexspinosus (Say, 1821)
Sozibius tuobukus (Chamberlin, 1911)
Strigamia fulva Sager, 1856
Theatops posticus (Say, 1821)

WISCONSIN:

Arenophilus bipuncticeps (H. C. Wood, 1862)
Bothropolys multidentatus (Newport, 1845)
*Cryptops hortensis (Donovan, 1810)
Escaryus urbicus (Meinert, 1886)
Geophilus oweni Bollman, 1887
Geophilus vittatus (Rafinesque, 1820)
*Lamyctes emarginatus (Newport, 1844)
Lamyctes pius Chamberlin, 1911
Lithobius celer Bollman, 1888
*Lithobius forficatus (Linnaeus, 1758)
*Lithobius melanops Newport, 1845
Lithobius minnesotae Bollman, 1887
Nadabius eigenmanni (Bollman, 1887)
Nadabius holzingeri (Bollman, 1887)
Nadabius jowensis (Meinert, 1886)
Neolithobius mordax (L. Koch, 1862)
Neolithobius voracior (Chamberlin, 1912)
*Pachymerium ferrugineum (C. L. Koch, 1835)
Paitobius exiguus (Meinert, 1886)
Paitobius juventus (Bollman, 1887)
Pokabius bilabiatus (Wood, 1867)
*Schendyla nemorensis (C. L. Koch, 1837)
Scolopocryptops rubiginosus Koch, 1878
Scolopocryptops sexspinosus (Say, 1821)
*Scutigera coleoptrata (Linnaeus, 1758)
Sonibius politus (McNeill in Bollman, 1887)
Strigamia bothriopa Wood, 1862

Strigamia chionophila Wood, 1862
Taiyuna opita Chamberlin, 1912
Tidabius opiphilus Chamberlin, 1913
Tidabius tivius (Chamberlin, 1909)

WYOMING:
Arctogeophilus atopus (Chamberlin, 1902)
Arctogeophilus umbraticus (McNeill, 1887)
Arebius fremontus Chamberlin, 1943
Arebius tetonus Chamberlin, 1943
Arenophilus bipuncticeps (H. C. Wood, 1862)
*Lithobius forficatus (Linnaeus, 1758)
Nadabius coloradensis (Cockerell, 1893)
Nadabius jowensis (Meinert, 1886)
Nadabius vaquens Chamberlin & Wang, 1952
Scolopendra polymorpha H. C. Wood, 1861
Strigamia chionophila Wood, 1862

Appendix IV

List of Cavernicolous Centipedes

The following list includes species that are troglobites, troglophiles, trogloxenes or accidental occurrences. Further information about each species may be found in the main catalog.

LITHOBIOMORPHA

Bothropolys multidentatus (Newport, 1845)—Holsinger & Culver, 1988: 57 (USA: Virginia). [**Note:** Reported as an accidental occurrence in a cave].

Ethopolys (Archethopolys) bipunctatus insulatus Chamberlin, 1955—Chamberlin, 1955a: 180-181 (USA: Utah). [**Note:** The locality of this species had "Spider Cave" which may or may not be an actual cave but this is unknown to the author].

Eulithobius sphactes Crabill, 1958—Crabill, 1958d: 260-262 (USA: Oklahoma).

Nampabius major Chamberlin, 1925—Chamberlin, 1925f: 291 (USA: Tennessee).

Nampabius parienus Chamberlin, 1913—Holsinger & Culver, 1988: 57 (USA: Virginia). [**Note:** Reported as either a trogloxene or accidental occurrence in a cave. Holsinger & Culver also report a *Nampabius* sp. from Virginia: Montgomery Co.: Erharts Cave but do not indicate why they could not identify it to species; maybe due to immaturity or possibly a new species].

Nampabius turbator Crabill, 1952—Crabill, 1952c: 203-206 (USA: Virginia). [**Note:** This species is strongly indicative of being cavernicolous].

Neolithobius **sp.**—Reddell, 1965: 166 (USA: Texas).
—Reddell, 1994: ? (*Neolithobius* sp.; notes; USA: Texas).

Nuevobius cottus Crabill, 1960—Crabill, 1960f: 122-127; Figs. 1-4, 6 (USA: Tennessee). [**Note:** This species is cavernicolous and shows some resemblance to *Sozibius*].

Tidabius tivius (Chamberlin, 1909)—Dearolf, 1953: 231 (USA: Kentucky). [**Note:** Cave record].

SCOLOPENDROMORPHA

Cryptops (*Cryptops*) *hortensis* (Donovan, 1810)—Holsinger & Culver, 1988: 57 (USA: Virginia). [**Note:** Reported as an accidental occurrence in a cave].

Cryptops (*Cryptops*) *leucopodus* (Rafinesque, 1820)—Dearolf, 1953: 231 (USA: West Virginia). [**Note:** Cave record].
—Holsinger & Culver, 1988: 57 (USA: Virginia). [**Note:** Reported as an accidental occurrence in a cave].

Theatops phanus Chamberlin, 1951—Chamberlin, 1951a: 101 (USA: Texas).

Theatops posticus (Say, 1821)—Holsinger & Culver, 1988: 57 (USA: Virginia). [**Note:** Reported as an accidental occurrence in a cave].

Thalkethops grallatrix Crabill, 1960—Crabill, 1960a: 3-8; Figs. 1-5, 7, 9-11, 13-16 (USA: New Mexico).

Scolopocryptops sexspinosus (Say, 1821)—Dearolf, 1953: 231 (USA: Tennessee). [**Note:** Cave records].
—Holsinger & Culver, 1988: 57 (*Scolopocryptops*; USA: Tennessee). [**Note:** Reported as an accidental occurrence in a cave].

GEOPHILOMORPHA

[**Note:** Reddell (1965) reports of two geophilomorph specimens in the family Himantariidae. Undetermined species possibly from the same genus. The first is reported from USA: Texas: San Saba County: Harrell's

Cave. The second is reported from USA: Texas: Williamson County: Laubach Cave. Reddell (1994) again mentions the specimens above].

Arctogeophilus umbraticus (McNeill, 1887)—Holsinger & Culver, 1988: 57 (USA: Virginia). [**Note:** Reported as an accidental occurrence in a cave].

Arenophilus psednus Crabill, 1969—Crabill, 1969a: 10-11; Figs. 3-5 (USA: Kentucky).

Ephemerozaster antaeus Crabill, 1959—Crabill, 1959a: 120-126; Figs. 1-9; p. 155 (USA: New Mexico).

Stenophilus californicus (Chamberlin, 1930)—Chamberlin, 1930b: 297-299; Figs. 1-6 (USA: California).

Nyctunguis pholeter Crabill, 1958—Crabill, 1958c: 154-159; Figs. 1-14 (USA: Tennessee).

Appendix V

Additional References

This appendix lists other references on centipedes, most of which are not listed in the bibliography that are useful for anatomy, behavior, ecology, evolution, molecular biology, physiology, taxonomy, ultrastructure, venom and more. The taxonomic references listed will typically have something to do with the familial or generic level in some significant way to North American fauna, but may also only refer to higher classification of centipedes. Some of these papers may deal with ecological studies done in foreign countries but have application in North America. There are a number of papers by J.G.E. Lewis, and others in reference to morphological variation in the Geophilomorphs, Lithobiomorphs, Scolopendromorphs and Scutigeromorphs that may help with the taxonomy of these groups in North America and the world over.

Andersson, G. 1976. Post-embryonic development of *Lithobius forficatus* (L.), (Chilopoda: Lithobiidae). Entomologica Scandinavica 7: 161-168.

Applegarth, A.G. 1952. The anatomy of the cephalic region of a centipede *Pseudolithobius megaloporus* (Stuxberg) (Chilopoda). Microentomology. 17: 127-171.

Baerg, W.J. 1924. The effect of the venom of some supposedly poisonous arthropods. Annals of the Entomological Society of America 17: 343-352.

Bähr, R. 1974. Contribution to the morphology of compound eyes. Symposia of the Zoological Society of London. No. 32: 388-404.

Banks, N. 1892. Notes on the mouth parts of insects and chilopoda. American Naturalist 27: 400-401.

Beniouri, R. 1985. Etude cytologique comparée des spermatozoids chez quelques Géophilomorphes (Chilopoda). Bijdragen tot de Dierkunde 55(1): 25-35.

Beniouri, R. and M. Descamps. 1985. Ultrastructural study of the spermatozoa in *Cryptops hortensis* Leach (Myriapoda, Chilopoda). *Arch. Biol.* 96(2): 195-208.

Beniouri, R., M. Descamps and G. Torpier. 1985. The spermatocyte membrane in *Lithobius forficatus* L. (Myriapoda, Chilopoda). Changes induced by hormonal actions. Preliminary results. Reproduction Nutrition Development 25(1a): 83-92.

Bishop, S.C. 1941. The salamanders of New York. New York State Museum Bulletin, no. 324, 365 p. [reporting of salamanders consuming centipedes]

Blackman, M.W. 1905. The spermatogenesis of *Scolopendra heros*. Bulletin of the Museum of Comparative Zoology 48: 1-138.

Blackman, M.W. 1911. The spermatogenesis of the Myriopods. 6. An analysis of the chromosome group of *Scolopendra heros*. Biol. Bull. Woods Hole Mass. 19: 138-160, pls. i-ii.

Blower, J.G. 1952. Epidermal glands in centipedes. Nature, Lond., 170: 166-167.

Blower, J.G. 1955. Millipedes and centipedes as soil animals: 138-151. *In*: D. K. M. Kevan (ed.). Soil zoology; proceedings of the University of Nottingham 2d Easter School in Agricultural Science. Academic Press, New York.

Blower, J.G. (Ed.) 1974. Myriapoda. Symposia of the Zoological Society of London 32. 712 pp. [Also publ. Academic Press, London, New York, San Francisco - a general work including some taxonomy of Chilopoda and synopsis of classification.]

Borucki, H. 1996. Evolution und Phylogenetisches System der Chilopoda (Mandibulata, Tracheata). *Verh. Naturwiss. Ver.* Hamburg, 35: 95-226.

Britten, H. 1920. Food hunting habits of *Lithobius forficatus* L. *Lancs. Chesh. Nat.*, 13: 118.

Bush, S.P., MD; B.O. King, MD; R.L. Norris, MD; and S.A. Stockwell, PhD. 2001. Centipede envenomation. Wilderness and Environmental Medicine, 12: 93-99.

Campbell, M.I., 1932. Some observations on Scolopendra heros under laboratory conditions. Transactions of the Kansas Academy of Science 35: 80-81.

Case, S. 1920. General Reactions of a centipede. Journal of Entomology and Zoology 12: 79-81.

Castellani-Ceresa, L., G. Berruti and E. Bezzi, 1979. Cytochemical localization of glucose-6-phosphatase and thiamine-pyrophosphtase in Chilopoda spermatazoa. *In*: Myriapod Biology. Ch. 12, (Camatini Marina ed.) p. 113-119.

Chamberlin, R.V. (1918). Report on the Chilopoda and Diplopoda. University of Iowa Studies in Natural History 10(4): 41-44.

Chamberlin, R.V. 1920a. (misspelled "Chamberlain, R. V."). Note on the writings of H. C. Wood. Entomological News 31(4): 117-118.

Chao, J.L. and H.W. Chang 2003. The Scolopendromorph centipedes (Chilopoda) of Taiwan. African Invertebrates 44(1): 1-11.

Cochran, M.E. 1911. The biology of the red-backed salamander (*Plethodon cinereus* Green). Biology Bulletin, 20: 332-349. [reporting of salamanders consuming centipedes]

Cloudsley-Thompson, J. L. 1958. Spiders, Scorpions, Centipedes and Mites. Pergamon Press, London, 228 pp.

Crabill, R. E. Jr. 1960. Clear plastic spray for label restoration and preservation. Turtox News, 38(4): 110-111.

Curry, A., 1974. The Spiracle Structure and Resistance to Desiccation of Centipedes. Symposia of the Zoological Society of London. 32: 365-382.

Descamps, M. 1985. Hormonal control of spermatogenesis in the Myriapod *Lithobius forficatus*. *In*: Current trends in comparative endocrinology. Hong Kong University Press, Hong Kong, Loft a. Holes ed., 317-320.

Descamps, M. and R. Joly. 1985. Ultrastructure of the cerebral glands in *Scolopendra cingulata* Latr., *Cryptops savignyi* Leach and *C. hortensis* Leach (Myriapoda, Scolopendromorpha) *Int.* Journal of Insect Morphology and Embryology 14(2): 105-113.

Descamps, M., R. Joly and C. Jamault-Navarro 1985. Autoradigraphie localization of 5-hyroxytryptamine and noradrenaline in the central nervous system of *Lithobius forficatus* L. (Myriapoda, Chilopoda). Bijdragen tot de Dierkunde 55(1): 47-54.

Dobroruka, L.J. 1961. Hundertfüssler. Neue Brehm-Bücherei, A. Siemsen Verlag, Wittenberg Lutherstadt (now Kosmos-Verlag, Stuttgart), 281. 49 pp. [A key to all families and subfamilies.]

Dohle, W. 1985. Phylogenetic pathways in the Chilopoda. Bijdragen tot de Dierkunde 55: 55-66.

Drift, J. van der. 1951. Analysis of the animal community in a beech forest floor. Mededeling / Instituut voor Toegepast Biologisch Onderzoek in de Natuur. 9: 1-168.

Eason, E.H. 1964. Centipedes of the British Isles. Frederick Warne & Co., Ltd., London, 294 pp. + vi-x.

Eason, E.H. 1982. A review of the north-west European species of Lithobiomorpha with a revised key to their identification. Zoological Journal of the Linnean Society. 74: 9-33.

Eason, E.H. 1986. On the geographic distribution of *Lithobius variegatus* Leach, 1814, and the identity of *Lithobius rubriceps* Newport, 1845 (Chilopoda: Lithobiomorpha). Journal of Natural History 20: 23-29.

Edgecombe, G.D., S. Richter and G.D.F. Wilson, 2003. The mandibular edges: Homologous structures throughout Mandibulata? African Invertebrates 44(1): 115-135.

Edgecombe, G.D. and G. Giribet 2002. Myriapod Phylogeny and the Relationships of Chilopoda. In: Biodiversidad Taxonomía y Biogeographía de Arthrópodos de México: Hacia una síntesis de su conocimiento. Vol. 3, Ch. 6: 143-168.

Edgecombe, G.D. and G. Giribet 2004. Adding Mitochondrial Sequence Data (16S rRNA and Cytochrome *c* Oxidase Subunit I) to the Phylogeny of Centipedes (Myriapoda: Chilopoda): an Analysis of Morphology and Four Molecular Loci. Journal of Zoological Systematics and Evolutionary Research 42: 89-134.

Edgecombe G. & G. Giribet 2006. Evolutionary biology of centipedes (Myriapoda: Chilopoda) Annual Review of Entomology 52: 151-170.

Edgecombe G. & G. Giribet 2006. A century later - a total evidence re-evaluation of the phylogeny of scutigeromorph centipedes (Myriapoda: Chilopoda). Invertebrate Systematics 20: 503-525.

Enghoff, H. 1985. Millipedes and centipedes in urban areas. *Niche. Arhus.*, 5: 244-249.

Ernst, A. and J. Rosengerg, 2003. Structure and distribution of sensilla coeloconica on the maxillipedes of Chilopoda. African Invertebrates 44(1): 155-168.

Fahlander, K. 1938. Beitrage zur Anatomie und systematischen Einteilung der Chilopoden., Zool. Bidrag fran Uppsala, 17: 1-150.

Formanowicz, D.R., Jr. and P.J. Bradley 1987. Fluctuations in prey density: effects on the foraging tactics of scolopendrid centipedes. Animal Behavior 35: 453-461.

Frank, J.H., S. Sreenivasan, J. Benshoff, M.A. Deyrup, G.B. Edwards, S.E. Halbert, A.B. Hamon, M.D. Lowman, E.L. Mockford, R.H. Scheffrahn, G.J. Steck, M.C. Thomas, T.J. Walker, and W.C. Welbourn, 2004. Invertebrate animals extracted from native Tillandsia (Bromeliales: Bromeliaceae) in Sarasota county, Florida. Florida Entomologist 87(2): 176-185.

Friedrich, M. and D. Tautz, 1995. Ribosomal DNA phylogeny of the major extant arthropod classes and the evolution of myriapods. Nature 376: 165-167.

Fusco, G. and A. Minelli, 2000. Measuring morphological complexity of segmented animals: centipedes as model systems. Journal of Evolutionary Biology 13: 38-46.

Giribet, G., S. Carranza, M. Riutort, J. Baguñà and C. Ribera, 1999. Internal phylogeny of the Chilopoda (Myriapoda, Arthropoda) using complete 18S rDNA and partial 28S rDNA sequences. Philosophical Transactions of the Royal Society of London B 354: 215-222.

Giribet, G., G.D. Edgecombe and W.C. Wheeler, 2001. Arthropod phylogeny based on eight molecular loci and morphology. Nature, 413: 157-161.

Gopalakrishnakone P., 1992. Light and electron microscopic features of the venom apparatus of the centipede *Scolopendra morsitans*. Toxicon 30(5/6): 514.

Gupta, A.P. 1979. Arthropod Phylogeny. Van Nostrand Reinhold Company, New York, 762 pp.

Hamilton, W.J., Jr. and J.A. Pollack 1955. The food of some crotalid snakes from Fort Benning, Georgia. Chicago Academy of Science Nautral History Misc. 140: 1-4.

Harada, K., Asa, K., Imachi, T., Yamaguchi, Y. and Yoshida, K., 1999. Centipede Inflicted Postmortem Injury. Journal of Forensic Sciences 44(4): 849-850.

Haupt, J. 1979. Phylogenetic aspects of recent studies on myriapod sense organs. *In*: Myriapod Biology. Ch. 36, (Camatini Marina ed.) p. 391-406.

Helbing, G. 1985. Changes in the hemolymph protein spectra of *Lithobius forficatus* L. (Chilopoda). *Zool. Jahrb. (Zool. Physiol.)*, 89 (1): 99-106.

Hoffman, R.L. 1969. Myriapoda, exclusive of insecta. *In*: Treatise on Invertebrate Paleontology. Part R, Arthropoda 4, (Moore, R. C. ed.), 2: R572-R606.

Hoffman, R.L. 1982. Chilopoda. *In*: Parker, S. P., ed., *Synopsis and classification of living organisms*, 2: 681-688, McGraw-Hill Book Co., New York.

Hopkin, S.P., K. Watson, M.H. Martin, and M.L. Mould. 1985. The assimilation of heavy metals by *Lithobius variegatus* L. and *Glomeris marginata* (Chilopoda, Diplopoda). Bijdragen tot de Dierkunde 55 (1): 88-94.

Hosey, G.R., M. Wood, R.J. Thompson, and P.L. Druck. 1985. Social facilitation in a "non social" animal the centipede *Lithobius forficatus*. Beh. Processes, 10(1-2): 123-130.

Joly, R. 1979. Neurosecretion and endocrine glands in Chilopoda. *In*: Myriapod Biology. Ch. 26, (Camatini Marina ed.) p. 263-272.

Kaestner, A. 1968. Invertebrate Zoology. Interscience Publishers, A division of John Wiley & Sons, New York, London, Sydney. 472 pp.

Kendeigh, S.C. 1979. Invertebrate populations of the deciduous forest: fluctuations and relations to weather. Illinois Biological Monographs 50: 1-103.

Koch, M. 2003. Monophyly of the Myriapoda? Reliability of current arguments. African Invertebrates 44(1): 137-153.

Kraus, O. 2001. "Myriapoda" and the ancestry of the hexapoda. Annals *of the Soc. Entomol.* Fr. (N.S.), 37 (1-2): 105-127.

Lawrence, T.C. 1934 Notes on the feeding habits of *Scolopendra subspinipes* Leach (Myriapoda) Proceedings of the Hawaiian Entomological Society 8: 497-498.

Léger, L. and O. Duboscq 1903. Recherches sur les myriapodes de corse et eurs parasites. [Brölemann, H. W. Avec la description des diplopods.] Archives de Zoologie Expérimentale, ser. 4, 1: 307-358.

Leidy, J. 1853. Flora and fauna within living animals. Smithsonian contributions to knowledge. 5(2): 1-67 with 10 plates. (see page 21).

Levi, H.W. and L.R. Levi, 1990. Spiders and their kin. Golden Press, New York p. 1-160.

Lewis, J.G.E. 1963. On the spiracle structure and resistance to desiccation of four species of geophilomorph centipede. *Ent. Exp. & Appl.* 6: 89-94.

Lewis, J.G.E. 1965. The food and reproductive cycles of the centipedes *Lithobius variegatus* and *Lithobius forficatus* in a Yorkshire woodland. Proceedings of the Zoological Society of London 144: 269-283.

Lewis, J.G.E. 1981. The Biology of Centipedes. Cambridge: Cambridge University Press, 476pp.

Lewis, J.G.E. 1985. Possible species isolation mechanisms in some scolopendrid centipedes (Chilopoda; Scolopendridae). *Bijdragen tot de Dierkunde* 55: 125-130.

Lewis, J.G.E. 1985. Centipedes entering houses with the particular reference to *Geophilus carpophagus* Leach. The Entomologist's Monthly Magazine 121: 257-259.

Lewis, J.G.E. 1987. On some structural abnormalities in *Lithobius* and *Cryptops* (Chilopoda) and their possible significance. Bulletin of the British Myriapod Group 4: 3-6.

Lewis, J.G.E. 2003. The problems involved in the characterization of scolopendromorph species (Chilopoda: Scolopendromorpha). African Invertebrates 44(1): 61-69.

Lewis, J.G.E., A. Minelli and R.M. Shelley 2006. Taxonomic and nomenclatorial notes on scolopendrid centipedes (Chilopoda: Scolopendromorpha: Scolopendridae). Zootaxa 1155: 35-40.

Littlewood, P.M.H., J.G. Blower, 1987. The chemosensory behavior of *Lithobius forficatus*: evidence for a pheromone released by the coxal organs (Myriapoda: Chilopoda). Journal of Zoology 211:65-82.

Lopez, P., J. Martin, 2001. Chemosensory predator recognition induces specific defensive behaviors in a fossorial amphisbaenian. Animal Behavior 62: 259-264.

Manton, S.M. 1969. Introduction to classification of arthropoda. *In*: Treatise on Invertebrate Paleontology. Part R, Arthropoda 4, (Moore, R. C. ed.), (1): R3-R15.

Manton, S.M. 1969. Evolution and affinities of Onychophora, Myriapoda, Hexapoda, and Crustacea. *In*: Treatise on Invertebrate Paleontology. Part R, Arthropoda 4, (Moore, R. C. ed.), (1) R15-R56.

Manton, S.M. 1977. The Arthropoda, habits, functional morphology and evolution. Oxford University Press: ?pp.

McCormick, S and G.A. Polis 1982. Arthropods that Prey on Vertebrates. Biological Reviews of the Cambridge Philosophical Society 57: 29-58.

Meinert, F. 1883. Caput Scolopendrae. Copenhagen: H. Hagerups Forlag. p. 1-71 + 3 plates. [**Note:** Excellent diagrams of the musculature of the head and mouth parts of *Scolopendra subspinipes*.]

Mesibov, B. 1986. A guide to Tasmanian centipedes. 62 pp.

Minelli, A. 1978. Secretions of Centipedes. *In*: Bettini, S. *Arthropod venoms: handbook of experimental pharmacology*. Vol. 48, Heidelberg, Springer-Verlag p. 73-85

Minelli, A. 1993. Chilopoda. Ch. 3, *In*: Microscopic Anatomy of Invertebrates, Vol. 12: Onycophora, Chilopoda, and Lesser Protostomata: 57-114.

Minelli, A., D. Foddai, L.A. Pereira, and J.G.E. Lewis, 2000. The Evolution of Segmentation of Centipede Trunk and Appendages. Journal of Zoological Systematics and Evolutionary Research 38(2): 103-117.

Mundel: 1990. Chilopoda. Ch. 25, *In*: Soil Biology Guide, (Dindhal, D.L. ed.) John Wiley & Sons, New York: 819-833.

Negrisolo, E., A. Minelli, and G. Valle 2004a. Extensive gene order rearrangment in the mitochondrial genome of the centipede *Scutigera coleoptrata*. Journal of Molecular Evolution 58: 413-423.

Negrisolo, E., A. Minelli, and G. Valle 2004b. The Mitochondrial Genome of the House Centipede Scutigera and the Monophyly Versus Paraphyly of Myriapods. Molecular Biology and Evolution 21(4): 770-780.

Palmén, E. and M. Rantala. 1954. Life history and ecology of Pachymerium ferrugineum (Chilopoda: Geophilidae). Ann. Zool. Soc. Zool.-Bot. Fenn. "Vanamo," 16: 1-44.

Porter, R.N. 1973. Centipede feeding on field mouse. The Lammergeyer 19: 31.

Quistad, G.B., P.A. Dennis and S. Skinner, 1992. Insecticidal Activity ofSpider (Araneae), Centipede (Chilopoda), Scorpion (Scorpionida_, and Snake (Serpentes) Venoms. Journal of Economic Entomology 85(1): 33-39.

Regier, J.C. and J.W. Shultz, 1998. Molecular phylogeny of Arthropods and the Significance of the Cambrian "Explosion" for Molecular Systematics. American Zoology 38: 918-928.

Regier, J.C. and J.W. Schultz. 2001. A Phylogenetic Analysis of Myriapoda (Arthropoda) using two nuclear protein-encoding genes. Zoological J. Linnean Soc. 132: 469-486.

Ribaut, H. 1921. L'armement des pattes chez les Lithobies., Soc. D'Hist. Nat. Toul., 49: 312-319.

Roberts, R.B., 1956. An ecological study of the arthropods of a mixed beech-oak woodland, with particular reference to Lithobiidae (PhD Dissertation). Highfield: University of Southampton.

Rosenberg, J. and G. Seifert 1977. The coxal glands of Geophilomorpha (Chilopoda): Organs of osmoregulation. Cell Tissue Research. 182: p. 247-251.

Rosenberg, J. 1979. Fine structure of the "Lyphatic Tissue" of *Cryptops hortensis* (Chilopoda, Scolopendromorpha): General organization and intercellular junctions. *In*: Myriapod Biology. Ch. 29, (Camatini Marina ed.) p. 287-294.

Sinclair, F.G. 1895. Myriapoda in the Cambridge Natural History: 29-80.

Spelda, J., J. Rosenberg and K. Voigtländer 2003. The German myriapod project (GerMyLit). African Invertebrates 44(1): 325-330.

Stahnke, H.L. and R.H. Larson 1968. Obtaining Venom from Centipedes. Tutox News 46(5): 172-173.

Tobias, D. 1974. New criteria for the differentiation of species within the Lithobiidae. Symposia of the Zoological Society of London 32: 75-87.

Traniello, J.F.A. 1982. Population structure and social organization in the primitive ant Amblyopone pallipes (Hymenopter: Formicidae). Psyche 89(1-2): 65-80.

Tuf, I.H. 2003. Four-year development of a centipede (Chilopoda) community after a summer flood. African Invertebrates 44(1): 265-276.

Vandenbulcke, F., C. Grelle, M.C. Fabre and M. Descamps, 1998. Ultrastructural and Autometallographic Studies of the Nephrocytes of *Lithobius forficatus* L. (Myriapoda, Chilopoda): Role in Detoxification of Cadmium andLead. International Journal of Insect Morphology and Embryology 27(2): 111-120.

Verhoeff, K.W. 1906. Vergleichend morphologische studien über die coxo-pleuralen Körperteile der Chilopoden, mit besonderer

Berucksichtigung der Scolopendromorpha, ein Beitrag zur Anatomie und Systematik derselben, nebst physiologischen und phylogenetischen Mitteilungen und Ausblicken auf die Insekten. Nova acta Academiae Caesareae Leopoldino-Carolinae Germanicae Naturae Curiosorum. 86: 349-502.

Walls, J.G. 2000. The Guide to Owning Millipedes and Centipedes. T. F. H. Pubs., Inc., Neptune City, New Jersey, 64 pp.

Wang, T.H. and H.W. Wu 1948. On the structure of the Malpighian tubes of centipedes and their excretion of uric acid. Sinesia 18: 1-11.

Wells-Cole, H. 1898. A voracious centipede. The Journal of the Bombay Natural History Society 12: 214.

White, M.J.D. 1979. The present status of Myriapod Cytogentics. *In*: Myriapod Biology. Ch. 1, (Camatini Marina ed.) p. 3-8.

Wood, D.M. and A.G. Wheeler, 1972. First record on North America of the centipede (Myriap. Chil.) parasite *Loewia foeda* (Dipt. Tachinidae). Canadian Entomologist 104: 1363-1367.

Woznicki, P., Jolanta Wytwer and Malgorzata Kulesza. 2003. Chromosome Study of *Lithobius forficatus* (Lithobiomorpha, Chilopods). Folia Biologica (Kraków), 51(3-4): 147-150.

Würmli, M. 1974. Systematic criteria in the Scutigeromorpha. Symposia of the Zoological Society of London 32: 89-98.

Würmli, M. 1978. Biometrical studies on the taxonomy and the post-embryonic development of some species of *Scolopendra* Linnaeus (Chilopoda). *Abh. Verh. Naturwiss. Ver.* Hamburg (NF), 21/22: 51-54.

Index of Scientific Names

A

Abatobius 92
abboti 429
aberrans
  Nothembius 160
  Tidabius 207
aberrantus 89
acuminata 289
acuminatus. *See* chionophila
adelus 174
Adenoschendyla. *See* Pectiniunguis;
adjacens 162
adlatus 67
admarinus 336
aethes 182
africanus 36
agamus 68
Agathothus 280
aiolus 170
ajonus 162
alabamae
  Garibius 83
  Watophilus 318
alamedanus 331
alaska 311
alaskanus
  Ethopolys integer 62
  Geophilus 336
  Oabius 162
Alaskobius 67
alastus 284
albertanus 178
albus. *See* chionophila;
aletes 181
alicea 333

Allobothropolys
  subgenus of Bothropolys.
     See Ethopolys;
Allothereua 27
alternans 246
ameles
  Gosibius 89
  Nadabius 131
americanum. See umbraticus;
americanus
  Lithobius. *See* forficatus;
  Pectiniunguis 408. *See* also
    halirrhytus;
Amplobius 93
amplus
  Garriscaphus 372
  Nothembius 160
ampyx 336
Anamorpha 27, 36
anderis 207
anderisus 215
Anethops. *See* Scolopocryptops;
angelicus 92
angelus 205
annectus 123
Anobius 188
anomalans 218
anomalus. *See* carniolensis;
anonyx 337
antaeus 378
apachum 319
apachus
  Lophobius 129
  Nyctunguis 404
apheles 105
Apunguis 394

D

helvola. *See* sexspinosus;
Hemilithobius. *See* Lithobius;
  subgenus of Lithobius. *See* Lithobius;
Hemiscolopendra 240
Henicopidae 36
Henicopinae 36
Henicopini 36
heros 249
  heros. *See* heros;
  Scolopendra heros 251
hespera 362
Hesperobius. *See* Arebius;
hesperus
  Guambius 96
  Stenophilus 381
heterotarsus 42
Himantariidae 372
Himantariinae 373
hirsutipes
  Scolioplanes. *See* acuminata;
  Scolopendra. *See* alternans;
hodites 365
hoffmani 385
Holitys 400. *See* also Marsikomerus;
holzingeri 134
homa 33
Honuaphilus. See Tuoba;
hopianus 185
hoples 50
Horonia 281
hortensis 220
howei. *See* Neolithobius mordax;
hubrichti 99
huestoni 44
huronicus. *See* flavus;
    See bipuncticeps;
hyalinus. See leucopodus; *See* also
    hortensis;
hypogeus 82

I

idahoana 315
idahona 282
Idiona. *See* Arctogeophilus;

idium 322
iginus 191
imaharensis 51
imperialis. *See* parviceps;
inaequidens. *See* marginata;
incerta. *See* alternans;
indianae 344
ineptus 169
inermis 300
inexpectus 227
inimicus 143
insignis. *See* africanus;
insulae 160
insulare. *See* spinicaudus;
insularis 307
insulatus
  Ethopolys bipunctatus 61
integer 62
integrior 70
intermedius
  Gosibius 95
  Tygarrup 385
intermontanus 108
interrupta. *See* forficatus;
investigans 102
iosemiteus 192
isabella 331
isantus 315
Ityphilus 304
iugans 327

J

japonicus. *See* ethopus;
javanicus 386
josephus 67
jowensis 134
Juanobius 100
juventus 174

K

kaibabus 49
kansensis 208
Kauabius. *See* Lithobius;
kebus 214

S

U

uleorus 168
umbraticus 309
unaster 328
underwoodi 156
unicus 206
urania 285
urbanus 196
urbicus 398
utahana. *See* viridis;
utahensis. *See* also Oabius tiganus;
  Kethops 264
  Pokabius 192
uticola 216
utus 319

V

vagrans 180. *See* also crassipes;
valida. *See* barbarica;
vallis 407
vaquens 140
vaquero 187
variabilis. *See* acuminata;
varians 355
vector
  Cryptops 227
  Tidabius 213
venezuelae 227
verdescens 187
verus 78
vesuviana 283
vians 77
victorianus 57
virginiensis
  Geophilus 357. *See* also Geophilus
    mordax;
  Nampabius 147
viridilimbata. *See* polymorpha;
viridis 259. *See* also marginata;
vittatus 357
voracior 157
vorax 158

W

waccamanus 140
waikaneus 391
waipaheenas 392
walkeri. *See* bidens;
wamus 168
Watobius 214
Watophilus 317
  subgenus of Watophilus 317
watovius 110
watsingus
  Arenophilus 329
  Cryptops 227
watsuitus 176
weberi 228
whitei 361
winnetui 361
woodii. *See* marginata;
wurmanus. *See* Lithobius borealis;

X

xanti 64
xenoporus. *See* umbraticus;
xenopus 159

Y

yamashinai. *See* rapax;
yanikans 420
yavapainus 361
Yobius 45
yukus 420

Z

Zanobius 170
Zantaenia 282
Zantethobius. *See* Zygethobius;
Zelanion. *See* Steneurytion;
Zinapolys 215
zinus 176
zionicus 213
zipius 215
Zygethobiinae 43
Zygethobiini 43

CPSIA information can be obtained at www.ICGtesting.com
Printed in the USA
LVOW101941300413

331652LV00001B/384/P